Derrick de Kerckhove
Charles J. Lumsden (Eds.)

The Alphabet and the Brain

The Lateralization of Writing

With 59 Figures

Springer-Verlag Berlin Heidelberg GmbH

Associate Professor Derrick de Kerckhove
Co-Director
The McLuhan Program in Culture and Technology
University of Toronto
39A Queen's Park Circle
Toronto, Ontario M5S 1A1, Canada

Associate Professor Charles J. Lumsden
Sociobiology Research Group
Department of Medicine
University of Toronto
Medical Sciences Building, Room 7313
Toronto, Ontario M5S 1A8, Canada

ISBN 978-3-662-01095-2 ISBN 978-3-662-01093-8 (eBook)
DOI 10.1007/978-3-662-01093-8

Library of Congress Cataloging-in-Publication Data. The Alphabet and the Brain. 1. Alphabet. 2. Writing. 3. Neuropsychology. 4. Laterality. 5. Cerebral Dominance. I. De Kerckhove, Derrick. II. Lumsden, Charles J., 1949–
QP399.A435 1988 152.3′35 87-23351
ISBN 978-3-662-01095-2

Originally published by Springer-Verlag Berlin Heidelberg 1988
Softcover reprint of the hardcover 1st edition 1988

2126/3130-543210

Preface

This book is a consequence of the suggestion that a major key toward understanding cognition in any advanced culture is to be found in the relationships between processing orthographies, language, and thought. In this book, the contributors attempt to take only the first step, namely to ascertain that there are reliable constancies among the interactions between a given type of writing and specific brain processes. And, among the possible brain processes that could be investigated, only one apparently simple issue is being explored: namely, whether the lateralization of reading and writing to the right in fully phonemic alphabets is the result of formalized but essentially random occurrences, or whether some physiological determinants are at play.

The original project was much more complicated. It began with Derrick de Kerckhove's attempt to establish a connection between the rise of the alphabetic culture in Athens and the development of a theatrical tradition in that city from around the end of the 6th century B.C. to the Roman conquest. The underlying assumption, first proposed in a conversation with Marshall McLuhan, was that the Greek alphabet was responsible for a fundamental change in the psychology of the Athenians and that the creation of the great tragedies of Greek theatre was a kind of cultural response to a condition of deep psychological crisis. Thus, if a causal connection could be shown between alphabetic literacy and the development of theatrical phenomena, then theatre could be understood as a privileged cultural medium to work out cognitive and emotional solutions to the reorganization of the Greek and later the whole Western mind set. It should be noted that the subsequent Western theatrical tradition ranges all the way from the late Middle Ages and the Renaissance to our present "audio-visual era."

The editors hope to return eventually to this line of investigation. However, at the time, it appeared impossible to deal adequately with such vast issues without first establishing a scientific basis for the hypothesis. It became necessary to narrow the focus of the research to a single core issue, that of the lateralization of the Greek alphabet. Even so, as simple and straightforward as this question may appear, it does not automatically evoke uncontroversial evidence or answers.

Obviously, it was difficult to be informed and persuasive in the exact sciences with a background in literature and the sociology of art. A first attempt at bringing scientists and humanities scholars to bear upon the issue took the form of a day colloquium on "neuro-cultural research," that is, research focusing on the interactions between the human nervous system and technological or cultural environments, at St. Michael's College (University of Toronto), in May 1982. The second effort was made during a colloquium on McLuhan in December 1983 at the Canadian Cultural Centre in Paris. A special section of the colloquium was devoted to neuro-cultural research, and specifically to the issue of literacy and the brain, with papers from Professors Jean-Pierre Changeux, Michel Imbert, André Roch Lecours, Derrick de Kerckhove, Lynd Forguson, and Anthony Wilden. The proceedings of the colloquium were eventually published by the Canadian Commission for Unesco. Though the publication was well received, it did not attract much attention from the scientific community because it stood somewhere between the humanities and the exact sciences.

The solution came from joining forces with Charles Lumsden, who took interest in the project because he saw in the specific issue under discussion a possible indicator of a co-evolutionary pattern between genetic and cultural trends and constraints. Indeed, the area of interface between the human body and mind and the social environments created by human technologies may be best represented by writing systems, if only because writing is a formal code which represents language outside the human body, presumably along some of the lines of linguistic representations within the body – i.e., thought.

It is surprising that it has taken so long to address the issue of formal codes in terms of evolutionary biology. It is even more surprising that in the heyday of talks about the "two sides of the brain," when scientists and media would vie to find ways to divide the world in left and right, few people seemed to have been very concerned with the fact that in the West, people write from left to right, as if it were a natural thing. Regarding the particular impression that Western cultures were prone to favour what were deemed to be "left-hemisphere" approaches to general information processing, problem solving, and social organization, there were a lot of conjectures, but precious little firm evidence to support claims. This is partly why both editors felt to an equal degree the need to explore the issue by concentrating their attention on the structural features of the single most important system of processing information in their own cultural environment.

We have called upon scholars from different fields to ask their opinion on the principal issues relevant to the basic hypothesis. The key areas were provisions in the human body for different levels of biological adaptation, historical development of orthographies, relationships between the structure of specific languages and the structure of their writings systems, relevant neuropsychological investigations to date, and finally, the working out of the basic hypothesis from different angles.

Thus, what began as an unpublished monograph on theatre and the alphabet in Athens has now become a collection of scientific essays. Our first words of thanks must go to the contributors, who devoted time, effort, and attention to an issue that was presented to them often as a challenge outside the immediate focus of their specialty. And before anything else, the editors would like to pay tribute to many people who, for editorial or other reasons, do not appear among the contributors, but who have helped the project along at various stages. Two workshops, one in Paris at the Canadian Cultural Centre and the other in Toronto at the McLuhan Program were held in April and July 1985 during the planning stages of the book. Our thanks go to Professors Michel Imbert, Lynd Forguson, Brian Stock, Sandra Witelson, as well as to the regretted Paul Kolers, all of whom attended these workshops and have subsequently spent valuable time with us to guide the project along the way. Other supporters and collaborators of the project have included, at different times, in different places, sometimes in person, sometimes by mail, Professors Jean Saint-Cyr, Morris Moscovitch, Jacques Mehler, Louis Holtz, Denise Schmandt-Besserat, and Alfonso Caramazza. The editors have also benefited from suggestions and comments from Professors Karl Pribram, Diane McGuinness, Anthony Wilden, Marcel Kinsbourne, E. B. Hunt, and Daniel Schacter. A special mention should go to Sally Grande, who, as an informal volunteer researcher in the literature on the brain, sharing a fascination for literacy in early Greek culture with the editors, has provided much needed support and information especially in the early stages of neuro-cultural research. Some of the contributors themselves have gone extra lengths. Professor Insup Taylor, Jean-Luc Nespoulous, and David Olson, for example, carefully reread manuscripts pertaining to their fields in order to offer valuable suggestions. Professor Bhatt kindly provided translations for the papers of Jean-Pierre Changeux, Claude Hagège, Robert Lafont, Baudouin Jurdant, Jean-Luc Nespoulous, André Roch Lecours, and Collette Sirat.

In addition to the diligent work of our publishing team at Springer including Thomas Thiekötter, Janet Hamilton, and Susan Kentner, the production of the book required considerable preparation and revisions. For these, we are especially thankful to Ann Stilman, our copy editor, Cassie Rivers, who helped with proofreading and indexing and Sylvia Wookey, who, along with the Membrane Biology Group and Ann Hansen at the McLuhan Program did much of the time-consuming administrative and office work.

Another aspect of book production today is the rising cost of pre-production. We want to thank last, and certainly not least, institutions and individuals who have supported the project financially and without whom it could not have been carried to term.

Of the institutional support received, we would like to acknowledge that of the University of Toronto's General Research Grant and the Connaught Development Fund Grant for general editorial expenses. A travel grant from the Canadian Ministry of External Affairs also contributed to the Paris meeting. Some expenses for supplies and secretarial help were borne by the McLuhan Program in Culture and Technology, the Canadian Cultural Centre in Paris, and the Membrane Biology Group at the University of Toronto. Let them be thanked for this. We are also truly indebted to the people who have taken enough personal interest in the project to help support it financially. Among these, may we thank in closing Ms. Catherine Harris, Professor Robin Harris, Mrs. Dorothy Dunlop, and the late Janet Underwood.

Toronto, May 1988

DERRICK DE KERCKHOVE
CHARLES J. LUMSDEN

Contents

List of Contributors

PARTH M. BHATT is Assistant Professor in the Department of French and an associate researcher at the Experimental Phonetics Laboratory at the University of Toronto. The focus of his work is on applied phonetics and neurolinguistics. Among his publications, *Prosodie et lésions cérébrales* is to appear shortly.
Experimental Phonetics Laboratory, New College, University of Toronto, Toronto, Ontario M5S 1A1, Canada

JEAN-PIERRE CHANGEUX is Professor at the Collège de France, Director of Research at the Neurobiology Laboratory at the Institut Pasteur and a member of the Centre National de la Recherche Scientifique. Along with colleagues A. Danchin and P. Courrège, he has developed the theory of selective stabilization of developing synapses. Among his many publications, *L'Homme neuronal* has achieved international acclaim.
Institut Pasteur, 28, rue du Docteur Roux, Paris 75724, France

DERRICK DE KERCKHOVE is Associate Professor of French and Co-Director of the McLuhan Program at the University of Toronto. As an associate researcher at the Centre for Culture and Technology with Marshall McLuhan, he began exploring relationships between cultural artifacts and neurology in the early 1970s. He is the author of several papers on the alphabet and on the neurocultural effects of literacy and the new media.
McLuhan Program, 39A Queen's Park Crescent, Toronto, Ontario M5S 1A1, Canada

LEIF H. FINKEL is Assistant Professor at the Rockefeller University and a researcher at the Neurosciences Research Institute. His principal scientific interest is in elaborating models of selective mechanisms in the nervous system.
The Neurosciences Research Institute, The Rockefeller University, 1230 York Avenue, New York, New York 10021, USA

PATRICIA ELLEN GRANT is a research fellow in the Departments of Physics and Medicine at the University of Toronto. Her research in-

terests include the mathematical modeling of neural systems and cognitive processes.
Membrane Biology Group, Room 7313, Medical Sciences Building, University of Toronto, Toronto, Ontario M5S 1A8, Canada

CLAUDE HAGÈGE is Directeur d'études at the Ecole Pratique des Hautes Etudes, Professor of Linguistics at the University of Paris (Sorbonne), and a researcher at the Centre National de la Recherche Scientifique. He is the author of several books on linguistics and the anthropology and sociology of language, including *La structure de langues* (1982) and the recent *L'homme de paroles* (1985).
102, Boulevard Kellermann, 75013 Paris, France

EDWARD A. JONES is Associate Professor in the Department of Psychology at Labouré College in Boston. His extensive research in the culture, psychology, and literacy of the Japanese (often in collaboration with Chisato Aoki, doctoral candidate in psychology) is to appear in a book, *The Human Reaction,* in 1988.
72 Fells Avenue, Medford, Massachusetts 02155, USA

BAUDOUIN JURDANT is Director of the Groupe d'études et de recherches sur la science at the Université Louis Pasteur in Strasbourg. He is also Executive Co-Editor of the international multidisciplinary journal, *Fundamenta Scientiae.* His principal research area is in the relationships between cognitive processes, money, and literacy in classical antiquity.
GERSULP, 4, rue Blaise Pascal, 67451 Strasbourg, France

ROBERT LAFONT is Professor in the Faculté des Arts et des Lettres at the University Paul-Valéry (Montpellier III) and Director of Research at the Centre National de la Recherche Scientifique. In that capacity, he directed the work and publication of *Anthropologie de l'écriture* (1984). An earlier publication in language studies was *Le travail et la langue* (1978).
Arts et Lettres, Université Paul Valéry Montpellier III, Route de Mende, B.P. 5043, 34032 Montpellier CEDEX, France

ANDRÉ ROCH LECOURS is Professor in the Faculty of Medicine at the University of Montreal and the Director of the Centre de Recherches at the Centre Hospitalier Côte-des-neiges, also in Montreal. Among his principal publications are *L'aphasie, Aphasiology, Biological Perspectives on Language* and, with J. L. Nespoulous, *Biological Foundations of Gestures.*
Centre Hospitalier Côte-des-neiges, 4563, rue Queen Mary, Montréal, P.Q. H3W 1W5, Canada

CHARLES J. LUMSDEN is Associate Professor of Medicine at the University of Toronto and a member of the Membrane Biology Group at the same university. Approaching the theme of human evolution from the vantage points of several disciplines ranging from physics to population genetics, he is the author (with E. O. Wilson) of *Genes, Mind and Culture: The Coevolutionary Process* (1981) and *Promethean Fire: Reflections on the Origin of Mind* (1983).
Sociobiology Research Group Department of Medicine, Room 7313, Medical Sciences Bldg, University of Toronto, Toronto, Ontario M5S 1AB, Canada

JOSEPH NAVEH is Professor at the Hebrew University of Jerusalem where he works on ancient Hebraic and Aramean epigraphs and inscriptions. He is the author of several books on ancient Semitic writings, including *Early History of the Alphabet: An Introduction to West Semitic Epigraphy and Palaeography* (1982).
Hebrew University of Jerusalem, Jerusalem, Israel

JEAN-LUC NESPOULOUS is Professor in the Department of Linguistics and Philology at the University of Montreal and a senior researcher at the Centre Hospitalier Côte-des-neiges. Among his publications are *Etudes neurolinguistiques, Neuropsychologie de l'expression orale,* and, with A. R. Lecours, *Biological Foundations of Gestures.*
Centre Hospitalier Côte-des-neiges, 4563, rue Queen Mary, Montréal, P.Q. H3W 1W5, Canada

DAVID R. OLSON is Professor of Applied Psychology in the Centre for Applied Cognitive Science at the Ontario Institute for Studies in Education, as well as Co-Director of the McLuhan Program in Culture and Technology at the University of Toronto. His major research concerns the relation between oral conversational language of pre-school children and formalized language of written texts. Recent books include *Spatial Cognition* (with E. Bialystok, 1983).
McLuhan Program, 39A Queen's Park Crescent, Toronto, Ontario, M5S 1A1, Canada

COLETTE SIRAT is Directeur d'études at the Ecole Pratique des Hautes Etudes and Research Director at the Institut de recherche et d'histoire des textes and the Centre National de la Recherche Scientifique. She has published many books on ancient Semitic writings as well as on the methodology of deciphering and classifying ancient orthographies. Among her principal publications are *Les papyrus en caractères hébraïques* (1986) and (with M. Beit-Arié) *Les manuscrits en caractères hébraïques portant des indications de date jusqu'à 1540* (3 vols, 1972, 1979, 1986).
Institut de recherche et d'histoire des textes, Centre Félix Grat, 40, Avenue d'Iéna, 75116 Paris, France

JOHN R. SKOYLES did his graduate work in philosophy, logic and scientific method at the London School of Economics and has conducted independent research on the development of the alphabet and other writing systems during the past decade. He has published research papers and reviews in *Nature, New Ideas in Psychology, Principia Scientiae* and other journals.
6 Denning Road, London NW3 1SV, United Kingdom

INSUP TAYLOR is a faculty member at the McLuhan Program in Culture and Technology at the University of Toronto. Her research focuses on the relationships between literacy and psychology, with a special emphasis on Eastern and Far Eastern orthographies. She is the author, with M. M. Taylor, of *The Psychology of Reading* (1983).
McLuhan Program, University of Toronto, 39A Queen's Park Crescent, Toronto, Ontario, M5S 1A1, Canada

M. MARTIN TAYLOR is Senior Experimental Psychologist at the Defence and Civil Institute of Environmental Medicine, Toronto. He specializes in perception, human-computer interaction, and automatic speech recognition. Co-editor (with F. Néel and D.G. Bouwhuis) of *Structure of Multimodal Dialogue* (1988), he is the author, with I. Taylor, of *The Psychology of Reading* (1983).
DCIEM, Box 2000, Downsview, Ontario, M3M 3B9, Canada

OVID J. L. TZENG is Professor in the Department of Psychology at the University of California at Riverside, where he also directs the Center for Orthography, Reading and Dyslexic Studies. A senior scientist at the Salk Institute for Biological Studies, he is the author of many research articles on orthography and brain processes (several in collaboration with Daisy Hung, also a research associate at the Salk Institute). He has co-edited (with H. Singer) *Perception of Print* (1981).
Department of Psychology, University of California at Riverside, Riverside, California 92502, USA

WILLIAM C. WATT is Associate Professor in the School of Social Sciences at the University of California (Irvine). Author of several research papers in semiotics, he focuses his attention on the history and development of ancient writings, tracing the evolution of writing systems in their present form.
School of Social Sciences, University of California at Irvine, Irvine California 92717, USA

General Introduction

DERRICK DE KERCKHOVE and CHARLES J. LUMSDEN

I invented for them the art of numbering, the basis of all sciences, and the art of combining letters, memory of all things, mother of the Muses and source of all the other arts.

Æschylus, *Prometheus Bound* (1, 459 – 461, trans. D. de Kerckhove)

Literacy and the Cognitive Revolution of Ancient Greece

The Greeks were not unaware of the importance of the role that writing played in their civilization. Æschylus, the presumed author of *Prometheus Bound,* put the writing of numbers and letters at the source of all human inventions. The story of Prometheus, who was the object of a special cult in Athens as the god of intelligence and technical skill, can be viewed as a founding myth for the western world. It appears that Æschylus may even have had a fair idea of the implications of literacy upon cognitive styles. His interpretation of the myth suggests that Prometheus rescued humankind from the wrath of Zeus by giving them a mind to defend themselves against the hardships of nature:

... they were witless erst and I made them to have sense and be endowed with reason. [... Previously,] though they had eyes to see, they saw to no avail; they had ears, but understood not; but, like to shapes in dreams, throughout their length of days, without purpose they wrought all things in confusion.

Prometheus Bound (1, 443 – 4/447 – 450, trans. H. Weir Smith)

The relationship between writing and intellectual drive is more than plausible. In *Prometheus Bound,* the alphabet is called μονσομητοϱ (i.e., the "mother of the Muses," 1, 461), a term that by itself implies an interesting secularization of mythical figures, but which also indicates that writing, in the opinion of the author, was the source of all the arts of thought. The nine muses each represented one of the following domains of human understanding: philosophy, history, geography, music, comedy, tragedy, astronomy, dance, and lyric poetry. In the same speech, Prometheus goes on to list all the inventions he made available to humankind to protect it against its own frailty. Among them were agriculture and ploughing, architecture and bricklaying, divining, sailing, horse-riding, and many others. However, the preeminence given to thinking and to the principal tools of thought, writing, and counting may indicate that in the mind of the author, all these other inventions had followed from them.

Formal philosophy, as it was institutionalized in western thinking by the pre-Socratic philosophers and by Plato and Aristotle, was one among several innovations in the Greek world that point to a cognitive revolution. Other developments that were not exclusive to the Greek culture, but were to achieve a higher degree of formal articulation in Greece than in comparable situations in Babylonia or Egypt, included medical science and a systematic description of the human body; the creation of a legal constitution and a body of laws; the invention of the idea and the study of history; the formalization of geographical knowledge; the structuring of an educational system; the order of architecture; several important theories in physics that would only fully mature in the twentieth century, among which was the theory of the atom; Euclidean physics; and the invention and application of geometry. In Greek drama and lyric poetry, as Bruno Snell (1953), Zebedei Barbu (1972), and many other classical scholars have pointed out, a new notion of the "self" seems to have developed. In Plato, notably in some of his better-known dialogues, *Cratylus, Critias, Protagoras,* the *Symposium,* and also the *Laws,* one can see the development of a theory of language that is supported by many references to the alphabet. Thus, we find therein more evidence of a fundamental aspect of the cognitive revolution that might be deemed to have followed the development of writing.

There is no evidence from the ancients that they compared the merits of writing hieroglyphically as opposed to alphabetically, although it is clear that Æschylus and Plato were both aware of the distinctive structural features of the alphabet. Note that Æschylus' wording γραμματων τε συνθέσεισ (literally, the "combination of letter," *Prometheus Bound* 1, 460), gives as succinctly as possible an accurate description of the main structural feature of the alphabet: the fact that it is produced by combining characters. To literacy in general, Plato did attribute cognitive consequences, although not all positive ones, as is evident from reservations he expressed in the *Seventh Letter,* and more specifically in the following legend of origin whereby King Thamous of Egypt reproves Theuth, the inventor god, for having invented writing:

> If men learn this, it will implant forgetfulness in their souls; they will cease to exercise memory because they rely on that which is written, calling things to remembrance no longer from within themselves, but by means of external marks. What you have discovered is the recipe not for memory, but for reminder. And it is not true wisdom that you offer your disciples, but only its semblance, for by telling them of many things without teaching them you will make them seem to know much, while for the most part they know nothing, and as men filled, not with wisdom, but with the conceit of wisdom, they will be a burden to their fellows.
>
> *Phædrus* (275 a–b, trans. R. Hackforth)

The argument is very much along the lines of those of people who claim that pocket calculators will make children mentally lazy because they give them access to mechanized methods of calculation and complex forms of knowledge without effort. On the surface this argument is simply based on common sense, but the fact that Plato recounts it through myth gives it more solemnity and depth.

The *Dialogues* and the *Letters* give much evidence of a careful and systematic exploration of the mind and of the way it works. In particular, as a document of the mental framework of the first members, presumably, of western civilization, Plato's work stands unchallenged. It is not that generations of classicists and philosophers have failed to probe his thought, but rather that most have assumed that the human mind is universally thus, and only a few, notably Eric Havelock (1963), fully realized how novel was Plato's thought, not only in the Greek world, but among human cultures generally.

Approaches to the Question of Literacy

Both stories, that of Prometheus and that of Theuth, are myths of the origin of writing that imply a fundamental change in the psychology of humankind. It is therefore surprising that the cognitive consequences of literacy have received attention only recently. Since Havelock's work, research into literacy has become a major industry, but it has not yet found a foundation. In his review of the work done to date, David Olson (1986) reminds us that Jean-Jacques Rousseau (1749) gave us one of the first theories of writing and the development of civilization. It is predictably couched in the Europocentric terms of eighteenth century philosophy. Following another typically Gallic bias, Jacques Derrida, in *Of Grammatology* (1976) and many other essays on the subject of writing, proposes an esoteric theory of textuality that seems to put writing at the very principle of creation, at the core of reality and consciousness.

Another vein of research, less heady and better grounded in historical fact, was provided by Havelock (1963), who raised much controversy in classical scholarship with his theories of the impact of writing on cognition. Although his work has been criticized with some vigor (Woodbury, 1983; Larsen, 1986), one of Havelock's greatest achievements is to have suggested that the structure of the Greek alphabet, rather than just any kind of literacy, might be responsible for much of the cognitive change of Greek culture. His argument, as if inspired by a reaction to King Thamous' answer to Theuth, seems to take the opposite stance: namely, that the simplicity of the alphabet's structure enabled the learner to release the mind from the burden of memorizing objects of knowledge, making it available for speculation and critical thought. This laid the foundation for a new, more technical and factual attitude toward knowledge.

If ever there was an elephant described by blind experts, it is the matter of literacy, but one thing for sure is that this elephant is not white. Alphabetic literacy may be responsible for some of humankind's greatest leaps in the evolutionary processing of knowledge, and that is why it deserves a closer examination.

There are several levels at which literacy effects can be examined, and opinions differ as to which level is the most relevant. Quite apart from important historical and linguistic investigations on the history and development of writing itself (Cohen, 1958; Diringer, 1948; Driver, 1954; Février, 1959; Gelb, 1963; Jackson, 1981; Sampson, 1985), some studies concentrate on the social and cultural conditions accompanying and favoring literacy (Febvre & Martin, 1958; Graff, 1981, 1986; Innis, 1951, 1972; Martin, 1968–1970; Ong, 1967, 1971, 1982; Stock, 1983; Stone, 1969); others are concerned with schooling (Cole & Griffin, 1980; Scribner & Cole, 1981; Vygotsky, 1978), with epistemological and cognitive correlations (Cole & Bruner, 1971; de Kerckhove, 1984; Goody, 1977, 1986; Havelock, 1963, 1982, 1986; Olson, 1977, 1986; Ong, 1967; Vygotsky, 1978); there are historical studies of effects (Eisenstein, 1979; Innis, 1951, 1972; McLuhan, 1962), philosophical investigations (Derrida, 1976; Etiemble, 1973; Rousseau 1749) and experimental psychological research (Humphreys & Evett, 1985; Sinatra & Stahl-Gemake, 1983; Taylor and Taylor, 1983). Several annotated bibliographies have been produced on various aspects of literacy (e.g., Liggett, 1984; Scollon, 1985).

There are at least two ways of considering the relationship between writing and cognition. Either any form of literacy will have a similar effect on the cognitive style of the culture adopting it, which is the position adopted by several contributors to this volume (Sirat, I. Taylor, and Olson), or it takes a specific writing form, the highly abstract phonological system called the alphabet, to yield the kind of effects associated with Greco-Roman civilization, which is the fundamental hypothesis that gave rise to this book. This problem can be approached in different ways.

The conventional approaches range from the study of the sociocultural conditions favoring the development of literacy and literate cognition generally, to the cognitive operations associated with the script itself. The larger realm, that of sociocultural conditions, is best represented by the work of Scribner and Cole (1981), who, in their field study of the Vai in Liberia, concluded that neither the fact nor the nature of literacy could be deemed to carry the weight of the assumed consequences, but rather that literacy was generally associated with a schooling system which alone or principally made the difference between the educated and the uneducated cognition. In his own work supported by extensive field study, also in Africa, another anthropologist, Jack Goody (Goody and Watt, 1963; Goody, 1977), attributes the cognitive changes more precisely to literacy itself; claiming, for instance, that writing enables people to make lists and categories and that in itself it can generate a better control on mental classifications. One of the more probing arguments of Goody and Watt is the suggestion that at least two important cognitive features might be correlated with the development of Greek literacy, namely, the creation of a taxonomy of scientific categories, which we still consider adequate today, and the discovery of an epistemological dimension, that is, "knowledge about knowledge systems."

Initially, Goody and Watt (1963) had based their claim on the invention of the Greek alphabet, thus giving a special emphasis to the properties of Greek literacy to effect social change. Later, however, Goody (1977) returned to a more conservative opinion regarding the specificity of the effects of the alphabet itself. In *The Domestication of the Savage Mind*, the author attributes to a critical mass of literate concentration the psychological effects he had associated earlier with the orthography itself. The same caution characterizes Olson's work in developmental psychology. Olson (1977, 1986), however, follows a different path than Goody in investigating precisely what the major cognitive effect of literacy might be. Olson suggests in this volume that literacy, by making statements external, changes the reader's (and presumably the writer's) attitude toward meaning. A written document both provides a factual statement and requires interpretation. This situation of knowledge leads to several cognitive consequences, among which the principal ones are the development of hermeneutics and critical analysis, the distinction between objectivity and subjectivity, and the gradual development of a strong personal identity. In another vein, combining sociocultural data and intuitive cognitive assumptions, Brian Stock's (1983) research into medieval literacy and its consequences also supports Olson's approach.

In view of this impressive body of theory, why should one consider adding yet another level of investigation to the matter of literacy? The answer, for us at least, is first, that unless one turns to the brain, there may be no way to properly distinguish what is attributable to the alphabet from what is attributable to any writing system; and second, that given the progress in cognitive studies today, there appears to be a need for brain-related studies to begin to unify the available approaches, and to explore more deeply the cognitive differences between western cultures and others like the Arabic, the Indic, the Japanese, or the Chinese, which use different writing systems.

Another reason to turn to the brain is that there is a piece of evidence that has not been examined before and which may carry enough weight to require such a detour. This is the observation that all orthographies are spatially organized in definite and fairly constant patterns, and that these patterns differ for different orthographies in reliable ways. A quick survey of world scripts reveals, for instance, that almost all pictographic writing systems favor a vertical layout (later adaptations, particularly under the recent influence of western styles, have sometime adopted the horizontal orientation). Another striking feature is that practically all systems of writing that depend exclusively on the visual rendition of the phonological features of language are horizontally laid out. Even more surprising is the fact that 95% of phonological orthographies that include markings for vocalic sounds, whether in syllabaries, as isolated signs or parts of a syllabic character, or in alphabets, as individual vowel-letters, are written toward the right, whereas almost all the systems that do not include letters for vowels are written toward the left, and have been rendered so almost from the beginning, for over three millenia. Although there may be other constancies, such as distribution and comparable features of character-shapes or relationships between individual signs and the sounds they represent, the issue of writing direction has the merit of being immediately evident, testable over long periods of time, and can be related to known facts about the visual system and its organization in the brain. Directionality therefore can be selected to provide a benchmark for the study of relationships between reading/writing and neurological considerations.

Organization of the Book

This book brings together the work of historians, epigraphists, linguists, psychologists, biologists, and neuropsychologists to approach a single issue from their respective disciplines. The interdisciplinary contributions are focused on the core hypothesis that a specific writing system requires a specific set of neurological responses and decipherment strategies. The precision evidenced by a given writing system is thus a reflection of the kind of selective precision required from the brain.

The chosen case study revolves around the changes that followed the adoption of the Phoenician alphabet by the Greeks. The book is divided into five sections, each of which addresses a dominant issue that can be framed as a question.

What Are the Main Biological Provisions Within the Organism to Adapt to its Environmental and/or Cultural Determinants?

Going from the general conditions to the presentation of models and hypotheses, the book begins by establishing the biological foundations of adaptability. The hypothesis that writing might affect and bias selectivities in the human organism is indeed predicated on general conditions of adaptability of the biological and neurological systems. To get a firm ground in the physiological processes underlying human cultural development, two lines of research, genetic biology and developmental neurobiology, are presented. In biology, the gene-culture coevolution theory is making headway toward finding the causal relationships between cultural and biological phenomena, trying to provide concrete evidence in an area that until recently was open only to philosophical speculation. The idea is to study the impact of cultural change on the spectrum of the human genotype.

In his chapter, Charles Lumsden suggests that cultural conditions provide for a process complementary to Darwin's natural selection and add to it a cultural selection, which eventually, over generations of culturally conditioned populations, will find its way into the genotype. This is not the surreptitious return of a Lamarckian heresy, principally because there is no need to hypothesize a direct retroactive impact of the phenotype on the genotype. It is the study of indirect influences of a biocultural conditioning on certain genetic selectivities. Writing, as a significant modifier of the cultural domain, can be expected to affect culture sufficiently to create conditions of gene-cultural selection.

However, gene-culture coevolution theory stresses the fact that the nervous system of the organism is itself an area of interchange and adaptation. In terms of human evolution, there are four principal interactive domains of mutual regulation; the genetic, the cellular, the neuropsychological, and the cultural. For the promotion of conditions favoring a gradual process of selection in the genetic material over many generations, it may not be enough to be exposed preferentially to given environmental conditions in the culture. A more intimate relationship between culture, mind, and body is required for any consistent interaction to be established effectively. There is a need to find a common ground or a bridge between the cultural impact and the human body. This common ground is provided by the nervous system, which is an intermediary level of human adaptability, and especially the brain, which is known to evidence an impressive degree of plasticity.

It is suggested that writing can also affect selectivities at the level of the phenotype by teaching the brain to accommodate to its special requirements. This is where the approaches of Jean-Pierre Changeux and Leif Finkel are relevant. Both explore the neurological conditions whereby a given element in the environment can modify the brain's response to favor appropriate strategies, just as the exercising of specific muscles is known to improve the performance of athletes.

This work suggests that in the process of developing themselves within the strictly assigned parameters of genetic determinants, the human brain and the nervous system generally organize themselves selectively according to the kind of tasks to which the organism is exposed. As Changeux and Finkel explain in their contributions to this volume, it is not that more or fewer neurons grow or develop under the influence of behavior, but rather that the synaptic connections between them are selectively made to accommodate to the different requirements and differing combinations of requirements imposed by the environment. Although this is especially true of organisms while they are in the process of developing, it remains a condition of adaptation throughout the lifespan.

In Changeux' contribution to epigenetic theory, what he calls the "selective stabilization of synapses" (1983; also this volume) is the process whereby interactive pathways between and among neurons are created according to genetic constraints that are guided and modified by the exposure of the organism to its specific environmental conditions during the period of its development. What we have here are the neurological conditions whereby a given task would evoke a reliable strategy on the part of the brain. It is not a matter of simple "hardwiring" of neurological programs, but of different levels and different intensities of synaptic facilitation. A professional pianist, to maintain the necessary degree of digital flexibility, must practice an average of 3 – 5 hours a day. There must be, however, some measure of constancy in the neurological interpretation of the required program. Indeed, even years after abandoning practice, the same professional pianist can play many pieces, even complex ones.

Surely one task demanding long and continued practice is that of reading and writing. It is also one which, in western urban environments at least, most people practice daily, either over long periods, as, presumably, academics do, or at the very least in short spurts, such as casual reading of advertising panels in the subway. Thus, teaching a brain to read, a task that must be exceptionally stringent considering the length of time required for its acquisition, would require a measure of synaptic stabilization during the formative years, to endow an effortless capability to read as long as one's eyesight can support it. The training must be extremely precise and specifically adapted to the requirements, not only of the language of the user, but also of the specific structure of the code that is used. It is obvious that learning to read the Latin alphabet does not automatically confer the ability to read Cyrillic, let alone Minoan Linear A. A different processing strategy underlying and preceding the cognitive assumptions of the reader must be posited and eventually demonstrated for each and every scriptform. This is the debate that unfolds in the following sections.

How Did the Alphabet Develop and How Does It Compare with Other Writing Systems?

The first step is to get a historical perspective on writing and perceive the larger patterns of development, all the while avoiding the Europocentrism that led earlier students of literacy to see a sort of teleological process in the achievement of the Greek alphabet. If linguistic, logistic, and cultural constraints have had unquestionable effects on the shaping of orthographies, the dominant relationship tends to be that which binds the characteristics of the script to the characteristics of the language it represents.

Claude Hagège presents an introduction to writing systems in general, and focuses on their relationships to the languages and the cultural grounds that gave rise to them. One begins to see that different languages obey inherent structural modalities that may play an important role in selecting an appropriate orthography. Hagège also reviews some ancient and modern attitudes toward writing and opinions about their civilizing effects. His approach is historical, and sets the stage for Joseph Naveh's controversial theory about the origin of the Greek alphabet.

Although most students of the history of Greek writing agree to place the invention around the eighth century B.C., Naveh suggests that "the *argumentum a silentio*", (the fact that no Greek inscription predating the eighth century has yet been discovered), is not conclusive. He offers evidence to date the invention earlier, around the eleventh century B.C. As Naveh's research bears on the correlations between the Phoenician models and the shapes of the individual Greek letters, his paper presents an up-to-date compendium of the complex issue of the development of a complete set of Greek characters.

As many scholars (e.g., Cohen, 1958; Février, 1959; Havelock, 1982; Sampson, 1985) have pointed out, the word "alphabet" presents enough ambiguity to warrant a category distinction between *consonantal* and *vocalic* types of alphabetic systems. The Greek alphabet developed in the high-density areas of different Mediterranean cultures, and its lineage brings together features of the Sumerian and the Egyptian scripts. However, it was worked out as an adaptation of the Phoenician script to the specific needs of the Greek language. Robert Lafont's contribution brings out the differences, the main one being the addition of vowels, and compares the Phoenician and the Greek inventions to each other and to other ancient writing systems, notably syllabaries. Taking the consequences of linguistic constraints on graphemic representation to their logical conclusions, his comparisons also lead to historical observations on the impact of different kinds of writings on social attitudes toward language and knowledge.

To conclude this section from the point of view of phonetics, Parth Bhatt focuses on a variation of the Greek alphabet, the French version of the Latin script, to show the later developments of the relationships between graphic and phonic systems. He also offers a review of the current state of research in the relationships between speech and writing. His paper underscores the relative autonomy of each system, reminding us, for instance, that there are many words that are written the same in French and in English, but pronounced very differently. Such variations, of course, do not invalidate the structural principles that correlate speech and writing in a larger sense. The correlations, rather than the divergences, between the two systems are emphasized in the next section of the book.

Are There Structural Relationships Between Types of Orthographies and Their Layout on Writing Materials?

This section explores the relationships between the structure of orthographies and the way they are laid out spatially, in order to develop the hypothesis that writing systems are subjected to logical criteria that evidence consistent developments.

To begin, William Watt's theory approaches the development of letter shapes from a logical and a developmental viewpoint. Introducing considerations about the logistics of graphic design, he shows that the alphabet, over the course of its development, responded to two complementary constraints: the gradual definition of the configuration of each character, and the physical gestures required to write the letters. He makes suggestions about the direction of individual letters, which are particularly interesting when considered alongside complementary viewpoints such as Parth Bhatt's paper in the previous section, or Derrick de Kerckhove's and Colette Sirat's contributions to this one, and the models in the final section. He goes on to relate these developments in terms of evolutionary processes and cognitive tendencies.

Derrick de Kerckhove's point in this section is to emphasize a critical comparison between the Phoenician and the Greek alphabets: the Phoenician and Greek systems are the sources of two main streams of phonological representations that are reliably distinguished by the opposite directions of their respective scripts. Almost all consonantal variations of the Phoenician models are written horizontally to the left, while all vocalic alphabets are written to the right. The suggestion is that this constancy is not merely circumstantial: each type may require different processing strategies. Another surprising discovery is that fewer than 10 of over 200 world syllabaries are written to the left. De Kerckhove presents a set of structural principles to account for constancies in the layout of orthographies. This paper is intended to serve as a basis for the theory presented in the final section on the issue of brain processes and writing direction.

However, to explain such regularities, one should not be content to invoke internal causes alone. Colette Sirat's paper points out material and circumstantial explanations for the choice of direction and for the constancies adopted by writing systems, with a special emphasis on Semitic scriptforms. Her emphasis is on the *ductus,* the guiding force moving the hand along the path of writing. Several considerations are in order, notably the nature of the surface, the tool of writing, the posture of the scribe and other constraints presented by environmental conditions attending the writing process.

By comparing Eastern and Western orthographies from the vantage point of the psychology of literacy, Insup Taylor places the debate at the level, not of mere decipherment, but of cognition and learning. Her systematic review of different orthographies and of the conditions surrounding the process of learning them echoes and complements Hagège's introduction as it completes the survey of writings systems. Her concluding sections introduce the question of reading and hemispheric specialization, the object of the next part of the book.

Is There Evidence That the Alphabet May Have Imposed Specific Constraints on the Selectivities of Brain Organization?

This section samples some of the recent work done by neuropsychologists on the relationships between reading and the brain. Its aim is to identify the key issues, among which is the question of hemispheric specialization. It begins with a historical review by André Roch Lecours and Jean-Luc Nespoulous of earlier opinions about brain functions and later investigations in anatomical asymmetries as they relate to reading and writing.

Ellen Grant's paper maps out the complex neuroanatomical conditions underlying linguistic processes in the brain. She discusses a neurological model for reading and writing that can serve as a point of reference for the other papers in this section. The paper also reviews some of the recent observations about hemispheric specialization and includes valuable suggestions about language processing, hemispheric specialization, and scriptform.

Ovid Tzeng and Daisy Hung take the processing of Chinese orthography as a basis for comparison with the other principal writing systems. To search for evidence of different cognitive processes associated with these various scripts, they review the most important areas of research in phonological and visual recoding. By comparing Chinese with non-Chinese readers, investigators have shown that the specificity of Chinese logographs affects positively both direct access to meaning from characters and visual recall. Conversely, phonographic systems give a better access to naming. Experiments conducted with readers of Japanese Kanji (logographic) and Kana (phonographic) characters have given evidence for different lateralization processes. The authors report that they tried to replicate these results with Chinese readers, but unsuccessfully. They conclude that the evidence of right-hemispheric reading for Chinese characters is nonexistent, but that more work with better controls and a finer appreciation for the different levels of processing should be carried out.

The following contribution, by André Roch Lecours et al., is a review of the work done to date to establish evidence for a functional lateralization of cortical processes in reading and writing. It is based on test comparisons carried out by the authors on literate and illiterate peoples. It discusses results obtained through dichotic listening and aphasiology. Reporting on the results of the tests, the study concludes that there is a difference, albeit not very dramatic, in the extent of left hemispheric dominance for language for both groups, the illiterate being less evidently lateralized.

Approaching the same issue of dominant lateralization from a completely different angle. Edward Jones and Chisato Aoki assemble data on Japanese literacy to indicate that within the same cultural group making simultaneous use of two highly contrastive systems of writing, ideographic and syllabic, there is evidence of neuropsychological processing differentiations. They suggest that Japanese readers might be less lateralized than English ones because they use two complementary systems of writing, each requiring different strategies of deciphering involving both hemispheres in differing proportions. The paper concludes with important suggestions about the possible implications of the processing systems for cultural and cognitive significance.

What Would Be the Models of Interaction Between the Alphabet and the Brain That Are Pertinent to Features Examined and Presented in This Book?

The global hypothesis governing this last section is that rightward reading and writing might be an important promoter of the left hemisphere's (or, as Martin Taylor more modestly suggests, the "left track's") involvement in orthographic recoding, and maybe in cognitive activities as well. However, the most important task is to propose models to account for the mechanisms that appear to bind rightward reading to the structure of the Western alphabets.

Martin Taylor's Bilateral Cooperative Model of reading is an ambitious synthesis of what is known and what is believed about brain specialization, with the aim of arriving at a functional model of collaboration between the two hemispheres. Taylor does not emphasize the physiological or anatomical dimension as much as the logical interaction obtained at different levels of processing. Rather than talk about hemispheres specifically, Taylor prefers to attribute the processes to a right and a left "track," leaving to future research the necessary sorting out of anatomical localization. Taylor's paper also tests the model on the principal orthographies that have been considered so far and concludes that there may be some ground for suggesting not only that certain types of script reflect or favor one track over another, but also that different types of cognition might be associated with the preferential track.

Taylor's paper sets the general framework for two avenues of neuropsychological modeling. The first one investigates the possible correlation that could be established between the direction of writing and visual hemifield preferences (Skoyles, de Kerckhove), and the other (Jurdant) posits a functional differentiation in the cortical status of consonants and vowels.

In his contribution, John Skoyles defends the hypothesis that the right hemisphere might have been more involved in deciphering ancient orthographies than it is in decoding our present Latin alphabet. He begins by discussing the accepted notion that all reading is normally processed, as all linguistic functions, in the left hemisphere. He then proceeds to describe the evolution of writing direction, proposing an explanation that brings together neuropsychological and circumstantial evidence for different patterns of orthographic layout.

Baudouin Jurdant's theory is probably the most daring, with a proposed category distinction between vowels and consonants at the sensorimotor and cortical levels. Basing his argument on prenatal and postnatal observations of infants' differing reactions to vocalic and consonantal sounds, Jurdant suggests that the cortical treatment of both types of sounds, and consequently of their scriptorial representation, is lateralized and is partly responsible for the direction of writing.

Derrick de Kerckhove's argument can be stated in simple terms. He suggests that the relationship between reading and the brain was changed when the Greeks introduced fixed characters to represent the sound of vowels in the Phoenician system. The deciphering process of Phoenician – and of all subsequent variations on the West Semitic model – relied on the oral context to supply the missing vocalic elements. By supplying these signs in the line of writing, the Greeks eliminated the need to contextualize the script orally, but reinforced the need to rely on the proper sequencing of the phonemes. Because the emphasis changed from feature detection to sequence detection, the visual field preference changed. The change of visual field preference forced the re-lateralization of a script which, at first, was written like its model, the Phoenician system, right-to-left.

The last chapter, by David Olson, opens the way for further investigations by exploring some of the cognitive and epistemological changes that can be correlated with the spread of literacy in the culture. His argument is that literacy brings about a cognitively operant category distinction between what is given and what is interpreted. "Writing," suggests Olson, "creates the problem of meaning," which in turn creates conditions for thinking. After showing how writing affects the form and the use of memory, Olson expands his observations to the realm of modern communication media, including the computer.

Olson's paper is not a conclusion to this book, but suggests a new beginning. Indeed, the next stage of investigations in line with the hypothesis is to find out precisely what cognitive operations that are dependent on writing are biased by the characteristics of the particular orthography being considered. This investigation may require a new team of interdisciplinary researchers.

References

Æschylus (1973). *Complete works* (2 vols). Cambridge, MA: Harvard University Press (Loeb Classical Library).

Barbu, Z. (1972). *Aspects of psychological history.* London: Basil Blackwell.

Changeux, J.-P. (1983). *L'homme neuronal.* Paris: Fayard.

Cohen, M. (1958). *La grande invention de l'écriture et son évolution.* Paris: Imprimerie Nationale.

Cole, M., & Bruner, J.S. (1971). Cultural differences and inferences about psychological processes. *American Psychologist, 26,* 867–876.

Cole, M., & Griffin, P. (1980). Cultural amplifiers reconsidered. In D.R. Olson (Ed.). *Social foundations of language and thought: Essays in honour of J.S. Bruner.* New York: Norton.

Derrida, J. (1976). *Of Grammatology,* (G.C. Spivak, Trans.). Baltimore: John Hopkins Press.

Diringer, D. (1948). *The alphabet: A key to the history of mankind.* London: Hutchinson.

Driver, G.R. (1954). *Semitic writing: from pictograph to alphabet.* London: Oxford University Press.

Eisenstein, E. (1979). *The printing press as an agent of change* (2 vols). Cambridge: Cambridge University Press.

Etiemble, R. (1973). *L'écriture.* Paris: Gallimard.

Febvre, L., & Martin, H.J. (1958). *L'apparition du livre.* Paris: Albin Michel.

Février, J.G. (1959). *Histoire de l'écriture* (2nd ed.). Paris: Payot.

Gelb, I.G. (1963). *A study of writing: The foundations of grammatology* (2nd ed.). Chicago, IL: Chicago University Press.

Goody, J. (1977). *The domestication of the savage mind.* Cambridge: Cambridge University Press.

Goody, J. (1986). *La logique de l'écriture: aux origines des sociétés humaines.* Paris: Armand Colin.

Goody, J., & Watt, I. (1963). The consequences of literacy. *Comparative Studies in Society and History, 5,* 304–345.

Goody, J., & Watt, I. (1968). *Literacy in traditional societies.* Cambridge: Cambridge University Press.

Graff, H.J. (1981). *Literacy and social development in the West.* Cambridge: Cambridge University Press.

Graff, H.J. (1986). The history of literacy: towards the third generation. *Interchange, 17,* 122–134.

Hamilton, E., & Cairns, H. (Eds.), (1980). *Plato. The collected dialogues.* (Trans. R. Hackforth et al.). Bollinger series. Princeton, NJ: Princeton University Press.
Havelock, E.A. (1963). *Preface to Plato.* Cambridge, MA: Harvard University Press.
Havelock, E.A. (1982). *The literate revolution in Greece and its cultural consequences.* Princeton, NJ: Princeton University Press.
Havelock, E.A. (1986). The alphabetic mind: a gift of Greece to the modern world. *Oral Tradition, 1,* 134–150.
Humphreys, G.W., & Evett, L.J. (1985). Are there independent lexical and non-lexical routes in word processing? An evaluation of the dual-route theory of reading. *The Behavioral and Brain Sciences, 8,* 689–740.
Innis, H.A. (1951). *The bias of communications.* Toronto: Toronto University Press.
Innis, H.A. (1972). *Empire and communications* (2nd ed.). Oxford: Oxford University Press, Toronto: Toronto University Press.
Jackson, D. (1981). *The story of writing.* New York: Taplinger.
de Kerckhove, D. (1984). Effets cognitifs de l'alphabet. In D. de Kerckhove & D. Jutras (Eds.), *Pour comprendre 1984* (Occasional Paper N49, pp. 112–129). Ottawa: UNESCO.
Lafont, R. (Ed.) (1984). *Anthropologie de l'écriture.* Paris: Centre Georges Pompidou.
Larsen, M.T. (1986). Writing on clay: from pictograph to alphabet. *The Quarterly Newsletter of the Laboratory of Comparative Human Cognition, 8,* 3–9.
Liggett, S. (1984). The relationship between speaking and writing: an annotated bibliography. *College Composition and Communication, 35,* 334–344.
Martin, H.-J. (1968–1970). Le livre et la civilisation écrite (3 vols.). Paris: Ecole nationale supérieure des bibliothèques.
McLuhan, H.M. (1962). *The Gutenberg galaxy: the making of typographic man.* Toronto: Toronto University Press.
Olson, D.R. (1977). From utterance to text: the bias of language in speech and writing. *Harvard Educational Review, 47,* 257–281.
Olson, D.R. (1986). The cognitive consequences of literacy. *Canadian Psychology/Psychologie canadienne, 27,*109–121.
Ong, W.J. (1967). *The presence of the word.* New Haven: Yale University Press.
Ong, W.J. (1971). *Rhetoric, romance and technology: studies in the interaction of expression and culture.* Ithaca: Cornell University Press.
Ong, W.J. (1982). *Orality and literacy: the technologizing of the word.* London: Methuen.
Plato (1961). In E. Hamilton & H. Cairns (Eds.), *The collected dialogues of Plato.* Princeton: Princeton University Press.
Rousseau, J.J. (1749). Essay on the origin of language (Trans. J. Moran). New York: Frederic Ungar, 1966.
Sampson, G. (1985). *Writing systems: a linguistic introduction.* Stanford, CA: Stanford University Press.
Scollon, R. (1985). Language, literacy and learning: An annotated bibliography. In D.R. Olson, N. Torrance, & A. Hildyard (Eds.), *Literacy, language and learning* (pp. 412–426). Cambridge: Cambridge University Press.
Scribner, S., & Cole, M. (1981). *The psychology of literacy.* Cambridge: Harvard University Press.
Sinatra, R., & Stahl-Gemake, J. (1983). *Using the right brain in the language arts.* Springfield, IL: Charles Thomas.
Snell, B. (1953). *The discovery of mind.* London: Basil Blackwell.
Stock, B. (1983). *The implications of literacy.* Princeton: Princeton University Press.
Stone, L. (1969). Literacy and education in England, 1640–1900. *Past & Present, 42,* 69–139.
Taylor, I., & Taylor, M. (1983). *The psychology of reading.* New York: Academic Press.
Vygotsky, L.S. (1978). *Mind in society: The development of higher psychological processes.* Cambridge, MA: Harvard University Press.
Woodbury, L. (1983). The literate revolution: a review article. *Classical Views/Echos du monde classique, 27,* 329–352.

Part 1 Biological Foundations

Introductory Remarks

We begin, perhaps properly, with beginnings. Recent studies have given a new view of the Darwinian mechanisms that created mind and culture in their idiosyncratically human form. The process, termed gene-culture coevolution, is reviewed by Lumsden and placed in relation to studies of brain and language. Somewhat surprisingly, the vocabulary of Darwinian thought, with its stress on competition and selection, has also proven highly productive in the neurosciences, where it is providing new models of what goes on in the brain during learning and thinking. Changeux and Finkel, representing the two leading approaches to the brain as a selectional system, highlight the data and ideas and focus these on basic cognitive actions like learning and category formation, which appear to be fundamental to a detailed understanding of the relationships between alphabet and brain. The implication we draw from these chapters is that the time has come to begin placing studies of literacy and writing on the foundations provided by theoretical neuroscience and population biology.

CHAPTER 1

Gene-Culture Coevolution: Culture and Biology in Darwinian Perspective *

CHARLES J. LUMSDEN[1]

Introduction

It has become clear that the idiosyncratic properties of the human mind have a powerful effect on the evolution of culture. In some reciprocal manner as yet less easily grasped, culture has influenced the genetic evolution of the brain structures underlying the mind. Because this coevolution concerns both the biological and social sciences, efforts to construct provisional models and unify the still fragmentary evidence to explain it are of more than ordinary interest. With respect to the genesis of our own species, evolutionary biologists have begun to grapple with the fact that human beings were not created by the purely Darwinian evolution so relevant to other species. For the past several million years our ancestors have been shaped by biological evolution and cultural evolution proceeding together, in a manner still little understood. In each of 200000 or so generations two tracks of heritable information, one genetic and the other cultural, have met in the events of individual socialization. During this time biology does not appear to have overwhelmed culture and culture has not granted biology total independence from human affairs. The relationship resembles one of reciprocating interaction, in which culture is generated and shaped by biological imperatives while the course of genetic evolution shifts in response to cultural innovations.

Most of biology is concerned with "how" questions: how cells divide, how action potentials are generated, how genes prescribe information. Evolutionary biology concentrates on "why" questions: why action potentials are generated in a certain way or why the rules of language acquisition have a certain form rather than some other. The scientific query "why" can be answered only by the study of history, and the history of biological processes is by definition evolution. The creative process of genetic evolution is natural selection, the differential transmission across generations of genes that affect

---

* Research support was provided in part through the Natural Sciences and Engineering Research Council of Canada (NSERC), grant number A0393, and the Medical Research Council of Canada (MRC), grant number MA-8635. The author is a career Scholar of the MRC.

[1] Sociobiology Research Group, Department of Medicine, University of Toronto, Room 7313, Medical Sciences Building, Toronto, Ontario, M5S 1A8, Canada.

survival and reproduction. Evolution at times proceeds by means other than natural selection. Mutations can occur at such high frequency as to push up the percentage of mutants in a population without the aid of natural selection. Alternatively, immigrants can bring new genes into a population at a high enough rate to change a population's overall genetic composition. If the population is a small one, random sampling effects caused by mating choices and genetic recombination can change the gene frequency in an unpredictable fashion from generation to generation. It appears, however, that these various forces may in general be less potent than natural selection in directing evolution over long periods of time. Present evidence suggests that natural selection is a significant force shaping the coevolution of biology and culture (Lumsden & Wilson, 1981 a, b).

Language, with its extensions to diverse systems of reading and writing, ranks high on the list of prime evolutionary innovations that distinguish the genus *Homo.* Its creation allowed the maturation of *protoculture,* the social transmission of behavior through observational learning, into true culture: the networks of knowledge, meaning, and value that constitute our most remarkable means of adaptation. The sustained attempt, which we undertake in this book, to study the relations between mental development, brain operation, and historical changes in writing forms therefore begins naturally with a discussion of the basic gene-culture process.

The Fabric of Memory

Human memory tends to organize both continuous and discontinuous impressions into discrete clusters. Experimental studies have shown that these cuts are made around objects or abstractions that have the most attributes in common and share the fewest attributes with other objects or abstractions. They appear to be of a size that maximizes efficiency in storage and transfer of information (Brunswick, 1956; Rosch, Mervis, Gray, Johnson & Boyes-Braem, 1976). Hence while categories such as "fruit", "fish", and "furniture" do not exist in the real world, they comprise recognizable collections of objects that share an unusually large number of stimuli most easily processed by the mind. Children move naturally into this mode of memory formation, performing equally well on tasks involving objects and those involving collections of objects. Thus they organize the diagnostic stimuli into ensembles (such as "cookies" vs "crackers" and "bunch" vs "pile") that are as sharply distinguished as the objects themselves (Markman & Seibert, 1976).

The brain speeds processing further by compounding the clusters hierarchically into larger assemblages. A convenient classification of the levels of clustering is the following (Wickelgren, 1979; Horton & Mills, 1984). The units of memory, which are experienced as objects or abstractions, are appropriately called *nodes,* aligning the description to the nodes and links between nodes envisioned in spreading-activation models of memory storage and recall (Anderson, 1983). At least three levels of nodes can then be recognized. Concepts, the most elementary clusters, are frequently tagged by words or phrases (such as "dogs" and "hunt"). Vocalizable propositions are signaled by phrases, clauses, or sentences expressing objects and relations ("Dogs hunt."). Finally, schemata having lexical equivalents are signaled by sentences and larger units of text ("the technique of hunting with dogs"). Node-link structures were originally proposed by psychologists as theoretical representations, but they have gained considerable credence through methods that detect their organization. The main steps in their growth are not merely the accidents of personal lifeways, but general processes that possess regularity across cultures (Dasen, 1972). Hence, in a manner important for the entire relation of biology and culture, the semantic mechanisms of culture formation are robust and consistent, generally more so than the final products they generate.

For each concept the mind tends to dwell on a prototype pattern that constitutes the standard, such as a particular wavelength and intensity to form the idealized color red or a particular body shape and size to form the typical shark (Rosch, 1973, 1975; Medin & Smith, 1984). Given an array of similar variants, the mind can construct a standard near the average of the variants and use it as the prototype even without having perceived any example of it directly (Posner & Keele, 1968; Kagan, 1984). The most important result for gene-culture coevolution is that the divisions, however "fuzzy," are created and labeled, even when the stimuli being processed vary continuously. In short, the mental development in humans tends to impose a semidiscrete, hierarchical order upon the world.

Most of the concepts comprising the basic entries within memory are subject to purely phenotypic variation arising from the particularities of cultural history. There is nevertheless a tendency for those belonging to at least a few categories to occur consistently across cultures. As Rosch (1975) has shown, such categories include elementary geometric forms (square, circle, equilateral triangle), the facial expressions of six basic emotions (happiness, sadness, anger, fear, surprise, disgust), and the basic colors (red, yellow, green, blue).

The Logic of Mental Epigenesis

Directed Cognition

Epigenesis is a biological term that refers to the total process of interaction between genes and the environment during development. During *mental* epigenesis, information acquired through socialization is used in the development of the mind and its contents, including the knowledge structures accessed through memory recall. Information encoded in the genes guides and shapes this development. The logic of the way this genetic information operates is often expressible in the form of rules which constrain various options or alternative pathways in psychological development (Simon, 1979; Wexler & Culicover, 1980; Berwick, 1982). It has been proposed that the term "epigenetic rules" be used to refer to these interacting components of the developmental logic (Lumsden & Wilson, 1981 a, 1983). Connection with evolutionary genetics is made by noting the theoretical possibility that changes in a gene can alter one or more epigenetic rules and the relations among them. In physiological terms, the epigenetic rules of cognitive and behavioral development comprise one or more elements in a complex sequence of events occurring at various sites in the nervous system. Our analysis of the available data (Lumsden & Wilson, 1981 a, 1982, 1983; Lumsden, 1984 a) has led us to conclude that these elements are crudely but usefully separated into two main categories: primary epigenetic rules, which regulate the development of systems ranging from the peripheral sensory filters to perception, and secondary epigenetic rules, which assemble the inner mental processes (Fodor, 1983), including the procedures of consciously deliberated valuation and decision making.

The epigenetic rules embody an innate part of the individual's strategy for learning culture. It is theoretically possible for this culture learning to belong to one of three main classes of evolutionary strategies (Lumsden & Wilson, 1981 a; Lumsden, 1984 a): *pure genetic transmission,* in which innate epigenetic rules prescribe essentially one developmental response to any culture trait or array of traits (hence, an entirely genetic culture is a theoretical possibility); *pure cultural transmission,* in which the epigenetic rules prescribe genetically unbiased use of any culture trait in competition with others in organizing mental development (this is the traditional viewpoint of cultural determinism of tabula rasa individuals, widely applied in the social sciences; e.g., Harris, 1968; Freeman, 1983 for review); and *gene-culture transmission,* in which the innate epigenetic rules discriminate multiple culture traits and are more likely to use some rather than others. The term "gene-culture" in this context is not meant to reiterate the truism that both genes and culture somehow influence human development. Rather, it describes the presence of epigenetic rules that predispose mental development to take certain specific directions in the presence of certain kinds of cultural information.

On the basis of present evidence it appears that much if not most of human culture is sustained by gene-culture transmission rather than by pure cultural transmission. Whenever detailed studies have been conducted of development as mediated by choice among or directedness toward empirically distinguishable culture traits, they have almost always revealed an innate bias favoring some over others. Examples include a neonate preference for sugar combined with an active aversion to salty and bitter flavors (Maller & Desor, 1974; Chiva, 1979), affecting the evolution of cuisine; the innate discrimination for four basic colors (red, yellow, green, blue) (Bornstein, Kessen & Weiskopf, 1976a, b) and a greater ease of learning color classifications clustered on these color modes (Rosch, 1973, 1975), affecting the evolution of color-term systems (Berlin & Kay, 1969; Lumsden, 1985; see further discussion below); infant phoneme discrimination, affecting later speech structure (Eilers, Wilson & Moore, 1977); infant preference for certain kinds of visual patterns (Hershenson, Munsinger & Kessen, 1965; Fantz, Fagan & Miranda, 1975) regulating attention and arousal; neonate preference for normally composed facial features (a bias manifested within 10 minutes following birth) (Freedman, 1974), and locomotor patterns (Fox & McDaniel, 1982), orienting the infant learner toward human sources of information; smiling and other specific forms of nonverbal communication (Eibl-Eibesfeldt, 1979), facilitating the development of bonding, reciprocity, and communication; nonverbal signals used in mother-infant bonding, inducing long-lasting affects in later maternal care (Klaus, Jerauld, Kreger, McAlpine, Steffa & Kennel, 1972; DeCasper & Fifer, 1980); sexual differences in the carrying of infants and other larger objects (Salk, 1973; Lockhard, Daley & Gunderson, 1979); the fear-of-stranger response (Morgan & Ricciuti, 1973); the predisposition to acquire phobias against certain dangerous objects, such as heights, running water, and snakes, but not other dangerous objects, including electric sockets and guns (Marks, 1969); the development of sexual preferences within the family (Shepher, 1971, 1983; Wolf, 1968; Wolf & Huang, 1980; van den Berghe, 1980, 1983) affecting adult mating behavior and social structure; the size and operating speeds of long-term memory and short-term memory, affecting the choice of strategies in conscious deliberation and problem-solving (Simon, 1979); the development of linguistic knowledge (Chomsky, 1980; Wexler & Culicover, 1980), ontological knowledge, and knowledge about numerosity (Keil, 1979, 1981).

That the epigenetic rules have a genetic basis is indicated by several lines of evidence. Some appear during early childhood and are relatively inflexible. In addition, pedigree analysis and standard comparisons of fraternal and identical twins, in some instances strengthened by longitudinal studies of development, have yielded evidence of genetic variance in virtually every category of cognition and behavior investigated by these means, including some that either constitute epigenetic rules or share components with them. These categories include color vision, hearing acuity, odor and taste discrimination, number ability, word fluency, spatial ability, memory, timing of language acquisition, spelling, sentence construction, perceptual skills, psychomotor skill, extroversion/introversion, homosexuality, proneness to alcoholism, age of first sexual activity, timing of Piagetian developmental stages, some phobias, certain forms of neuroses and psychoses, and others (for reviews see McClearn & Defries, 1973; Loehlin & Nichols, 1976; Wilson, 1978; Lumsden & Wilson, 1981a, 1983). Single gene variants have been identified that affect certain cognitive abilities selectively (Ashton, Polovina & Vandenberg, 1979), as well as the ability to discriminate certain odorants

(Amoore, 1977). It has also become apparent that mutations at a single locus can result in profound but highly specific changes in the architecture and operation of brain tissues such as mammalian neocortex (Caviness & Rakic, 1978; Rakic, 1979). Not only do these alterations modify behavior at the locomotory and perceptual levels, they also introduce changes into such higher level functions as choice and decision (e.g., Bliss & Errington, 1977).

Gene-Culture Linkage

The conclusions about the existence of innate biasing epigenetic rules for gene-culture transmission are consistent with the recent theoretical finding that the tabula rasa state of pure cultural transmission tends to be unstable in evolutionary time (Lumsden & Wilson, 1981a). During the process of gene-culture coevolution a population of tabula rasa organisms, even if present initially, is very likely to evolve rapidly into a condition where the ancestral phenotypes have been replaced by organisms equipped for gene-culture transmission. In a cultural species the genetic fitness of an organism is affected not only by its genotype but also by its cultural heritage as expressed by a subset of cultural information that is allowed to affect development. The genetic fitness is influenced by the pathway of enculturation that the organism follows, and is enhanced by any tendency of mental epigenesis to use culturally transmitted information that confers greater relative genetic fitness. The innate epigenetic rules of gene-culture transmission provide this capability, guiding the organism to incorporate or respond to sets of relatively advantageous information more often than sets that are relatively deleterious.

Consider, in contrast, a population of tabula rasa organisms which alter the degree of influence over development exercised by cultural information without reference to the consequences for genetic fitness. The population exists in an environment that will in general contain both adaptive and less advantageous culture traits, but it is unable to distinguish between them. Moreover, its members are susceptible to cultural programming that could at whim shape their preferences to favor deleterious behavior. Over a period of generations the population is unstable against invasion by genetic mutants that set innate biases toward adaptive culture traits (i.e., traits which facilitate survival and reproduction). Because of their particular enculturation, such developmentally biased organisms outcompete the tabula rasa design, leaving more offspring generation after generation, until eventually they constitute virtually the entire population. This inference can be termed a principle of gene-culture linkage: genetic evolution operates in such a way as to influence both culture and individual cognitive development. A relation between individual genetic fitness and a choice of behaviors, predicted on the basis of this leash principle, has been explicitly documented in a wide array of real behavioral categories, including diet (Gajdusek, 1970), body marking (Blumberg & Hesser, 1975), sexual conventions (Daly & Wilson, 1978), marital customs (Daly & Wilson, 1978), economic practice (Irons, 1979), achieved socioeconomic status (Mealey, 1985), and others.

Some culture traits undeniably provide superior genetic fitness over others, but how is this possible? Survival and reproduction are not the products of an artifact lying in a campsite or an idea circulating in the recesses of long-term memory. They are fixed by explicit behavior, by muscular contraction and motion of parts of the body. The human mind intervenes to pose new strata of process and transformation between enculturation and explicit behavior. Mental activity and outward behavior are based on the knowledge structures that make up the contents of the various cognitive domains. But if it is the case that culture and the development of the human mind are sustained by gene-culture transmission, then the knowledge structures are psychological entities built up in forms governed by epigenetic rules. When organisms are predisposed to form certain mental structures and operations as opposed to others, the result is directed cognition.

In principle, the achievement of directed cognition through a gene-culture linkage can be the result of one or more processes. First, it could result in part from sensory screening, which limits perception to narrow windows opening on the vast arrays of physical stimuli impinging on the body. Second, it could follow from a tendency for certain node-link structures to take form and link preferentially with others in semantic memory, including those related more directly to activities in the limbic and brain reward systems, so that they are more likely to become differentially associated with particular informational, valuational, and emotive constructs. For example, bonding results from the virtually automatic positive linking of mother and infant during their initial contacts, whereas snakes, heights, and other typical subjects of human phobias are likely to acquire negative valuation and become tagged as objects of avoidance behavior. Third, directed cognition could be induced by constraints on achievable cognitive design, biasing development toward certain parameters of information processing capacity rather than others. Thus the symbol capacity of short-term memory is on the order of three to seven elements or chunks, while the comparatively infinite store of long-term memory admits new elements much more slowly than it allows them to be retrieved (Newell & Simon, 1972; Simon, 1979). The effects of these characteristics on the selection of search and computational strategies have been documented in many areas of human reasoning and problem solving (Larkin, McDermott, Simon & Simon, 1980; Newell & Simon, 1972; Simon, 1979, 1981). Overall, the canalization in cognitive development leads to a substantial convergence of the forms of mental activity among the members of societies and even among peoples belonging to different cultures (for reviews see Hallpike, 1979; Lumsden & Wilson, 1981 a; Williams, 1972). Idiosyncracies in concept formation and all other aspects of development obviously distinguish one human being from another. But the epigenetic rules of gene-culture transmission appear to be sufficiently specific to produce a broad overlap in mental activity and behavior of all individuals, and hence a convergence powerful enough to be labeled human nature.

The Structure of Culture

In gene-culture transmission behaviors and patterns of thought are not determined genetically. The human mind develops them from information it acquires during socialization. This information, combined with its shifting patterns of meaning and significance, comprises culture in the sense I shall consider here. Although the details vary (Kroeber & Kluckhohn, 1963), there is some agreement among anthropologists that culture is an ordered system of socially transmissible information encoding behavioral, affective, and cognitive characteristics of a social group, including such aspects as skills, knowledge, attitudes, beliefs, myths, and rituals (Kroeber & Kluckhohn, 1963; Schwartz & Ewald, 1968; Schneider & Bonjean, 1973; Keesing, 1976). This description of culture is often extended to include the actual lifeways and material artifacts that are the overt expressions of the transmissible information. In this chapter I will be concerned mainly with the informational aspects per se.

An extensive literature summarizes evidence that human cultures, as information systems, possess an inner order (for overviews see Murdock, 1949, 1967; Naroll & Cohen, 1970; Taylor, 1973). The discourse and body movements experienced in tasks, rituals, demonstrations, and other means of transmission can be hierarchically clustered as sets of phonemes, words, sentences, stories, and analogous components of motor patterns (Hutchinson, 1970; Laban, 1975). Considerable attention has focused on the elements of these various sets as natural units or building blocks of culture. In culture theory previous authors have referred to units of culturally transmitted information in diverse ways, such as mnemotype, idea, idene, sociogene, instruction, culture type, meme, and concept. (The history of the terminology is discussed more fully in Lumsden & Wilson, 1981 a.) This discussion has been further stimulated by the findings of Shannon (1948) and later sudents of information theory (Goldman, 1953; Brillouin, 1962; Khinchin, 1957; Gatlin, 1972; Chaitin, 1975), in which systems of transmissible information are objectified and ultimately ordered into quantifiable patterns composed of basic units.

This important problem of the existence of units can be resolved at least in part by considering the mechanics of gene-culture transmission. In order to understand the necessary steps, consider first an analogy from molecular biology. The genome is commonly envisaged as consisting of parts, the genes, that have an objective existence. However, inspection of an actual nucleotide sequence reveals only a continuous genetic text lacking obvious breaks or pauses (Fig. 1). Human observers are free to develop any number of useful decompositions or classifications for such sequences of nucleotide letters. But there is one decomposition whose biological meaning is known to be of the greatest significance, namely, that corresponding to the genetic code. By following this code, the cell's transcription-translation apparatus reads certain triplets of adjacent nucleotide letters as "start" (of gene) and "stop" (end of gene). (The others it associates with specific amino acids.) This decoding operation, which sustains the flow of genetic information from DNA through RNA to protein, in effect parses the genetic text into natural units. These are the gene sequences that code for distinct polypeptides within the living cell.

The relation between the epigenetic rules and culture appears to be similar to that between the cellular transcription-translation apparatus and DNA. During individual mental development, culture is scanned by the epigenetic rules. The rules respond to specific cues or patterns within the culture, and then only in specific ways. These cues and patterns act as detectable signals, influencing development. The particulate nature of their coding activity is revealed partly in the discrete node-link elements of memory, described earlier. Thus while culture might be decomposed in a great many different, academically useful ways, there is one *natural* decomposition of culture sustaining mental development and cultural evolution: that produced by its interaction with epigenetic rules. It is important to note that this characterization does *not* state that culture consists a priori of atomic units or symbols. Rather, such units are the emergent result of culture experienced as a whole by the individual.

CTCTGTGACGATGACCCGCCAGAGATCCCACACGCCACATTCAAAGCCATGG
CCTACAAGGAAGGAACCATGTTGAACTGTGAATGCAAGAGAGGTTTCCGCAG

Fig. 1. *Top:* Part of the genetic text for the gene encoding the interleukin-2 receptor in humans. The receptor molecule is situated in the external membrane of certain cells participating in the immune response. When interleukin-2 is present, cells with interleukin-2 receptors proliferate and can take part in the body's immune defense system. Away from the context of the genetic code, the interpretation instantiating the beginning, end, and order of the receptor's sequence of individual amino acids is not evident (nor, on a larger scale, is the sequence of distinct genes within the genome as a whole). The genetic text looks continuous. The nucleotide sequence shown here is part of a longer encoding reported in Leonhard et al. (1985). *Bottom:* Segment of the hieroglyphic text of the decree of Canopus, ca. 237 B.C., drawn after Budge (1929). The decree was also recorded in demotic and Greek versions. The lines shown record the decision of the Egyptian priesthood to honour the deceased Princess Berenice with rites like those offered to the daughter of RA, the sun-god. Away from the context of a decoding semantics, the sequence of distinct ideas and events recorded in the text is not evident. The stream of symbols appears more or less continuous

It would seem that the activity of epigenetic rules in mental development makes it possible to take the crucial steps delineated by Rosenberg (1980) for building more powerful middle-range theory in the social sciences: identification of natural units and the linkage of explanations across disciplines through the relation of these elements. It also encourages us to make a careful distinction between two conceivable kinds of units: *structural* units, which are phenotypic traits detectable within larger patterns, and *generative* units, which are structural units that act through a coding process to generate other structures. Just as genes are generative units for the RNA and (ultimately) protein, epigenetic rules extract generative units from culture and turn them into knowledge structures and mental representations, culture's inward manifestations.

The question of greatest immediate interest is therefore the nature of the generative unit. It appears possible that this element, which in earlier studies we have called the "culturgen" (Lumsden & Wilson, 1981 a, b, 1983), can be equated to the node of semantic memory. The level of the node, whether concept (the most elementary recognizable unit), proposition, or schema, determines the complexity of the generated behavior or artifact maintained in the culture. For example, the differentiation of letters in an alphabet is at the level of the concept, the initial verbal reaction to a stranger is a proposition, and the lexical expression of an incest taboo is a schema.

Although a direct correspondence between nodes and generative units of culture appears feasible at lower levels of organization, there is no reason to expect the more complex constructions of culture to be mapped onto semantic nodes in a one-to-one fashion. Subjects of cultural anthropology such as marriage ceremonies and the architecture of temples are the outcomes of numerous interlocking behaviors that result from cognitive activity accessing multiple knowledge structures. These in turn vary according to the particularities of local history. Nevertheless, each can be realistically interpreted as the outcome of human mental development, which involves the assembly of knowledge structures and their processing mechanisms. Cultural evolution then involves the shifting of the outer phenotypes of behavior and artifacts through the insertion and combination of their basic generative structures into human thinking.

The epigenetic rules of mental development determine the manner in which the nodes are created and linked to form semantic networks. These physiological processes impose a strict filtering of stimuli from the environment and alter each event of mental life thereafter, from short-term memory and storage in long-term memory to recall, feeling, reverie, and decision-making. As a consequence the difference between genetic and cultural inheritance can be expressed in terms of two strategies of information flow. In genetic inheritance the basic information, encoded in DNA, is directly replicated and transcribed unidirectionally out through RNA and protein. The information in cultural inheritance, encoded in the semantic networks, cannot replicate itself directly and must generate culture in order to reproduce. The process is the logical equivalent of the reverse transcription forbidden in genetic inheritance. The two systems are linked in the following manner: because of the intervention of the epigenetic rules, which have a genetic basis, culture is based not just on learning but also, ultimately, on the particular structure of the DNA.

Coevolutionary Mechanisms

Modern theories of gene-culture interaction can be sorted on the basis of degree to which the intervening psychological mechanisms are described and Darwinian factors involving survival and reproduction explicitly considered. In my view evolutionary theories that do not explicitly deal with thinking and mental development cannot produce adequate descriptions of the human case. For humankind, genetic determinism shapes (at most) the epigenetic rules of mental development, not finished thoughts or behaviors. The feasibility of producing useful formal theories based on this perspective has been investigated in our laboratory (e.g., Lumsden & Wilson, 1981 a, b, 1983; Lumsden 1983, 1984 a, b; Lumsden & Gushurst, 1985), with encouraging initial results. Our findings extend and in some cases revise inferences obtained from earlier methods of behavioral genetics and sociobiology. Comparison of possible approaches (e.g., Lumsden & Wilson, 1981 a) indicates that between genes and culture two particular intermediate levels of organization are especially pertinent to the correct formal delineation of the linkage. They are also especially accessible in terms of available empirical data and theoretical representation (Fig. 2). The first is cellular development in the brain, leading to the formation of nerve cell circuits. The second is mental development, in which the activity of these circuits builds mental structures and the capacity for overt behavior.

The transition from genes to culture can therefore be represented in initial approximation by a progression around a circuit of interactions rather than by steps up a ladder of independent levels of organization. The feedforward from genes to culture, via individual development and behavior, combines with the feedback from culture to genes: acts of overt behavior in particular settings are the beginnings of large-scale social phenomena. Culture and macrosocial patterns in turn form a principal part of the environment in which human genetic evolution takes place (Holloway, 1966; Tobias, 1981; Lumsden & Wilson, 1983; Lumsden, 1983). One of the advantages of such an approach to analyzing gene-culture interaction is its avoidance of a reductionistic strategy, and its replacement with a formal treatment of stepwise emergence, from genes through the mind to culture. The resulting task of theory construction is challenging, and only partially completed at the present time. To describe social as well as biological events one must account for the correlated activity of many individuals doing dif-

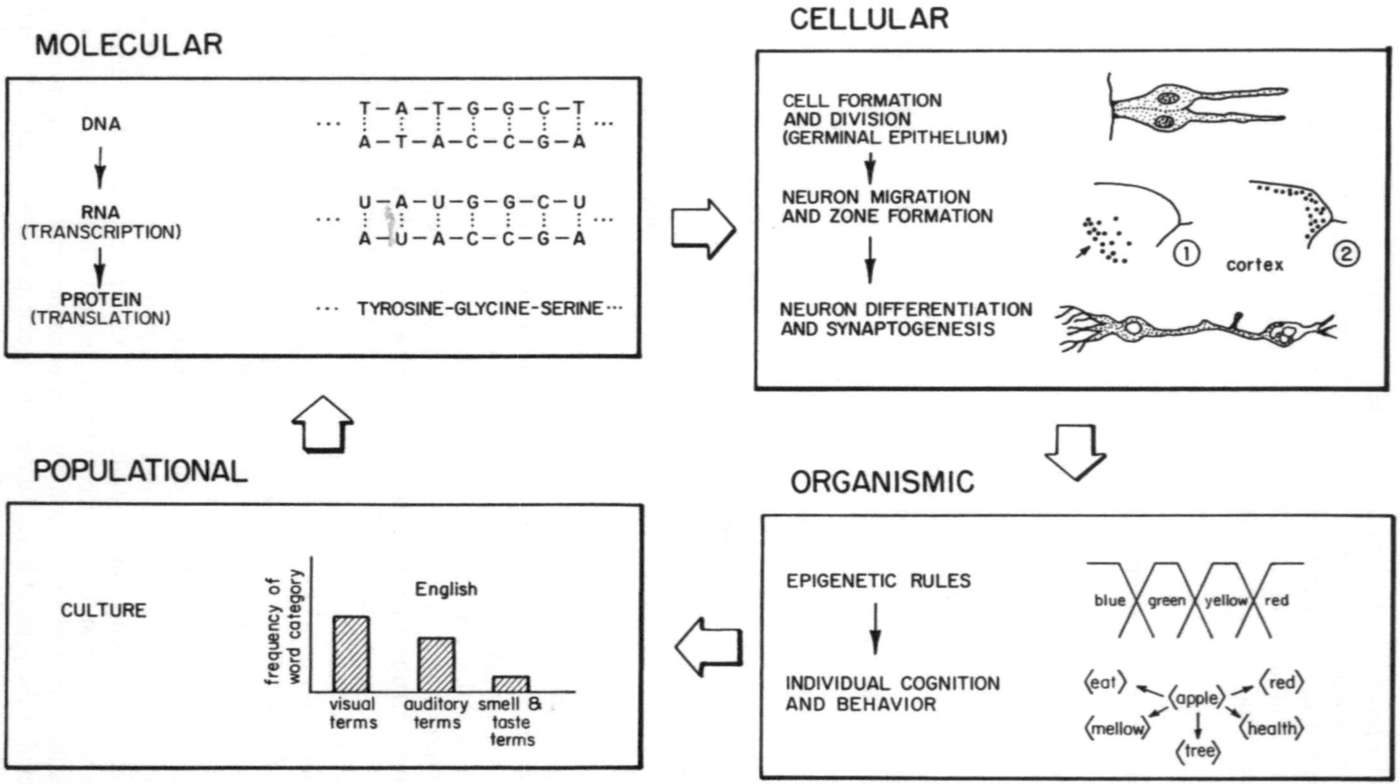

Fig. 2. The circuit of gene-culture coevolution showing the principal levels of interaction. (From Lumsden and Wilson, 1981 a; reprinted by permission)

ferent things in the presence of multiple traits of culture, for the prevailing social patterns, and for the feedback of these patterns to cognition. In evolutionary time, on the scale of societal history and genetic change, multiple genotypes with different epigenetic rules and cognitive properties interact with shifting cultures. The effects of fitness and other evolutionary forces can be obtained only if life histories embodying predictions about individual survival and reproduction are evaluated for each genotype in the interacting assembly.

A number of general findings have emerged from studies of such theoretical models, involving the entire coevolutionary circuit. The first is that even small innate biases in the epigenetic rules can be strongly amplified in the emergent patterns of sociocultural diversity. Analysis of the available data on infant development (e.g., Maller & Desor, 1974; Freedman, 1974; Lockhard et al., 1979) suggests that human epigenetic rules are in fact strong enough to create this type of effect. A second finding is the complement of the first: even when the options allowed by the epigenetic rules are rigidly constrained by genetic prescription, they can still generate wide cultural diversity. This variation is further enhanced by the cultural dependence of the epigenetic rules and is reinforced by the probabilistic nature of individual thought and behavior. In cases tied quite directly to survival and reproduction, such as incest, color vision, and certain instances of agricultural production, it has been possible to estimate the strength of the effects that genetic and cultural evolution exert on one another (e.g., Lumsden & Wilson, 1981a, 1983; Durham, 1982a).

A third result is a generalization of gene-culture linkage: the tabula rasa strategy of mental development, leading to cultural determination of individual thought and behavior, is generally found to be unstable. Under conditions that entail differential fitness and no transitions to allelic polymorphism under frequency-dependent selection (see below), genotypes prescribing a perfect lack of bias toward Darwinian-advantageous culture traits are rapidly replaced by genotypes prescribing a favorable bias and the maintenance of culture via gene-culture transmission. This result is consistent with the directedness observed to date in all classes of human epigenetic rules known to us. Although cultural evolution toward civilizations and other forms of complex societal organization (e.g., Lenski & Lenski, 1970) was previously thought to weaken such effects of biological evolution, this may not be the case. When a number of properties characteristic of later human social orders, such as stratification and increasing cultural complexity, are considered, they are found to favor the increase, rather than the decrease, of the number and complexity of epigenetic rules coding for gene-culture transmission (Lumsden, 1983).

A fourth finding is that the genetic evolution of human nature can proceed much more quickly than implied in earlier discussions. Even when selectional forces are only moderate and the innate directedness provided by the epigenetic rules is mild in comparison with that generally observed in human developmental studies, the rate of change of gene frequency can be high enough to achieve the near-replacement of one allele by another within as few as 20 – 50 generations, or about 1000 years of human history. Finally, because people use the decisions and behaviors of others as part of their own deliberations, natural selection of genotypes during gene-culture coevolution is in general frequency-dependent (Lumsden & Wilson, 1981 a, b; Lewontin, 1981): its direction and intensity are determined in part by the number of individuals using alternative culture traits and thus by the relative frequencies of the competing gene variants. Classical sociobiological arguments have usually assumed that some measure of fitness (such as inclusive fitness) is optimized during human evolution. However, if this finding of frequency dependence turns out to have wide validity, the basic assumption of genetic fitness optimization and the many models based on it may require extensive re-examination. Frequency-dependent selection may not lead to the optimization of a fitness measure at all. The pathways of both genetic and cultural evolution may exhibit behavior *much more complex* than an approach to a coevolutionarily stable optimum. A diversity of genetic and culturgenic variants can persist through time, even changing from generation to generation in a chaotic fashion (e.g., Lumsden & Wilson, 1981 a, pp. 298 – 302). For such systems arguments from optimization are wrong and must be replaced by methods using more basic principles of population biology.

Language

This is a book about language in neurocultural perspective. Our particular focus is a remarkable event, the reversal of writing direction, that occurred in the history of its written form among the early Greeks. It is of great interest that of all the cognitive domains, linguistic knowledge has most often been singled out as a paradigm of the manner in which the human mind differs from that of its primate ancestors (e.g., Hockett & Ascher, 1964). The older evolutionary literature abounds in stories about how a facility for language would have been helpful to earlier hominids. These well-known presentations are somewhat too numerous, and, at the present time, unfortunately, rather too "just so" in their content to warrant coverage here.

Traditional social science models have generally taken language to be the instrument through which culture imposes its idiosyncratic form on the human mind (e.g., Berger & Luckmann, 1966; Gleason, 1961; White, 1949; Whorf, 1956), but the combined evidence of psycholinguistics, neurobiology, and human genetics suggests that this extreme position should be approached with caution. Instead, it appears likely that the ontogeny of linguistic knowledge structures is the result of mental operations carried out on cultural cues under the guidance of a rich set of innate developmental instructions.

The signs of epigenetic rules operating in the domain of linguistic knowledge have been detected piecemeal by a number of investigators. Infants possess innate rules of speech perception that are adultlike and facilitate the development of language (Eimas, Siqueland, Jusczyk & Vigorito, 1971; Liberman, Cooper, Shankweiler & Studdert-Kennedy, 1967). Whereas variations in pitch are perceived as arrayed along a smoothly varying continuum, distinctions of voicing are automatically classified into categories – in this case, into phoneme clusters. For example, sounds ranging between /ba/ and /ga/ and /s/ and /v/ are clustered on the basis of perceived similarity into one or the other of these paired units. A principal component of the discrimination process is the voice onset time (VOT) for phonemes, which is the timing of the acoustical formants or energy bands relative to one another. The recognition of stop and fricative consonants, for instance, depends on the extent of the first formant and the direction of the second formant. This development of phoneme discrimination is channeled, moving through more than one stage during the first year of life or longer (Eilers et al., 1977). In the 11 languages surveyed by Lisker and Abramson (1964) one or more values along each VOT continuum served as reference points, dividing the continuum into two or three phonetic clusters. Ultimately a repertory of from 20 to 60 phonemes develops around the reference points, the total number varying according to culture.

Psycholinguists have pointed out the impressive speed and precision with which linguistic knowledge develops in children of all cultures. Some have contrasted this state of affairs with the seeming absence from childhood linguistic environments of the order and completeness required for a developmentally unconstrained inductive learner to achieve competence (e.g., Wexler & Culicover, 1980 and references therein). Put briefly, there is a "paradox of the poverty of the stimulus" (Chomsky, 1980): how can linguistic knowledge develop in children when their environments apparently lack information in the requisite quantity and form?

A possible answer is that their genes provide the missing clues by way of innate developmental constraints specialized for language acquisition. These channel the assembly of linguistic knowledge to successful completion (Chomsky, 1980; Keil, 1981; Pinker, 1981), and as a corollary have partially shaped language and its cultural evolution. Wexler and Culicover (1980) present empirical and theoretical evidence of the existence and universality of certain such constraints in human language systems. In important related work Berwick (1982) has built an acquisition front-end for the English-parsing software system PARSIFAL and demonstrated that, given similar types of constraints, it can acquire rules for parsing English syntax.

Linguistic theory is in a phase of rapid development, and it is possible that specific epigenetic rules will have to be rethought in the light of new developments. But already the work of Wexler, Culicover, Berwick, and their associates has provided a first glimpse of the epigenetic field in one of the most complex human cognitive domains shaped by gene-culture coevolution.

Language and Neurocultural Analysis: The Case of Basic Color Terms

It is one thing to warm to the task of defining epigenetic rules and neurocultural linkages, but is it realistic to envisage a mode of analysis of language systems in which the causal connections among genes, mind, and culture could be worked out in specific instances? The complexity of human language and its history requires a cautious response pending actual case studies. In this regard a recent report by Lumsden (1985) begins to demonstrate the method's efficacy.

The phenomenon considered was that of basic color terms. The categorization and naming of colors is known to be a cultural universal (Berlin & Kay, 1969; Kay & McDaniel, 1978). Although lexicon size varies, all human societies so considered have terms for colors or degrees of light and dark (Witkowski & Brown, 1977; Kay & McDaniel, 1978). The explanation of the ethnographic distribution of color terms and their evolution has been labeled an important anthropological issue (Berlin & Kay, 1969; Ratliff, 1976). The neurophysiology of color perception has also been extensively studied using human subjects and primate models (Mollon, 1982; Lennie, 1984). Discussion of ethnographic distributions of color naming has therefore taken on an increasingly biological tone (Witkowski & Brown, 1977; Kay & McDaniel, 1978; Ratliff, 1976; Bornstein, 1973; Whitfield, 1980). Moreover, the study of color categorization has recently been extended to very young infants prior to the onset of language acquisition (Bornstein, 1975, 1979, 1981; Bornstein and Marks, 1982). It is possible, therefore, to begin to compare the performance of humans who have not yet begun color-term socialization through language with that of fully enculturated adults. Qualitative correspondences between the color categories perceived by young infants and by the adult populations examined by psychologists have been previously noted (Bornstein et al., 1976a, b; Lumsden & Wilson, 1981a).

The study reported by Lumsden used the specifications of infant color categorization obtained by Bornstein and his coworkers (Bornstein, 1975, 1979, 1981). Sixteen-week-old infants were found to respond to variation in wavelength as though four basic color categories were being discriminated. Conventional terms for these categories would be *red*, *yellow*, *green*, and *blue*. The categorization was detected by measuring the span of the infants' attention to monochromatic lights ordered across the visible spectrum. Within a short time the infants habituated to the repeated presentation of the stimulus light. Recovery from habituation was strong only when changes in wavelength crossed certain wavelength values, which were judged to be the boundary regions between perceptual categories of color. Boundaries between the categories were mapped, together with the perceptual responses near the category centers.

To quantify the relationship between the infant data and the ethnographic observations, boundary values between infant categories were identified (Lumsden, 1985): *red-yellow* at 600 nm, *yellow-green* at 560 nm, and *green-blue* at 480 nm. The values given correspond to the peaks of maximum wavelength discriminability in the boundary regions (Bornstein, 1981) and to the midrange crossover points between the respective categories as measured by Bornstein et al. (1976 a, b). The width of the boundary regions is in all cases narrow: 20 nm for *red-yellow,* 10 nm for *yellow-green,* and 20 nm for *green-blue* (Bornstein et al., 1976 a, b).

The ethnographic observations were drawn from the data on color naming developed by Berlin and Kay (1969). In the Berlin-Kay study, native speakers of 20 languages were shown arrays of patches ordered in color and brigthness by the Munsell system. A list of basic color terms (terms operationally defined to include such characteristics as monolexemic structure and broad applicability) was elicited from each subject. Subjects were then asked to place each basic color term of their native language on the two-dimensional color chart by picking the color patch or patches that best represented each term. Except for the 40-member Tzeltal group, the subjects were bilingual in their native language and American English. The Tzeltal subjects varied from monolingual fluency in Tzeltal to Tzeltal-Spanish bilingualism. The close accord observed between the response patterns of the Tzeltal speakers and those of the other informants suggested that biases arising from bilingual ties to American English were not substantial. The center of gravity of the best exemplars was plotted by Berlin and Kay for each basic color term in each language.

The Berlin-Kay (BK) map was analyzed by Lumsden (1985) in quantitative terms. The Munsell brightness axis holds the dominant wavelength of the color patches approximately constant. To allow tests of correspondence with the four infant spectral categories, the BK color groups were clustered on the basis of vertical order to bring together all sample points in a given dominant wavelength range. This procedure led to BK red forming RED (R), BK brown, orange, and yellow forming YELLOW, (Y), BK green forming GREEN (G), and BK blue forming BLUE (B) (Fig. 3). Lateral boundaries for the clusters were those reported by Berlin and Kay for their original color groups (Berlin & Kay, 1969). Two data points from BK pink fell in the Munsell patch (R 5, 8) vertically aligned with BK red and were included in RED.

Values of the dominant wavelength and excitation purity (saturation of the color patch emission by the dominant wavelength) were determined for each data point observable on the BK map. So transformed into wavelength equivalents, the BK data were compared to psychological studies of infant categorization. The qualitative correspondence between the infant category structure and the BK ethnographic distribution proved to be quite marked (Fig. 4). The boundaries of the ethnographic clusters align closely with those of the infant categories. In Fig. 4 the histograms of sample points represent

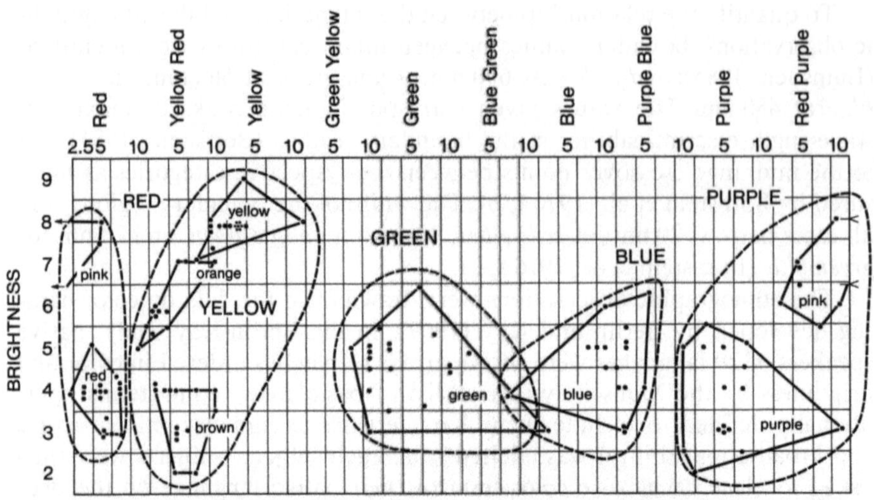

Fig. 3. Clustering of the Berlin-Kay color-term groups into four major spectral hue categories *Red, Yellow, Green, Blue, Purple* corresponds to a nonspectral purple and was omitted from the analysis (see text). *Solid polygons,* original groups according to Berlin and Kay (1969); *dashed lines,* proposed clusterings. (From Lumsden, 1985; reprinted by permission)

distributions of wavelengths that speakers of different languages judge to be best examples of particular colors. Thus this figure presents wavelength ranges that infants group as similar in hue against wavelength ranges that adults from different linguistic communities each label the same way.

To quantify the relationship, each of the data points on the BK map was assigned to an infant category on the basis of its dominant wavelength compared to the values selected for the boundaries between the infant categories. The results were organized into a 4 × 4 cross-tabulation of the infant categories and the categories of the BK ethnographic distribution. In this form the data showed a striking diagonal order, with most of the ethnographic BLUE points falling in the cell corresponding to infant *blue,* and similarly through GREEN, YELLOW, and RED. The strength of this association was highly significant: $\chi^2 = 246.5$, d.f. $= 9$, $p < 0.001$; Spearman's *rho* $= 0.88$; Kendall's *tau* B $= 0.86$. The measures of category association obtained in this and subsequent procedures indicate that between 75% and 95% of the variance in the membership in the ethnographic color categories is accounted for solely by the prelinguistic infant color categories.

These results suggest that the wavelength clustering of the BK ethnographic color term distribution can be parsimoniously explained on the basis of an epigenetic rule (Lumsden & Wilson, 1981a) serving color category development. This epigenetic rule involves procedures of color discrimination present since early infancy. Other findings also support such a conclusion. Bornstein has related the centers of the BK clusters to response

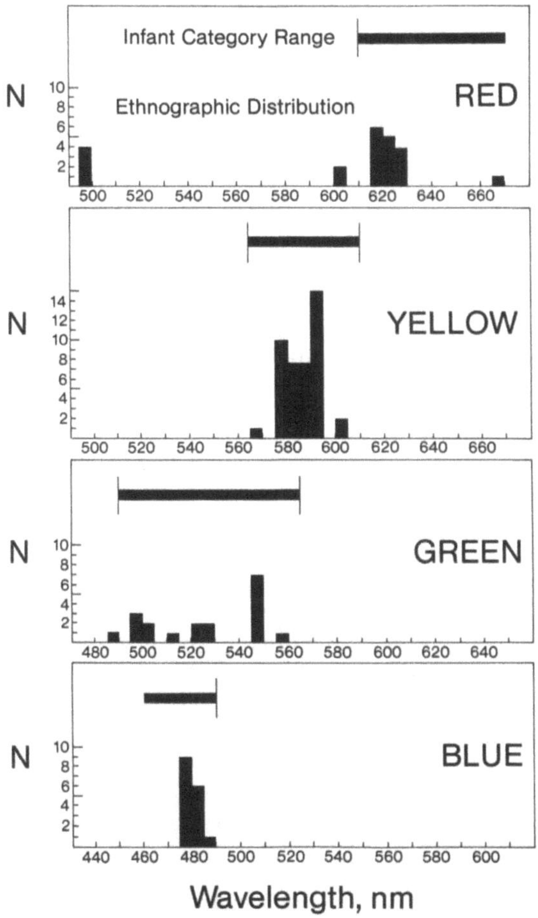

Fig. 4. Alignment of the infant color categories (Bornstein et al. 1976 a, b) and the Berlin-Kay ethnographic distribution of basic color terms. (From Lumsden, 1985; reprinted by permission)

peaks of neurons in the noncortical visual systems serving color processing (Bornstein, 1973). Similar correspondences have been documented at length by Kay and McDaniel (1978). Investigation of color categorization has also been extended to nonhuman primates. The macaque, whose visual system appears neurophysiologically similar to the human case, categorizes the spectrum into similar basic categories of *red, yellow, green,* and *blue* (Sandell, Gross & Bornstein, 1979). Whether these properties are shared on the basis of parallel adaptation or descent from a common ancestor remains to be determined.

Of course, the fact that the basic hue categories of infants, which resemble those of the macaque, account for most of the variance in the basic color categories of the ethnographic distribution does not imply that culture is irrelevant or plays no role. Color terms themselves are part of the language system and are transmitted culturally. Human societies differ greatly in the complexity of their color lexica and in the ways in which color terms are used in social exchange (Berlin & Kay, 1969; Witkowski & Brown, 1977; Ratliff, 1976). Identification of the epigenetic rule does not address the determinants of this complexity or the associated systems of conventions. The social evolution of color-term systems is also an involved process whose regularities might be explained by a combination of cultural and biological mechanisms (Witkowski & Brown, 1977; Kay & McDaniel, 1978; Ratliff, 1976; Bornstein, 1973; Whitfield, 1980). However, an epigenetic rule for basic color categories would indicate that these important processes of language evolution have been occurring in a manner consistent with innate constraints operating since early infancy. It would also suggest the usefulness of approaching problems of language evolution with neurocultural methods based in evolutionary biology.

Discussion

In addition to the case of color-term evolution, the mapping of gene-culture connections has made progress in a half-dozen or so other cases (Table 1). Testable hypotheses that relate writing direction, alphabet composition, and innate constraints on text comprehension will therefore add to the growing list of sociocultural phenomena being illuminated by modern evolutionary biology. Inevitably, some social scientists have suggested that such explicit

Table 1. Cases of gene-culture interaction in which the key linkages have been at least partly resolved

Phenomenon	References
Basic color terms	Berlin & Kay (1969) Borstein et al. (1976) Lumsden (1985)
Sibling incest avoidance	Sheper (1983) Wolf & Huang (1980) Lumsden & Wilson (1981 a)
Lactose intolerance and animal husbandry	Durham (1982 a) Aoki (1986)
Sickle cell anemia and slash/burn horticulture	Durham (1982 a, b)
Color blindness and Asiatic caste hierarchies	Lumsden (1983)
Neanderthal transition to modern *Homo sapiens sapiens*	Lumsden (1983)
Alcohol usage patterns	Jones & Aoki (1986)
Greek writing direction	This volume, chapters 9, 13, 20

concern for biology in studies of culture is unnecessary (Sahlins, 1976; Caplan, 1978; Harris, 1979; Bock, 1980; Montague, 1980; Trigg, 1982; Marks, Staski and Schiffer, 1983). They argue that biological underpinnings or even constraints may exist, but they are common from culture to culture and form a universal raw material that culture proceeds to shape. The differences among cultures, together with the essential nature of social diversity, can be understood by reference to sociocultural patterns and processes, taking biology as a common denominator among humans. Why one culture differs from another is thus a question that does not concern biology (e.g., Goodenough, 1983; for a history of this position together with an opposing view by a cultural anthropologist, see Freeman, 1983).

The evidence summarized in this chapter indicates that such views are at the very least wrong. In the presence of neurocultural linkages sustained by epigenetic rules for gene-culture transmission, "explanations" of cultural differences or social diversity by reference only to the variations between information contents of societies are, however artful (cf. Geertz, 1973), not explanations at all. They are primarily clarifications of the boundary conditions under which these differences developed. Without simultaneous reference to biology, one is at a loss to explain why certain environmental or sociocultural events led to their specific consequences for individuals and societies, rather than causing some other conceivable set of outcomes. The opportunities for hermeneutical improvisation also sadly diminish. The explanation of diversity and difference among humans and their societies appears to require the explicit methodological consideration of both biology and culture.

Acknowledgements. It is a pleasure to thank Christopher Scott Findlay and Christine Littlefield for discussion, and Edward O. Wilson for the continuing collaboration that produced the basis for the work reported here.

References

Amoore, J.E. (1977). Specific anosmia and the concept of primary odors. *Chemical Senses and Flavour, 2,* 267–281.

Anderson, J.R. (1983). *The architecture of cognition.* Cambridge, MA: Harvard University Press.

Aoki, K. (1986). A stochastic model of gene-culture coevolution suggested by the "culture historical hypothesis" for the evolution of adult lactose absorption in man. *Proceedings of the National Academy of Sciences of the United States of America, 83,* 2929–2933.

Ashton, G.C., Polovina, J.J., & Vandenberg, S.G. (1979). Segregation analysis of family data for 15 tests of cognitive ability. *Behavior Genetics, 9,* 329–347.

Berger, P.L., & Luckmann, T. (1966). *The social construction of reality. A treatise in the sociology of knowledge.* Garden City, New York: Doubleday.

Berlin, B., & Kay, P. (1969). *Basic color terms: their universality and evolution.* Berkeley, CA: University of California Press.

Berwick, R.C. (1982). *Locality principles and the acquisition of syntactic knowledge.* Unpublished doctoral dissertation, Department of Electrical Engineering and Computer Science, MIT, Cambridge, MA.

Bliss, T. V. P., & Errington, M. L. (1977). "Reeler" mutant mice fail to show spontaneous alternation. *Brain Research, 124,* 168–170.
Blumberg, B. S., & Hesser, J. E. (1975). Anthropology and infectious disease. In A. Damon (Ed.), *Physiological anthropology.* New York: Oxford University Press, pp. 260–294.
Bock, K. (1980). *Human nature and history: a response to sociobiology.* New York: Columbia University Press.
Bornstein, M. H. (1973). Color vision and color naming: a psychophysiological hypothesis. *Psychological Bulletin, 80,* 257–285.
Bornstein, M. H. (1975). Qualities of color vision in infancy. *Journal of Experimental Child Psychology, 19,* 401–419.
Bornstein, M. H. (1979). Perceptual development: stability and change in feature perception. In M. H. Bornstein, & W. Kessen (Eds.), *Psychological development from infancy: image to intention* (pp. 37–81). Hillsdale, NJ: Lawrence Erlbaum.
Bornstein, M. H. (1981). "Human infant color vision and color perception" reviewed and reassessed: a critique of Werner and Wooten (1979 a). *Infant Behavior and Development, 4,* 119–150.
Bornstein, M. H., & Marks, L. E. (1982). Color revisionism. *Psychology Today, 16,* 64–88, 68–70, 73.
Bornstein, M. W., Kessen, W., & Weiskopf, S. (1976 a). Color vision and hue categorization in young human infants. *Journal of Experimental Psychology and Human Perception and Performance, 2,* 115–129.
Bornstein, M. H., Kessen, W., & Weiskopf, S. (1976 b). The categories of hue in infancy. *Science, 191,* 201–202.
Bouchard, T. J., Jr., & McGue, M. (1981). Familial studies of intelligence: a review. *Science, 212,* 1055–1059.
Brillouin, L. (1962). *Science and information theory.* New York: Academic.
Brunswick, E. (1956). *Perception and the representative design of experiments.* Berkeley, CA: University of California Press.
Budge, E. A. W. (1929). *The Rosetta stone in the British Museum.* London: The Religious Tract Society.
Caplan, A. L. (Ed.) (1978). *The sociobiology debate: readings on ethical and scientific issues.* New York: Harper and Rowe.
Caviness, V. S., Jr., & Rakic, P. (1978). Mechanisms of cortical development: a view from mutations in mice. *Annual Review of Neuroscience, 1,* 297–326.
Chaitin, E. (1975). Randomness and mathematical proof. *Scientific American, 232,* 47–52.
Chiva, M. (1979). Comment la personne se construit en mangeant. *Communications* (École des Hautes Études en Sciences Sociales – Centre d'Études Transdisciplinaires, Paris), *31,* 107–118.
Chomsky, N. (1980). *Rules and representations.* New York: Columbia University Press.
Daly, M., & Wilson, M. (1978). *Sex, evolution, and behavior.* North Scituate, MA: Duxburg Press.
Dasen, P. R. (1972). Cross-cultural Piagetian research: a summary. *Journal of Cross-Cultural Psychology, 3,* 23–29.
DeCasper, A. J., & Fifer, W. P. (1980). Of human bonding: newborns prefer their mothers' voices. *Science, 208,* 1174–1176.
Durham, W. H. (1982 a). Interactions of genetic and cultural evolution: models and examples. *Human Ecology, 10,* 289–323.
Durham, W. H. (1982 b). The coevolution of yam festivals and sickle-cell frequencies in West Africa. *American Journal of Physical Anthropology, 57,* 183.
Eibl-Eibesfeldt, I. (1979). Human ethology: concepts and implications for the sciences of man. *The Behavioral and Brain Sciences, 2,* 1–57.
Eilers, R. E., Wilson, W. R., Moore, J. M. (1977). Developmental changes in speech discrimination in infants. *Journal of Speech and Hearing Research, 20,* 766–780.
Eimas, P. D., Siqueland, E. R., Jusczyk, P., & Vigorito, J. (1971). Speech perception in infants. *Science, 171,* 303–306.

Fantz, R.L., Fagan III, J.F., & Miranda, S.B. (1975). Early visual selectivity: as a function of pattern variables, previous exposure, age from birth and conception, and expected cognitive deficit. In L.B. Cohen & P. Salapatek (Eds.), *Infant perception: from sensation to cognition: basic visual processes* (Vol. 1, pp. 249–345). New York: Academic.

Fodor, J.A. (1983). *The modularity of mind: an essay on faculty psychology.* Cambridge, MA: MIT Press.

Fox, R., & McDaniel, C. (1982). The perception of biological motion by human infants. *Science, 218,* 486–487.

Freedman, D.G. (1974). *Human infancy: an evolutionary perspective.* Hillsdale, NJ: Lawrence Erlbaum.

Freeman, D. (1983). *Margaret Mead and Samoa: the making and unmaking of an anthropological myth.* Cambridge, MA: Harvard University Press.

Gajdusek, D.C. (1970). Physiological and psychological characteristics of stone age man. *Science and Technology, 33,* 26–62.

Gatlin, L.L. (1972). *Information theory and the living system.* New York: Columbia University Press.

Geertz, C. (1973). *The interpretation of cultures.* New York: Basic Books.

Gleason, H.A. (1961). *An introduction to descriptive linguistics.* New York: Holt, Rinehart & Winston.

Goldman, S. (1953). *Information theory.* New York: Dover.

Goodenough, W. (1983). Margaret Mead and cultural anthropology. *Science, 220,* 906, 908.

Hallpike, C.R. (1979). *The foundations of primitive thought.* New York: Oxford University (Clarendon) Press.

Harris, M. (1968). *The rise of anthropological theory: a history of theories of culture.* New York: Harper and Rowe.

Harris, M. (1979). *Cultural materialism: the struggle for a science of culture.* New York: Random House.

Hershenson, M., Munsinger, H., & Kessen, W. (1965). Preference for shapes of intermediate variability in the newborn human. *Science, 147,* 630–631.

Hockett, C.F., & Ascher, R. (1964). The human revolution. Current Anthropology, 5, 135–147, 166–168.

Holloway, R.L. (1966). Cranial capacity, neural organization, and hominid evolution: a search for more suitable parameters. *American Anthropologist, 68,* 103–121.

Horton, D.L., Mills, C.B. (1984). Human learning and memory. *Annual Review of Psychology, 35,* 361–394.

Hutchinson, A. (1970). *Labanotation or kinetography Laban: the system of analyzing and recording movement.* New York: Theatre Arts Books.

Irons, W. (1979). Cultural and biological success. In N.A. Chagnon & W. Irons (Eds.), *Evolutionary biology and human social behavior: an anthropological perspective* (pp 257–272). North Scituate, MA: Duxbury Press.

Jones, E., Aoki, C. (1986, to be submitted). The gene-culture coevolution of alcohol use. *Journal of Social & Biological Structures.*

Kagan, J. (1984). *The nature of the child.* New York: Basic Books.

Kay, P., & McDaniel, C.K. (1978). The linguistic significance of the meanings of basic color terms. *Language, 54,* 610–646.

Keesing, R.M. (1976). *Cultural anthropology: a contemporary perspective.* New York: Holt, Rinehart and Winston.

Keil, F.C. (1979). Semantic and cognitive development: an ontological perspective. Cambridge, MA: Harvard University Press.

Keil, F.C. (1981). Constraints on knowledge and cognitive development. *Physiological Reviews, 88,* 197–227.

Khinchin, A.I. (1957). *Mathematical foundations of information theory.* (Silverman, R.A. & Friedman, M.D., Trans). New York: Dover.

Klaus, M.H., Jerauld, R., Kreger, N.C., McAlpine, W., Steffa, M., & Kennel, J.H. (1972). Maternal attachment: importance of the first post-partum days. *New England Journal of Medicine, 286,* 460–463.

Kroeber, A.L., & Kluckhohn, C. (1963). *Culture: a critical review of concepts and definitions.* New York: Random House.
Laban, R. (1975). *Laban's principles of dance and movement notation* (2nd ed.). London: MacDonald & Evans.
Larkin, J., McDermott J., Simon, D.P., & Simon, H.A. (1980). Expert and novice performance in solving physics problems. *Science, 208,* 1335–1342.
Lennie, P. (1984). Recent developments in the physiology of color vision. *Trends in Neurosciences, 7,* 243–248.
Lenski, G., & Lenski, J. (1970). *Human societies: a macrolevel introduction to sociology.* New York: McGraw-Hill Book Company.
Leonard, W.J., Depper, J.M., Kanehisa, M., Kronke, M., Peffer, N.J., Svetlik, P.B., Sullivan, M., & Greene, W.C. (1985). Structure of the interleukin-2 receptor gene. *Science, 230,* 633–639.
Lewontin, R.C. (1981). Sleight of hand. *The Sciences, 21,* 23–26.
Liberman, A.M., Cooper, F.S., Shankweiler, D.P., & Studdert-Kennedy, M. (1967). Perception of speech code. *Psychological Review, 74,* 431–461.
Lisker, L., & Abramson, A.S. (1964). A cross-language study of voicing in initial stops: acoustical measurements. *Word, 20,* 384–422.
Lockhard, J.S., Daley, P.C., Gunderson, V.M. (1979). Maternal and paternal differences in infant carry: U.S. and African data. *American Naturalist, 113,* 235–246.
Loehlin, J.C., & Nichols, R.C. (1976). *Heredity, environment, and personality.* Austin, TX: University of Texas Press.
Lumsden, C.J. (1983). Gene-culture coevolution and the devolution of tabula rasa. *Journal of Social and Biological Structures, 6,* 101–114.
Lumsden, C.J. (1984a). Parent-offspring conflict over the transmission of culture. *Ethology and Sociobiology, 5,* 111–130.
Lumsden, C.J. (1984b). Dual inheritance in haploid organisms: a model of magnetotactic bacteria. Journal of Theoretical Biology, 111, 11–16.
Lumsden, C.J. (1985). Color categorization: a possible concordance between genes and culture. *Proceedings of the National Academy of Sciences of the United States of America, 82,* 5805–5808.
Lumsden, C.J., & Gushurst, A. (1985). Gene-culture coevolution: humankind in the making. In J.H. Fetzer (Ed.), *Sociobiology and Epistemology* (pp. 3–28). Boston, MA: D. Reidel.
Lumsden, C.J., & Wilson, E.O. (1980). Translation of epigenetic rules of individual behavior into ethnographic patterns. *Proceedings of the National Academy of Sciences of the United States of America, 77,* 4382–4386.
Lumsden, C.J., & Wilson, E.O. (1981a). *Genes, mind, and culture: the coevolutionary process.* Cambridge, MA: Harvard University Press.
Lumsden, C.J., & Wilson, E.O. (1981b). Genes, mind, and ideology. *The Sciences, 21,* 6–8.
Lumsden, C.J., Wilson, E.O. (1982). Mind and the linkage between genes and culture: a précis of *Genes, mind, and culture. Behavioral and Brain Sciences, 5,* 1–7.
Lumsden, C.J., & Wilson, E.O. (1983). *Promethean fire: reflections on the origin of mind.* Cambridge, MA: Harvard University Press.
Maller, O., & Desor, J.A. (1974). Effect of taste on ingestion by human newborns. In J. Bosma (Ed.), *Fourth symposium on oral sensation and perception: development in the fetus and infant,* pp. 279–311. Washington DC: Government Printing House.
Markman, E.M., & Seibert, J. (1976). Classes and collections: internal organization and resulting holistic properties. *Cognitive Psychology, 8,* 561–577.
Marks, I.M. (1969). *Fears and phobias.* New York: Academic.
Marks, J., Staski, E., & Schiffer, M.B. (1983). A return to basics. *Nature, 302,* 15–16.
McClearn, G.E., & Defries, J.C. (1973). *Introduction to behavioral genetics.* San Francisco: Freeman.

Mealey, L. (1985). The relationship between social status and biological success: a case study of Mormon religious hierarchy. *Ethology & Sociobiology, 6,* 249–257.
Medin, D.L., & Smith, E.E. (1984). Concepts and concept formation. *Annual Review of Psychology, 35,* 113–138.
Mollon, J.D. (1982). Color vision. *Annual Review of Psychology, 33,* 41–85.
Montague, A. (Ed.) (1980). *Sociobiology examined.* New York: Oxford University Press.
Morgan, G.A., & Riccuiti, H.N. (1973). Infants' response to strangers during the first year. In L.J. Stone, H.T. Smith & L.B. Murphy (Eds.), *The competent infant: research and commentary,* pp. 1128–1138, New York: Basic Books.
Murdock, G.P. (1949). *Social structure:* New York: Macmillan.
Murdock, G.P. (1967). Ethnographic atlas: a summary. *Ethnology, 6,* 109–236.
Naroll, R., & Cohen, R. (Eds.) (1970). *A handbook of method in cultural anthropology.* Garden City, NY.: The Natural History Press.
Newell, A., & Simon, H.A. (1972). *Human problem solving.* Englewood Cliffs, NJ.: Prentice-Hall.
Pinker, S. (1981). What is language, that a child may learn it, and a child, that he may learn language? *Journal of Mathematical Psychology, 23,* 90–97.
Posner, M.I., & Keele, S.W. (1968). On the genesis of abstract ideas. *Journal of Experimental Psychology, 77,* 353–363.
Rakic, P. (1979). Genetic and epigenetic determinants of local neuronal circuits in the mammalian central nervous system. In F.O. Schmitt & F.G. Worden (Eds.), *The neurosciences: fourth study program* (pp 109–127). Cambridge, MA: MIT Press.
Ratliff, F. (1976). On the psychophysiological basis of universal color terms. *Proceedings of the American Philosophical Society, 120,* 311–330.
Rosch, E. (1973). Natural categories. *Cognitive Psychology, 4,* 328–350.
Rosch, E. (1975). Universals and cultural specifics in human categorization. In R.W. Brishin, S. Bochner & W.J. Lonner (Eds.), *Cross-cultural perspectives on learning* (pp 177–206). New York: Halsted Press, Wiley.
Rosch, E., Mervis, C.B., Gray, W.D., Johnson, D.M., & Boyes-Braem, P. (1976). Basic objects in natural categories. *Cognitive Psychology, 8,* 382–439.
Rosenberg, A. (1980). *Sociobiology and the preemption of social science.* Baltimore: Johns Hopkins University Press.
Sahlins, M. (1976). *The use and abuse of biology: an anthropological critique of sociobiology.* Ann Arbor, MI: University of Michigan Press.
Salk, L. (1973). The role of the heartbeat in the relations between mother and infant. *Scientific American, 228,* 24–29.
Sandell, J.H., Gross, C.G., & Bornstein, M.H. (1979). Color categories in macaques. *Journal of Comparative and Physiological Psychology, 93,* 626–635.
Schneider, L., & Bonjean, C.M. (1973). *The idea of culture in the social sciences.* Cambridge: Cambridge University Press.
Schwartz, B.M., & Ewald, R.H. (1968). *Culture and society.* New York: Ronald.
Shannon, C.E. (1948). The mathematical theory of communication. *Bell System Technical Journal, 27,* 379–423, 623–656.
Shepher, J. (1971). Mate selection among second-generation kibbutz adolescents and adults: incest avoidance and negative imprinting. *Archives of Sexual Behavior, 1,* 293–307.
Shepher, J. (1973). *Incest: a biosocial view.* New York: Academic.
Simon, H.A. (1979). *Models of thought.* New Haven, CT: Yale University Press.
Simon, H.A. (1981). *The sciences of the artificial* (2nd ed.). Cambridge, MA: MIT Press.
Taylor, R.B. (1973). *Introduction to cultural anthropology.* Boston, MA: Allyn & Bacon.
Tobias, P.V. (1981). *The evolution of the human brain, intellect and spirit. 1st Abbie Memorial Lecture, University of Adelaide, South Australia, 12th October, 1979.* Adelaide: University of Adelaide Information Office.
Trigg, R. (1982). *The shaping of man: philosophical aspects of sociobiology.* Oxford: Basil Blackwell.

van den Berghe, P. L. (1980). Royal incest and inclusive fitness. *American Ethnologist, 7,* 300–317.
van den Berghe, P. L. (1983). Human inbreeding avoidance: culture in nature. *Behavioral and Brain Sciences, 6,* 91–123.
Wexler, K., & Culicover, P. W. (1980). *Formal principles of language acquisition.* Cambridge, MA: MIT Press.
White, L. A. (1949). Ethnological theory. In: R. W. Sellars, V. J. McGill & M. Farber (Eds.), *Philosophy for the future* (pp 357–384). New York: Macmillan.
Whitfield, T. W. A. (1980). Salient features of color space. *Perception and Psychophysics, 29,* 87–90.
Whorf, B. L. (1956). *Language, thought, and reality.* Cambridge, MA: MIT Press.
Wickelgren, W. A. (1979). *Cognitive psychology.* Englewood Cliffs, NJ: Prentice Hall.
Williams, T. R. (1972). *Introduction to socialization: human culture transmitted.* St. Louis, MO: C. V. Mosby.
Wilson, R. S. (1978). Synchronies in mental development: an epigenetic perspective. *Science, 202,* 939–948.
Witkowski, S. R., & Brown, C. H. (1977). An explanation for color nomenclature universals. *American Anthropologist, 79,* 50–57.
Wolf, A. P. (1968). Adopt a daughter-in-law, marry a sister: a Chinese solution to the problem of incest taboo. *American Anthropologist, 70,* 864–874.
Wolf, A. P., & Huang, C. S. (1980). *Marriage and adoption in China, 1845–1945.* Stanford, CA: Stanford University Press.

CHAPTER 2

Learning and Selection in the Nervous System

JEAN-PIERRE CHANGEUX[1]

The need to strengthen the links between the neurosciences and the humanities, and sometimes even to create these links in the context of "neuro-cultural research" (de Kerckhove, 1984), would not have to be emphasized if these disciplines had not traditionally evolved in an independent and even divergent fashion. Spinoza's aphorism – "men judge according to the dispositions of their brains" – has too long been neglected by researchers in both areas, who are mainly concerned with analyses within the contexts of their own disciplines. In the field of neurolinguistics, for example, there is a wide gap separating the genetic code and language, and to span it presents a number of risks. It is a fact that some attempts at applying biological models to the humanities have ended in spectacular failure.

Should we therefore abandon any attempt to bridge the gap between these two approaches? Of course the answer is no; however, we must accept that many proposals in this direction will only be *models,* that is, hypothetical representations of external reality. In many cases it will be possible to test these models through experimentation, which will then either confirm or invalidate them. Thus, insofar as models can contribute to the process of experimentation and generate new research, they can prove themselves useful despite their hypothetical nature and (it must be said) despite their ultimate inadequacy at representing real objects. Any considerations discussed herein must be interpreted in this context. All living beings are "open" thermodynamic systems (Glansdorff & Prigogine, 1971), with internal structures that constitute privileged states of organization of matter in time and space. The question can then be asked: Where does this organization come from? Does it come from inside the organism, from its environment, or from both? There are two extreme views, or models, to account for this development of order:

1. The *instructive* model: the environment imposes an order that is transmitted directly to the organism.
2. The *selective* model: this is the Darwinian approach, according to which the effect of the environment is indirect. The organism spontaneously generates multiple internal variations *prior* to any interaction with the outside world, and the environment then *selects,* or *selectively stabilizes,* one or another of these endogeneous variations.

---

[1] Institut Pasteur, 28, Rue du Dr. Roux, F-75724 Paris, France.

The instructive model is the one that comes naturally to the scientist's mind and historically was the first to be proposed. However, we do not know of any actual mechanisms based upon it. In contrast, there are a number of selective mechanisms whose validity has been unequivocally confirmed.

Two conditions are necessary for the functioning of a selective machine: (1) It must have an *internal generator of diversity;* that is, it must be able to produce variations by means of some endogenous combinatorial mechanism. (2) It must have a *system of selection,* in order to retain specific combinations while rejecting others, on the basis of some exchange of coded signals with the outside world. If both these conditions are met, it is possible to study the development of order of a biological system on a tangible basis.

Over the course of the history of scientific epistemology, selective models have been proposed in three distinct areas: the evolution of species, the biosynthesis of antibodies, and developmental neurobiology: It may be instructive to examine each of these in turn, in order to clarify how a selective mechanism may function.

Selective Theories in Evolution

In Greece, in the fifth century B.C., Empedocles of Agrigentus wrote in Fragment 57 of his work *Peri Phuseos* (Of Nature):

> There grew on earth many heads without necks; individual arms without shoulders wandered about; and eyes existed that were not attached to any forehead. On their own these parts of the body wandered about here and there, but when a divine presence linked one to the other, they adjusted through chance contact, and even more were born that added themselves to those that already existed. Beings were born with legs that turned, and innumerable hands. There were many beings with two faces and two chests, cows with faces of men and men with heads of bulls, and hermaphrodites with delicate limbs. Now listen how it was that fire produced the race of men and women with abundant tears. Listen because this speech is neither irrelevant nor frivolous. In the beginning these forms had their origin in a piece of ground that had equal parts of heat and humidity. Fire tried to make contact with a similar element....

As quaint as this text is, it is uncannily close to modern views about evolution, and totally different from, say, Creation as the "instructive" molding of a clay statue into the image of humanity. The concept of a generator of diversity is present through limbs "wandering about here and there" and joining together at random, as is a mechanism of selection (albeit obscurely) through the effects of "fire and humidity".

By the eighteenth century, the concept of the biological evolution of species was developed, along with the concept of evolution by natural selection. The latter can be found in Diderot's writings. However, Charles Darwin was the first to explicitly formulate this theory and illustrate it with examples drawn from the animal kingdom. From then on, the variations that are a prerequisite to natural selection and upon which this selection operates became understood. Such variations arise from mutations in DNA molecules, or from larger-scale reorganisations of chromosomes such as changes in number, translocations, transpositions, inversions, deletions, duplications, or gene conversions. DNA thus serves as the internal generator of diversity. On the other hand, the mechanisms of selective stabilization of these variations, of their segregation in natural populations, have not as yet been firmly established.

Selective Theories in Biochemistry

In the field of molecular biology, the biosynthesis of antibodies posed a formidable problem for decades. It is well established that vertebrate organisms react to invasions of foreign molecules by synthesizing antibodies specifically designed to attack them. Antibodies are proteins whose sequences of amino acids, like any protein, are coded by DNA. However, the variety of known antibodies is so large that at first it seemed impossible to encode them in the genome: the number of genes necessary to code for each one individually would exceed by far the total quantity of DNA in the fertilized egg! This initially led to the proposal of instructive models to account for such extreme structural diversity. In recent years, however, there has been a reassessment of the genetic determinants involved. It now appears that the DNA structures (and mRNA molecules that result from their transcription) that determine the sequence of each antibody do not exist as discrete units, but as families of segments, with each segment encoding for only some fragments of the antibody molecule. There are only about 300 such coding units, but they have the ability to recombine, thereby enabling a gigantic number of combinations to be generated (approximately 18 billion!). This internal combinatorial mechanism (the generator of diversity) can thus produce billions of antibody molecules with different sequences, and out of this vast population there will be one that is complementary to almost any foreign antigen. The formation of an antigen-antibody complex will then cause the multiplication of the lymphocyte that carries it, and as a result that antibody is "selected". The general framework of this selective model has been confirmed by experimentation. It is thus no longer simply a model, but the established description of a molecular mechanism.

Selective Theories in the Neurosciences

In psychology, the first pre-Darwinian references to selective models are found in the writings of the early British associationists, Locke and Hume.

Darwin himself, in his treatise on *The Expression of the Emotions in Animals and Man,* relied rather curiously on instructive models inspired by Lamarck rather than on his own selective models. However, Taine (1870), in his work *On Intelligence,* writes: "In the struggle for life (Darwin) which, at any given time, takes place between all our images, the one which, at the origin, possesses a higher energy, keeps at each conflict, by the law of repetition upon which it is based, the capacity to repel its rivals." Also, James (1909) states explicity that "to think is to make selections."

The revival of selective models for use in the neurosciences has been mainly due to the recent convergence of two lines of research: (1) studies on the diversification of nerve cells and their synapses during pre- and postnatal development, and the effects of the environment upon this diversification (Changeux, 1972, 1983a; Changeux, Courrège & Danchin 1973; Changeux & Danchin, 1976), and (2) extrapolation of the selective model of the biosynthesis of antibodies to the higher functions of the human brain (Edelman & Mountcastle, 1978; Edelman & Finkel, 1984). These two lines of work will now be discussed.

The Selective Stabilization of Synapses in Development

Unlike cells of the immune system, neurons stop reproducing at a very early stage of development, well before the formation of synaptic contacts and thus before the interaction with the outside world can make a mark on them. As a model based on processes of recombination and selection requires that the system be able to generate new combinations, it would thus seem that it could only be applied to the embryonic stage. However, after the cells have stopped dividing, their synaptic contacts continue to multiply. It is this feature that provides each neuron with its functional characteristics, its "singularity." Even in adults, nerve cells retain the ability to form and regenerate new synaptic contacts. Thus, if selection does exist, we must look for it mainly at the level of the synapse.

The Genetic Envelope

At both the anatomical and functional levels, the physical organization of the nervous system is species-specific and largely independent of the environment. However, genetic mutations and/or chromosomic alterations can profoundly modify this organization and are transmitted by heredity. In mice, for example, the *nervous* mutation results in the death of Purkinje cells in the cerebellum, the *reeler* mutation destroys granular cells, and the *staggerer* mutation interferes with the formation of synapses between the granular cells and Purkinje cells. These mutated genes affect particular types of cerebellar neurons and synapses, but all of them can attack other targets as well. For instance, the *nervous* mutation may affect certain retinal cells and even cells of the male reproductive system. Thus, there is *divergence* of the action of the same gene upon many different targets.

Reciprocally, when looking at the percentage of genes in the mouse's genome that code for the brain, it can be seen that most genes in some way act upon the nervous system. We can say that the brain "contains" the essential elements of the genetic information present in the chromosomes; that is, there exists a *convergence* of multiple genetic actions upon the brain. There is therefore no simple relationship between genes and cerebral function; one cannot speak of "a language gene" or "an intelligence gene." These integrated functions develop in the context of a complex neural organization to which multiple genetic activities contribute. The set of genes that participates in the development of cerebral organization is defined as its *genetic envelope.*

Variability of Cerebral Phenotype

Does the existence of a genetic envelope coding for brain structure therefore set limits to the variability of the cerebral phenotype, including the variability of its synaptic connections? To answer this question, Levinthal and his colleagues (1976) studied in great detail the morphology of well-defined, identifiable neurons in genetically identical individuals, using a species of crustacean (*Daphnia magna*) and a parthogenetic fish (*Poecilostoma*). They found that within an individual there was a variation in the branching patterns of axons and dendrites between right and left, across the symmetrical division of the body, and when comparing two isogenic individuals this variability was larger (although the major lines of connection might be the same). In humans, a striking example of phenotypic variability between isogenic individuals is that of manual preference: in a pair of monozygotic twins, one may be right-handed and the other left-handed (references in Changeux, 1983 b). Other examples could be mentioned. Undoubtedly, there is a significant potential for variability of the cerebral phenotype that escapes the constraints of the genetic envelope.

Epigenesis Through Synaptic Stabilization of Synapses in Development

The selective model as applied here suggests that as synapses develop, a "Darwinian" selection takes place upon the variability allowed by the genetic envelope. Once all cell divisions have taken place, the neurons make contact with each other through their axonal and dendritic extensions. These branches keep multiplying until they reach a "critical" stage of maximal connectivity (limited by the genetic envelope). Each neuron sends out and receives a far greater number of synaptic contacts than will be used functionally in adulthood; i.e., there is "redundancy" but also "maximal diversity." It is at this point, according to the model, that "selection" intervenes. Each synapse of the developing nervous system is assumed to exist in at least three stages: labile, stable, and degenerative. As the network becomes active, either spontaneously or through induction, it differentially regulates the evolution of the labile synapses, stabilizing some and eliminating others. Consequently, each neuron acquires its own connectivity, its singularity.

This general model has been put into mathematical form (Changeux et al., 1973). One of its major predictions is the variability of the stabilized network following the epigenetic experience of the critical stage. The same relationship between input and output can be obtained after the experience via the stabilization of different networks. The model thus predicts the phenotypic variability that is observed in isogenic individuals and a fortiori in individuals who are not isogenic.

A similar mathematical model of "local competition" has been developed to account for the innervation of striated muscle (Gouzé, Lasry & Changeux, 1983). On the basis of very simple reasoning it is possible to predict the outcome of competition between nerve endings reaching the same muscle fiber. The uptake of some retrograde factor produced in limited quantity by each muscle fiber would lead to the activity-dependent stabilization of endings that have accumulated and transformed a sufficient quantity of some factors to the exclusion of the others.

It has been experimentally determined in several systems (e.g., neuromuscular junction, cerebellum) that in the course of normal development, important regressive events may occur that affect synaptic contacts and even nerve cells (Hamburger, 1970). At the neuromuscular junction, it has been shown that the state of activity regulates this evolution: whereas lack of stimulation maintains the multiple innervation, proper stimulations accelerate synapse elimination. Contrary to what is generally assumed, the process of learning may be accompanied not by an increase, but by a *reduction* in the number of synaptic contacts. To learn is to eliminate.

Learning Through Selection in the Adult

The adult brain retains the capacity to learn, but it does not appear that this capacity can be explained solely through the processes of synaptic growth and regression seen in the embryo and young subject. A model that is currently being developed (Heidmann & Changeux, 1982; Toulouse, Dehaene & Changeux, 1986) and has been presented in its main outlines in *Neuronal Man* (Changeux, 1983b; see also Edelman & Mountcastle, 1978; Von der Malsburg & Bienenstock, 1987; Hopfield & Tank, 1986), is based on (1) the possibility that the *efficacy* rather than the number of synapses is altered during learning, and (2) that selection may concern *global units*, resulting from the activation of *cooperative* sets of nerve cells. These functional units have long been proposed by psychologists under such terms as "mental representations" (Hebb, 1949).

A model to account for how these mental representations are entered into memory through selection obviously requires the two indispensable elements of any selective model. The generator of diversity would be the combinatorial mechanism that allows groups of neurons to split up, associate, and recombine amongst themselves. These dynamics would lead to the spontaneous development of multiple "prerepresentations", and a restricted number of these would then be selected as a result of interaction with the outside world. Selection might result, for example, through a *resonance* or "matching" between the percept evoked by this interaction, and the prerepresentation that is spontaneously formed in the brain at the same time (or after an adequate period of time). Molecular models of the modification of synaptic efficacy based on the allosteric transitions of postsynaptic receptors are currently being developed (Heidmann & Changeux, 1982; Changeux & Heidmann, 1987).

One of the major advantages of a selective model of learning is its capacity to *reject* external information that is not pertinent for the subject at a given time. Memory input occurs only on the basis of a *congruence* between internal activity and that activity evoked by the interaction with the environment. The instructive model does not simply predict the elimination of nonpertinent elements: in fact, this model may lead to rapid, nonselective saturation by all incoming information. The selective model, on the other hand, leads to a storage of fragmentary but "relevant" features. Another of its advantages is that the stored states are not simply packed but are organized into a hierarchical "ultrametric" structure (Toulouse et al., 1986).

Conclusion

The shaping of the adult brain, in particular the acquisition of its ability to learn and utilize the alphabet, may be interpreted in terms of Darwinian selection mechanisms operating at three distinct levels with different time scales: (1) The main features of its neural organization would be determined by a genetic envelope resulting from evolution on geological time scales. (2) The detailed shaping of its cortical maps, in particular those involved in written language processing, would be imprinted during postnatal development. (3) The actual transfer from short- to long-term memory of novel "objects" would occur on a time scale of fractions of seconds to longer periods. Finally, mechanisms of selection may also take place at the cultural level for the genesis, propagation, utilization, and storage of "communication symbols" no longer within a single brain but between "populations of brains." Accordingly, the differentiation of the alphabet from ideograms might be viewed as resulting from such a Darwinian selection at the cultural level (Changeux, 1983).

References

Changeux, J.P. (1972). Le cerveau et l'évenement. *Communications, 18,* 37–47.

Changeux, J.P. (1983a). Remarks about the singularity of nerve cells and its ontogenesis, *Progress Brain Research, 58,* 465–478.

Changeux, J.P. (1983b). *Neuronal man.* New York: Pantheon.

Changeux, J.P., & Danchin, A. (1976). The selective stabilization of developing synapses: a plausible mechanism for the specification of neuronal networks. *Nature, 264,* 705–712.

Changeux, J.P., & Heidmann, Th. (1987). Allosteric receptors and molecular models of learning. In G. Edelman, W. Gall & W.M. Cowan (Eds.). *New insights into synaptic function.* New York: John Wiley.

Changeux, J.P., Courrège, P., & Danchin, A. (1973). A theory of the epigenesis of neural network by selective stabilization of synapses. *Proceedings of the National Academy of Sciences, USA, 70,* 2974–2978.

Edelman, G., & Finkel, C. (1984). Neuronal group selection in the cerebral cortex. In G. Edelman, W. E. Goll & M. W. Cowan (Eds.). *Dynamic aspects of neocortical function* (pp. 653–695). New York: Wiley.

Edelman, G., & Mountcastle, V. (1978). *The mindful brain.* Cambridge, MA: MIT Press.

Glansdorff, P., & Prigogine, I. (1971). *Structure, stabilité et fluctuations.* Paris: Masson.

Gouzé, J.L., Lasry, J.M., & Changeux, J.P. (1983). Selective stabilization of muscle innervation during development: a mathematical model. *Biological Cybernetics* (Berlin) *46,* 207–215.

Hamburger, V. (1970). Embryonic Motility in Vertebrates. In F.O. Schmidt (Ed.) *Neuroscience: Second study program* (pp 141–151). New York: Rockefeller University Press.

Hebb, D.O. (1949). *Organization in behavior.* New York: John Wiley.

Heidmann, T., & Changeux, J.P. (1982). Un modèle moléculaire de régulation d'éfficacité d'une synapse chimique au niveau post-synaptique. *Compte Rendus de l'Académie des Sciences, 295,* 665–670.

Hopfield, J., & Tank, D.W. (1986). Computing with neural circuits: a model. *Science, 233,* 625–635.

James, W. (1909). *The meaning of truth: A sequel to pragmatism.* London: Longmans.

de Kerckhove, D. (1984). Neuro-cultural research. *Understanding 1984/Pour comprendre 1984,* Occasional Paper, Canadian Commission for Unesco, *48,* 129–139.

Levinthal, F., Macagno, E., Levinthal, C. (1976). Anatomy and development of identified cells in isogenic organisms. *Cold Spring Harbour Symposium on Quantitative Biology, 40,* 321–332.

Taine, H. (1870). *De l'intelligence,* Paris.

Toulouse, G., Dehaene, S.F., Changeux, J.P. (1986). Spin glass model of learning by selection. *Proceedings of the National Academy of Science, USA, 83,* 1695–1698.

Von der Malsburg, Ch., & Bienenstock, E. (1987). Statistical coding and short term synaptic plasticity: Scheme for knowledge representation in the brain. In E. Bienenstock, S. Fogelman & G. Wesbuch (Eds.) *Disordered systems and biological organization.* Berlin: Springer.

CHAPTER 3

Neuronal Group Selection: A Basis for Categorization by the Nervous System *

LEIF H. FINKEL [1]

Introduction

The history of neuroscience, like that of all of biology, has been a process of relating structure and function. Three major periods in neuroscience can be characterized by attempts to pitch this relationship at various levels in the nervous system. In the mid-nineteenth century, a series of neurological observations showed that particular functions – notably, the ability to understand and generate speech – appear to be localized to specific regions of the cerebral cortex. Around the turn of the century, the discrete nature of neurons as cells was recognized, and initial discoveries were made about the synaptic connections between them. Finally, in the last quarter-century has come the discovery of the cortical column: a modular structure repeated throughout much of the cortex that is thought to serve as an input/output unit, and in which all cells share at least some functional properties, (e.g., receptive fields).

None of these three levels of structure-function relations has led to a full understanding of the higher brain functions. With regard to the first case (cerebral localization), the behavioral functions are well defined, but the anatomical areas involved comprise an appreciable fraction of the cortex. In the second case, the functions of individual neurons are at best uncertain except for a few specialized types (e.g. the Mauthner neuron in fish, which mediates the tail flick used in escape swimming). The case of the cortical column comes closest to rigorously relating structure and function, but it is generally recognized that there are too few cortical columns for them to represent the basic units of the nervous system (Mountcastle, 1978).

The theory of neuronal group selection (Edelman, 1978, 1981) proposes that the structure-function relationship is best posed at the level of *neuronal groups.* Neuronal groups are small populations (hundreds to thousands) of strongly coupled cells. Neurons within the cortex and other neural areas are partitioned into groups during embryonic development. The large repertoire of groups thus created manifest a tremendous variability in the fine structure

---

* This work was supported by a grant from the International Business Machines Corporation.

[1] The Rockefeller University, 1230 York Avenue, New York, NY 10021, USA.

of their anatomical connections. The theory proposes that the nervous system operates as a selective system, picking particular neuronal groups from the vast repertoire of possible candidates according to a selective process akin to the natural selection of species in evolution. Selected groups are then dynamically maintained in the mature animal through modifications of the synapses linking the cells.

The theory puts forward a population approach to neurobiology emphasizing variance within the population, and stresses the relation of mature function to development and evolution. It holds that the primary problem faced by the nervous system is how to *categorize* the world. This problem arises because the world is "unlabeled," and the nervous system itself must create the categories, through a subjective, individual, adaptive process. Most theories of higher brain functions implicitly assume that categories somehow intrinsically exist in the external world – hence, direct perception and other nativist notions (together with learning) are all that is required. The present theory maintains that the fundamental problem is perceptual categorization prior to learning.

An adequate theory of categorization must provide a somatic means of categorization based on initial encounter, and also must account for the relating of the process to development and evolution. An analysis of the situation (Edelman & Finkel, 1984) indicates that only selective models can satisfy these requirements in a way that avoids the vicious circle of the homunculus, a brain inside the brain. However, an acceptable selective theory must be rigorous; it cannot be based on purely eliminative selection (Changeux & Danchin, 1976; Young, 1971) without the generation of new variance in the system, and the level of selection within the system must be clearly articulated with definite units and mechanisms.

In this paper, I will give a brief outline of the theory as it is presently formulated. This will entail a discussion of the basic concepts of selection, an identification of the units and mechanisms of selection, and an indication of the formalism behind our models. I will then review some of the evidence that has been found to support the theory. Finally, I will speculate on the relationship of selective processes in the nervous system to some higher brain functions, most particularly categorization, and I will use this relationship as a hopeful bridge from neurobiology to learning and culture.

Overview of the Theory

There are three minimal requirements in order for a selective system to function adaptively: (1) a large population of variant individuals must exist or be created; (2) there must be a mechanism of selection operating according to independent criteria set by boundary conditions in the outside world; and (3) there must be some form of heritability of selected forms. In natural selection, variability among individuals is produced by a number of genetic and developmental processes; the mechanism of selection is the differential reproduction of viable offspring; and heritability occurs through genetics, sexual or asexual. In the only other known example of a selective system, the immune system, variant antibody types are generated through a mechanism of somatic recombination of genes; the mechanism of selection here involves the suprathreshold binding of antigens by antibodies on those lymphocytes in the population with matching complementarities: and heritability occurs through clonal selection – the repeated mitoses of selected cells (Edelman, 1975).

The theory of neuronal group selection proposes that the units of selection in the nervous system are groups of neurons, each with a different structure at the finest level of wiring. There are two phases of selection – a developmental phase in which the mechanism involves the primary processes of development, and a secondary, experiential phase in which connections between neurons are no longer added or removed but selection occurs through modifications in synaptic efficacies. A form of heritability exists in the effects of synaptic changes, but here the analogy is less obvious, as will be discussed later.

A concept that is central to the theory is that of degeneracy. The primary repertoire of neuronal groups is "degenerate" in the sense that each group can potentially be selected for several different functions; and conversely, there are many structurally different groups that can be selected for the same function. This property follows, as will be discussed below, from the large degree of variability found in the anatomical fine structure of the nervous system and in the even greater degree of variability found in the physiological properties of the anatomical networks. Thus, degeneracy describes the nature of the map between structure and function and indicates that this map is not one-to-one.

Degeneracy is required by any finite system that must categorize an open world of stimuli in continuum (Edelman, 1981). For example, our ability to perceive color depends upon three color pigments in the cones of our retinas. The spectral absorption curves of these pigments overlap considerably, so that most wavelengths of light stimulate either two or all three pigments to some degree, and conversely, each pigment is sensitive to a wide range of wavelengths. The point has been made that the overlap of these absorption curves maximizes the efficiency of light detection (Barlow, 1982; Buchsbaum & Gottschalk, 1983). From our point of view, this overlap is symptomatic of the degeneracy found in selective systems, from the overlap of receptive fields to the overlap of niches in an ecology (May & MacArthur, 1972).

Degeneracy can act in a similar fashion to redundancy, in that it assures the reliable operation of a system in a stochastic and unpredictable environment. In this last sense, the reduction of degeneracy in the system corresponds to a measure of information accrued in the system *ex post facto,* i.e., after experience.

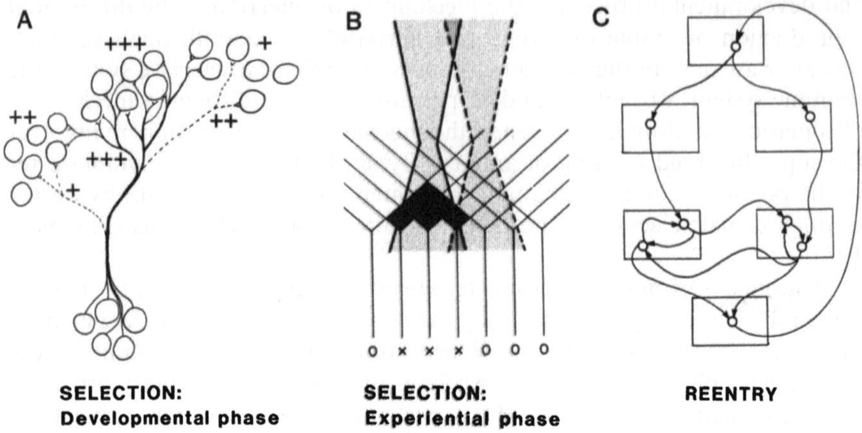

Fig. 1A–C. Neuronal group selection. **A** Developmental phase – Connections formed between cells (*circles*) depend upon the modulation of CAMs on their surfaces. As growing neurites make contact with other cells, the amount, distribution, or chemical state of the CAMs on the presynaptic and postsynaptic processes (indicated by the number of + signs), determine whether a connection will be stabilized. CAM modulation depends on the state of the cells at the time, and on the previous history of contacts by the cell. In the case depicted, only the middle branches are stabilized. **B** Experiential phase. *Hourglass figures* represent groups, *branching fibers* represent overlapped extrinsic inputs (synapses within branching region are not shown). Coactivated inputs (×) from overlapped fibers lead to intense firing within a small portion of the network (darkened area). If stimulation is repeated, groups in this area are selected by the synaptic modification rules (see text). Uncorrelated inputs (o) can lead to a decrease in the strength of synaptic connections. **C** Reentry. Schematic of the flow of activity from a common source to groups (*circles*) in several repertoires (*large rectangles*). The particular anatomical configuration considered allows cross-connections only between the second-level repertoires. *Large arrow* running up the extreme right side indicates that the ouput of a final common repertoire (e.g., a motor repertoire) can influence the common input, either directly or through a movement of the animal in space that changes the sensory stimulation received by peripheral receptors

Figure 1 portrays the two phases of selection (developmental and experiential), which will be discussed below, as well as the process of phasic reentry, which will be discussed at the end of this paper. Reentry links the output of different repertoires of neuronal groups in such a way as to create and preserve unity of perception despite the multimodal and temporally extended nature of the stimulus. For example, as shown in the figure, inputs can reach a given repertoire from several sources, at several different times, and activate many groups in the repertoire (for simplicity, only one group is shown activated in the figure). Reentry deals with the coordination of these multiple inputs and the resulting patterns of activity in the various repertoires.

The primary phase of selection includes all of the developmental interactions leading to the establishment of a "primary" repertoire of neuronal groups. The mechanisms of neural development can only be touched on: this is a period of intense competitive and cooperative interactions between developing neurites, characterized by cell proliferation, neurite outgrowth, and also by reduction of polyinnervation and massive amounts of cell death. There is substantial evidence for the role of cell-adhesion molecules (CAMs), notably N-CAM and Ng-CAM, in the development of ordered neural maps (see Fig. 1 A), as well as in the relationship between histogenesis and morphogenesis (Edelman, 1984a, b). This evidence seriously questions the prevalent view that order in the nervous system arises from a set of genetically coded neuronal markers, a specific addressing system that allows each axon to find its way during development to a unique target with the complementary molecular address. The existence of CAMs argues for a more variable and dynamic view of development in which there are no "addresses," and connections are made by modulating the prevalence or binding strength of the adhesion molecules present on the surfaces of the neurons involved. These modulations can be local biochemical changes in the CAMs themselves, or global changes in the cytoskeleton over the whole cell. The main point is that the genes controlling the expression of CAMs are proposed to be independently regulated from those controlling the expression of structural proteins (Edelman, 1985). Thus, cell differentiation is independent of changes in the adhesion properties. This principle allows collectives of cells to come together by adhesion and to interact with other collectives, leading to modulations of the CAMs and differentiation of the cells. Differentiation changes the rules for further modulations of the CAMs and further aggregations and interactions. Indeed, changes in the CAMs have been found to be correlated with the process of embryonic induction (Edelman, 1984a, b). It has long been believed that induction occurs between groups of cells (Spemann, 1938), and in fact, Weiss' (1939) descriptions of induction bear a striking resemblance to those interactions I shall propose to occur between neuronal groups in the mature nervous system. Thus, the idea of selection amongst groups through modifications in populations of synapses allows developmental mechanisms to persist, in transmogrified form, throughout the lifetime of the animal. Since evolution acts on the outcome of development (i.e., the phenotype) through changes in the process of development, this ties adult function into a mechanism of adaptation.

The interaction of CAMs with the primary processes of development also provides a mechanism for introducing variability into the developmental program. It provides, for example, a natural mechanism for allometric growth of various cell groups, and more generally, it provides a mechanism for heterochrony (the speeding up or slowing down of various developmental processes, particularly the onset of sexual maturation). An obligate variability in the fine structure of the nervous system follows, particularly at the level of individual synaptic contacts. Indeed, such anatomical variability has recently begun to be documented. Pearson and Goodman (1979) compared the branching patterns of an identified neuron (the descending contralateral movement detector) in different individuals of the locust *L. migratoria.* They found that the pattern of branches varied so greatly that no "normal" pattern could be said to exist. Furthermore, the corresponding axons on the left and right sides of each individual varied greatly. Of course, all individuals differed genetically; however, similar results were found in optic neurons from genetically *identical* (parthenogenetically reproduced) individuals of the water flea *Daphnia magna* (Macagno, Lopresti & Levinthal, 1973). In another example, Stent and colleagues (Kramer, Golden & Stent, 1985) studied neural innervation patterns in the leech *Haementeria,* and concluded that while certain neurites characteristically appear to innervate the same region, the corresponding innervation patterns of other neurites varied greatly.

Further examples at the same level of detail in higher species are not currently available. However, these preliminary examples support the requirement for neuronal group selection that there exist a degenerate anatomical repertoire upon which selection can operate.

The second phase of selection operates upon the repertoire of variant neuronal groups created during development. The mechanism of this selection involves the modification of synaptic strengths of the connections between cells in the same and/or different groups. These modifications change the properties of the affected groups: for example, by sharpening receptive fields or by strengthening the associational links between groups. In order to induce synaptic modifications the stimuli must, in general, generate multiple coactivated inputs (Fig. 1B). However, groups will differ in the exact patterns of stimuli most effective in producing modifications.

The dimensions and anatomical constituents of neuronal groups may vary in different regions of the nervous system and between species; a specific example from the somatosensory cortex of the owl monkey has been previously proposed (Edelman & Finkel, 1984). Groups in this region are proposed to extend vertically through all six cortical layers and to cover 50 – 150 μm horizontally. Based on current estimates of cortical cell densities (Rockel, Hiorns & Powell, 1980), they thus include roughly 500 – 1500 neurons.

Neuronal groups are proposed to arise from the combined action of three processes: confinement, selection, and competition. Confinement results from the intrinsic balance between excitatory and inhibitory inputs in the cerebral cortex. Group confinement is a dynamic constraint upon the horizontal spread of a group. We have speculated on the roles of various cell types in the process of group confinement (Edelman & Finkel, 1984); suffice it to say that some cell types act to spread excitation radially outward, others act to dampen the sources of excitation, while still others link processes in different cortical layers, so that together a dynamic balance is maintained. A similar view of the flow of cortical activation is shared by several other researchers (Szentagothai, 1978; Eccles, 1984). Neuronal groups thus form as a result of normal developmental processes. During the experiential phase of selection, some of these groups are selected, others are weakened or completely dissolved. The mechanism of group selection depends on the rules of synaptic modification, which are discussed below.

Neuronal group selection, like natural selection, has a large stochastic component. A particular selected group may actually be less adapted than another group which does not get selected, for any of various historical, random, or accidental reasons. On the average, groups that are intensely stimulated in a correlated or coactivated manner (see Fig. 1B) will be selected, and will thenceforth be particularly responsive to the set of stimuli that selected them. Due to the divergence of axonal arbors over many group diameters, nearby groups will receive very similar sets of anatomical inputs. The receptive fields of nearby groups will therefore be significantly overlapped, with exceptions near representational borders. Cells within the same group will have much more highly overlapped receptive fields than cells in different groups.

The receptive field of a group reflects the fraction of anatomic inputs with synaptic efficacies strong enough to fire that group. The degree of overlap of the receptive fields of nearby groups is controlled by the process of group competition. A cell may belong to only one group at a time; however, through group competition a cell in one group can be "captured" by another group, thereby changing its own receptive field and perhaps that of the capturing group as well. As groups are strengthened (through activation of their receptive fields) they tend to expand, capturing cells from nearby groups, until the restraining forces of group confinement and group competition limit further expansion. The rules of the competition are determined by the dynamics of activation of the groups in question, and are mediated by the synaptic modification mechanisms present.

The existence of neuronal groups and predictions as to their properties have been given extensive support from recent physiological experiments on changes in topographic maps in the CNS. These findings have been mainly from the somatosensory cortex of monkeys, but similar results have been found at other CNS levels, such as thalamus and dorsal column nuclei, and in other species (for review see Edelman and Finkel, 1984). The best evidence comes from Merzenich and colleagues (Kaas, Merzenich & Killackey, 1983), who found that there is a great deal of intrinsic variability in the receptive field maps of areas 3b and 1 of owl monkeys. They performed a series of perturbative experiments which support the conclusion that the physiological maps found in cortex are the result of a dynamic competitive process, a selective process that selects the map from a degenerate manifold of possible physiological maps under the existing anatomy. Their studies also suggested a set of rules regulating reorganization of cortical maps due to trauma or normal experience, including rules on the overlap of receptive fields, the magnification factor of representations, and the shift of representations across cortex. It has previously been shown that selection acting on neuronal groups can account for these rules (Edelman & Finkel, 1984).

Neuronal group selection in cortical maps serves as a paradigm for considering experience-driven perceptual and even conceptual changes. Note that the groups compete not only for control of cell firing, but also for the repertoire of responses allowed to each group. In other words, the outcome of group competition determines not only the adaptedness, but also to some degree the *adaptability* of a group. I turn now to the synaptic modifications that are the major mechanism of selection in somatic time in the behaving animal.

Rules for Synaptic Modifications

Shortly after the first anatomical descriptions of the synapse were made (at the turn of the century), the idea was proposed that synaptic modifications could provide the neural basis for learning and memory. Subsequently, there have been many formal models of the modification of synaptic efficacy – which is usually defined as the voltage produced in a postsynaptic cell due to an action potential in the presynaptic cell. Perhaps the most influential of these models was the proposal by Hebb (1949) that synapses are strengthened (synaptic efficacy is increased) when the pre- and postsynaptic cells fire in a correlated manner. Recent physiological and cell biological evidence has rendered many of these formal models simplistic or implausible. A set of formal rules has therefore been proposed (Finkel & Edelman, 1985) that are consistent with recent experimental findings. These rules have been formalized since the biological mechanisms involved are too complicated to model in detail, particularly if the ultimate aim is to model the dynamics of large populations of synapses in various structurally different neural networks. Other synaptic mechanisms could be equally compatible with selection in the nervous system, and in this sense these particular rules are only exemplary. However, they do generate several properties necessary for any selective system and are "realistic," i.e., they are based on the available experimental evidence.

The main assumption here is that the presynaptic and postsynaptic components of synaptic efficacy are separately and independently regulated. Thus, two independent rules are proposed. This reflects the physical independence of the presynaptic process of transmitter release from the postsynaptic process of transmitter binding and generation of a potential across the postsynaptic membrane. Of course, greater realism could be achieved at the price of complicating the model by considering larger numbers of independent or coupled rules.

The presynaptic rule deals with predominantly long-term modifications. They occur at all presynaptic terminals of a neuron, and thus are widely distributed according to the ramifications of the particular axon. The postsynaptic rule involves modifications that are predominantly short-term in nature and are exquisitely localized to those synapses receiving the appropriate spatiotemporal combination of homosynaptic and heterosynaptic inputs (the homosynaptic input is the local input to the synapse in question; heterosynaptic inputs are those to other postsynaptic sites on the same neuron). The postsynaptic rule thus establishes a sensitivity to the pattern of inputs to a cell. Moreover, it provides an explanation for the experimental observation that heterosynaptic inputs can modulate local (homosynaptic) synaptic efficacy.

It has been shown that these two rules can operate concurrently and in parallel in a neural network, and interact such that the long-term presynaptic changes can "recall" appropriate short-term postsynaptic changes and thereby maintain specificity on the long time scale. There is no intrinsic reason why long-term changes could not also occur postsynaptically; but it is argued that long-term changes cannot be specific for individual synapses – they must affect large numbers of synapses on the same neuron. This is because the only mechanism for maintaining a cellular change for a period of many years (such as is required for long-term memory) is to invoke a change in gene expression. However, there is no way to selectively route gene products from the nucleus to appropriate synapses; therefore, the long-term change will be manifested at many of the synapses of a neuron and will thus be spatially distributed. Although several schemes have been proposed to get around this problem (Lynch & Baudry, 1984; Crick, 1984; Lisman, 1985), usually by invoking structural changes in the cytoskeleton, in the absence of experimental evidence for truly long-term local changes the distributed nature of long-term modifications will be taken as a basic constraint.

The presynaptic rule (Finkel & Edelman, 1985) deals with changes in presynaptic efficacy, which is defined as the amount of transmitter released due to depolarization of the terminal. The presynaptic efficacy fluctuates around a set baseline level with transient facilitations and depressions in transmitter release due to the recent pattern of stimulation. For example, short, high-frequency bursts generally cause facilitation, probably due to accumulation of intracellular calcium. Prolonged trains of stimulation can lead to depression of transmitter release due to depletion of transmitter reserves, inactivation of release sites, or some related factors. It is assumed that a long-term average is kept of the fluctuations of efficacy around its baseline value, and that if this average exceeds a threshold, then the baseline efficacy is permanently reset to a new value. The reset baseline efficacy alters the dynamic response of the cell. For example, increasing the baseline efficacy can increase total transmitter release for short stimulus trains and decrease release for longer trains.

The short-term postsynaptic modifications are postulated to result from biochemical modifications of transmitter receptors or voltage-gated channels in the postsynaptic membrane. There is a family of related mechanisms of which I will mention only one. Suppose that local homosynaptic inputs initiate a cascade of local biochemical events (e.g., second messenger-mediated phosphorylations of channel proteins) leading to modifications of local voltage-sensitive channels. Such mechanisms are known to alter the current-voltage relations of channels, thus altering the local potential change produced by a bolus of transmitter. These modifications are intrinsically short-term in nature, since the biochemical reactions are reversible and the channels themselves have a short lifetime.

The main assumption of the postsynaptic rule is that the local modifications are "state-dependent,", i.e., the probability of modifying a channel depends upon what functional state it is in at the time (channels can be in a variety of kinetic states, e.g., open, closed, active, inactive). Heterosynaptic inputs generate voltages that are electrotonically conducted through the dendritic tree. When, after a time delay, the conducted voltages reach the local synapse, they alter the state of some of the voltage-sensitive channels. The critical point here is that if the heterosynaptic voltages increase the fraction of channels in the preferentially modifiable channel state (according to the state-dependent assumption) *at a time* when the concentration of the locally produced modifying substance (for example, cAMP) is still high, then the heterosynaptic inputs will have increased the amount of local modification. In this way, synaptic inputs distributed widely over a neuron can interact to modulate local postsynaptic modifications. The postsynaptic rule critically depends upon the relative timing of the inputs, their anatomical relation to each other on the cell, and on the types of transmitters, receptors, channels, and modifying substances present. In fact, a kind of "transmitter logic" can be imagined in which particular combinations of transmitters are able to selectively interact with each other within defined anatomical and temporal windows. This allows the biochemical diversity of transmitters, etc., to be used for increasing the specificity of synaptic modifications in a combinatorial manner.

To recapitulate, the two independent synaptic rules are a short-term, highly specific postsynaptic rule, and a long-term, anatomically diffuse presynaptic rule. A major prediction of the model is that if these two rules are allowed to operate together in a neural network, their mutual interactions will depend critically upon the anatomical wiring of the net (Finkel & Edelman, 1985). If the connections are randomly organized, with any neuron equally likely to contact any other, then there is no orderly relation between the two rules. However, it has been shown both analytically and through computer simulations that if the network is organized into neuronal groups, then the rules interact to give three useful properties: (1) Short-term changes in a group preferentially generate long-term changes in the same group. (2) Long-term changes generated in a group preferentially enhance short-term changes within the same group. This follows from the stronger connections within the group (versus between groups) that serve to magnify nonlinearly the effect of the long-term changes. The long-term and short-term changes are thus linked through group structure. (3) Long-term changes also generate variability in subsequent patterns of short-term changes – both in the af-

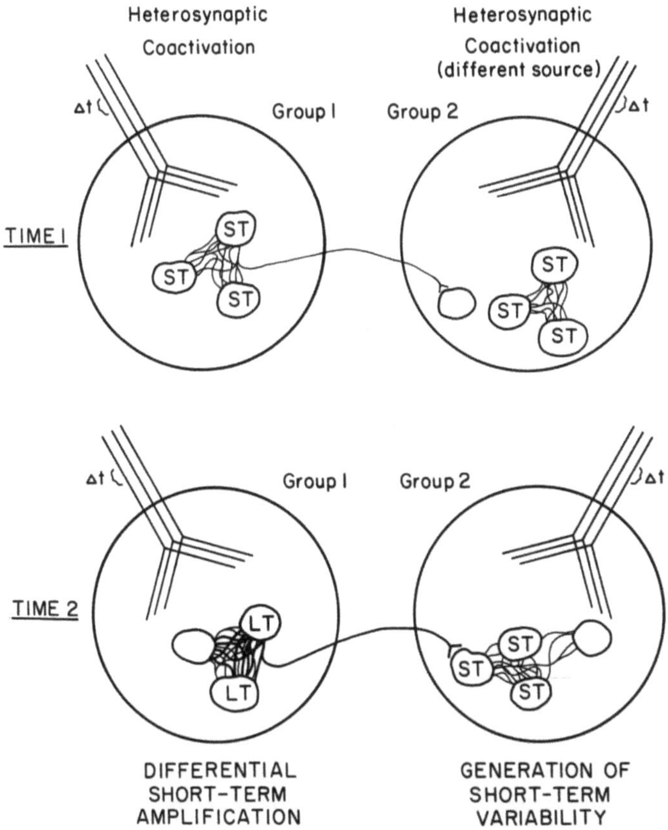

Fig. 2. Interaction of the synaptic rules. Schematic of two neuronal groups (*large circles*) at two different times; only a few cells (*small circles*) in each group are shown. *Top:* Group 1 receives coactivated stimulation from external inputs which induce a pattern of short-term changes (*ST*) in the connections drawn. Group 2 receives another pattern of coactivated inputs resulting in different patterns of short-term changes. The inputs to group 1 are then repeated so as to induce long-term changes (*LT*) in some of the cells of group 1 via the presynaptic rule. *Bottom:* At subsequent times, due to the LT changes, re-presentation of the same external heterosynaptic inputs more easily induce the original ST pattern in group 1. In addition, the divergent fibers from group 1 (which are now strengthened by the presynaptic changes) increase the variability of the ST patterns in group 2

fected group and in other nearby groups due to the wide ramifications of the presynaptic arbors. This generation of variability is a necessary property of all selective systems, and in this system it ensures that groups will always be able to respond to novel situations without becoming fixed or sterotyped in their responses. As shown in Fig. 2, the long-term changes have two effects: they differentially enhance particular short-term changes, and they generate variability in these short-term changes. The synaptic rules lead to competition at the group level wherein long-term changes within each group select particular patterns of short-term changes, but these choices are countered by the effects of long-term changes in other groups. The synaptic changes serve both to increase adaptedness (in that they improve the response to prevalent environmental conditions), and they maintain adaptability to new environmental situations. The synaptic rules, in this sense, serve the same function the rules of genetics perform in natural selection. And thus the mechanism of neuronal group selection also has a type of "heredity" associated with it.

Of Cows, Categories, and Culture

The paradigm that has conditioned most thinking about information storage in the nervous system has been that of associative memory and stimulus-response conditioning. In this view, each input stimulus alters the neural structure/function relation such that re-presentation of a similar stimulus leads to an accentuated response of just that neural construct previously stimulated, i.e., without degeneracy. Hebb (1949) arrived at his synaptic model by interpreting this paradigm in terms of instructionist models of cell assemblies and thus making the reductionist assumption that these stimulus-response associations are localized to the individual synapses. Every current model of synaptic modification has implicitly maintained a paradigm of stimulus-controlled synaptic modifications to yield conditioned associations.

Such a stimulus-response paradigm takes the naive classical and "essentialist" view of categorization, i.e., it assumes that stimulus categories are given and that the problem faced by the nervous system is association of these categories. There can be little doubt that associational processes play a major role in memory. But the missing step in associational paradigms is a clear-cut description of how to categorize the stimulus. The ability to measure whether a particular stimulus is the same as or sufficiently similar to a previous stimulus to warrant the conditioned response, implies an a priori classification scheme. But categories must to some extent be defined by the organism in a context-dependent fashion that is adaptive for that organism in its eco-niche. Therefore, context-dependent categorization is the basic required function of both memory and perception; associations mainly serve to enrich and extend the taxonomy of categories.

The theory of neuronal group selection argues that context-dependent categorization is the result of selection during somatic time in groups of neurons which, in their degenerate patterns of response, will bias the performance of the system upon confrontation with the identical or similar stimulus (recategorization). According to the theory, the environment acts upon preexisting neuronal groups and induces short-term changes in synaptic efficacy. If these changes occur repeatedly and in conjunction with other compatible patterns of activity in the network, they are translated into long-term changes that result in the relative strengthening of the involved groups and relative weakening of others. The neural map formed by group membership determines the ongoing classificatory scheme of the network. Whereas categorization reflects long-term structure, context-dependent effects arise mainly from the short-term modifications. "Context" as used here means the set of all contemporaneous inputs received by the network. In the postsynaptic rule, conducted voltages provide the context for local synaptic modifications. The criterion of state-dependent modification in the postsynaptic rule is the mechanism that enforces context dependency, inasmuch as only those synapses that experience the appropriate conjunction of events will be modified. The existence of both presynaptic and postsynaptic mechanisms in a population model allows the same network to carry out both context-dependent and context-independent operations simultaneously.

One of the tenets of the theory of neuronal group selection is that categorization is central to many higher brain functions, including perception and memory. Indeed, our implicit categorization schemes completely dominate our recognition abilities. It is not possible to review here the literature on preattentive perception in humans (Treisman & Gelade, 1980; Julesz, 1981): suffice it to mention that we are particularly adept at perceiving certain categories and embarrassingly inept at others. These categories seem to be species-dependent, but the ability to categorize and to learn categories is remarkably general across taxa. Herrnstein and colleagues (Cerella, 1979; Herrnstein, 1982) have shown that pigeons, for example, demonstrate an uncanny ability to categorize various objects such as leaves of different species, people, and even fish. Furthermore, these pigeons seem able to generalize upon the categories they learn. Learning, in some senses, can be defined as a process of adaptive categorization. But categorization in this sense obligately involves an output function; a response indicating the result of the categorization. How the nervous system generates this response is perhaps its deepest mystery – it involves the coordination of sensory and motor maps around an object that is defined by that very coordination. Before considering such matters, however, I should state how I define a category.

The Platonic definition of a category had to do with absolute "essential" ideals that are reflected only imperfectly in worldly examples. Thus the spots on one cow differ from those on the next in an insignificant way; the ideal cow is ideally spotted. The variation among members of a category is therefore of no intrinsic importance. Together with the concepts of a hierarchical relationship between species and the great chain of being, such Platonic ideals motivated taxonomic schemes of species classification (Mayr, 1982). The Darwinian theory of evolution provided the empirical answer to the problem of taxonomy by giving the mechanism of generating the taxa, and in this sense can be seen as a theory of categorization. Darwinism proposes that biological categories (species) originate through selective processes. Under selection, the coloration of a cow can be critical for its survival. The definition of a species is still hotly debated by evolutionists (Sober, 1984); our point here is merely that selective systems are natural categorization machines.

Artificial machines constructed to perform pattern recognition almost universally depend upon a Platonic classification scheme. Objects are described by a set of attributes (color, shape, velocity, etc.) usually determined by low-level feature detectors, and class boundaries are defined in the large dimensional hyperspace of attributes. Thus, the "ideal" example of each class can be imagined to sit in the middle of its N-dimensional hypercube and all other members of the class imperfectly resemble the ideal. Parenthetically, the recent development of so-called "Boltzmann machines" (Hopfield, 1982) represents a similar type of view in which, via the formalism of attractors in dynamical systems, a stored memory (of an object) is modeled by the minimum of a complex energy function.

Bongard (1970) recognized that the classification problem could never be solved by a "hyperspace" approach – he compared the classification boundaries in the hyperspace to the boundaries surrounding water in a wet sponge. The problem was articulated by Wittgenstein (1968), who observed that membership in a category need have no singly necessary or jointly sufficient conditions. Taking his example of family resemblances, faces in a family tend to resemble each other, but I may have my mother's eyes, my father's nose, etc., and there is clearly no list of defining attributes (e.g., "you must have black hair" or "long neck and small head mean you are in the family"). Ryle (1951) has formalized these notions in the concept of a polymorphous set. A polymorphous set is defined as those elements that have *any* M out of N possible properties ($M < N$). In other words, given a set of N properties or attributes, any object that possesses a threshold number (M) of the attributes is in the set. Figure 3 shows an example of a polymorphous set whose criterion for membership is that elements must have at least two of the three following properties: dark, round, double-edged. Note that members of the set need share no properties in common, there are no necessary properties, and no unique combination of properties guarantees membership. Thus, in some sense, this definition approaches the kind of categorization sought – a partition of disjunctions of features. It is also tempting to think how naturally suited a neuron or a group of neurons would be to making these threshold decisions.

The selective view of the nervous system is a radical departure from the view of the brain as an information storage and recall device. Memory is seen as a process of recategorization, as a change in the way the event, object, or relationship is categorized. A categorization machine (Edelman & Reeke, 1982; Reeke & Edelman, 1984) based on this concept of categorization has been built. This automaton, called Darwin II, works on selective principles: neuronal groups in various repertoires are selected by synaptic modifications that result from activation of the groups by other groups or by stimulation of peripheral receptors. Darwin II receives no outside in-

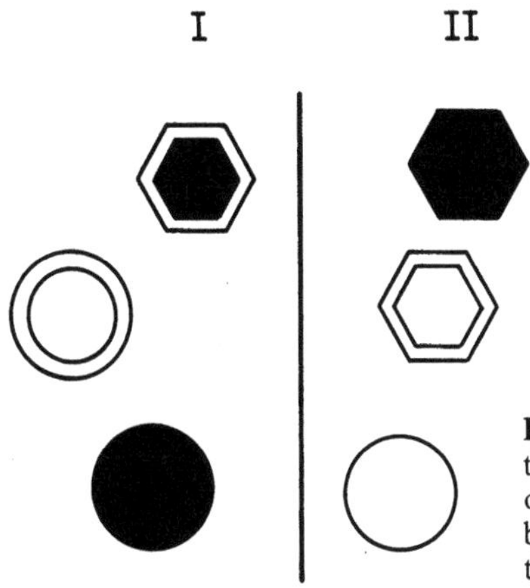

Fig. 3. Example of a polymorphous set. Objects on the left (class *I*) are in the polymorphous set, those on the right (class *II*) are not. The rule for set membership is that the object must have at least two of the properties: dark center, round, or double edge

struction as to how to categorize the geometrical objects presented to it, yet, after repeated presentations, it does respond differently to objects in (what we also would usually recognize as) different classes and it responds similarly to objects in the same class. The successful operation of Darwin II supports the self-consistency of a selective theory. However, the result of Darwin II's categorization must be interpreted from the pattern of activity in its various repertoires – a functioning output device is currently being implemented.

As mentioned above, the output of the system involves the coordination of many maps or repertoires. One of the original principles of the theory was that this coordination occurs through a phasic reentry (see Fig. 1 C) of signals between the maps (Edelman, 1978). The anatomical substrate for such reentry exists, for example, in the reciprocal connections between thalamus and cortex, in the connective loops between cortex, basal ganglia, and cerebellum, and in callosal connections. The basic idea of reentry is that since a neuron is ignorant of the ultimate source of its inputs, it is the context in which these inputs arrive that defines their only "meaning." Given the degeneracy of the network, many such contexts are equivalent. Nevertheless, multiple repertoires operating simultaneously and in parallel can define a common object by the mutual context of their outputs. There are constraints on the temporal delays in the signaling process, but the postsynaptic rule discussed above allows some leniency. The phasic nature of reentry arises from the duration of the temporal window within which inputs can arrive and still be considered equivalent. As a group in a repertoire undergoes synaptic changes due to reentrant inputs, the response of the group changes, and various associations with other groups are allowed or disallowed. In this way, a representation can be generated in an output repertoire which can then drive a degenerate output device such as the musculoskeletal system. Movements of the animal change the stimuli received, and thus provide the final global reentrant pathway (Fig. 1 C).

The idea of the degenerate outputs of a set of muscles, for example the voice-producing muscles, has prompted Liberman (1982) to speak of "gestures" as the categories of related vocal motions. Similar sets of sounds can be generated by many different patterns of activation of the vocal muscles; conversely, very similar movements, when they occur in different contexts, can produce radically different sounds. Plausibly, the generation of these sounds may be intimately related to the neurobiological bases of the words themselves. Thus, the degenerate relations between output patterns in a selective system and their relation to the generated categorical scheme may have some pertinence in the study of the neurobiological basis of language.

The degeneracy of motor responses was first recognized by Bernstein (1967), who noted that the same action can be performed by numerous different muscle groups. For example, he showed that as a child matures there is a dramatic change in the sequence of activation of his or her limbs during the act of running. It is as if the nervous system coordinates (selects) its output with the output device available. This has been shown at a more peripheral level by examining the gait patterns of centipedes as various numbers of legs are successively amputated – with each removal, the animal immediately switches to the most efficient remaining gait (von Holst, 1973). Such coordination is achieved through reentry not only within the system, but also through the global feedback loop involving the behavioral result of the action. This functional loop makes sense from a developmental and evolutionary point of view. Rapid and dramatic changes occur in body morphology and the CNS itself both during normal development and over the course of evolution, and these changes require an ability for a coordination of the mapping between periphery and CNS. Lumsden (1983) has invoked selective arguments to explain how rapid increases in the size of the brain may have come about. The point is that as the environment changes (sometimes due to antecedent changes in the organism or its niche, such as the change from an arboreal to a terrestrial existence), the selection pressures on the animal are altered, and this is compensated for by internal selective changes during development and in the mature nervous system. In summary, a nervous system operating on selective principles can adapt to the external environment as well as to the internal constraints of its body, and can do so in a continuing developmental fashion throughout the lifetime of the individual.

This last point suggests that at higher functional levels, a selective system can act as an instructive or information-processing system. This is equivalent to the emergence of Lamarckian behavior in societies of Darwinian individuals. There is no doubt that culture has Lamarckian components, and indeed this is of tremendous adaptive advantage. The interaction of genes, individuals, and culture has been studied in detail elsewhere (Boyd and Richerson, 1985; Lumsden, this volume). The inference is that culture is adaptive – through imitation and other forms of social learning, individual learning can be rapidly and efficiently propagated to both peers and progeny. Various external rewards or punishments acting through internal value schemes associated with emotions and affect can then serve as the environmental component of selection.

An understanding of the link between percept and concept is still not at hand. However, one can perhaps dimly see how the multimodal nature of the sensory inputs and motor outputs of the nervous system may allow repertoires of selected neuronal groups to reflect the associations and generalizations of perceptual categories. In this sense, selective processes acting on groups of neurons may underlie the higher functions of the nervous system just as they tie these functions to development and evolution.

Acknowledgements. The ideas of the author expressed here developed after a series of critical discussions with Dr. Gerald Edelman. I would also like to thank Drs. George Reeke, Jr. and John Pearson for their insights.

References

Barlow, H.B. (1982). What causes trichromacy – a theoretical analysis using comb-filtered spectra. *Vision Research, 22,* 635–643.

Bernstein, N.A. (1967). *The coordination and regulation of movement.* London: Pergamon Press.

Bongard, M.M. (1970). *Pattern recognition.* New York: Spartan Books.

Boyd, R., & Richerson, P.J. (1985). *Culture and the evolutionary process.* Chicago: University of Chicago Press.

Buchsbaum, G., & Gottschalk, A. (1983). Trichromacy, opponent colors coding, and optimum color information transmission in the retina. *Proceedings of the Royal Society of London. Series B: Biological Sciences (London), 220,* 89–113.

Cerella, J. (1979). Visual classes and natural categories in the pigeon. *Journal of Experimental Psychology: Human Perception and Performance (Washington), 5,* 68–77.

Changeux, J.P., & Danchin, A. (1976). Selective stabilization of developing synapses as a mechanism for the specification of neuronal networks. *Nature, 264,* 705–711.

Crick, F. (1984). Memory and molecular turnover. *Nature, 312,* 101.

Eccles, J.C. (1984). The cerebral neocortex: a theory of its operation. In E.G. Jones, & A. Peters (Eds.), *Cerebral cortex* (vol. 2, pp. 1–38). New York: Plenum.

Edelman, G.M. (1975). The shock of molecular recognition. In E.E. Smith (Ed.), *Molecular approaches to immunology.* New York: Academic.

Edelman, G.M. (1978). Group selection and phasic reentrant signalling: a theory of higher brain function. In G.M. Edelman, & V.B. Mountcastle (Eds.), *The mindful brain: cortical organization and the group-selective theory of higher brain function* (pp. 55–100). Cambridge: MIT Press.

Edelman, G.M. (1981). Group selection as the basis for higher brain function. In F.O. Schmitt, F.G. Worden, G. Adelman, & S.G. Dennis (Eds.), *Organization of the cerebral cortex* (pp. 535–563). Cambridge: MIT Press.

Edelman, G.M. (1983). Cell adhesion molecules. *Science, 219,* 450–57.

Edelman, G.M. (1984a). Modulation of cell adhesion during induction, histogenesis, and perinatal development of the nervous system. *Annual Review of Neuroscience, 7,* 339–377.

Edelman, G.M. (1984b). Cell adhesion and morphogenesis: the regulator hypothesis. *Proceedings of the National Academy of Sciences of the USA, 81,* 1460–1464.

Edelman, G.M. (1985). Cell adhesion and the molecular processes of morphogenesis. *Annual Review of Biochemistry, 54,* 135–169.

Edelman, G.M., & Finkel, L.H. (1984). Neuronal group selection in the cerebral cortex. In G.M. Edelman, W.E. Gall, & W.M. Cowan (Eds.), *Dynamic aspects of neocortical function* (pp. 653–695). New York: Wiley.

Edelman, G.M., & Reeke, G.N. Jr. (1982). Selective networks capable of representative transformations, limited generalizations, and associative memory. *Proceedings of the National Academy of Sciences of the USA, 79,* 2091–2095.

Finkel, L.H., & Edelman, G.M. (1985). Interaction of synaptic modification rules within populations of neurons. *Proceedings of the National Academy of Sciences of the USA, 82,* 1291–1295.

Hebb, D.O. (1949). *The organization of behavior.* New York: Wiley.

Herrnstein, R.J. (1982). Stimuli and the texture of experience. *Neuroscience and Biobehavioral Reviews, 6,* 105–117.

Hopfield, J.J. (1982). Neural networks and physical systems with emergent collective computational abilities. *Proceedings of the National Academy of Sciences of the USA, 79,* 2554–2558.

Julesz, B. (1981). Textons, the elements of texture perception and their interactions. *Nature, 290,* 91–97.

Kaas, J.H., Merzenich, M.M., & Killackey, H.P. (1983). The reorganization of somatosensory cortex following peripheral-nerve damage in adult and developing mammals. *Annual Review of Neuroscience, 6,* 325–356.

Kramer, A.P., Golden, J.R., & Stent, G.S. (1985). Developmental arborization of sensory neurons in the leech *Haementeria ghilani,* 1. Origins of natural variations in the branching pattern. *Journal of Neuroscience, 5,* 759–767.

Liberman, A.M. (1982). On finding that speech is special. *American Psychologist, 37,* 148–167.

Lisman, J.E. (1985). A mechanism for memory storage insensitive to molecular turnover – a bistable autophosphorylating kinase. *Proceedings of the National Academy of Sciences of the USA, 82,* 3055–3057.

Lumsden, C.J. (1983). Neuronal group selection and the evolution of hominid cranial capacity. *Journal of Human Evolution, 12,* 169–184.

Lynch, G.S., & Baudry, M. (1984). The biochemistry of memory: a new and specific hypothesis. *Science, 224,* 1057–1063.

Macagno, E.R., Lopresti, V., & Levinthal, C. (1973). Structure and development of neuronal connections in isogenic organisms: variations and similarities in the optic system of *Daphnia magna. Proceedings of the National Academy of Sciences of the USA, 70,* 57–61.

May, R.M., & MacArthur, R.H. (1972). Niche overlap as a function of environmental variability. *Proceedings of the National Academy of Sciences of the USA, 69,* 1109–1113.

Mayr, E. (1982). *The growth of biological thought: diversity, evolution, and inheritance.* Cambridge, MA: Harvard University Press.

Mountcastle, V.B. (1978). An organizing principle for cerebral function: the unit module and the distributed system. In G.M. Edelman & V.B. Mountcastle, (Eds.), *The mindful brain: cortical organization and the group-selective theory of higher brain function* (pp. 7–50). Cambridge, MA: MIT Press.

Pearson, K.G., & Goodman, C.S. (1979). Correlation of variability in structure with variability in synaptic connections of an identified interneuron in locusts. *Journal of Comparative Neurology, 184,* 141–165.

Reeke, G.N., & Edelman, G.M. (1984). Selective networks and recognition automata. *Annals of the New York Academy of Sciences, 426,* 181–201.

Rockel, A.J., Hiorns, R.W., & Powell, T.P.S. (1980). The basic uniformity in structure of the neocortex. *Brain, 103,* 221–244.

Ryle, G. (1951). *The concept of mind.* London: Hutchinson.

Sober, E. (1984). *The nature of selection.* Cambridge, MA: MIT Press.

Spemann, H. (1938). *Embryonic development and induction.* New Haven: Yale University Press.

Szentagothai, J. (1978). The neuron network of the cerebral cortex: a functional interpretation. *Proceedings of the Royal Society of London. Series B: Biological Sciences, 201,* 219–248.
Treisman, A., & Gelade, G. (1980). A feature-integration theory of attention. *Cognitive Psychology, 12,* 97–136.
von Holst, E. (1973). *The behavioral physiology of animal and man: the collected papers of Erich von Holst.* London: Methuen.
Weiss, P. (1939). *Principles of development.* New York: Henry Holt.
Wittgenstein, L. (1968). *Philosophical investigations* (3rd ed.). Oxford: Blackewell.
Young, T. Z. (1979). Learning as a process of selection and amplification. *J. Roy. Soc. Med.* (Lond.) 72, 801–814.

Part 2 The Evolution of Writing Systems

Introductory Remarks

This section brings together four papers on the history and development of orthographies, with a special emphasis on the Greek alphabet. The intent is to identify the kinds of relationships existing between the structure of languages and their graphemic representations. The section goes from a general introduction to writings systems (Hagège) to the particular study of one of today's descendents of the Greek system, the French alphabet (Bhatt). Joseph Naveh and Robert Lafont's papers help to place the invention of the alphabet in its historical and its structural contexts.

CHAPTER 4

Writing: The Invention and the Dream

CLAUDE HAGÈGE[1]

The term "writing" can have many different meanings. In its broadest sense it might be taken to include such examples as the cave-writing mythograms of the superior Paleolithic period depicting hunting scenes. However, if the definition is limited to the modern meaning, according to which it is a technique of representing speech by a durable trace, one can then speak (in a fairly broad sense) of an "invention," and attribute a historical role to it. After all, this invention was a remarkable leap forward for that portion of humanity able to take advantage of it; a leap that must have been comparable to the discovery, much further back in the depths of time, of the uses of fire. Through it, our species began to have at its disposal a durable means of concretizing speech and of holding historical knowledge back from the edge of the abyss threatening to engulf it, as collective oral memory, stretched across thousands of years of transmission, could no longer suffice to do.

Thus, for ancient civilizations the birth of writing was also the birth of history. Like any revolutionary innovation it brought both positive and negative effects. The intangibility of its contents and its dissimilarity to oral language altered many of the normal circumstances of discourse, creating long-distance dialogues where the usual proximity of the communicators was lacking. Yet precisely because of this, knowledge could become accessible to a far greater number of recipients – writing possessed the advantages of both longevity and range. In spreading to different areas and societies it allowed for all the changes, input, and variations that any culture would require, permitting the encoding of new words as well as of already existing ones.

Writing has the power to initiate reflection and possibly to encourage the higher processes of analysis and abstraction. Humans in nonliterate societies in no way lack these abilities, but they must develop them by means that are without doubt less generally available. Moreover, at least one intellectual activity is not conceivable at all without writing: sequential numbering, which presupposes a system of number signs and a written order of arithmetical succession.

In prehistoric times, the inclination for social living and the aptitude for language came, in an increasingly decisive way over hundreds of thousands of years, to distinguish a new species. However, as far as is presently known,

[1] École Pratique des Hautes Études, Paris, France.

it was only in a small number of societies that writing appeared. In all cases its development seems to have been linked to a particularly complex order of human relations and finely-woven network of hierarchies, characteristic of urban societies with highly structured economies. Thus, it was not the inevitable result of a "natural" development, much less a defining property of the human species.

To grasp the importance of the historical role of writing and its influence over the fate of our species, a digression is necessary. The three centers of civilization where the phenomenon appeared were all established agricultural societies, partially urbanized, with relatively large populations and well-developed systems of exchange. The two situated in the Middle East, Sumeria and Ancient Egypt, invented writing at almost exactly the same time, 200 years apart: Sumeria in approximately 3300 B.C. (the Uruk inscriptions), and Egypt in approximately 3100 B.C. It is not clearly known whether one was in fact the model for the other: the relationship between the two civilizations was certainly close, but the likelihood of a direct influence is questionable when the differences between their techniques are noted.

In Sumeria, which arose on the silty flooded earth of Lower Mesopotamia, the writing base employed was a block of fresh clay molded into tablets. A *calamus,* made of a reed capped with the heads of nails, was used as the writing instrument. This imprinted lines in the form of wedge-shaped figures, which led to the term "cuneiform" (from the Latin *cuneus* meaning corner). This technique passed through two classical stages: the *pictogram,* or simple representation of an object, and the *ideogram,* or schema of an idea, a character corresponding to a word in the language. Due to ever-increasing stylization, any resemblance between the characters and the objects they represented was soon abolished. (These historical forms, despite their antiquity, regain familiarity when one thinks how the modern world is increasingly discovering the advantages of ideograms: tourist guidebooks, public notices, street signs, all kinds of advertising, boxes and packages of objects printed with unambiguous schemas indicating the top, bottom, fragility, volumetric capacity, etc. – see Eco, 1964). It was only after the stage of pictograms that *phonograms* appeared in Sumeria (André-Leicknam, 1982), a phonogram being a sign which, because it represented a common word whose pronunciation contained a particular sound, came to be used as the written version of that sound for all other words or parts of words that contained it.

In comparison, Egyptian scribes used the stems of rushes as the writing instrument, chewing the extremities to form brushes, which were then dipped in lamp-black ink. They would run these over sheets of papyrus, a writing material made from a cyperaceous plant that in those days grew in abundance on the banks of the Nile. The stems of this plant were cut into sections and sheets of it pressed one against the other in order to produce, after a process of drying, smoothing, and assembling, a supple yet solid roll (Beyer, 1982, p. 351). Materials were not the only difference between the techniques of Egypt and Sumeria. A far more fundamental distinction is that

Egyptian writing appears to have always existed in a single form, as far back as the documentation goes. In fact, the hieroglyphs of even the most ancient texts already include not only both pictograms and ideograms, but also a complete system of phonograms, functioning in the same way as that of cuneiform; i.e., on the same principle as the rebus. They even contain a series of special signs called *determinatives,* which, when placed next to those signs that corresponded to homonymic words – i.e., words possessing the same consonants, the only letters there were – allowed one to resolve ambiguities by specifying, as do the keys of homophonic Chinese characters, the semantic or syntactic category to which the word belonged.

The degree of refinement shown by this system of writing, considering its antiquity, has for a long time been poorly understood. Its interpretation has led to many misunderstandings. According to Rousseau:

> The more the writing is unrefined, the older the language is. The first way of writing is not to paint the sounds but the objects themselves, either directly, as did the Mexicans, or by allegorical figures, as did the Egyptians. This state relates to passionate language and presupposes a form of society and the needs that the passions have led to . . . The painting of the objects is appropriate for savage peoples (Rousseau, 1826, chap. V).

Champollion deciphered the hieroglyphs in 1822, but a full 6 years later C. Nodier still claimed: "The names of things when spoken have been the imitation of their sounds, and the names of things when written the imitation of their forms. Onomatopoeia thus belongs to spoken languages and hieroglyphs belong to written languages" (Nodier, 1828, p. 11). Thus Nodier, whose name is associated in literature with fantasy and illuminism, sought the key to the mystery of language in neo-Cratylic speculations at the very time of the rise of comparative grammar. The view he takes here is not surprising when one further reads in his *Notions elementaires de linguistique:*

"The names of created beings were . . . their true names in Adam's language . . . who . . . formed them according to his sensation, that is, by reason of the most salient aspect in which the thing appeared to him" (Nodier, 1834, cited by Yaguello, 1984, p. 182).

These mystical visions obviously failed to recognize the complexity of the cultures that invented cuneiform and hieroglyphic writing. In both cases, despite the differences mentioned, the birth of writing seems to be have been linked to the development of increasingly complex social systems, necessitating the management of accumulated riches. Just as money resulted from the substitution of signs for things, writing too was (in the Near East) the invention of merchants. Hermes, the Greek god of craftiness, thieves, and merchants, corresponded in Egyptian mythology to Thot, the god of sciences and technology, who was also the god of writing (Plato, in *Phaedrus,* considers him to be its inventor). By all appearances, it was the peripheral users of writing – foreigners, travellers, merchants of all the regions surrounding the two great central empires – who were responsible for a decisive step forward: stylization.

This is the first step on the path leading to true writing, detaching it from pictographic representations and bringing about the development and eventually the systematization of phonetic syllables. Another instrumental group may have been the scribes. The extreme specialization required by this profession, which required a long apprenticeship and consequently the appropriate financial means, made the knowledge of writing a privilege. It is quite possible that the scribes, invested with official functions, and the priests, who appropriated their services, were themselves the developers of stylized writing. Perhaps, having come to monopolize an invention that had been collectively conceived, they diverted it for their own profit. After all, writing is an instrument of power: it enables the sending of orders to far-off fiefdoms and can determine which laws will prevail. And if it is filled with mysteries, it is all the more effective. One can believe that "esotericism, far from being the first form of knowledge, is in fact only a perversion of this knowledge" (Foucault, 1966, p. 103). Pure hypothesis, perhaps, but ancient Egypt was certainly not the only place where elite classes were careful to preserve their privileges and little inclined to share them. To, mention only one comparable example from an entirely different geographic and cultural setting, the knowledge of Aztec writing, another complex system, was also restricted to priests and eminent members of the community: "Halfway between pictography and phonetism, passing through ideography. Aztec writing remains esoteric, as does knowledge itself in a society that is highly hierarchized" (Soustelle, 1975, p. 173).

However, contact with other societies cannot occur without eventual exchanges that profoundly affect the established orders. In Mesopotamia, beginning in the first half of the third millenium B.C., the Semitic language that coexisted with Sumerian also used cuneiform. In the study of cuneiform, one notes (in a somewhat similar way as in Japanese, which uses a special syllabary called Katakana) the numerous borrowings by Sumerian from the Semitic language, including names (Bottéro, 1982, p. 32). This situation gave rise to two fundamental consequences. In Akkadian writing, the official language of the Akkad Empire from 2340 B.C., and as a consequence in Sumerian itself (since at that date the Semitic Akkad Empire conquered the Sumerians and thereby both inherited and influenced their writing system), phonograms flourished to the detriment of ideograms, after a period of mixed usage. This eventually led to a system that encoded the actual sounds of language, with signs representing, unit by unit, the phonemes as they were pronounced by the users. Ultimately, this situation led to an invention of incalculable importance: the alphabet. Its first expression then, dated at approximately 1500 B.C., was cuneiform rather than hieroglyphic, despite the numerous contacts that its creators, the Semitic inhabitants of the Kingdom of Ougarit (today Ras Shamra, in Syria) had with the Egyptians.

This invention, as impressive as it was, was destined to remain imperfect. One notes in all languages a fairly rapid modification of pronounciation, which can make an initially faithful graphism obsolete. This is the cause of the complexity of modern French orthography, for example. It must be noted, however, that even though alphabetic notations may become more difficult due to their retaining traces of ancient pronunciations, they can also thereby serve as a stabilizing factor. Conversely, widespread ignorance of the alphabet can be a factor in the *increase* of changes. It was during the Middle Ages, a period that preceded printing and when illiteracy was very widespread, that the French language underwent its most significant evolution.

In any case, from the moment the alphabet was born its advantages far outweighed its disadvantages. Very soon it came to encode many languages, both Semitic and non-Semitic (Février, 1959, pp. 173–179) – as did a later alphabet, the linear writing of the merchants of Phoenicia (modern Lebanon). The Phoenician letters, in straight or curved style traced on papyrus, were destined for a brilliant future. It is a form of this alphabet, having evolved through different stages that included the adding of vowels by the Greeks (consonants were the only elements up to that point), that has survived right up to the present time in the Western world. It was not mere chance that the alphabet was invented by people who spoke a Semitic language. *Writing is a linguistic analysis at various degrees of consciousness.* Given the nature of the language they spoke, the Semites could not end written segmentation at the level of the word, as do, say, Chinese ideograms, which encode a monosyllabic language with unchanging words. In Semitic languages a very large number of words are comprised of more than one syllable, and the changes of consonants and vowels (alternations) play a grammatical role – for example, they distinguish the singular and plural of nouns, as well as different verb forms. It was the more or less conscious understanding of phonemes, linked to this type of language, that favored the blossoming of the alphabet.

Reciprocally, alphabetic writing has nourished a semiotic tradition that belongs to the West. Alphabetic letters transcribe (though imperfectly, due to phonetic changes) the sounds that constitute the words; hence, to linguists in the Greek and Latin tradition, the meanings of the words themselves appear to be linked by a univocal relationship. This is not the case with ideograms, as for example in modern Chinese writing or the ideographic part of Japanese writing (the other part of Japanese is a syllabary). A word denoted by an ideogram is divorced from its attachments to sound, and consequently is conceived of separately from the relationship between phonetic structure and content. Thus, this type of writing does not favor the formation of a univocal link between the signs and the words.

With this in mind, Sumeria and Egypt, the earliest centres of writing, must be considered in separate terms from the history of the alphabet. It is a posteriori, and because the Middle East and the West are the seats of literate civilizations, that there has been a tendency to assign to prealphabetic writing systems a teleology destining them to become alphabets. The nature of Egyptian writing proves that there is no element of necessity in this evolution. An old Europocentric preoccupation has led to the seeking of the origin of alphabetic writing in all stages of the history of writing, whereas it is strictly the "reciprocal role of the words and the signs" (Leclant, 1975) that sould be the focus of interest.

The third major writing system contributes to a better understanding of this role. Chinese characters certainly show some traits in common with those of Sumeria and Egypt. One is their antiquity [although there is not full agreement as to this; some scholars (Février, 1959, p. 69) say they date back to the end of the second millenium B.C., and others (Tsung-I, 1982, pp. 272ff) to 4000 B.C.]. Another is their spread over an entire cultural region, in this case the Far East. They were used until the 17th century in Vietnam, and are still in use in Japan, where they are associated with the syllabary signs, and on a smaller scale in Korea, where a very precise semi-alphabetic code is in use (Hagège & Haudricourt, 1978, pp. 31–37).

However, at this point the similarity to the Sumerian and Egyptian systems ends. The origin of Chinese writing appears to have been magicoreligious and divinatory rather than economic and mercantile. Moreover, even though some stylization of the pictograms did take place, it was not very generalizable, and traces of the direct representations of the words are still visible today. But above all, despite the introduction of a phonetic principle into the majority of the characters, thus combining notations of sound and meaning to create "ideophonograms", this system never resulted in a syllabic form of writing. And this is even less so since the rebuses which are the foundation of this practice have never been systematized: neither in extension, since there is no character having a fixed phonetic value for all elements of the language that are phonetically identical to the one that the character originally notes, nor in comprehension, since, most often, the phonetic portion, in those words where it appears, only notes certain features, not the exact pronunciation of the corresponding word. Furthermore, as in all languages, pronunciation has been modified over the course of time, so the inexactitude is all the greater. Since in any case the nonphonetic portion of the ideophonogram represents only the meaning and not the sound, Chinese characters give no indication of the important phonetic changes that have marked the history of this language.

This system of pictograms and ideophonograms, which has existed in a more or less fixed form since ancient times, has been preserved right up to the present day. It is of great interest that it has long held a degree of power over the imagination of Westerners. The speculations of poets and philosophers demonstrate the recurrent temptation of those who feel both master and slave to words to escape this prison-house of language. Here, it is through writing, toward and against speech, that one hopes to open a pathway.

As long as an exact understanding was not available, ignorance fed speculation. P. A. Kircher was fascinated by hieroglyphs, completely excluding the possibility of their decipherment and being content to see in them "a more excellent system of writing, more sublime and closer to abstraction which, by a more ingenious linking of the symbols, or their equivalent, proposes directly to the intelligence of the wise complex reasoning, high ideas or some remarkable mystery hidden in the bosom of nature or of divinity" (Kircher, 1636, cited by Derrida, 1967, p. 120). In the 18th century, a number of important literary figures could not resist the utopian quest for a *pasigraphism,* a universal system of writing that would be comprehensible to all, whatever one's country and whatever one's language. Leibniz wished to take, with a few modifications, Chinese writing as the model, which he admired as he judged it to be more "philosophical" than Egyptian writing. The ideal system would be "a universal type of writing that would have the advantages of the Chinese system, since each could understand it in his own language, but it would infinitely surpass Chinese since it could be learned in a few weeks, since its characters are well related according to the order and the connection of things" (Leibniz, 1703, p. 25).

The truth is that what was known of the Chinese writing system at the time was largely through Jesuit missionaries, and partially erroneous. It was not until 1836 that P. S. Du Ponceau, a sinologist and Americanist, showed in his *Dissertation on the Nature and Character of the Chinese System of Writing* that this system represented the Chinese language only, and not a universal system of ideas (Hagège, 1982, pp. 4–9).

As for poets, Chinese writing, which says things beyond the shell of words, is for many an object of fascination. The ideographical dream (see Formentelli, 1982, pp. 209–233) abolishes the prisons of language and seeks to rediscover the harmony of worlds buried within the drawing, where history is also inscribed. For whatever effort one makes to imagine the vocal articulations of early humans during the infancy of language, the mythograms of the anthropologist, faraway ancestors of pictograms, cover only the walls of the caves. Voices have not left fossils.

This "sacralization" of a nonphonetic writing system can only be conceived of at the expense of speech. It is not accidental that the scholarly study of speech, as it has emerged over the centuries to become one of the major preoccupations of linguistics today, originated with Westerners who use a writing system that reflects sound:

> Writing, which in China did not result in a phonetic analysis of language, was never felt to be a more or less faithful imitation of speech and it is for this reason that the graphic sign, symbol of a reality that is as unique and singular as itself, has kept much of its primitive prestige. There is no reason to believe that speech did not have in early times in China the same effectiveness as writing, but its power was in part eclipsed by that of writing. On the contrary, in civilizations where writing evolved very early towards a syllabary or an alphabet, it is the word which concentrated in itself all the powers of magic and religious creation. And in fact, it is noteworthy that one does not find in China this remarkable importance attributed to speech, to the word, to the syllable or to the vowel that one finds in all the great ancient civilizations from the Mediterranean basin to India (Gernet, 1963).

Even if alphabetic writing appears to be closer to speech and pronunciation, the distance, as we shall see, remains great between the activities of writing and speaking, and also the cultural attitudes and conceptions of language that underlie each of these activities.

Writing as an End in Itself

Whatever the virtues of spoken language, it is through writing that humanity is best able to express an age-old dream: the dream of a release from nature, from the material tissue, from one's own constraining existence. The divergence between written and spoken language can go very far. In Chinese, for example, it resulted very early on in a shorthand language wherein most of the words can, according to the context, fulfill quite diverse functions: this shorthand language, called *wenyan,* never seems to have had a correspondence with true spoken usage (Hagège, 1975, pp. 21–22), let alone that for at least 1000 years Chinese writing was used for only ritual and magical purposes. However, it is true that the resistance of Chinese to the romanization of writing cannot be explained by tradition alone. Only ideographic characters can serve to distinguish homophones, which this language contains in rare abundance. At any rate Chinese is an extreme case, since *wenyan* constitutes a third register of linguistic processing in addition to written and spoken Chinese.

Such a divergence does not simply indicate two systems representing the same level of meaning in different ways. In reality it implies a divergence between their actual purposes, with speech being easier and more spontaneous, and writing more prestigious and endowed with greater power. For when one writes, even if only to a single recipient and particularly to an unfamiliar one, a more solemn function is conferred upon the message and more formality taken with its construction. It has often been noted that in a given language the written and spoken styles call upon different grammatical resources. In English for example, written communications contain a greater number of nominalizations, participles, and qualifying adjectives than their oral counterparts (Chafe, 1982, pp. 35–53). In some cases the prestige given to writing extends to using archaic forms of language and outdated sentence structures derived from scholarly writings, as is often seen in liturgy. Thus, modern tongues such as the Romance languages, Indo-Aryan languages, Bulgarian, Burmese, modern Arabic, Amharic, and modern Mongol, still utilize, respectively, Latin, Sanskrit, Old Slavic, Pali, Koranic Arabic, Gueze, and classical Mongol. (It must be noted that the use of an old religious language is not unknown in oral societies as well; Hawaiian is an illustration of this, although on a fairly restricted scale.)

The autonomy of the written word consecrates it as an end in itself. In literate civilizations, literary pleasure is first of all the pleasure of style. All elements contribute to individualize writing as a style in itself. What writing expresses above all is the abolition of linearity, this inescapable law of oral expression, which for so long has been at the center of reflection on language. Upon the horizontal plane of the writing surface, scripts have the possibility of being written in various directions: vertically or horizontally, going to the right or to the left or even both, as in boustrophedon. The hieroglyphs of Ancient Egypt were an extreme example of counterpoint, but other manifestations of escape from the constraints of linearity are found everywhere, in all ages. Palindromes can only be conceived of in the written form, as they are sets of words or sentences that are identical if read from left to right and from right to left. The so-called "concrete" and "spatializing" poetry of today are not imprisoned, as is oral poetry, within the constraints of a single dimension. Calligrams, iconosyntax, toposyntax and other techniques dating back to *Coup de dés* (Throw of the Dice) by Mallarmé, can give to the text a picture-like contour that reflects its content. Other procedures that confer autonomy on writing as an end in itself include typographical techniques such as indentations, blanks, chapters, capital letters, titles, and subtitles. By freezing speech in time through spatializing it, these techniques transform it into an object in two dimensions on the page and three dimensions in a volume. By adding new components they impose (if imperfectly) rhythm and life (Butor, 1982).

The interpretation (reading) of alphabetic writing itself, which implies highly complex cerebral mechanisms (Husson, 1967, pp. 23–28), does not necessarily address the represented phonemes, even though it is true that this writing, which is analytic, represents them with relative exactness. If this were so, deaf-mutes would only be able to read those words that they have learned to articulate, but if correctly educated, they are in fact able to read and write a far greater number. In those cases where their abilities are limited to those words that they can articulate, the fault lies in poorly conducted training, based on the scriptophobic premise that a direct relation between written words and referents is impossible. Such an approach fails to appreciate the relative autonomy of the written code in relation to oral language (Alisedo, 1972).

However, writing is not autonomous in relation to culture. For instance, Japanese writing is a complex combination of two systems that use Chinese characters. There are roughly 850 of these characters, many possessing one or two Sino-Japanese meanings in addition to the Japanese one. This writing system is quite poorly adapted to the language it encodes. However, when it was borrowed from China in the fourth century A.D., it allowed the recording of a language that had up to that point been without any writing system, and so became deeply integrated into Japanese civilization. The ideograms are also a means of artistic expression, and attempts to increase the use of syllabaries have only resulted in the establishment of a limited number of officially recognized characters. Turkey provides another example: here, it was because Arabic writing was intimately linked to Islam and denoted Arabic words belonging to a philosophical religious, and political vocabulary that were abundant in the Turkish lexicon, that Mustapha Kemal, wishing to secularize his nation, had the Latin alphabet adopted in 1928. His aim was not a simple spelling reform but a cultural revolution.

If writing is only slightly independent of culture, it is much more independent of spoken language. Writing has a surprising ability to instill meaning into an object. It tends to become what its nature carried in seed at its inception: an aesthetic. Egyptian hieroglyphs very early on become part of the overall decor, and their plastic arrangements can only be interpreted as a love for the written sign. Chinese calligraphy is intimately linked to poetry and to painting, which it always accompanies and of which it is in fact a component. Certain complex Chinese characters, made from combinations of many simple ones, permit graphic games: by juxtaposing the complex and the simple, one can sometimes end up with interpretable sentences (Alleton, 1970, pp. 63–66). Arabesques transmit in stone aesthetic messages as well as verses from the Koran. Devanagari writing and its numerous variants and derivatives in the countries of South East Asia speak to the eye by proposing different decorations to it, through the types of drawings it enables one to make.

Orality, Literacy and Society

Is it the wish for entry into the economic structure of the contemporary world, or simply a consequence of colonialism, that leads so many countries today, African ones in particular, to adopt the alphabet to encode languages that up to now have been purely oral? Or is this due to pressure by the media, who unambiguously decry the notion of illiteracy? Now is certainly not the time to revive the tone of neo-Rousseaulian elegies. It is absurd to claim that writing can be an instrument of oppression because it allows the transmission and enforcement of orders: laws need not be equated with oppression. It is doubtful that it was really just because he attempted to found his authority on the semblance of a writing system that the chief of the Nambikwara was abandoned by his followers (Lévi-Strauss, 1955, pp. 337–349; see also Derrida, 1967, p. 191; Calvet, 1984, pp. 105–111). However, the introduction of writing into the heart of an oral society does require certain precautions. Writing has been a progressive rather than a spontaneous development, and important cultural differences separate societies that are literate from those that are not. The latter have over many years developed on the basis of oral language their own modes of expression, their own systems of exchange and balance. To avoid the risks of a dangerous intervention of writing into an oral milieu, these societies must design for themselves the paths through which they hope eventually to accede to the rewards of literacy. No one can deny them this goal. It should also be recognized, however, that the notion of illiteracy in oral societies need not have the stigmatizing and privative implications that it does in those Europocentric parts of the world where languages have been written for many years (Hagège, 1975, pp. 251–266). The depositories of the history of oral societies are their scholars and their poets.

"Writing veils one's view of language: it is not a cloth but a travesty," taught F. de Saussure (1972, pp. 51–52). And long before him, Rousseau stated: "Languages are made to be spoken, writing is simply a supplement to speech" (1826, chap. 8). Jacques Derrida (1967, chaps. 1 & 2), a modern advocate of writing, reproaches these illustrious scriptophobes for their phonocentrism: he suggests that by giving priority to the oral discourse they have ignored the trace, which, since it does not require presence, is more powerful because it represents.

With regard to the relationships between written and oral discourse, despite the problems posed by their popularity, the new mechanical means of preserving speech (electronic recording devices) are not without relevance for the field of linguistics. It was, a very long time ago, the invention of alphabetic writing that no doubt gave the first decisive impulse to grammatical research. From the moment that a single sign was consistently used to note the numerous regional or individual variations of a *p*, an *a*, or an *r*, an important phenomenon emerged: despite the immensity of these variations, members of the same linguistic community could understand one another. Invariants must exist, and what is linguistics but the definition of these invariants, in the domain of pronunciation as well as in lexicon or syntax? However, the reason why change is not impossible in the near future is that machines which record speech do the opposite of what the linguist does: they retain only variation. Linguistics cannot remain indifferent to this evolution in technology: in fact, it has taken advantage of it as an opportunity for renewal. Variations in speech were being studied well before the advent of machines that could accurately reproduce them, but the availability of such machines has advanced what had already begun. Born of the discovery of invariants, linguistics is largely becoming a science of variation on the basis of invariance – a science that no longer studies sameness in itself, but subsumes it under the thousand faces of otherness.

But is writing, for all its power, really assured of such a brilliant future that those deprived of it should be forced to covet it so greatly? The last few decades of technological advance have seriously eroded its power and endangered its reign. From public officials to advertisers, from poets to reporters, the professions are steadily increasing in which all effective action, whether it be to inform, to entertain, or to persuade, can no longer content itself with written messages but must be transmitted through speech. The tape recorder, the computer, the video recorder all have profound potential for changing, or are already in the midst of changing, the relationship between the spoken and the written word. We do not know what their exact effect on the basic essence of language will be, but they definitely exert a negative influence. Despite the essential role it still plays and the prestige it still preserves, writing is finding itself in an increasing position of exteriority.

References

Alisedo, G. M. (1972). *Las languas graficas.* Unpublished doctoral dissertation, University of Cordoba, Argentina.

Alleton, V. (1970). *L'écriture chinoise* (pp. 63–66). Colloque "Que sais-je?" Paris: Presses Universitaires de France.

André-Leicknam, B. (1982). Cunéiformes et hiéroglyphes. *Naissance de l'écriture.* Paris: Editions de la Réunion des musées nationaux.

Beyer, D. (1982). *Naissance de l'écriture* (p. 351). Paris: Editions de la Réunion des musées nationaux.

Bottéro, J. (1982). De l'aide-mémoire à l'écriture. *Ecritures, systèmes idéographiques et pratiques expressives* (p. 32). Actes du Colloque international de l'Université Paris VII. Paris: Le Sycomore.

Butor, M. (1982). Le livre comme objet. *Répertoires II.* Paris: Minuit.

Calvet, L. J. (1984). *La tradition orale* (pp. 105–111). Colloque "Que sais-je?" Paris: Presses Universitaires de France.

Chafe, W. L. (1982). Integration and involvement in speaking, writing, and oral literature. In D. Tannen (Ed.), *Spoken and written language* (pp. 35–53). Norwood, NJ: Ablex.

Derrida, J. (1967). *De la grammatologie,* Paris: Minuit.

Eco, U. (1964). *Apocalittici e integrati.* Milano: Fabbri-Bompiani.

Février, J. G. (1959). *Histoire de l'écriture.* Paris: Payot.

Formentelli, E. (1982). Rêver l'idéogramme: Mallarmé, Segalen, Michaux, Macé. *Ecritures, systèmes idéographiques et pratiques expressives* (pp. 209–233). Actes du Colloque international de l'Université Paris VII. Paris: Le Sycomore.

Foucault, M. (1966). *Les mots et les choses* (p. 103). Paris: Gallimard.

Gernet, J. (1963). Aspects et fonctions psychologiques de l'écriture. *L'écriture et la psychologie des peuples.* Actes du colloque de Paris. Paris: Armand Colin.

Hagège, C. (1975). *Le problème linguistique des prépositions et la solution chinoise.* Louvain: Peeters.

Hagège, C. (1982). *La structure des langues* (pp. 4–9). Colloque "Que sais-je?" Paris: Presses Universitaires de France.

Hagège, C., & Haudricourt, A. G. (1978). *La phonologie panchronique* (pp. 31–37). Paris: Presses Universitaires de France.

Husson, R. (1967). Mécanismes cérébraux du langage oral, de la lecture et de l'écriture. *Les Cahiers du Collège de Médecine, 1,* 2.

Kircher, A. (1636). *Prodromus coptus sive aegyptiacus.* Rome.

Leclant, J. (1975). Présentation. *Le déchiffrement des écritures et des langues.* In J. Leclant (Ed.) *Actes du colloque du XXIXe congrès international des orientalistes.* Paris: L'Asiathèque.

Leibniz (1703). Letter to Reverend Bouvet. *Philosophische Schriften,* Gerhardt edition.

Lévi-Strauss, C. (1955). *Tristes tropiques* (pp. 337–349). Paris: Gallimard.

Nodier, C. (1828). *Dictionnaire raisonné des onomatopées françaises* (p. 11). Paris.

Nodier, C. (1834). Langue organique. *Notions élémentaires de linguistique,* Paris.

Rousseau, J. J. (1826). De l'écriture. *Essai sur l'origine des langues,* Oeuvres, (Vol. 1).

Saussure, F. de (1972). In T. de Mauro (Ed.), *Cours de linguistique générale* (pp. 51–52). Paris: Payot, (1st ed.: Geneva, 1916).

Soustelle, J. (1975). De la pictographie au phonétisme dans l'écriture aztèque. In J. Leclant (Ed.) *Le déchiffrement des écritures et des langues.* (p. 173). Actes du Colloque du XXIXe Congrès international des Orientalistes. Paris: L'Asiathèque.

Tsung-I, J. (1982). Caractères chinois et poétique. *Ecritures, systèmes idéographiques et pratiques expressives* (p. 272). Actes du Colloque international de l'Université Paris VII. Paris: Le Sycomore.

Yaguello, M. (1984). *Les fous du langage* (p. 182). Paris: Le Seuil.

CHAPTER 5

The Origin of the Greek Alphabet *

JOSEPH NAVEH[1]

The common ancestor of all alphabetic writing systems existing today is the so-called Proto-Canaanite script, which was introduced by the Canaanites, presumably under the inspiration of the Egyptian uniconsonantal hieroglyphic signs, in the first half of the second millennium B.C. Had the Egyptians used only the uniconsonantal signs and let their bi- and triconsonantal pictographic symbols fall into disuse, their writing would have been alphabetic like that of the Semites. However, adhering conservatively to their writing tradition, they were not able to reduce the number of the signs in their script. The revolutionary innovation of reducing the number of signs was made by the Canaanites in ca. 1700 B.C. The alphabetic writing invented in Canaan was a pictographic acrophonic script: thus, the pictures of "house", "palm of the hand", and "water", for example, did not stand for the respective Canaanite words, *bet, kaf,* and *mem,* but designated the first consonant of each word: *b, k, m.* The number of these pictographs was presumably 27. By the thirteenth century B.C. the Proto-Canaanite signs had been reduced to 22, but the pictographic conception still permitted the flexibility of the stances and the writing in any direction: from left to right, from right to left, in vertical columns, and even in horizontal or vertical boustrophedon. Vertical writing disappeared ca. 1100 B.C. At this stage the symbols became more and more linear. Until the middle of the eleventh century B.C. there were still pictographic forms (e.g., the *'ayin* depicting an eye with its pupil), and the letters could have different stances. From the middle of the eleventh century B.C., when all the letters had become linear, most of them had stabilized stances, and they were written from right to left, our terminology changes: the script is no longer called Proto-Canaanite (or Canaanite), but rather Phoenician.

From the beginning of the first millennium B.C. cursive letters evolved in the Phoenician alphabet which began to affect their lapidary counterparts. By the mid-eighth century B.C. the Phoenician alphabet had developed a uniform script with cursive and lapidary styles. The Hebrew script branched

* The main part of the paper is based on the last chapter of a previous publication (Naveh, 1982); it is reproduced here with the kind permission of the director of Magnes Press, The Hebrew University of Jerusalem.

[1] Department of Ancient Semitic Languages, Faculty of Humanities, The Hebrew University, Jerusalem, Israel.

off from the Phoenician in the middle of the ninth century B.C., the Aramaic script in the mid-eighth century B.C. The three national scripts – Phoenician, Hebrew and Aramaic – albeit in three different ways, continued to develop cursive letter forms, which had already begun to evolve in the Phoenician script in the early centuries of the first millennium B.C. (Naveh, 1982, pp. 23–42, 53–89).

There is a consensus among scholars regarding the West Semitic origin of the Greek alphabet; however, its earliest use among the Greeks is still a subject of controversy. The consensus is based on four points:

1. According to Greek legend, the alphabetic characters – named *phoinikeia grammata* (Phoenician letters) or *kadmeia grammata* (the letters of Kadmos) – were introduced together with other arts by the Phoenicians who came to Greece with a leader named Kadmos.
2. The names of the letters, *alpha, beta, gamma, delta,* etc. have no meaning in Greek, but most of their Semitic equivalents, *alef, bet, gimel, dalet,* etc. are Semitic words.
3. The letter sequence in the Greek alphabet is basically identical to the Phoenician (= Hebrew and Aramaic) alphabetical order.
4. The earliest Greek letter forms are very similar, and some even identical, to the equivalent West Semitic letters.

"Obviously the Greek alphabet must have branched off from the Semitic at the point where the chronologically contemporary resemblances are strongest, i.e. where the two sets of alphabets most nearly agree in the forms of letters." This generally accepted statement was made by Rhys Carpenter in 1933 (p. 10). Its application, however, is a matter of scholarly controversy. Moreover, there are certain other epigraphic traits which have to be taken into consideration. The main characteristics of the archaic Greek alphabet can be summarized in five points:

1. The earliest Greek inscriptions known today belong to the eighth century B.C.
2. The archaic Greek script used the 22 West Semitic letters, some of which designated vowels, and gradually introduced five supplementary letters: Y, then Φ, X, Ψ, and finally Ω.
3. The archaic Greek script was not uniform – it had some local variations.
4. It was lapidary in style.
5. The direction of writing and the letter stances were not stable. The archaic Greeks wrote in horizontal lines, either from right to left, or from left to right, or in horizontal boustrophedon.

Some scholars maintain that the deviations from the West Semitic prototype, mainly the introduction of vowel letters and supplementary letters, and the alteration of some letter forms, must have taken place over a considerable period. It is assumed that at a former stage, prior to the eighth century B.C., the Greek alphabet was closer to the West Semitic and that the archaic

Greek script in the earliest known inscriptions was itself the result of an evolution. Nowadays there is little support for such a theory of a standard Greek "Uralphabet." Carpenter, who dates the Greek adoption of the alphabet to the second half of the eighth century B.C., states that "the *argumentum a silentio* [i.e., the fact that no Greek inscription predating the eighth century has yet been discovered] grows every year more formidable and more conclusive" (Carpenter, 1933, p. 27).

The adoption of the alphabet by the Greeks is a complex problem. In studying it, scholars also take into consideration the Homeric question and the first witnessed date of the Olympic games (776 B.C.); in addition, they seek to determine the actual place at which the Greeks could have adopted the Semitic alphabet. The results seem to corroborate the assumption that the Greeks learned the Phoenician alphabet in the eighth century B.C. At that time, following 300 years of decay in the wake of the Dorian invasion (ca. 1100 B.C.), Greece once again began to flourish and there is evidence of commercial contacts between Phoenicians and Greeks. However, the problem centers mainly on epigraphic indications. Although we cannot demonstrate that Greek inscriptions existed earlier than the eighth century B.C., a comparative analysis of the characteristic traits of the West Semitic script and those of archaic Greek writing leads to the assumption that the Greek borrowing of the alphabet should be dated some 300 years earlier than the earliest known Greek inscriptions.

True, the "argument from silence" cannot be disregarded; but how conclusive is it? The Hebrews adopted the alphabet in the twelfth or eleventh century B.C., but only one Hebrew inscription – the Gezer Calendar (which may, in fact, be Phoenician) – (Naveh, 1982, pp. 65, 76, Fig. 54) is known to be older than the eighth century B.C. Although it is likely that the Hebrew script was widely used in the ninth century B.C., even by Israel's eastern neighbours (cf. the Mesha inscription; Naveh, 1982, pp. 65–67, Fig. 55), virtually no ninth-century Hebrew inscriptions are known to date. The Aramaeans began to use the alphabet slightly later than the Hebrews, but we know of only a few early Aramaic inscriptions. However, from the eighth century B.C. onward, the number of Hebrew and Aramaic inscriptions gradually increases, testifying to the spread of writing. The progress of literacy in Greece was probably very similar to that in the East.

It is commonly held that the Greeks adopted the Semitic practice of writing from right to left and that the earliest Greek inscriptions followed this practice; according to this view, the Greeks subsequently wrote in boustrophedon, and then from left to right. But Jeffery, who argues for an initial date in the mid-eighth century, has pointed out that all three systems existed concurrently even at the early stage of archaic Greek writing. She also maintains that boustrophedon writing "simply implies a pictorial conception of letters as outlined figures which can be turned in either direction according to need" (Jeffery, 1961, p. 46). This means that if the Greeks adopted the Phoenician alphabet in the eighth century, when it was already a systematic linear script, they neglected its achievements and turned it into a more primitive, almost pictographic script. Thus we have to look for a West Semitic model in which vertical writing had ceased to exist but the left-to-right direction of writing and horizontal boustrophedon were still in use. This stage of evolution is represented by the Proto-Canaanite script of the late twelfth century and first half of the eleventh century B.C., i.e. just before right-to-left writing became the standard practice, around 1050 B.C.

The Greek letters are considerably less cursive than the eighth- or even ninth-century Phoenician letters. This cannot be explained merely by the fact that most of the archaic Greek inscriptions known to date are graffiti and unofficial texts that were written on stone and pottery. The Greeks must have borrowed their script from a lapidary Semitic prototype. It is hardly conceivable that they did so in the eighth century and yet ignored most of the cursive achievements of the Phoenician script. A more plausible assumption is that the Greeks adopted a lapidary style of writing because it was the only existing model.

The most ancient Greek inscriptions known today, i.e. those from the eighth century, include inscribed vases from Athens and Mount Hymettos, inscribed sherds from Corinth, and rock-cut inscriptions from Thera (Jeffery, 1961, Pls. 1:1,3; 18:1; 61:1). In the scripts of these early texts and in the other archaic inscriptions down to the fifth century B.C., local variations are discernable. These local scripts were in use until the fourth century B.C., when the Ionian was adopted universally and became the classical Greek script. In all the archaic local scripts, *alpha, epsilon,* and *omicron* were used as signs for *a, e* and *o,* respectively, whereas *iota* before or after a consonant indicated *i. Upsilon* – the first supplementary letter – denoted the vowel *u.* The direction of writing and letter profiles were erratic in all the local scripts. Only from the fourth century B.C. onward, when the archaic local scripts were replaced by the uniform classical script, did left-to-right writing with fixed letter forms become common practice for all Greeks.

Deviation from the West Semitic letter forms and the introduction of vowel signs to improve the West Semitic consonantal system must have been a lengthy process. There were some archaic Greek letters, however, which preserved the original Proto-Canaanite prototypes. The *sigma* (ξ) has the shape of the thirteenth- and twelfth-century vertical *shin,* as it is inscribed on the ewer and the bowls from Lachish and Qubur el-Walaydah (Naveh, 1982, pp. 33–36, Figs. 28, 30). The *san* (Ϻ) does not seem to have evolved from Semitic *sade* [although in some abecedaries, e.g., the one from Marsiliana (Jeffery, 1961, Pl. 48:18), it is placed between *pi* and *qoppa*], but is another rotation of *shin-sigma.* The *mu* of five equal strokes (𐌌) as it appears in the local scripts of Crete, Melos, Euboia, and its colonies (Jeffery, 1961, p. 31, Fig. 13:4) resembles the pictographic *mem* designating water. The *omicron* with the dot in its center (⊙) is thought to be a result of cutting with a compass; but it seems more likely to represent the pictographic form of the *ʿayin,* i.e. an eye with the pupil, which still occurs in the eleventh-century Proto-

Canaanite inscriptions. To these examples we may add the box-shaped *heta* (𐌇), the tall, I-shaped *zeta* (I), as well as the forms of *delta* (Δ), *epsilon* (E), *nu* (𐌍), *ksi* (Ξ), *pi* (Γ), *qoppa* (Ϙ), and *rho* (P). All resemble their late Proto-Canaanite equivalents.

If the Greeks adopted the Proto-Canaanite alphabet when its letters were still in a process of evolution from pictographic to linear forms, this would explain the variations in the stances of certain letters. For example, the *lambda* was written with its crook at the top or at the base (Γ, L): this does not mean that one of the two versions was incorrect, but simply that the pictographic conception allowed for both forms. The archaic *gamma* could be written either Γ, or Λ, or <. On the eighth-century Dipylon vase from Athens, *alpha* has a 90° rotation (ᐳ), and on the painted aryballos of the Museum of Fine Arts at Boston, the *alpha* is "inverted" (∀). (Jeffery, 1961, Pls. 1:1, 6:22). These are not errors, but rather realizations of the pictorial conception of the letter forms, which were inherited from the Proto-Canaanite script. Although the tenth-century Phoenician script still had two *gimel* forms – 7 and Λ – from the ninth century B.C. onwards only one form prevailed. *Alef* and *lamed*, however, were stabilized ca. 1050 B.C.

In addition to letters which preserved the old Proto-Canaanite forms, there were some archaic Greek letters which underwent longer or shorter evolutions from their Semitic prototypes. While *beta* (B), *vau* (F), and the straight, vertical *iota* (I) are the product of a relatively long evolution, *tau* (T) and some local forms of *rho* (R) and *sigma* (ϟ, S) show shorter Greek developments.

The fact that the archaic Greek alphabet had not one set of letters, but various local forms, also poses a problem. Cook and Woodhead (1959) presumed that there were certain local Phoenician scripts, from which Greek individuals and groups learned their local scripts at some time in the second half of the eighth century B.C. However, we know that the Phoenician script was a uniform one, without regional variations. In the second half of the eighth century, the Hebrew and the Aramaic scripts had already developed independent traditions of writing, without any local variations. It is hardly feasible that the local archaic Greek scripts could have developed from more than one West Semitic script. Certainly, one Semitic tradition was adopted, and the evolution of the local scripts must be explained as a Greek phenomenon, a process which could not have been accomplished in one generation.

It has been suggested that the eighth-century *waw* as it appears in the Samaria ostraca (Naveh, 1982, p. 72, Fig. 65) may have been the prototype of the archaic Greek *vau* (Jeffery, 1961, pp. 24–25). However, as we have stated above, we cannot assume that the Greeks adopted Phoenician and Hebrew (or even Aramaic) letter forms at the same time. The *vau* must have been developed from the Proto-Canaanite *waw* of *c.* 1100 B.C., which was also the ancestor of the Hebrew *waw*. Therefore, if some forms of the *vau* resemble the eighth-century Hebrew *waw*, this should be seen only as the result of parallel developments. There is no reason to assume that there was any connection between the Hebrew and the archaic Greek scripts.

Another theory has it that the Greek alphabet evolved from Aramaic. This is based mainly on the assumption that the final -a in the names *alpha, beta, gamma, delta,* etc. was probably adopted from the Aramaeans; in Aramaic, final -a designates the determinative state of the nouns (Segert, 1963). However, the letter names *iota, pi,* and *rho* indicate that they were taken not from Aramaic, but from a Canaanite language: *yod* in Phoenician is "hand" (in Hebrew and Aramaic *yad*): *pe* means "mouth" (in Aramaic *pum*); *rosh* means "head" (in Aramaic *resh*). Moreover, if *alpha, beta,* etc., were Aramaic, they would appear in the names of the Syriac alphabet; that alphabet, however, has no names with final -a. Interestingly enough, the Aramaic name *resh* survives in Hebrew tradition, whereas the Canaanite form is preserved in the Greek *rho.*

The transition from the defective spelling of the Proto-Canaanite and Phoenician writing to the Greek vocalization system appears to pose a problem. In the earliest Aramaic and Hebrew inscriptions *matres lectionis* were used, a fact which scholars adduced as partial evidence for the argument that the Greeks adopted the alphabet from the Aramaeans. However, the idea of *matres lectionis* was known as early as the thirteenth century B.C. The North Canaanites in Ugarit used *yod* in certain instances to designate *i;* they also introduced the supplementary letters *'u* and *'i* and turned the original *'alef* into *'a* (Gordon, 1965).

There is, however, one archaic Greek letter which could not have been adopted from the eleventh-century Phoenician script. This is the *kappa,* which in all local scripts is written 𐌊 or K. Until the tenth century B.C., its West Semitic equivalent, the *kaf,* was written in the form of three fingers meeting at a common base (Ѵ). However, an identical reproduction of this early *kaf* can be found in the local western Greek scripts as a supplementary letter denoting *khi.* Thus it may well be that the letter shaped like three fingers served for both *k* and *kh* in the Greek alphabet of the eleventh and tenth centuries. Later, in the ninth century, wishing to differentiate between these two sounds, the Greeks borrowed the contemporary Phoenician *kaf,* which in the meantime had developed a downstroke (𐤊), and used it for *k;* the older form thenceforth denoted *kh* only.

Another Greek letter, *upsilon,* was adopted from the Phoenician script after the eleventh century B.C. In the alphabetic sequence, *upsilon* (Y) follows T. It was the first supplementary letter, and it served to represent the vowel *u.* When the Greeks adopted the Proto-Canaanite script, they presumably used the *waw* (= Greek *vau*) to write the consonant *w,* as did the Canaanites. Later, in the tenth and ninth centuries, the Greek *vau* underwent certain changes and its shape became Ϝ. When the Greeks invented their system of vocalization, the 22 letters supplied them with signs for *a* (*aleph-alpha*), *e* (*he-epsilon*), *i* (*yod-iota*) and *o* (*'ayin-omicron*), but they did not find a "free" letter to use for *u* (*waw-vau* was used in archaic Greek as a consonant). Since the Greeks were always aware of the origin of their alphabet (cf. their tradition), whenever they needed an additional letter, they looked primarily for Phoenician models. So they chose the *waw* (which by then had a form different from their *vau,* though it had evolved from the same letter) as suitable for *u.* This happened possibly around the tenth century B.C., but at any rate before the invention of the *kappa.* This much is indicated by the fact that *upsilon* is the first supplementary letter, while *khi* is the third (after *phi*).

The antiquity of the Greek alphabet is not a question of epigraphy alone; it is also, and primarily, a historical issue. It is widely maintained that, after the disappearance of the Mycenaean script together with the Late Helladic civilization, Greece knew a Dark Age of illiteracy which lasted until the eighth century; this view cannot be substantiated. It is a common belief that the Greek adoption of the alphabet took place in a bilingual environment, where Greeks and Semites lived as neighbors. Recent archeological excavations have yielded evidence of a Greek settlement at Tell Sukas in Phoenicia (today southern Lebanon) as early as the late ninth century B.C. (Riis, 1970, pp. 126–127, 159–162). However, ninth-century Phoenician inscriptions were found long ago in Cyprus and Sardinia. Since most of the features of the archaic Greek alphabet resemble those of the West Semitic script of ca. 1100 B.C., we have to give serious consideration to the theory of the early adoption of the alphabet by the Greeks. We suggest, therefore, that the Greeks learned the West Semitic writing at approximately the same time that the Hebrews and Aramaeans achieved literacy. These two peoples followed the practice of their Canaanite-Phoenician neighbors for some two centuries before they began to develop their own national scripts. Greeks living in more distant parts may perhaps have borrowed the Semitic alphabet after seeing it used by Canaanite merchants visiting the Aegean islands. It is also possible that in some areas, only a few Greeks might have used the new writing over a fairly long period. Since it was in Crete and Thera that the most archaic letter forms were preserved, it may well be that the inhabitants of these islands were the first Greeks to employ the alphabetic writing. Later it spread to the Greek mainland and to other islands.

The above-mentioned evidence formed the basis of observations published in 1973 (Naveh, 1973). Whereas classical epigraphists did not react, Semitic epigraphists, who could not ignore the evidence, tried to compromise between the conventional view and the new proposition. Millard's conclusion was: "Unsatisfactory though the position may be, no more precise date can be given for the adoption of the alphabet by the Greeks than the three centuries and a half, 1100 to 750 B.C." (Millard, 1976, p. 142). McCarter (1974) surmised that "the Greeks, though their script did not diverge as an independent tradition before *c.* 800, had experimented with the Semitic alphabet as early as *c.* 1100 . . . The memory of the earlier experimentation survived long enough . . . to exert a limited influence upon the final formulation of the Greek alphabet years later." (McCarter, 1974, p. 68). This impossible formula reflects the hesitation that McCarter shared with his teacher, Cross. However, after discussing the inscriptions found recently in Qubur el-Walaydah and Izbet Sartah, and, especially, the inscription from Crete (Naveh, 1982, pp. 36–37, 41, Figs. 30, 31, 36). Cross concludes as follows: "These new data must be said to give added support to the thesis of . . . the high antiquity of the earliest use by the Greeks of the alphabet, and remove obstacles to dating their borrowing to the time of transition from Old Canaanite to Linear Phoenician toward 1100 B.C.E." (Cross, 1980, p. 17).

References

Carpenter, R. (1933). The antiquity of the Greek alphabet. *American Journal of Archaeology, 37,* 8–29.

Cook R.M., & Woodhead, A.G. (1959). The diffusion of the Greek alphabet. *American Journal of Archaeology, 63,* 175–178.

Cross, F.M. (1980). Newly found inscriptions in old Canaanite and early Phoenician script. *Bulletin of the American Schools of Oriental Research, 238,* 1–20.

Gordon, C.H. (1965). *Ugaritic textbook.* Rome: Pontifical Biblical Institute.

Jeffery, L.H. (1961). *The local scripts of archaic Greece.* Oxford: Clarendon.

McCarter, F.K. (1974). The early diffusion of the alphabet. *Biblical Archaeologist, 37,* 54–68.

Millard, A.R. (1976). The Canaanite linear alphabet and its passage to the Greeks, *Kadmos, 15,* 130–144.

Naveh, J. (1973). Some Semitic epigraphical considerations on the antiquity of the Greek alphabet, *American Journal of Archaeology, 77,* 1–8.

Naveh, J. (1982). *Early history of the alphabet: an introduction to West Semitic epigraphy and palaeography,* Jerusalem-Leiden: Magnes-E. J. Brill.

Riis, P.J. (1970). *Sukas* I: *The north-east sanctuary and the first settling of Greeks in Syria and Palestine.* Copenhagen: The Royal Danish Academy of Sciences and Letters.

Segert, S. (1963). Altaramäische Schrift und Anfänge des griechischen Alphabets. *Klio, 41,* 38–57.

CHAPTER 6

Relationships Between Speech and Writing Systems in Ancient Alphabets and Syllabaries

ROBERT LAFONT[1]

The Canaanite Innovation

During the 14th century B.C., in the country of Canaan (Phoenicia to the Greeks, Byblos to be precise) there appeared an alphabet that was distinct from cuneiform. Its letters would come to be used, with some modifications, by the Hellenics, the Etruscans, the Oscs, the Umbrians, the Latins, and other cultures. With respect to writing systems, Canaan was certainly a privileged area in the whole Middle Eastern region. Two relatively simple writing systems were in use there, unfortunately still undeciphered today: the "pseudo-hieroglyphic" of Byblos, which had 114 characters, and the Proto-Sinaic of Palestine which had 35 (Cohen, 1958). In addition, there were the 22 signs used in the funerary inscription of Ahiram in Byblos. These characters had a long history behind them, and gave rise to the letters of our alphabet.

We must however comprehend the full meaning of the term "alphabet." The manner in which this Canaanite invention was eventually used by the Indo-European Greeks may prevent us from understanding the nature of the invention itself. Canaanite, Hebrew, Moabite, and Samarian were very close dialects, part of the broad family of Semitic languages. All of these, within a short period of time, came to use the new phonography. In this family, the lexemes – the roots of individual words – consist of consonants: i.e., consonantal characters alone are sufficient for the production of meaning. Vocalic sounds are essentially used only for verbal or nominal derivation, i.e., to create grammatical variables. Indo-European languages also used this convention to a certain extent: we find it in classical Greek, for instance, with three degrees of vocalic alternation: *naught*, ε, and *o*, as they are expressed, for example, in "gig*n*omai," "eg*é*nomen," and "gé*go*na." However, what was only an occasional occurrence in Indo-European languages was and still is a living function, perfectly understood and practiced, within the entire lexical inventory of present-day Semitic languages. It must be added that a general constraint exists in these languages, prohibiting vowels in the initial position: there are only glottal occlusions which are treated as phonemes, such as the

[1] Université Paul Valéry, Arts et Lettres, Montpellier III, Route de Mende, B.P. 5043, 34032 Montpellier Cedex, France.

laryngeal occlusive *alef.* It is precisely this phoneme that traditionally opens the alphabetic order.

Thus, the principle of acrophonic symbolization, originally developed by the Akkadians (and also practiced by the Egyptians to specify the reading of some of their hieroglyphs with phonological determinants), led to the establishment of a system of notation not of isolated consonants, as had been believed, but of a kind of syllabary made up of characters consisting of a syllable implying an undifferentiated vowel. For the sake of precision, the Phoenician letters should be transcribed not as: ɔ, b, g, d, etc., but as ɔ. . ., b . . ., g . . , d . . ., etc.

The true significance of the Canaanite invention was not the leap from syllabic to phonemic representation, but from a fully syllabic system to one in which the vocalic part was not differentiated within the syllable. Thus, this writing did not completely reflect vocalizations: it did not transcribe orality. Rather, it noted the elements that produce meaning itself. It was a kind of stenography that consisted of specifying the words by their lexical markings alone. This explains why Semitic languages to this day have been reluctant to adopt the use of vowels. What is valuable to them is not a complete phonography, as exists for example in special instances of vocalized Arabic or dotted Hebrew scripts, but the simplest system that can produce the desired meaning.

The Egyptians, whose language was of the same type, did not reflect vocalic articulation in writing (which is why we are not able to orally reconstitute their language). They took the same step forward, but did not make the new syllabic system autonomous – instead, they forced it back within pictographic symbolics. The true Canaanite breakthrough, grafted onto the invention of the consonantic syllable, was that the new system was adaptable for every Semitic language. Along with putting to rest the enormous and unwieldy tinkering with cuneiform, it escaped its cultural limitations. It also enabled language to escape the ideological binding that was characteristic of hieroglyphics. Paradoxically, one might say that once language was relieved from the sacredness of its own representation, writing could become universally available, and the power vested in scribes was abolished. This revolution was generally associated with the functioning of a commerce-based civilization; one not yet completely secular of course, but mobile. The kings of Persia appear to have been mindful of the merits of this innovation, as they came to administer their empire using Aramaic functionaries and Semitic script.

The Hellenic Innovation

In the fifth century B.C., in the same general region, more changes took place. A new wave of Semitic peoples were becoming urbanized and imposing their language, Aramaic, over the older, related idioms. Phoenician and Hebrew were no longer spoken locally, and were becoming literary languages. (Phoenician, however, through the establishment of Carthage, went on to exert its influence on Western and African cultures.) This Aramaic hegemony was social rather than political and permitted intellectual exchanges, which was the true advantage of the Canaanite innovation. The graphic system of the Semites became widespread: in fact, all the Aramaic alphabets, whether in Syria, Palestine, Egypt, Palmyria, or Nabatenia, differed only in certain subtleties of the characters: all of them used the 22 syllables with consonantal openings.

In addition, the transfer of the system to Indo-European languages had begun. Earlier, in the course of the second millenium B.C., vast migrations of these nomadic peoples had entered the urbanized and state-controlled territories of the Semites, troubling them with their raids and establishing new centralized powers. The first to arrive, the Hittites, settled in Anatolia, forming an empire modeled after the Akkadians. Alongside their own system they adopted Mesopotamian writing, introducing into it the new complication of yet another level of allography. However, the spread of Hittite cultural influence ceased in the eleventh century B.C. At about the same time, the Creto-Mycenaean culture, which had developed a regional script adapted from Minoan Linear B (used for archaic Greek, an Indo-European language), fell apart due to the Dorian invasion (Ventris & Chadwick, 1956).

During the first half of the first millenium B.C., the progress and expansion of writing in the Eastern Mediterranean region presented a challenge that could only have been taken up by the alphabetic, hence Semitic, solution. Quite simply, this challenge led to the adoption of the Canaanite system by the new Indo-Europeans of the West, the Cypriots, the Aeolians, the Ionians, and the Dorians themselves. The adoption was systematic and direct: not only did the letters remain the same in their form and value, but their order and their Semitic names did not change. Some inconsistencies in the adopted forms, such as the angle of inclination of the characters and the orientation of the lines, ended with the decision by Athens in 403 B.C. to standardize the Ionian style, which had developed in Miletus. Greek colonization had by this time covered the central and eastern part of the Mediterranean, and spread a network of modified Phoenician writing which even made inroads into the Punic empire. The Carthaginian and Hellenic armies fighting the battle of Himeria in Sicily in 490 B.C. were peoples of the same "alphabet," but the difference between their languages had created an important problem of adaptation. Eventually, the Canaanite innovation led to a second innovation: the notation of vowels.

This again arose out of an issue of syllabification. Words beginning with vocalic sounds were common in Greek, and the first Greek scriptors, as they began to adapt the Phoenician model to the needs of their own language, must have quickly perceived that what was missing were characters to indicate those sounds. It is perhaps due to early experimentation in adapting the initial occlusives of the Phoenician alphabet that we owe the two "breathings," hard and soft, that are characteristic of initial vowels (including diphthongs) in classical Greek. While the hard breathing sign marked the aspiration of an initial semiconsonant η (in French this character later became the letter H, which may or may not be pronounced depending upon whether or not the word is derived from a Greek root), the Greek grammarians during the Alexandrian era invented the soft breathing marker to differentiate the sound made by a weak consonant from the corresponding vocalic sound. The solution was in fact already contained in the Semitic script in its very economy. All that was needed was to take the character for glottal occlusion ɔ, which began the abecedarium, as a sign indicating a sound that did not fully mature into a consonant: one thus obtained the basic vocalic value.

What other antecedents could have served as solutions? In Ugaritic cuneiform there were three signs that corresponded to occlusive /ɔa/, occlusive /ɔi/, and occlusive /ɔu/. Indeed, the phonological system of any Semitic language, whatever the dialect might be, has three vowels that classical Arabic notes as /a/, /i/, and /u/. For /i/ and /u/ the notation is already in the system: these are the vocalic forms of the semivowels *yod* and *waw*. It was thus sufficient to use the symbols for *yod* and *waw* in cases where the vowels were not supported by an occlusive sonant. The Greeks thus obtained syllables which, instead of being of the type /bay/ and /baw/, were read as /b.y/ and /b.w/: that is, as /bi/ and /bu/. Finally, to note /a/, the ɔ was appropriate.

Another problem existed, that of noting vocalic duration. This was especially pertinent in Indo-European languages, because of the vocalization of their laryngeals. In Semitic languages, these laryngeals are represented as consonantal sounds marked by deeply throated aspirations. They already appeared in Indo-European languages in the Hittite script, functioning as sonants: /aH/ was equivalent to /ay/ or /aw/. However, probably because of the lack of aspiration in the majority of other dialects, they tended to disappear into the lengthening of the preceding vowel (as with s in the evolution of French *asne*>*aHne*>*âne*). Moreover, Greek contained the series of long /ā/, /ē/, and /ō/, three different laryngeals (an effect of coloration) which Hittite contained in only two different states: /hh/ corresponding to the coloration of long /ā/ of Greek, and /h/ corresponding to long /ē/ and /ō/. To write the Greek syllable one had to represent phonologically pertinent combinations of /.H1/, /.H2/, /.H3/ as one had to note *yod* and *waw*. Adding the fact that Greek also posed the problem of short /ă/, which derived from a laryngeal sound between two consonants (as in *pátēr* of *pHte+r*), we can see that the whole problem was to find a notation for the double series of syllabic peaks:

long /ā/ /ē/ /ī/ /ō/ /ū/

short /ă/ /ĕ/ /ĭ/ /ŏ/ /ŭ/

It is worth remarking that for /ă/ā/, /ĭ/ī, and /ŭ/ū the solutions were all provided by the Semitic model: the Greeks simply used the signs for *alef, yod,* and *waw* for their vocalic values. However, this implied that they could no longer mark the long/short opposition, whereas Semitic languages (as we shall see in Arabic) indirectly found a recourse by avoiding the need to mark the presence of the short vowels and by reserving the use of vocalized sonants for the long laryngeals. But in the case of /ĕ/ē and /ŏ/ō, the Greeks eventually resorted to innovations with oppositions ε/η and o/ω, which allowed differentiation.

It is remarkable that this necessary reworking of a borrowed alphabet searched for just those signs of Semitic laryngeals that were useless in Indo-European. From the weak unvoiced continuant /ε/ noted as O in Semitic, the Greeks obtained short /ŏ/; from the weak continuant /h/ noted as ϝ (also known in Greek as *digamma*), which they reversed into the uppercase E, they obtained the short /ĕ/, marked as ε; and from the strong unvoiced continuant /ḥ/, they obtained η, the sign for long /ē/. In fact, this sign, which became capitalized as H, was used in western Greek writing styles to mark initial aspiration (therefore leading to its present value in the Latin alphabet), and in the east, to fill in for the long /ē/.

These adaptations were completed by a reworking of the consonants. In order to denote their unvoiced aspirated consonants *th* and *ph,* the Greeks used the Phoenician notations – which were useless to them – of the emphatic /ṭ/ and /ḳ), written θ and φ, and they invented χ for /kh/. Finally, since they had palatalized the /u/ into /ü/, they adopted the digraph ou for /u/. And just as they used ou for /ọ̄/, they used the digraph ei for /ẹ̄/. For long open /ō/, ω was added to the alphabet.

Just as in the age-old symbolism, the inventory of the letters became an extension from A to Ω (Lafont, Boyer, Gardès-Madray, Marconot, Rieusset, 1984). The emphasis on vowels was certainly not only practical, but also ideological. It is often stated that the discovery of the vowel implied the discovery of the phoneme, which is certainly true. In this sense, the Hellenic innovation which completed the Canaanite one was just as important. It broke down another barrier to the domination of people by the scribes. Henceforth, the totality of the voice could be reflected according to an analysis that the Prague School (linguists working with R. Jakobson and L. Troubetskoy, who developed structural phonology and linguistics) would honor with a theory more than 2000 years later. The "mystery" of the production of meaning was brought to light and revealed by the alphabet. At the same time, grammatical structure was mastered through writing. The system acquired a quasi-perfection: the metaphors and the metonymies that constituted pictography became but a faint memory preserved in the acrophonic names of the letters. The era of the pictogram had come to an end. Just as the phoneme is simply an abstract sign with no meaning in itself, the notational system could henceforth be divorced from the objects and statements that the language referred to. Writing, like speech, had become the support for meaning, but sufficiently detached from it not to have to be dependent upon it.

This process of abstraction had another consequence. The fidelity of the representation of speech had to be exacting to the point where, as it transcribed its ebb and flow, the grapheme-to-phoneme correspondence ceased to be evident. The Greeks succeeded in transcribing their complex syntactic phonetics, but found it difficult to parse meaningful units as moveable or substitutable. In the beginning, at least, they had a rather poor sense of the word as a unit. Individual words had a much firmer place in the Semitic tradition, where they constituted the graphic frameworks for the consonantal roots. It was only during the Alexandrian era that the Greeks conquered the word by applying their energies to a systematic analysis of language. This analysis gave words and grammar their stability even as the tendency to rely on phonetics gave way to a more systematic control of syntax. Phonology itself and intonation were mastered graphically by the notation of the tone that identified it.

The Double Revenge of Syllabism

Vowels also appeared on the eastern side of the Chamito-Semitic world, in the land of Kush. [To be sure, much later, between the third century B.C. and the second century A.D., probably because of a distant Greek influence, the Kush language of the Kingdom of Meroe came to be written with 17 syllabic characters, all beginning with a consonant, borrowed from the Egyptian demotic, with three fully syllabic and three vocalic signs (Février, 1959, pp. 134–137).] Even as the Greek world was unfolding, a new wave of Indo-European nomads assaulted Mediterranean and Oriental Babylonia. Out of this vast migration of peoples of similar cultures speaking closely related dialects, the Medes and the Persians, who together founded the Achemenid Empire during the seventh century B.C., stand out particularly. As the Hittites had done before them, they adopted the civilization grounded in ancient Sumeria, and became people of writing by burrowing it from others. In fact, as soon as they established state control, they began to foster local traditions and languages through the Sumerian heritage. Thus, the memorial of Darius in Behistun is set in three languages: Classical Akkadian, Elamite Akkadian, and Persian, all of it in cuneiforms.

Old Persian seems to have been primarily a dynastic language (possibly in the same way that the household of the Great Inca had its own private language). The preserved epigraphs are all written on monuments, are all from the area of Persepolis, and all date from the time of the Achemenids. Their phonological structure presented a new challenge to writing. The Eastern dialects of Indo-European, namely Indo-Iranian, had but one vowel –

short /ă/, which included short /ă/, /ĕ/, and /ŏ/. They also had /i/ and /u/, which were the vocalic components of corresponding sonants. Short /ă/, statistically speaking, was the most commonly represented sound in Old Persian. Its role, somewhat similar to that of the French silent /e/, was "neutral" and optional. The Achemenid scribes used one character only, /ma/ or /m/, to specify any syllable that included the sound /a/ plus a single consonant. The guiding principle, of course, was syllabism. Additionally, since /mi/ and /mu/ were clearly differentiated in this language, the syllabic structure contained two characters for these sounds. Thanks to these letters a new solution was adopted (although its benefit was not extended to the whole system): the signs for /i/ and /u/ were used to modify the syllables containing /a/. Thus, "ma" + "i" would read as "ma-i", giving the syllable /mi/. To produce long /ā/, the character was simply repeated: i.e. it would read as "ma" + "a," effecting long /mā/.

In India, the notion of modifying a syllable by adding a vocalic component was to become the guiding principle for the creation of *Nagari* (the "writing of the city," more specifically, "of the capital"). The inventors of Nagari seem to have given the principle of syllabism a systematic and thorough examination. As much as one would like to be precise about dates, there are precious few clues about the relationships between historical landmarks and orthographies. Could there possibly have been a Semitic influence? Semitic models had been introduced long before, and had branched out into two widely spread adaptations, the *Kharoṣtri* (better known as Arameo-Indian) and the *Brahmi,* which was probably closer to South Semitic origins (Février, 1959, pp. 501–521; Cohen, 1958). Did Darius himself import the Achemenid system? Of course, nothing forbids one to speculate that Nagari was an exclusively Indian creation.

Whatever the historical antecedents, it was in the Indus valley – in that region where the old pictograms of Mohenjo-Daro and even the city-states of the third and the second millenia had been all but forgotten – that new Indo-European settlers approached the creation of a writing system with the utmost care and rigor. They introduced an unprecedented feature: the typical left-right horizontal line to which the characters seem "suspended", and which seems to indicate that linearity was the rule. The characters represented open syllables beginning with a consonant and systematically containing the expression of short /ă/. They numbered 33 in all, and covered the full range of consonantal possibilities. These included occlusives featuring variables of voicedness, aspiration and nasalization, combined with labials, dentals, cerebrals, and gutturals. In addition, they represented continuants and /h/, three vocalic phonemes, /a/, /i/ and /u/, and the vocalic sonants, the liquid /r̥/ and /l̥/.

The modification of the neutral /a/ contained in all syllables was effected by the same technique as that adapted by the Achemenids, but in a simpler and more complete manner. The reduction of /a/ and its replacement by /i/ and /u/ were specified by modifiers positioned to the right, the left, above, or below the syllabic character. Thus, we have a phonological analysis of the kind that underlies the Greek invention, but Nagari, instead of eliminating the syllabic system, actually reinforced it.

However, there was yet another problem; that of representing the duration of the vocalic sounds that existed both in short and long forms in the language. Here again we may see a connection with Indo-Iranian, inasmuch as Indic had also subsumed the pitch of /e/ and /o/ under the general value of short /ă/. However, these sounds were evident in the diphthongs /ay/ and /aw/, which had appeared during the evolution of the language. The solution was to add /i/ to /a/, "a" + "i", to obtain /e/; then to add another /a/, "a" + "a" + "i", to effect /ai/. Similarly, /u/ was added to /a/ to obtain /o/. An even more ambitious contraction of /e/ + /a/ + /u/ effected long /āu/. The first degree of modification was called "guṇa" by the Indian grammarians, a word that means "attribute" or "property." The second degree was called "vr$_0$ddhi" ("augmentation"). As in Old Persian long /ā/ was obtained by repeating the sign (i.e., "a" + "a"), while long /ī/ and /ū/ were given characters of their own (Whitney, 1889, pp. 82–84).

Paradoxically, then, Nagari was more rigorously "phonological" than Greek writing, yet it remained a syllabary. Although it was destined, under King Açoka, to be used to write down the abundant oral legacy of the vedas and the Buddhist literature, which was strongly biased towards oral speech, Nagari acquired a prominent position in world cultures, taking as large a place as the Chinese and the Semitic-Greco-Roman orthographies.

Nagari ran into the same problem of literal phonocentrism that had plagued the early Greek scribes: there was a degree of incompatibility between representing words as stable individual units of meaning, and representing the flow of oral speech as faithfully as possible. Consequently, there was a conflict of priorities in the adoption of either a lexical or a phonological syntax. The solution chosen by the Indians, under the influence of powerful ideological pressure, was to go for a grammar of sound structures rather than a grammar of words. Sanskrit was considered to be a "perfect" language, not only because it was supported by a very complete, although probably artificial linguistic system ("Saṃskr$_0$ta" means the same thing as the Latin "confectus," i.e. "made"), but especially because it was deemed to be a sacred language that had to be preserved.

The understanding of phonology by Indian scribes was therefore both enhanced by their finely tuned attention to subtle variations in oral speech-forms, and marred by their exacting rendering of these variations, including intonation and rhythm, often in opposition to the syntactical value of the sentence. Even as the roots of lexical morphemes were revealed by this analysis, they were buried under the extravagances of an obsessive need to reproduce phonology. Whereas the Greeks gave an ideal status to individual words by working out a strategy to normalize the notation of the voice, the Indians put the notation of the voice above that of the phonological signifier.

However, the survival of the Prakrits, the living Indic dialects, offers proof of the implicit efficiency of their orthography.

In any event, what we have here is a scriptural system that did not follow the Semitic-Hellenic model. Another return to syllabism was to occur much later, around the fifth century A.D., on the southwestern front of the Semitic expansion in Ethiopia. Learned people living in Aksoum, Ethiopia's christianized capital, made the decision to include the five vowels in long and short forms, /ā/, /ă/, /ē/, /ĕ/, ī/, /ĭ/, /ō/, /ŏ/, /ū/, /ŭ/, in the Sudarabic system which they had been using up until that time to write their Chamitic language. They accomplished this by a rational process that systematically defined the consonantal characters by additional curves. It is possible that the Greek evangelical influence was behind this vocalizing intent. In any case, the system became a new syllabary with 182 characters (Cohen, 1958, pp. 131–132).

Not too far away from Ethiopia, another development in line with the Semitic model occurred in what was to become the Arabic script. In their Yemenite kingdoms, the Arabs had created a brilliant civilization that spread over neighboring Africa. Ever since the seventh century B.C., they had been using a consonantal syllabary, which, although different in style, as it was probably derived from ancient pictograms, had exactly the same number and the same values as the Aramaeo-Canaanite system (Cohen, 1958).

The Aramaean script, following the caravan routes of Syrians, Jews, and Christians, was already widespread when a local reaction developed a little before the time of Islam. It began in Petra, a great intellectual and commercial center and a crossroads for the caravans of the Arabic Peninsula. This was the Nabatean cursive script, which evidenced a strong Syrian influence and which through the triumph of Islam would gain a secure position in the Arab world.

Refined during the era of Muslim expansion in the seventh and eighth centuries, this script was the outcome of a thorough linguistic analysis, comparable only to the Greek and Sanskrit theoretical works from which it also benefited. Yet another variant on the consonantal syllabic model (Lecomte, 1968; Fleisch, 1956), it based its structure on the linguistic system itself, which did not allow for initial vowels but rather, following the Semitic standards, turned glottal occlusions into phonemes. Nor were there any explosive groupings, which explains why borrowed words, particularly proper names, e.g., "Plato" or "Frank", were transformed into words such as ʾAflaṭūn(u) and ʾIfranǧ. Written syllables consisted of a group made of a consonant plus a vowel plus another potential consonant even in syntactic phonology: thus, "he left," from the verb *ṭlq*, when written separately, becomes *in*-ṭa-la-qa and as a composite word, becomes *tum-man-ṭa-la-qa* ("and then, he left").

As the development of the script progressed, it was not deemed necessary to indicate any vowels except long /ī/ and /ū/, which were included in the syllabic extensions of *yod* and *waw*. These did not call for any special characters, and were written exactly the same way as when they were utilized for their consonantal value. *Q.ī-l(a)* ("it was said") and *q(a)y-l(un),* a royal title, are thus written exactly the same way. It is a practical system, ruled by the principle of total economy of means. Nasalization, for example, being merely a matter of morphology, receives no more attention than the implicit vowel it is supposed to affect. Gemination, or repeated consonants, also a morphological happenstance, is not represented. This principle of economy is applied in any ordinary day-to-day writing, including and especially in the modern Arabic press. It is quite resolutely a choice for a lexical over a grammatical solution. Thanks to their form of writing, the Arabs have developed a strong and precise sense of the individual word, which is represented by its three-lettered – occasionally two-lettered – root. The words of the language are thus identified and supported by their written forms, which provide a lexical inventory without being complicated by grammatical categories.

However, there is a situation where the notation of vowels is useful: not because of the social use of the language, but to help to teach writing. There again, the principle is very simple (comparable to the diacritical signs of Hebrew): three markers are used to designate initial short /ă/, /ĭ/ and /ŭ/ (there is no need to indicate short /ĕ/, which only affects some dialects). Nasalization is specified by simply repeating the marker. The vocalic marker naught is the *sukun.* The *šadda* is added to a consonant to signify that it is geminated. And a special character, the *waṣla,* indicates that the initial occlusion, *hamza,* is cancelled. There is an internal logic to this refined system, ingenious in spite of its apparent complexity.

Vocalized Arabic script is the medium of a privileged learning: the word of God. The Koran must be recited as faithfully as possible to the form in which it was received by Mohammed himself, i.e., including vowels. The Holy Book therefore requires the supplementary notation, just as is the case for the notation of the samdhi in Sanskrit.

Such was the writing that would come to cover the Islamic world. Its adaptability was and still is exceptional. Fifteen characters out of 22 are discriminated by one, two, or even three dots. Adding or subtracting dots gives the basic characters enough scope to accommodate new phonemes in other linguistic groups, such as Indo-European languages influenced by Islam (Urdu, Persian, etc.) or those derived from Turkish (Usmanli, Uigur, etc.). And of course, the notation of vowels always remained an open option.

If some of its adaptations have yielded to romanization, such as in Turkey, or to Cyrillic, in present-day Russia, we should never forget that the evolution which began in Canaan over 3000 years ago has issued in two equally powerful though divergent phonological strains, Greco-Roman alphabetism and Semitic consonantal syllabism.

The Effects of the Alphabet

There is general agreement on a number of the effects of the invention of the alphabet and on their sociohistorical interpretation. Syllabic writing, a phonography that did not have to rely on any pictorial resemblance to what it represented, divested writing of its autonomy, easing the considerable efforts previously required to master it. It also posed a threat to the cultures of earlier systems, and for this reason the civilizations of hieroglyphs and cuneiforms, which gave birth to syllabic writing, avoided it. They absorbed some of it and relativized its use, but continued basically as they were. The kings of Persia may have submitted their own oral language to syllabic transcription, but they also maintained the Sumero-Akkadian graphic apparatus as a means of preserving the culture whose heritage they had imperially seized. In contrast, the Phoenicians, who did not seek a continental empire but a commercial conquest of sea routes, sacrificed this heritage.

Evolved Semitic writing – Canaanite, Aramaic, and later Arabic – heralded an agonizing change of civilizations. It was the writing of cities, created around the needs of urban societies. It was used by individuals for their personal archives; it was used for economic exchanges; and it was contemporary with the first monetary signs. It entrenched what had been the purpose of writing from the start: the eternalization of thought. The written trace lost its elaboration and became plain and unesthetic. It could also henceforth be erased: the tablet was no longer a brick baked in the oven or sun, but a temporary surface now called a slate. It was in this form that the system was passed on to the Greeks, the Ionians in particular.

What was transformed by this kind of writing was the contextual production of meaning. Phonographic coding highlights the word, but by the same token makes it transparent: it submerges the medium behind the meaning, emphasizing only the concept. In a purely oral situation one needs a situational context and especially a discursive context to be able to decode meaning as it develops. However when this oral flow is immobilized by writing, caught in the trap of space, meaning then becomes made up of stable elements. The code is fixed; the inscribed concept brings civilization into the era of the signified (*signifié*).

In its ideographic phase, writing had already created this absolute value of words and their meaning. Ideography represented the first necessary stage of the reification of meaning by writing, whose minimal unit is now considered to be the word. Phonological writing introduced the next stage, which allowed one, through directing inscribed speech and following the oral system of meaning-production in a linear fashion, to freeze the meaning in its course: to standardize the word–meaning relationships of words as they were being produced and exchanged.

For the written word was no longer being established at a distance from speech, but in successivity. The linear sequence – right to left, left to right, or boustrophedon – of nonideographical writing systems had the effect of graphically fractioning the cadence of the voice into moveable units of meaning. This effect, as we have said, was bound to the consonantal form of the Semitic vocabulary, and reached its high point in Arabic. It would have been overruled by grammatical considerations if the three-lettered roots had been taken not as representing words, but only as phonematic coordinates and the material support for the production of meaning.

The second phonological revolution, that of the Greeks (which Europocentrism takes as the ultimate goal of writing), was an adaptation to another systematic order where the vowel was lexically pertinent. At the same time it provoked an important reevaluation of the problem of meaning. By including vowels, the Greeks noted not only the lexical values of their vocabulary, but the totality of the voice. The consequence was that at first, despite their very intelligent analysis of how to produce meaning (the phonemes in a complete series), their efforts were directed toward a full phonological representation: their script attempted to emulate the flow of the human voice in speech. A great deal of work was still to be done. This evolution would run in a direction counter to that of Arabic and Sanskrit scripts, for in those two languages, which noted not practical or common but hieratic and divine speech, it was deemed necessary not only to note the vowels, but also to restore the full flow of speech.

In Greek one began at the level of human, not divine, speech, and in order to discover the word and its meaning it was necessary to proceed to its decipherment. It was necessary to discover the laws of λόγος under φονή. *Logos,* in Greek, implies the very process of discovery: first the order of the signified, then the unit meaning, the word. Since the analysis was carried out in writing and through writing, γϱάμμα was to give γϱαμματική (ἐπιστήμη), "science of the characters," more precisely, λογικη, "science of the logos." Thus began the reign of grammar: the institutionalization of an abstract model.

The Worldwide Spread of the Alphabet

The history of orthographies can be distinguished in different stages: pictography, syllabic phonology, and literal phonology. The last two systems have come to cover the world, in an expansion whose progress we can follow and understand. The Sanskrit syllabary was spread by Buddhism and came to dominante Asia. Historically and regionally modified, it developed along one line for the Mon-Khmers and the Burmese; along another for the Cambodians, the Laotians, and the Siamese; and along a third for the Tibetans. From Tibet it passed into Mongolia, where it remained official until the modern era. In China, where it remained official until 1368, it was used to transcribe Uigur Turkish as well as the Chinese language of Genghis Khan's empire.

The legacy of the Canaanite system has been even vaster, spreading to the west and the south: Sumerian Hebrew, Judaean Hebrew (from which Himyaritic of Southern Arabia, Ethiopian, and Libyco-Berber writing systems derive), and Iberian and Lihyanic or Northern Arabia. Aramaic expansion had the largest spread: Egyptian Aramaic and Palmyrean; biblical Square Hebrew and its Rabbinic Judeo-Persian, Judeo-Spanish and other derivations; Edessan Syriac, Malabar Jacobite, Karshuni, Arabic, Mandaean, Estranguelo; and Armazi from Tiflis.

Aramaic also followed two other routes: that taken up by the Indo-Europeans, which in India led to the Brahmi script, from which are derived not only the graphic forms of Tokharian and Koranic Persian, but also the scriptforms of Buddhist, Turkish, and above all Tamil. Through Pahlavian and Avestan Iranian, this branch led to Sogdian and Ienissei Turkish, Mongolian, Kalmuck, Manchu, etc. The line that began in Petra and spread through the Meccan Arabic brought by Islam covers all that is Muslim today: Uigur, Kazakh of China, Malay, Swahili, Hausa, Peul, the Urdu branch of Hindi, and Persian – which has also covered Berber, Malagasy, Spanish Mozarabic, and, until 1928, when it was officially romanized, Ottoman Turkish. Another route has only left a few allogeneous pockets: Coptic script in Egypt, the above-mentioned Karshuni, and Square Hebrew, used to denote Arabic by the Jewish communities that remained in Semitic countries (Cohen, 1958, pp. 321 – 337).

This inventory shows the determined effects that practical needs can have on the choice of a writing system (one can thus explain the development of the Aramaic model as it progressed through Persia and India to the Orient), especially the ideological and religious determinations that give form to culture. Because of this, it is difficult to separate the internal modeling of a writing system from the weight that the language it represents can bring to bear upon it. Lexical and morphological influences are exerted along with the scriptural influence. All languages of Islamic peoples show the influence of Arabic to some extent: this influence can go so far as to produce a true hybridization, as with Swahili and Mozarabic.

It is known that Greek script passed to the West through the Etruscans and the small Indo-European groups of Central Italy: the Latins, Osks, and Umbrians. The successive advent of the Alexandrian Empire, the Roman Empire, and the two ecumenical forms of the Christian Empire, the Western and the Eastern, explains the present divergence between Roman and Greek letters. The latter influenced the Coptic script and the Cyrillic script, which embraces part of the Slavic domain, while Greek Catholicism is at the foundation of Armenian and Georgian writing. In Europe, the boundaries of writing systems are the boundaries of religions, separating Pole and Czech from Russian, Croatian and Slovak from Serbian.

The economic and cultural imperialism of Western Europe in the nineteenth century initiated a new wave of romanization, this time on a world scale. The Latin alphabet served to bring writing to a diversity of languages that had not yet develped it. It modernized and secularized Turkish, de-Arabicized Berber, and denationalized Moroccan Arabic during a colonial phase. Russian imperialism within the USSR resisted this trend, and reconquered Siberia through the alphabetization of Turkish and Mongolian languages: thus the Mongolian of Outer Mongolia is divided by the Russian letters from the Mongolian of China, just as Moldavian is cut off from Romanian Latin by an authoritarian reimposition of Cyrillic. The ethnically diverse regions of Islamic culture also resisted this alphabet, as did the Sanskrit and Chinese worlds.

The worldwide expansion of the Latin alphabet, and of other writing systems for that matter, has essentially been a cultural expansion in linguistic form. The distinctive phonological features and nationalistic sentiments of German and Basque, for instance, should not lead us to forget that these two languages are profoundly Latinized in their lexical and writing systems, especially with regard to their articulations of meanings. It is fairly clear that German evidences the semantic distinctions of Latin, not to mention Greco-Roman intellectual production. Similarly, Israeli Hebrew is a modern language as far as its articulations of meanings are concerned: it has gone a great distance from biblical Hebrew, that dialect of the Canaanite culture which died out in the fourth century B.C. Biblical Hebrew was retained in the Diaspora as a sacred and erudite language, serving to unite the Jews separated by Romance or Yiddish Germanic languages into a European Weltanschauung, into which Jews with an Arabic language base could also be automatically integrated. The return to a Semitic script, as with the adoption of the Latin alphabet by modern Turkey, should not lead us to false assumptions: in both cases the underlying reason was nationalism. At most, universal romanization implies a standardization of alphabets, corresponding to the Whorfian school's description of Standard Average European (SAE) in linguistics (Lafont, 1978, pp. 103–104). What may be in the making is the standardization of the logosphere.

References

Cardona, G.R. (1981). *Antropologia della scrittura* Turin: Loescher.

Cohen, M. (1958). *La Grande invention de l'écriture et son évolution.* (3 vols.). Paris: Imprimerie Nationale.

Février, J.G. (1959). *Histoire de l'écriture.* Paris: Payot.

Fleisch, M. (1956). *L'arabe classique, esquisse d'une structure linguistique.* Beyrouth: Imprimerie catholique.

Lafont, R. (1978). *Le travail et la langue.* Paris: Flammarion.

Lafont, R., Boyer, H., Gardès-Madray, F., Marconot, J.M., & Rieusset, I. (1984). *Anthropologie de l'écriture.* Paris: Centre Georges Pompidou.

Lecomte, G. (1968). *Grammaire de l'arabe.* Paris: Presses Universitaires de France.

Ventris, M., & Chadwick, J. (1956). *Documents in Mycenian Greek* Cambridge: Cambridge University Press.

Whitney, W. (1889). *Sanskrit grammar* (2nd ed.). London: Oxford University Press (Reprint).

CHAPTER 7

Graphic Systems, Phonic Systems, and Linguistic Representations

PARTH M. BHATT[1]

Introduction

The study of writing and writing systems has received only minor attention from researchers developing theories of natural human languages. The majority of linguistic investigations have centered on:

- The status of written language as opposed to spoken language (Pulgram, 1951; Uldall, 1944; Vachek, 1945, 1973)
- The present status of orthographic systems and their relation to underlying linguistic systems (Catach, 1978; Léon, 1966)

The general consensus among linguists is to affirm the dominant status of the spoken word over the written word within theoretical linguistics (Bloomfield, 1933; Chomsky, 1970; De Saussure, 1915). Systems of graphic representation are seen as derivative, substitute systems (such as, among others, sign language, Braille, Morse code, and shorthand).

One of the probable causes for this closure of the linguistic paradigm can perhaps be found in the historical development of the study of language at the end of the nineteenth century. At this point in time, the primary focus of attention was the study of classical languages, the discovery of Sanskrit, its comparison to other known classical languages, and the subsequent attempts to reconstruct a common Indo-European language base (Pederson, 1931). Written records were perforce the single major source of language data.

The structural theoreticians of the late nineteenth and early twentieth century sought to reverse this tripartite interest in classical languages, written records, and diachronic studies and were perhaps influenced by two concurring theoretical developments:

1. The insistence (mostly on the part of structural linguists such as De Saussure, 1915) on the study of languages as abstract rule-governed value systems agreed upon through a complex social contract, rather than as individual organisms that evolved historically. This conceptual position linked the study of languages with contemporary currents of sociological and political thought, and provided an initial impetus for the analysis of "living," "modern" languages.

---

[1] Experimental Phonetics Laboratory, Department of French, University of Toronto, Toronto, Ontario, M5S 1A1 Canada.

2. The ensuing theoretical dominance of phonology within the linguistic paradigm (De Saussure, 1915; Troubetzkoy, 1939) with its reliance on abstract sound units and rule systems which compounded the tendency to the study of current languages, since written records give only indirect access to underlying phonological systems.

The study of written language as an autonomous system of expression was of secondary importance and this conception is still present in current linguistic analysis, although there has been a renewal of interest in written systems (Coulmas & Ehlich, 1983; Haas, 1970, 1976; Herrick, 1966; Llorach, 1968; Sampson, 1985). The task of analyzing graphic representations has been left to five major groups of researchers outside the main current of theoretical linguistics:

1. Anthropologists, archeologists, and historical linguists interested in the origin and evolution of the multiple forms of written representation found in ancient civilizations (Bange, 1971; Bouuaert, 1949; Christin, 1982; Cohen, 1953, 1958; Diringer, 1948, 1962, 1983; Driver, 1954; Fevrier, 1959; Friedrich, 1957; Gaur, 1984; Gelb, 1952, 1968; Gordon, 1968; Higounet, 1955; Jackson, 1963; Jensen, 1970; Naveh, 1982; Petrie, 1912; Rahi, 1977; Sprengling, 1931; Taylor, 1883)
2. Pedagogues and linguists motivated by the need to teach and preserve written forms or to convert written forms to spoken forms and vice versa (Catach, 1978; Léon, 1973)
3. Sociologists, social historians, and social psychologists concerned with the impact of the use of written communication on individuals and societies (De Kerckhove, 1985; Goodman, 1982; Goody, 1977; Olson, Torrance & Hildyard, 1985; Olson, this volume; Ong, 1982)
4. Experimental and cognitive psychologists interested in the psychological processes involved in reading and writing (Firth, 1980; Gibson & Levin, 1975; Gregg & Steinberg, 1980; Henderson, 1982, 1984; Huey, 1968; Kavanaugh & Mattingly, 1972; Kirk, 1983; Kolers, Wrolstad & Bouma, 1980; Laberge & Samuels, 1977; Lesgold & Perfetti, 1981; Levin & Addis, 1979; Martlew, 1983; Neissen, 1967; Pirozzolo & Wittrock, 1981; Pynte, 1983; Reber & Scarborough, 1977; Rosenberg, 1982; Sanford & Garrod, 1981; Scinto, 1986; Taylor & Taylor, 1983; Vygotsky, 1962, 1978; Weaver, 1977; Whiteman, 1981)
5. Clinicians and researchers on language pathology involved in describing and rehabilitating subjects with specific disorders of written representations (alexia, dyslexia, agraphia, dysgraphia etc.; Alajouanine, 1968; Alajouanine, Lhermitte and Ribaucourt-Ducarne, 1960; Albert, 1979; Benson, 1979; Charcot, 1890; Critchley, 1928; Déjerine, 1891, 1892; Dubois-Charlier, 1972, 1976; Hécaen and Kremin, 1977; Hécaen & Marcie, 1974; Kremin, 1976; Kussmaul, 1877; Leischner 1957; Marcé, 1856; Marcie, 1977; Marcie & Hécaen, 1979; Sasanuma, 1974, 1975; Sasanuma & Fujimura, 1971, 1972; Sasanuma & Monoi, 1975; Yamadori, 1975)

The tasks of examining the internal structural properties of past and present writing systems and analyzing the specific differences between spoken and written representations have not received systematic attention within the linguistic paradigm.

The aim of this article is to contribute to the study of these two problems by providing a detailed description and comparison of the graphological and phonological system of a specific language, modern French. It is hoped that the methodology and the theoretical concepts applied here will be of use in future studies of this type.

It is also important to note that the French graphological and phonological systems have strong Greco-Latin roots, notably the set of alphabetic characters. The observations made here on the basis of the particular example of modern French are thus relevant for a number of different alphabetic systems.

Differences Between Graphic Systems and Phonic Systems

It is essential to begin by distinguishing the internal properties of these systems from the actual physical channels involved in processing their output. In processing spoken language the distinction is between *phonological* and *phonetic* processes.

The phonology of a language includes the specification of the stock of abstract sound units, the rule systems, and the levels of representation that result in the segmental (essentially subphonemic, phonemic, and morphophonemic features) and the suprasegmental (essentially tonal, accentual, and intonational features) representation of an utterance.

It is fundamental to underline the fact that phonological representations and rule systems refer only to abstract, internal acoustic images or "signifiers" in Saussurian terminology. Phonetic processes, on the other hand, refer to all the mechanisms involved in what has been called the "speech chain" (Denes & Pinson, 1963): phonatory-articulatory production, speech signal, acoustic transmission, and auditory reception.

Similarly, it is possible to distinguish between graphological and graphetic processes. The graphology of a specific language would include the stock of abstract graphic units, the rule systems, and the levels of representation necessary for the generation of the graphic specification of an utterance. Graphetic processes would then refer to the set of mechanisms involved in the "graphic chain": brachiomanual production, graphic signal, visual transmission, and visual reception.

Given this distinction between phonological and phonetic systems on the one hand and between graphological and graphetic systems on the other, we intend to begin this discussion by analyzing the levels of representation that characterize phonological representations in modern French and then attempt to discover the corresponding levels of representation in graphological representations.

Distinctive Features

The existence of a set of subphonemic features (Jakobson, Fant & Halle, 1952) is well accepted in phonology, even though the exact nature and number of these features remains a subject of controversy (Martinet, 1960).

There have been few systematic attempts to specify the set of subgraphemic features of written systems (Herrick, 1966). An initial analysis of French capital letters presents the following sets of features:

1. A set of three binary features related to stroke mode:
   a) straight/curved
   b) horizontal/vertical
   c) concave/convex
2. A limited set of coordinates required to specify stroke trajectories on a proposed cube-like grid with 21 points (see Fig. 1)

The question of whether this set of subgraphemic features will be empirically adequate to describe all types of alphabetic writing is open to investigation, but the major purpose of this proposal is to suggest that such features do in fact exist in a limited set, in much the same way that distinctive sound features exist for spoken language.

Figure 2 provides a subgraphemic feature matrix, which allows the immediate formal description of systematic similarities and distinctive differences between the graphemes. Thus E and F are similar with the exception of the straight stroke 17–19, P and R are distinguished only by the straight stroke 9–19, and O and Q are distinguished by the straight stroke 11–21. This matrix also implies that graphemes result from specific combinations of these features.

It is also important to note that for both phonological and graphological systems the units used at this level of representation are abstract, phonic, or geometrical figures, that have no auditory or visual autonomy in themselves.

Graphemes/Phonemes

Specific combinations of distinctive sound features result in phoneme level units that exist in relatively circumscribed number for all human languages. Modern "standardized" French (Léon, 1966) has a total of 36 discrete, abstract sound units. Sixteen are vocalic phonemes /i, y, u, ə, e, ø, o, ɛ, œ, ɔ, a, ɑ, œ̃, õ, ɛ̃, ã/; 17 are consonantal phonemes /p, t, k, b, d, g, f, s, ʃ, v, z, ʒ, m, n, r, l, n/; three are semiconsonantal (or semivocalic) phonemes /w, j, ɥ/.

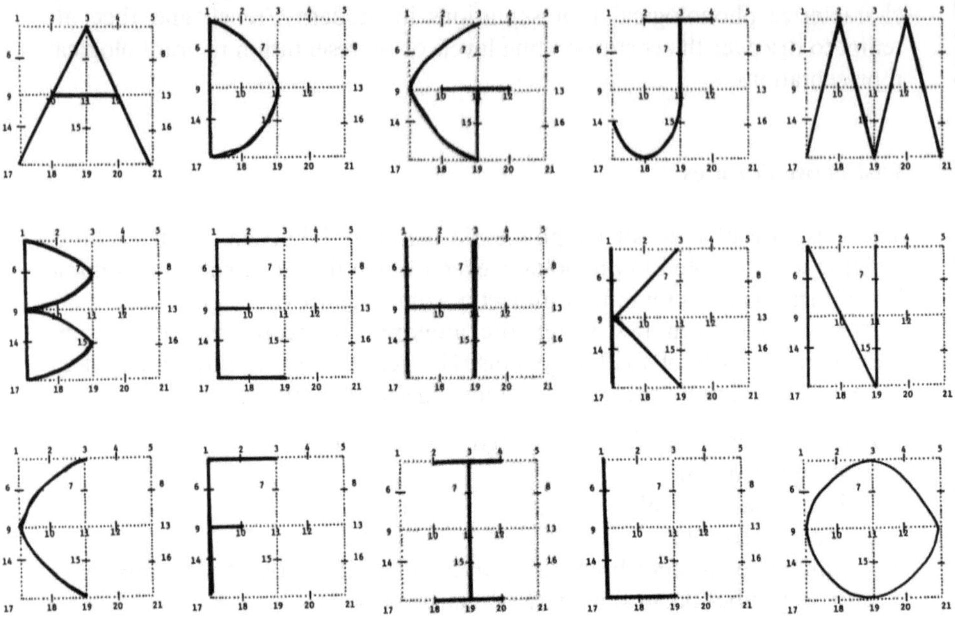

Fig. 1. Graphemic representation of the capital letters of the French alphabet

The modern French repertory of graphemes contains 26 letters and six diacritical marks giving a total of 38 graphemes, of which 17 are vocalic graphemes (a, à, â, e, é, è, ê, ë, i, î, ï, o, ô, u, ù, û, y) and 21 are consonantal or semiconsonantal graphemes (b, c, ç, d, f, g, h, j, k, l, m, n, p, q, r, s, t, v, w, x, z).

When compared in terms of overall numbers the two sets are similar, but this should not lead one to conclude that there is a systematic one-to-one correspondence between graphemes and phonemes.

According to Catach (1978), 92% of the occurrences of the archigrapheme A correspond to an underlying archiphoneme /A/. But the details of what seems quantitatively to be a very stable correspondence show the complexity of the underlying relationship. For example:

The grapheme a corresponds to the phoneme /a/ in the very large majority of cases:

- But the grapheme clusters "oi," "oî," "oie," "ois," "oit," "oix," as in *roi, foie, cloître, mois, voit, croix*, can also correspond to the phoneme cluster /wa/
- The posterior phoneme /a/ corresponds generally to the grapheme â
- But the phoneme /a/ can also correspond to the grapheme cluster "-as," as in *pas, las, tas*.

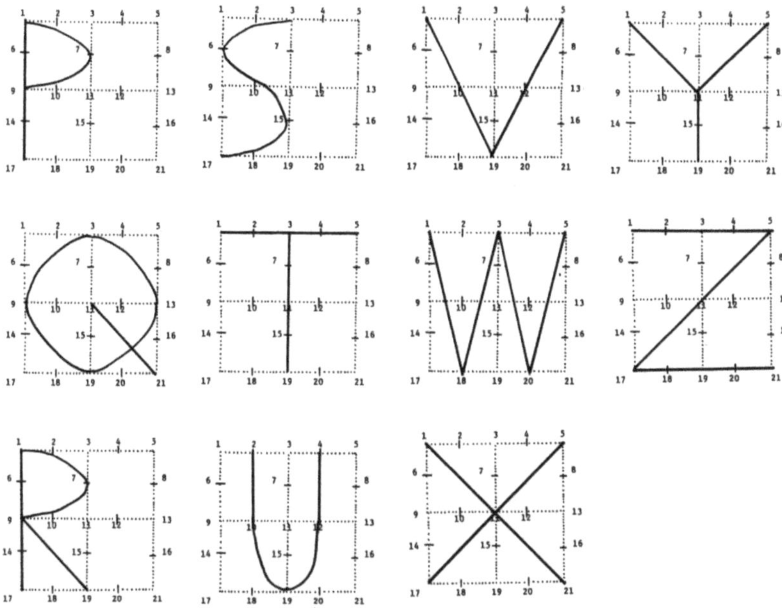

The example proposed here is based on a maximal standardized phonological system such as that proposed by Léon (1966), and does not take into account variation due to dialect differences or social differences.

Examples of this type of complex correspondence abound for both consonants and vowels:

- The grapheme j- in word initial position corresponds to the phoneme /ʒ/, as in *je, jeu, joue,*
- But the grapheme cluster "-ge-" also corresponds to the phoneme /ʒ/.

It can thus be said that even though graphic representations do have corresponding phonological representations, the relationship between the two types of representation is far from being simple and requires systematic apprenticeship on the part of the learner.

Once again it is essential to note that isolated single graphemes do not usually carry meaning (with certain exceptions such as the grapheme a), and that individual graphemes bear no intrinsic visual resemblance to the concepts they represent or to the sound units related to them. Their shapes are arbitrary, abstract constructs determined within a closed system of differential values.

These abstract figures can also be used to represent formal mathematical languages, as in algebra. In fact, one could suggest that written representations were a necessary preliminary stage for the construction of systems of formal mathematical or logical representation.

	Curved Stroke Trajectories								
	1-7-9	1-11-17	3-9-19	3-6-10	9-15-17	10-19-12	10-15-17	11-18-14	19-13-3
A									
B	+				+				
C			+						
D		+							
E									
F									
G			+						
H									
I									
J								+	
K									
L									
M									
N									
O			+						+
P	+								
Q			+						+
R	+								
S				+			+		
T									
U						+			
V									
W									
X									
Y									
Z									

	Straight Stroke Trajectories																
	1-3	1-5	1-11	1-17	1-18	1-19	1-21	2-4	2-10	2-17	2-19	3-9	3-11	3-17	3-18	3-19	3-20
A														+			
B				+													
C																	
D				+													
E	+			+													
F	+			+													
G																	
H				+												+	
I								+								+	
J								+					+				
K				+								+					
L				+													
M										+	+						
N				+		+										+	
O																	
P				+													
Q																	
R				+													
S																	
T		+														+	
U									+								
V						+											
W					+										+		+
X							+										
Y			+														
Z		+															

	Straight Stroke Trajectories (cont.)															
	3-21	4-12	4-19	4-21	5-11	5-17	5-19	5-20	9-11	9-19	10-12	11-19	11-21	17-19	17-21	18-20
A	+										+					
B																
C																
D																
E									+					+		
F									+							
G											+	+				
H									+							
I																+
J																
K										+						
L														+		
M			+	+												
N																
O																
P																
Q													+			
R										+						
S																
T																
U		+														
V							+									
W								+								
X						+										
Y					+							+				
Z						+									+	

Fig. 2. Subgraphemic feature matrix for French capital letters

Syllabic Clusters/Graphemic Clusters

The phonological status of the syllable is also a topic of debate within theoretical linguistics. Within the phonology of French, serious arguments can be made for including syllables as an independent level of phonological representation (Léon, 1966; Selkirk, 1978, 1980). One such argument is the conditioning of vocalic phoneme distribution by syllabic structure. Thus the vocalic phonemes /e/, /ɛ/, /ø/, /œ/, /o/, /ɔ/ are said to be in complementary distribution according to the accentual and syllabic properties of the chain of phonemes.

Direct phonological oppositions between the vocalic phonemes are rare, and the majority of such oppositions tend to be neutralized in modern French. The distribution attributes the series of open vocalic phonemes /ɛ/, /œ/, /ɔ/ to accentuated syllables containing a consonant in syllable final position, and the series of closed vocalic phonemes /e/, /ø/, /o/ to accentuated syllables containing a vocalic phoneme in syllable final position.

In terms of graphological representations, grapheme clusters exist for both vocalic and consonant graphemes. As mentioned above, "oi" corresponds to /wa/, "ge" corresponds to /ʒ/, "ch" corresponds to /ʃ/, s + consonant grapheme corresponds to /s/, "-ill-" corresponds to /j/ etc. Some grapheme clusters are ambiguous, such as "-gui-", as in "linguistique/droguiste," and "-gn-", as in "agneau/diagnostique." However, although local grapheme clusters do exist it is difficult to argue for grapheme clusters as an independent level of graphological representation.

Word Level

The existence of a word level representation in French phonology is controversial (Carton, 1974). French phonology seems to organize syllabic representations directly into higher levels of representation, bypassing the word level.

On the graphological level exactly the opposite holds true. In fact, the primary segmentation of graphological representations is at the word level. Each word unit is separated by blank spaces which immediately precede and follow it.

For some linguists, such as A. Martinet (1960, 1985), the word is not in fact a primary linguistic construct, but a written construct. It is written representations that have conferred conceptual autonomy and even conceptual dominance to the "word." Morphologically, the word unit is an informal, intuitive construct resulting from the combination of a lexical unit (lexeme) responsible for conceptual identity, and a grammatical unit (morpheme) responsible for indicating morphological and syntactic values.

Thus even graphically similar written representations can have widely divergent underlying syntactic and morphological values resulting in different phonological interpretations. For example, the graphic chain "portions" in French can correspond to a phonological representation /pɔRsjɔ̃/, in which case it is a conjugated form of the verb *porter*. In this case the lexeme "port-" is supplemented by the morpheme suffix "-ions" (first person plural, imperfect indicative tense). The same graphic chain can correspond to a different phonological representation /pɔRsjɔ̃/ in which case it is the plural form of the noun *portion*. The lexeme "portion" then carries the plural suffix "-s" (see Llorach, 1968, for further French examples). Similarly the graphic representations "sow," "row," "wind," in English, each have two independent phonological interpretations depending on their category.

Furthermore, graphic representations of morphological suffixes do not always correspond to underlying phonological representations. Thus in the above example there is no phonological difference between the singular *portion* and the plural *portions*. This is the case for all plural forms of nouns carrying the "-s" suffix in modern French. In modern French, graphic forms carry a far greater number of plural markers than do their phonological counterparts. Sentences such as *Il ne voit pas de solution* and *Ils ne voient pas de solutions* are phonologically identical but graphically well differentiated.

Phrase

Phrase boundaries are of crucial importance in French phonology. Liaison, accentuation, and intonation are all closely associated with syntactic boundaries (Carton, 1974; Léon, 1966; Martin, 1980, 1982; Selkirk, 1974, 1980). Liaison occurs only within phonological phrases and is blocked between phrases. Phrasal accentuation and intonation fall on the final syllable of the phonological phrase which generally correspond to the major semantico-syntactic units of the utterance. Phrasal units thus constitute one of the major levels of phonological representation in French and are of particular importance for prosodic systems.

On the other hand, phrase boundaries are not always indicated in graphological representations. Diacritical marks such as commas, semicolons, and colons can be used to indicate major sentence constituents, but the correspondence is not systematic.

Sentence

Sentences also constitute an autonomous level of representation in French phonology. Once again prosodic systems play a significant role, since major intonational movements are the principal phonological units associated with sentence boundaries (Carton, 1974; Léon, 1971; Martin, 1980, 1982).

In graphological representations, sentence units have a privileged status, being delimited by the use of capital letters in sentence initial position and by diacritical marks such as periods in sentence final position.

Expressive, Paralinguistic Levels of Representation

In addition to segmental and suprasegmental information necessary for the decoding of the linguistic message, the speech signal can also carry paralinguistic information that conveys a number of clues as to the speaker's attitude or emotion, dialectal origin, and individual characteristics such as sex and age (Léon, 1971). The majority of spoken communicative situations also have the advantage of the presence of a listener who shares common references and presuppositions (Ducrot, 1972). Oral discourse has the advantage of economy on this level.

Graphology has a limited set of procedures for conveying paralinguistic information, with the exception of written communication in comic books or in concrete poetry where graphic procedures can be used to convey paralinguistic information (Léon, 1971). Written representations therefore need to specify attitudes through the use of performative verbs or periphrases (Searle, 1969, 1979).

Semiological Status

When compared globally in terms of their semiological status, phonological and graphological representations share a number of common properties. In particular, the two systems have a common semiological function in being able to associate an abstract representation or signifier with an underlying concept.

The systems also share similar structural properties of construction by having stratified levels of representation wereby units from a lower level are integrated to form units at a higher level (Hjelmslev, 1954, 1961). In addition this confers a high degree of structural economy, since individual units can be recombined in a number of unrelated contexts in order to form different signifiers. Thus combination of the three graphemes, c, t, and a can form "cat," "act," "tac," "cta" (the last two combinations not forming actual words, but being used in combination with other graphemes within words).

Another common property is the arbitrary nature of the representations. As already stated, there is no necessary auditory or visual resemblance between the string of phonemes and graphemes and their underlying concepts. Both sets of representations are tacitly agreed upon by conventional contract between the members of a linguistic community.

In addition, it is essential to note that graphemes and phonemes are attributed differential, negative values within a given closed system. Thus the closed, bilabial, front vocalic phoneme /y/ exists in French phonology, but not in English. The consonantal phoneme /θ/ exists in English but not in French. Similarly, the grapheme string "lui" is interpretable as a single word in French, but not in English. In the same way, the corresponding phonological representation of a grapheme string is determined by the specific relations established between the orthography and phonology of a particular language. Thus the strings "rang," "fort," "vacation," "stage," which exist in both languages, have widely divergent phonological (as well as morphological and semantic) interpretations in English and in French. There is thus no universal phonetic or phonological motivation for the graphological characters.

Despite the fact that the two systems share a number of common structural characteristics and semiological properties, it is important to underline the fact that they are autonomous systems of representation. There are several major arguments for their independent status:

- Phonological systems and representations of a given language can change without corresponding changes in graphic representation; this can occur in synchrony due to dialectal variation or in diachrony, as was the case for "-oi" grapheme clusters and for almost all word final consonants in French.
- Graphological systems within a given language can change without a corresponding change in phonological representation; once again, this can occur in synchrony due to dialectal variation, for example the "-our" series and intervocalic "-s-" in British English as opposed to North American English, or in diachrony as was the case for the shift from the "-ei-" to the "-ai-" grapheme cluster in verb endings in French.
- A single set of graphemes can be used to correspond to a number of different phonological systems, as we have already seen above.
- Although the systems share a stratified structural construction, each chooses different levels of primary representation; for example the syllable, phrase, and sentence levels are crucial in French phonology, whereas it is the word and sentence levels that are primary in graphic representations.

The independence of the two systems is all the more evident on the graphetic and phonetic levels:

- The production of graphic representations involves brachiomanual and vestibular mechanisms and visual feedback, whereas speaking uses pulmonolaryngeal and buccolinguofacial mechanisms and auditory feedback.
- Writing produces a concrete signal that remains relatively constant over time, whereas the speech signal is transient and quickly degenerates after a short time span.
- Writing is transmitted to a reader at first in the form of visual stimuli processed by the external receptors of the eye, then transmitted through the optic nerve to primary projection areas in the occipital lobes (Alajouanine et al., 1960), whereas spoken language is transmitted as acoustic stimuli to the ears and is then passed along auditory pathways to the primary projection areas in the temporal lobes.

Despite these obvious physiological differences, one can still see a common principle at work in the overall structural status of both speaking and writing: both systems of higher cognitive functions are superimposed on sensorimotor systems the necessity of which for biological survival is more readily understandable. Speaking is grafted onto systems required for breathing and digestion, writing is superimposed onto systems involved in general manual skills such as use of tools, gathering of food, feeding etc. In both cases the cognitive system exploits and dominates an already existent sensorimotor circuit as its channel of expression.

It is tempting to suggest that the common structural properties and semiological functions of graphological and phonological systems are not the result of mere coincidence but reflect natural language-processing strategies. That is, meaning systems constructed from arbitrary figures are continually reorganized into autonomous levels of representation and reflect deep-rooted information-processing strategies in both sensorimotor and cognitive functional circuits.

A single string of graphological and phonological stimuli (or a corresponding pattern of neuronal activity) can be reorganized into multiple levels of independent information units which are then given a global interpretation (Bhatt, 1981). A parallel analogy in the visual system is to consider that a cube-like shape is formed of a particular concatenation of a series of dots or lines which can be analyzed into a given set of natural levels of representation and then interpreted as a single object. Thus one should not expect to find discrete, isomorphic cerebral representations of these types of mental constructs. The crucial task is to discover the rules and principles that allow us to create and interpret them.

Conclusion

The principal theme running throughout this discussion has been that graphology and phonology constitute two parallel systems of expression that share a similar semiological status and construction. It is essential to emphasize that the two systems also share a common underlying function, that their sole purpose is the expression of an underlying content (Hjelmslev, 1961). Written communication has played a key role in building economic, social, moral, and political structures (Goody, 1977); however, it is crucial to recognize that the major contribution of writing is not to be found in its usefulness for administrative, bureaucratic, or economic tasks. The principal advantage of the medium of the written word is the spatial and temporal transmissibility of the written signal, and the major function of written communication lies in its capacity to create *long-lasting mental structures* (collective scientific, philosophical, political, moral, and religious theories, as well as individual dreams and emotions) which the human organism *needs* and spontaneously creates for its psychological and biological survival.

Language was the tool that liberated thought from immediate experience, allowing us to speculate and create new, intangible societies and worlds. Written language was the tool which liberated thought from the immediacy of speech, from the bonds of time and space, allowing our mental constructs to be reincarnated in the minds and bodies of others, giving our ideas a life of their own.

References

Alajouanine, T. (1968). *L'aphasie et le langage pathologique.* Paris: Baillières.

Alajouanine, T., Lhermitte, F., & Ribaucourt-Ducarne, B. (1960). Les alexies agnosiques et aphasiques. In Alajounine, T. (Ed.), *Les grandes activités du lobe occipital* (pp. 235–265). Paris: Masson.

Albert, M. (1979). Alexia. In Heilman, K., & Valenstein, E. (Eds.), *Clinical neuropsychology* (pp. 59–91). Oxford: Oxford.

Bange, L. (1971). *A study of the use of vowel letters in consonantal writing.* Munich: University of Munich.

Benson, D.F. (1979). *Aphasia, alexia and agraphia.* London: Churchill Livingstone.

Bhatt, P. (1981). Fundamental frequency and the interhemispheric perception of time. In Martin, P. (Ed.), *Symposium prosodie/prosody symposium* (pp. 53–60). Toronto: Groupement des Acousticiens de Langue Française.

Bloomfield, L. (1933). *Language.* Chicago: University of Chicago Press.

Bouuaert, J. (1949). *Petite histoire de l'alphabet.* Bruxelles: Lebègue.

Carton, F. (1974). *Introduction à la phonétique du français.* Paris: Bordas.

Catach, N. (1978). *L'orthographe,* Paris: Presses Universitaires de France.

Charcot, J.M. (1890). *Leçons sur les maladies du système nerveux.* Paris: Lecrosnier et Babé.

Chomsky, N. (1970). Phonology and reading. In Levin, H., & Williams, J. (Eds.), *Basic writings on readings.* New York: Basic Books.

Christin, A.M. (Ed.) (1982). *Ecritures.* Paris: Le Sycomore.

Cohen, M. (1953). *L'écriture.* Paris: Editions sociales.

Cohen, M. (1958). *La grande invention de l'écriture et son évolution.* Paris: Imprimerie Nationale.

Coulmas, F., & Ehlich, K. (1983). *Writing in focus.* New York: Mouton.

Critchley, M. (1982). *Mirror writing.* London: Kegan Paul, Trench, Trubner.

Déjerine, J. (1891). Sur un cas de cécité verbale avec agraphie. *Mémoires de la Société de Biologie, 3,* 197–201.

Déjerine, J. (1892). Contribution a l'étude anatomopathologique et clinique des différentes variétés de cécité verbale. *Mémoires de la Société de Biologie, 4,* 61–90.

De Kerckhove, D. (1985). Effets cognitifs de l'alphabet. In De Kerckhove, D. (Ed.), *Understanding 1984.* Paris: UNESCO.

Denes, P., & Pinson, E. (1963). *The speech chain.* New York: Bell Telephone Laboratories.

De Saussure, F. (1915). *Cours de linguistique générale.* Paris: Payot.

Diringer, D. (1948). *The alphabet: a key to the history of mankind.* London: Hutchinsons.

Diringer, D. (1962). *Writing.* London: Thames and Hudson.

Diringer, D. (1983). *A history of the alphabet.* New York: Gresham.

Driver, G.R. (1954). *Semitic writing from pictograph to alphabet.* London: Oxford University Press.

Dubois-Charlier, F. (1972). A propos de l'alexie pure. *Langages, 25,* 76–94.

Dubois-Charlier, F. (1976). Les analyses neuropsychologiques et neurolinguistiques de l'alexie: 1838–1969. *Langages, 44,* 20–62.

Ducrot, O. (1972). *Dire et ne pas dire.* Paris: Hermann.

Fevrier, J. (1959). *Histoire de l'écriture.* Paris: Payot.

Firth, U. (Ed.) (1980). *Cognitive processes in spelling.* New York: Academic.

Friedrich, J. (1957). *Extinct languages.* New York: Philosophical Library.

Gaur, A. (1984). *A history of writing.* London: The British Library.

Gelb, I. (1952). *A study of writing.* Chicago: University of Chicago Press.

Gelb, I. (1968). Grammatology and graphemics. In *Papers from the fourth regional meeting, Chicago Linguistic Society.* Chicago: University of Chicago Press.

Gibson, E., & Levin, H. (1975). *The psychology of reading.* Cambridge: MIT Press.

Goodman, K. (1982). *Language and literacy.* (2 vols.). London: Routledge & Kegan Paul.

Goody, J. (1977). *The domestication of the savage mind.* Cambridge: Cambridge University Press.

Gordon, C. (1968). *Forgotten scripts.* Harmondsworth: Penguin.

Gregg, L., & Steinberg, E. (Eds.) (1980). *Cognitive processes in writing.* Hillsdale: Erlbaum.

Haas, W. (1970). *Phonographic translation.* Manchester: Manchester University Press.

Haas, W. (Ed.) (1976). *Writing without letters.* Manchester: Manchester University Press.

Hécaen, H., & Kremin, H. (1977). Reading disorders resulting from left hemisphere lesions: aphasic and "pure" alexias. In Whitaker, H., & Whitaker, H. (Eds.), *Studies in neurolinguistics,* II. New York: Academic.

Hécaen, H., & Marcie, P. (1974). Disorders of written language following right hemisphere lesions: spatial dysgraphia. In Diamond, S., & Beaumont, J. (Eds.), *Hemisphere function in the human brain.* London: Elek.

Henderson, L. (1982). *Orthography and word recognition in reading.* New York: Academic.

Henderson, L. (Ed.) (1984). *Orthographies and reading.* Hillsdale: Erlbaum.

Herrick, E. (1966). *A linguistic description of Roman alphabets.* Unpublished master's thesis, Hartford Seminary Foundation.

Higounet, C. (1955). *L'écriture.* Paris: Presses Universitaires de France.

Hjelmslev, L. (1954). La stratification du langage. *Word, 10,* 163–188.

Hjelmslev, L. (1961). *Prolegomena to a theory of language.* Madison: University of Wisconsin Press.

Huey, E. (1968). *The psychology and pedagogy of reading.* Cambridge: MIT Press.

Jackson, D. (1963). *The story of writing.* London: Macmillan.

Jakobson, R., Fant, G., & Halle, M. (1952). *Preliminaries to speech analysis.* Cambridge: MIT Press.

Jensen, H. (1970). *Sign, symbol and script.* London: Allen & Unwin.

Kavanaugh, J., & Mattingly, I. (Eds.) (1972). *Language by ear and by eye.* Cambridge: MIT Press.

Kirk, U. (Ed.) (1983). *Neuropsychology of spelling, reading and writing.* New York: Academic.

Kolers, P., Wrolstad, M., & Bouma, H. (Eds.) (1980). *Processing of visible language.* II. New York: Plenum.

Kremin, H. (1976). L'approche neurolinguistique des alexies: 1969–1976. *Langages, 44,* 63–81.

Kussmaul, A. (1877). *Die Störungen der Sprache.* Leipzig: Vogel.

Laberge, D., & Samuels, S. (1977). *Basic processes in reading.* Hillsdale: Erlbaum.

Leischner, A. (1957). *Die Störungen der Schriftsprache.* Stuttgart: Thieme.

Léon, P. (1966). *Prononciation du français standard.* Paris: Didier.

Léon, P. (1971). *Essais de phonostylistique.* Paris: Didier.

Lesgold, A., & Perfetti, C. (Eds.) (1981). *Interactive processes in reading.* Hillsdale: Erlbaum.

Levin, H., & Addis, A. (1979). *The eye-voice span.* Cambridge: MIT Press.

Llorach, E.A. (1968). Les representations graphiques du langage. In Martinet, A. (Ed.), *Le langage* (pp. 513–568). Paris: Pléiade.

Marcé, O. (1856). Mémoire sur quelques observations de physiologie pathologique tendant à démontrer l'existence d'un principe coordinateur de l'écriture. *Compte-rendu de la Société de Biologie, 3,* 93–115.

Marcie, P. (1977). L'agraphie: histoire neuropsychologique. *Langages, 47,* 70–85.

Marcie, P., & Hécaen, H. (1979). Agraphia: writing disorders associated with unilateral cortical lesions. In Heilman, K., & Valenstein, E. (Eds.), *Clinical neuropsychology.* Oxford: Oxford.

Martin, P. (1980). Sur les principes d'une théorie syntaxique de l'intonation. In Léon, P., & Rossi, M. (Eds.), *Problèmes de prosodie, I* (pp. 91–102) Paris: Didier.

Martin, P. (1982). L'intonation dans la description linguistique. *Recherches Sémiotiques, 2,* 63–85.

Martinet, A. (1960). *Eléments de linguistique générale.* Paris: A. Colin.

Martinet, A. (1985). *Syntaxe générale.* Paris: A. Colin.

Martlew, M. (1983). *The psychology of written language.* New York: Wiley.

Naveh, J. (1982). *Early history of the alphabet.* Jerusalem: Magnes.

Neisser, U. (1967). *Cognitive psychology.* New York: Appleton, Century, Crofts.

Olson, D. (this volume). Mind, media and memory: the archival and epistemic functions of written text. In Lumsden, C., & De Kerckhove, D. (Eds.), *Western literacy, brain and culture.* Berlin: Springer.
Olson, D., Torrance, N., & Hildyard, A. (Eds.) (1985). *Literacy, language and learning: the nature and consequences of reading and writing.* Cambridge: Cambridge University Press.
Ong, W. (1982). *Orality and literacy: the technologizing of the word.* London: Methuen.
Pedersen, H. (1931). *The discovery of language.* Bloomington: Indiana University Press.
Petrie, W. (1912). *The formation of the alphabet.* London: Macmillan.
Pirozzolo, F., & Wittrock, M. (Eds.) (1981). *Neuropsychological and cognitive processes in reading.* New York: Academic.
Pulgram, E. (1951). Phoneme and grapheme: a parallel. *Word, 7,* 15–20.
Pynte, J. (1983). *Lire, identifier, comprendre.* Lille: Presses Universitaires de Lille.
Rahi, I. (1977). *World alphabets: their origin and development.* Allahbad: Bhargava.
Reber, A., & Scarborough, D. (Eds.) (1977). *Towards a psychology of reading.* Hillsdale: Erlbaum.
Rosenberg, S. (Ed.) (1982). *Handbook of applied psycholinguistics. Hillsdale: Erlbaum.*
Sampson, G. (1985). *Writing systems: a linguistic introduction.* Stanford: Stanford University Press.
Sanford, A., & Garrod, S. (1981). *Understanding written language.* New York: Wiley.
Sasanuma, S. (1974). Kanji versus kana processing in alexia with transient agraphia: a case report. *Cortex, 10,* 88–97.
Sasanuma, S. (1975). Kana and kanji processing in Japanese aphasics. *Brain and Language, 2,* 369–383.
Sasanuma, S., & Fujimura, O. (1971). Selective impairment of phonetic and non-phonetic transcription of words in Japanese aphasic patients: kana and kanji in visual recognition and writing. *Cortex, 7,* 1–18.
Sasanuma, S., & Fujimura, O. (1972). An analysis of writing errors in Japanese aphasic patients: kanji versus kana words. *Cortex, 8,* 265–282.
Sasanuma, S., & Monoi, H. (1975). The syndrome of gogi (word-meaning) aphasia: selective impairment of kanji processing. *Neurology, 25,* 627–632.
Scinto, L. (1986). *Written language and psychological development.* New York: Academic.
Searle, J. (1969). *Speech acts.* Cambridge: Cambridge University Press.
Searle, J. (1979). *Expression and meaning.* Cambridge: Cambridge University Press.
Selkirk, E. (1974). Liaison and the X-bar notation. *Linguistic Inquiry, 5,* 573–590.
Selkirk, E. (1978). The French foot: on the status of mute e, *Studies in French Linguistics, 1,* 141–150.
Selkirk, E. (1980). *The phrase phonology of English and French.* New York: Garland.
Sprengling, M. (1931). *The alphabet: its rise and development from the Sinai inscriptions.* Chicago: University of Chicago Press.
Taylor, I. (1883). *The alphabet: an account of the origin and development of letters* (2 vols.). London: Kegan Paul, Trench.
Taylor, I., & Taylor, M. (1983). *The psychology of reading.* New York: Academic.
Troubetzkoy, N. (1939). *Principes de phonologie.* Paris: Klincksieck.
Uldall, H.J. (1944). Speech and writing. *Acta Linguistica, 4,* 11–16.
Vachek, J. (1945). Some remarks on writing and phonetic transcription. *Acta Linguistica, 5,* 86–93.
Vachek, J. (1973). *Written language: general problems and problems of English.* The Hague: Mouton.
Vygotsky, L. (1962). *Thought and language.* Cambridge: MIT Press.
Vygotsky, L. (1978). *Mind in society.* Cambridge: Harvard University Press.
Weaver, W. (1977). *Towards a psychology of reading and language.* Athens: University of Georgia Press.
Whiteman, M. (Ed.) (1981). *Writing: the nature development and teaching of written communication.* Hillsdale: Erlbaum.
Yamadori, A. (1975). Ideogram reading in alexia. *Brain, 98,* 231–238.

Part 3 Writing Right and Left

Introductory Remarks

The main issue selected for the hypothesis examined in this book is that of the direction of writing. This section therefore looks at this question from several angles before the hypothesis is examined at the neuropsychological level in the following sections. The first three papers bear on the question of the rightward direction of the Greek and all other vocalic alphabets. William Watt's approach is to examine each letter individually and estimate the lines of force that may have guided its fixation from the early experimentations to the present standards. Writing with his model theory in mind, Derrick de Kerckhove prepares the ground by suggesting that there may be structural constraints to guide the direction of writing systems. He proposes several logical principles of graphic layout to that effect. Colette Sirat tackles the problem from another angle, that of the circumstantial conditions guiding and supporting the hand movements of the scriptor. Finally, presenting a survey of world writing systems from the point of view of develomental and learning conditions, Insup Taylor introduces considerations concerning brain involvement and lateralization patterns in reading and writing.

CHAPTER 8

Canons of Alphabetic Change

WILLIAM C. WATT*

In all sciences we are being progressively relieved of the burden of singular instances, the tyranny of the particular.

Sir Peter Medawar

Introduction

It is well attested that between about 800 and 500 B.C. the Greeks inherited the Phoenician alphabet and set about modifying it in various ways. The present chapter is concerned with interpreting the fundamental forces that lay behind these modifications, with an ultimate view toward providing a new approach to understanding the human mind. Some of these cognitive implications are supported by the findings of contemporary psycholinguistics, but others are frankly speculative. However, all the arguments presented here are to some extent testable. In the aggregate, it is hoped that they will provide a picture of an historical and an ontogenetic process that in many ways mirror one another, both of a decidedly "Lamarckian" cast. This will be gone into in some detail.

The arguments will proceed as follows. The first section will define the basic characteristics of an alphabet, as a necessary prelude to examining the historical processes that can change it. Briefly, an alphabet can be viewed as two interrelated but distinct systems: one consisting of particular figures, or patterns, and the other of the programs needed to execute those figures manually, as writing. Both systems can be described in terms of their component "distinctive features." This bifurcation will be supported by references to specific "evolutionary processes," which can affect the two systems very differently. In the second section these evolutionary processes are further investigated, and are shown (not surprisingly given that their source is the human mind) to be cognitive tendencies that can manifest themselves in a number of ways. Finally, these processes will be firmly related to different phases of alphabetic evolution.

It may not be too much to claim that this analysis and the closely-related evolutionary model offer significant implications for the cognitive study of

* School of Social-Sciences, University of California, Irvine, CA 92717, USA.

the alphabet (as well as of other writing systems such as syllabaries[1]). This is plainest in the irrefragable emphasis placed on the bipartite division noted above: to the effect that there are really *two* alphabets at issue (e.g., in the English alphabet a set of 26 patterns and a set of 26 programs). Any study of the cognitive nature of the alphabet that fails to acknowledge this division is bound to miss the mutual influence of the programs and the patterns, and so must fail to winnow out related confounding factors. It would thus fall short of providing an adequate account of alphabetic evolution as a special case of cognitive evolution. Once understood, this point is likely to seem self-evident.

We now turn to examining the analysis in full.

The Alphabet in Iconic Perspective

We begin by contrasting two general approaches to studying the alphabet: one heavily influenced by linguistics and by present-day semiotics, which I will call the "iconic" account, the other attempting to describe the alphabet in such a way as to jibe with the results of psychological experiments, and perhaps best called the "task-oriented" account. Two examples of the iconic account are discussed here: that advanced by Eden and Halle in the early 1960s (e.g., in Eden & Halle, 1961) and that under development in my own work since 1975 (Watt 1975, 1980, 1981, in press). [A third important account, that of Herrick (1969) is omitted from consideration here as it is in 'stratificational' notation and so offers a less ready basis of comparison with the description on which the present paper rests.] A salient attribute that both iconic accounts share with many of the task-oriented ones is that they describe letters in terms of "distinctive features," factoring them into pictorial characteristics, such as "has a straight line." However, they differ importantly from the latter in assigning these features to *parts* of letters, whereas the task-oriented descriptions mostly assign them to letters as a whole. The two iconic accounts differ even between themselves in this respect: for Eden and Halle the distinctive features are essentially line segments per se, while in my own work they are far more abstract, being, in the analysis of the visual aspect of the letters, *attributes* of line segments (such as curvilinearity). My account also differs from that of Eden and Halle in the following respects: (1) it derives from what is known about the evolution of the alphabet

[1] Obviously, all ordinary writing systems have in common that they are executed by hand (when not being set in type), and are read by the eye. Recent research on syllabaries and ideographs (in Japanese, Kana and Kanji) is best exemplified by the sterling work of Hung and Tzeng (e.g., Tzeng and Hung, 1981). From the point of view adopted in the present paper, the evolution of syllabaries has yet to be studied systematically. It would be quite interesting, for example, to know whether the evolution of writing systems has been affected by the number of their symbols (50 for Japanese Katakana, 50 plus many variants for Japanese Hiragana, and 43 for the very large Cyrillic alphabet in its early days).

(largely in the distant past) and so is deeply influenced by developments in linguistic theory that came about only after Eden and Halle's work; (2) it treats the alphabet as two closely related but distinct systems: the "patterns" of the visual shapes and the "programs" of the manual performances needed to execute those shapes; and (3) owing to the abstraction of the distinctive features from the letters' constituent line segments (or, in the case of the letters-as-programs, from their constituent strokes or vectors), several features may occur simultaneously (or concurrently). Finally, (4) although the distinctive features in my analysis obviously differ from those of the psychology-based task-oriented approach (mine factor letter-components, rather than factoring the letters themselves directly, so that the two kinds of description are as it were on tracks of different gauge), they have definitely been influenced by certain psychological findings: for example, the early claim by Gibson, Osser, Schiff, and Smith (1963) that letters may partly be discriminated by reference to a feature of axial symmetry (see also Fehrer, 1935).

For readers familiar with linguistics, some striking if superficial differences between the various analyses may best be clarified by an analogy with phonology. One such difference is the manner in which letters are factored into distinctive features. In the task-oriented approach, they are treated like phonemes and are factored directly into features; in Eden and Halle's approach they are treated more like morphemes, and the "phonemes" into which they are segmented act, with very little further abstraction, as their distinctive features as well; in my own approach the letters again are analogous to morphemes, and are segmented into phoneme-like elements, but these are then factored further into co-occurrent features, much as in phonology itself. Another salient distinction from my analysis, the bifurcation into analyses of patterns and of programs, can be likened to a bifurcation of phonology into separate but related acoustic and articulatory accounts [as has been proposed (see, e.g., Fischer-Jørgensen, 1952), but little pursued]. Finally, my analysis recognizes that in many cases full specifications of the letters' attributes contain redundancies, i.e., some attributes may be calculable from neighboring ones, and all letters exhibit some attributes in common; thus, as in phonology, "redundancy rules" are provided to make these calculations. For example, whereas a typical task-oriented description would assign the whole letter P such features as + CURVED and + STRAIGHT – features which of course are mutually exclusive, but which must both be assigned because P contains both – my account first decomposes P into "I" and "ɔ" (actually, even further decomposition is accomplished, but this is not relevant at this juncture) and only then is the equivalent of + STRAIGHT assigned to "I" and the equivalent of + CURVED assigned to "ɔ"[2].

---

[2] Actually, as discussed in Watt (1981), the attributes of straightness and curvilinearity are assigned by means of different values for a single feature, CNCVE. This carries the value

This analysis, which I have put forward elsewhere (e.g., Watt 1980, 1981), may of course prove wrong in various ways – given the usual course of science, the probability is good that it will – but at least it is far more detailed than any other, thus providing many more potential points of comparison among the letters. This may well be an advantage; in any case, no other existing analysis can be reconciled satisfactorily with the relevant psychological evidence, although it must be admitted that this evidence embodies conflicts of its own (see, e.g., Townsend, 1971; Townsend, Hu, and Evans, 1984[3]).

Perhaps it should be noted at this point that scholarly interest in the alphabet hardly began in 1961: in fact, it may be as old as the alphabet itself. The sequence of letters or *abecedarium* (A, B, C, D . . .) dates roughly back to the alphabet's very origin as a Semitic borrowing from the Egyptian consonantal signs, and has recently been shown to result from a simple beginning-to-end reading of a matrix into which early scholars placed the letters so as to classify them by the phonological properties of their associated sounds (Watt, 1987)[4]. More broadly, attested interest in writing as such dates, appropriately enough, back to the Sumerians:

> Bring me my sister, my Geshtinanna,
> She understands letters;
> Bring me my little sister, my scribe,
> She is the singer
> Who understands the song . . .[5]

A few thousand years later the Greeks, who had borrowed the alphabet from the western Semitic nation we call the Phoenicians and were as curious about its origin as they were about everything, ascribed its importation to a number of legendary heroes, especially Cadmos, Prince of Tyre (Diringer, 1968, p. 358). The Romans, who got their letters from the Etruscans (who had gotten them from Greeks settling on the Bay of Naples), took an oc-

'+' for curves presenting concavity when read from left to right, '–' for curves presenting convexity when so read, and '∧' for straight lines.

[3] An aspect of adequately reported letter-confusion experiments that leaps to the eye from 26 × 26 matrices such as in Townsend (1971, pp. 44–45) is that if one letter is partly composed of another, the first is more apt to be confused with the second than vice versa; for instance, R is more often confused with P, which is included in R, than P is with R. As Townsend, Hu, and Evans put it, "... the probability of falsely sampling ghost features is lower than the probability of *losing* features in the stimulus" (Townsend et al. 1984, p. 43; italics added). If wholes are more like their parts than parts are like wholes, one cannot say that simple "similarity" is being judged, unless, with Tversky (1977), we allow that similarity is asymmetrical.

[4] This means that the letter order of the alphabet is about 3500 years old, and is rational.

[5] Translated by N.J. Sandars in *Poems of Heaven and Hell from Ancient Mesopotamia* (1971); quoted by N. Hall (1980, pp. 192–193).

casional interest in improving them[6]. Today the shelves of any well-stocked library groan beneath the volumes devoted to the story of how the alphabet, a "key to the history of mankind" as one writer puts it[7], has been handed from generation to generation and from culture to culture, in the process altering its forms and to some extent coming to stand for new sounds[8]. Many of these books are scarcely scholarly, and not a few have some of their facts wrong; on the other hand, some are works of the highest scholarship [e.g., Lillian H. Jeffery's monumental *Local Scripts of Archaic Greece* (1961) and the late Sir Godfrey Driver's comprehensive *Semitic Writing* (1976)]. In any event, the true history of the alphabet is still in the process of being unraveled. For instance, it was only comparatively recently that the runic alphabets were persuasively shown to have derived from northern Etruscan forms, via the letters used by the Etruscans' Venetic or other Alpine neighbors (Diringer, 1968, pp. 393, 402–403), and only relatively recently that the ultimate source of our letters, hence of the idea of the alphabet itself, was convincingly traced back to the Egyptian penalphabet of one-consonant signs (Driver, 1976, pp. 136–139, 161–171)[9]. Indeed, the early history of the Semitic acquisition of these Egyptian notions is still having fresh details added to it (Albright, 1950; Cross and Lambdin, 1960; Segert, 1983).

Nor, turning another scholarly page, has psychology neglected the rich opportunities afforded by the study of alphabetic letters (as well as syllabaries). Studies have been done on how people learn (and mislearn) a relatively small set of simple figures; how they recognize them in reading and learn to skip or "clump" them; how they discriminate (or confuse) them in the laboratory; where in the brain (in a rough neuroarchitectonic sense) they process them; how they handprint and handwrite them; how their visual systems, becoming satiated after fixation, selectively eliminate and so appear to disclose unconscious components of them: and in general, how these signs

[6] For example, Emperor Claudius I (who ruled from 41 to 54 A.D., between Caligula and Nero) added three new letters, all of which died at the end of his reign. In 312 B.C., one Appius Claudius Censor, using C as his model, added G. Thanks to the Etruscans, from whom the Romans had inherited their alphabet, C (a curvilinearized Γ) had come to stand for /k/, so that the Romans, using C for both /k/ and /g/, could not distinguish between the two sounds (Diringer, 1968, p. 419). The new G, being a needed improvement, remained.

[7] This is the subtitle of Diringer's *The Alphabet* (1968).

[8] For example, the letter F, in Phoenician *waw* [for /w/ or perhaps /β/, and for vocalic /u/ (Jeffery, 1961, pp. 24–25)] has been handed down to the modern western European alphabet as no fewer than five distinct letters, F, Y, U, V, and W, standing for as many sounds and then some.

[9] An extremely useful table exhibiting the Egyptian one-consonant symbols and their Semitic counterparts may be found in Driver (1976, p. 169). A set of one-consonant signs is of course a genuine alphabet of sorts, or "betagamma" at least. I have called the Egyptian system a "penalphabet" as a reminder that for the Egyptians these signs were used together with two-consonant signs and hieroglyphs. By the same token the Japanese Kanas are genuine syllabaries even though the Japanese commonly use them together with ideographs derived from Chinese characters (Diringer, 1968, p. 126).

can serve as a "window" into human cognition[10]. The present volume, indeed, explores and extends this vein of research.

A lot, then, is known about the alphabet (and other writing systems), from both the historical and the psychological points of view. Oddly, however, the modern linguistic (iconic) approach, in which one would expect both these viewpoints to be featly combined, in fact (outside the work cited here) lies mostly unexplored[11]. At first glance this is surprising. Given that the facts of historical change are at least as well known for the letters as for the sounds of language, and that therefore essentially the same sort of phenomena are to be explained and the same pleasure and profit to be gained thereby, one would think that study of the letters would long ago have attracted independent interest. This neglect is all the more surprising in view of the fact that since the work of Gibson et al. in 1963 an increasing number of "confusion" studies have been carried out on the alphabetic letters, similar to those that have been performed on the sounds of language. These studies provide psychological confirmation that the intuitive judgments of letter similarities are comparable to judgments of sound similarities and suggest that letters (like sounds) can be factored into distinctive features. (The basic reasoning is that with either sounds or letters, a single difference on any one feature-characterization – i.e., one being + X whereas the other is – X – should suffice to distinguish them, and the more features shared relative to the number of features not shared, the more similar they are, and hence the more confusable with each other.)[12] Yet despite these sur-

---

[10] Valuable collections of recent investigations into the cognitive dimensions of alphabets and other writing-systems are Kolers, Wrolstad, & Bouma, (1979, 1980). In other areas, the following papers are recommended: for children's mislearning, Frith (1971); for reading, Samuels, LaBerge, & Bremer (1978) and Terry, Samuels, & LaBerge (1976) and the papers in Reber & Scarborough (1977), Pirozzolo & Wittrock (1981), and Tzeng & Singer (1981); for discrimination and confusion, Townsend (1971) and Townsend et al. (1984) and the papers referenced therein; for neuropsychological implications, Tzeng & Hung (1981; pp. 246–248) and their references to the rich literature on Japanese Kana and Kanji; for reading abilities as affected by reversing and reorienting the letters, Kolers and Perkins (1969, 1975); for selective elimination of components after satiety (under retinal fixation), Pritchard, Heron, & Hebb (1960); for phonological confusion studies, Brown and Hildum (1956; this publication introduced such studies); for an early study of simple figures *other than* letters, Fehrer (1935).

[11] Linguists have, however, analyzed alphabetic letters and similar symbols, and analyzed them well (Eden & Halle, 1961; Rankin, Sillars, & Hsu, 1965; Rankin, Siegel, McClelland, & Tan, 1966; Herrick, 1969). For the best current work in the linguistic analysis of writing, including the search for "universals," see the analyses of J.S. Justeson and L.D. Stephens (e.g., Justeson, Norman, Campbell, & Kaufman, 1985; Stephens & Justeson, 1978).

[12] The process just described of drawing features from confusion matrices depends on the notion that features are attributes of letters as wholes. As I will argue below, this notion is mistaken. On the other hand, like many mistakes in science, it has been very fruitful, and its rectification consists of deriving letter attributes as amalgams of part attributes. (This is comparable to deriving morpheme attributes as amalgams of phoneme attributes.) For an initial suggestion as to how this is to be done, see (Watt, in press).

face similarities between sounds and letters, the latter have drawn little interest from students of language.

This neglect is explained, I think, partly by tradition and partly by circumstances. One such circumstance is that there is not always a correlation between the phonological changes that are of interest to linguists and the alphabetic changes whose neglect was just lamented. The same letters may be used in related languages – for instance, Latin and French – even though the sounds they represent may be very different. Conversely, the same sounds may be represented by very different letters: the most extreme instance of this is the Serbo-Croatian language, which can be written in either the Cyrillic or Roman alphabet and is called "Serbian" or "Croatian" accordingly. Of course, phonological changes are often reflected in a redistribution of the letters, and once in a great while some phonological factor even motivates the introduction of a new letter [such as the Romans' adoption of G in 312 B.C. (Diringer, 1968; pp. 419–420)] or the disappearance of an old one [as in the steady reduction of the Etruscan alphabet to fit that language's lack of voiced consonants (Diringer, 1968, pp. 389–390)]. In some of these cases, going against the near-universal rule that letter shapes bear a wholly arbitrary relationship to the sounds they represent, the shape of a new letter may derive from that of another which represents a similar sound, so that the expanded alphabet now contains a pair of letters whose structural similarity mirrors the similarity between their sounds: Roman G, for /g/, derived from C, for both /g/ and /k/ then reverting to /k/, supplies a ready instance. Of course such cases are few: moreover, they are clouded by the fact that in most instances pairs of very similar letters – P and R; O and Q – cover sounds altogether dissimilar[13].

A second circumstance, and one probably just as telling, is that whereas language itself guides the inquiring scholar into seeking out the components of sounds (i.e., their distinctive features or minimal differences), it does nothing comparable for letters. That is, one cannot examine the morphophonemics of a language without being constrained to begin factoring the sounds into their phonological attributes (if only to group them as labials and so on), for otherwise the language's rules of phonological combination would remain an ungeneralizable list of individual cases. The fact that one

[13] O has always existed much in its present form, while Q descends from ϙ. That P and R look so much alike is even more accidental, since P descends from Γ and R from a P (*rho*) that was contemporaneous with archaic Γ (Diringer, 1968, pp. 419–421). When one letter's shape is influenced by another, this is likely to be without regard to the sounds they each represent: for instance, it is likely that Ͱ became F under the influence of its neighbor in the abecedarium, E. Of course it is not imperative that the symbols of a writing system be unmotivated; for instance, those of the Korean system are apparently modeled in part on intuitive transverse X-rays of the tongue's articulatory positions (see, e.g., Taylor 1980); and it was at one time seriously proposed that the Hebrew alphabet – taken to be the *Urschrift* – had the same character [Van Helmont 1667, quoted in Mendelson, Siger, Kubzansky & Solomon (1964)]. And there is no reason, practicality apart, why oscillographs could not be used for letters.

says "*in*visible" but "*im*permeable," "*in*spire" but "*im*brue," "*in*credible" but "*im*modest," would suggest even to an untutored observer that 'in-' becomes 'im-' when it precedes another labial consonant. So a feature like LABIAL – a primitive first approximation to a more sophisticated analysis, perhaps – stands out immediately on even a cursory look at the morphophonemics of English (and many other languages). But nothing of the sort stands out when looking at how the letters co-occur. The printed letter 'n,' which has a feature we can call ROUNDED, does not change its shape depending on whether the letter following it is Rounded or not (e.g., 'b' vs 'v': "u*n*balanced," "u*n*verifiable"). The ordinary concern of linguists – the analysis of language – does not extend to their taking an interest in the shapes of the letters. This leads us, having considered two sorts of circumstance militating against such an interest, to the "tradition" to which these circumstances gave rise. "Sounds are primary; letters, parasitic" would be one way of putting it. Naturally, this and similar dicta were particularly stressed at a time when linguistics was struggling to free itself from philology, which had largely devoted itself to the study of written languages (chiefly Greek and Latin); but they have remained as hard dogma ever since. And with justice, for they are quite correct.

It is only when linguists begin to consider unusual aspects of language, or better yet when they begin to train their insights and techniques on something other than language *per se,* that they can find themselves drawn to a study of the language's 26 letters comparable to that of the study of its 30-odd segmental phonemes. In other words, setting aside for the moment the relatively minor fact that the letters must be kept distinct by their users if they are to continue to be discriminable as representing different sounds, it is likely to prove, even for the linguist, an uninteresting happenstance that the letters have anything to do with language at all. But they can be considered interesting in their own right: as a semiotic system whose elements share many attributes but differ on others; as a system that has undergone a "natural" evolution (as will be discussed below); and so on. In fact, it follows from the preceding discussion that linguists undertaking such a study will know from the outset to look elsewhere than to the usual linguistic sources for confirmation or rebuttal of their analyses: to their description's ability to "predict" the past, for example, to its explanatory power relative to children's learning and performative mistakes, to the difficulties experienced by victims of agraphia and to their ad hoc circumvention of such deficits, or to the findings of "confusion" experiments: in short, sources of confirmation quite outside the normal purviews. In large part, then, even though one may be aided by past experience and a linguistic armamentarium, this will be *terra incognita.*

This is all the more true because, although it is convenient to begin by comparing letters and sounds (segmental phonemes) and by suggesting that distinctive features can be assigned to both, on closer inspection the two sets of elements exhibit glaring differences. First, as noted above, they differ markedly in how they are best factored into features, since letters are analogous to morphemes (phoneme sequences), not phonemes. No other single issue has so misled investigators venturing into this area. Assigning to letter-wholes features properly assigned to letter-parts means that for a letter such as P a description of + STRAIGHT and + CURVED or the like, as above, would (quite predictably, if you think of it) apply just as well to ꟼ: i.e., the very description that should minimally distinguish P from its nearest neighbor cannot distinguish it at all. Of course ꟼ is not a letter, but this is not the point: reversing P to ꟼ is a mistake nearly all schoolchildren make (as reversing N to И is a mistake even many adults make), so obviously no analysis that is not capable of making this distinction can hope to determine what it is that is neglected when such a mistake is made, or what it is that is learned when it ceases to be made. Naturally, additional features could be added – e.g., + RIGHT-FACING to distinguish P from ꟼ and + TOP-HEAVY to distinguish P from b – but to make all the appropriate distinctions such ad hoc features would, in principle, have to be added indefinitely. Indeed, it is easily seen that failure to distinguish elements from their reversals or other permutations is precisely what should be expected from any solution in which phonological features are assigned to morphemes: it would be an automatic consequence, for instance, of assigning such features as + LABIAL and + LIQUID to the morpheme 'mar' (for 'mar' and 'ram' would be identically described)[14].

The second main difference, also alluded to above, is that in the plane of their physical realization – in their "phorology," to generalize over phonology and all other semiotic means of realization[15] – letters are best analyzed as members of two related systems: that of patterns and that of programs. The difference is immediately conveyed by Fig. 1. In Fig. 1a the letter A is represented and described as a pattern, and as such it is analyzed as a sequence of "phanemes," each of which is in turn factored into phanemic distinctive features. In Fig. 1b the same letter is represented and described as a program of execution or handprinting, and here it is analyzed as a sequence of "kinemes", each of which is in turn factored into kinemic distinctive features. (It should be noted in Fig. 1a that two features, those for vertical and horizontal axial symmetry, are assigned the letter as a whole. These could indeed be regarded as morphemic features, or better as "phanemic long components.[16]) Of course, the morpheme of Fig. 1a and that of Fig. 1b – of the phanology and the kinology respectively – are closely relat-

---

[14] Essentially this is what Dante did when assigning such features as + SHAGGY to whole morphemes (Alighieri ca. 1304).

[15] The term "phorology," suggested especially for use in reference to semiotic systems other than language per se – such as alphabets – avoids the etymological limitations of "phonology" (Watt, 1984; p. 103).

[16] The idea of phanemic long components is of course borrowed from Harris' "phonological long components" (Harris, 1961; pp. 125–136).

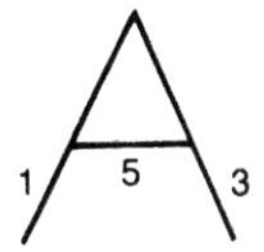

The visual or phanemic characterization of the letter 'A' is in terms of three line-segments, or phanemes, joined by two concatenators, as follows:

[1]	[2]	[3]	[4]	[5]
− VRTCL	− VRTCL	+ VRTCL	^ VRTCL	− VRTCL
− HRZTL	+ HRZTL	+ HRZTL	^ HRZTL	+ HRZTL
+ TRACE	− TRACE	+ TRACE	− TRACE	+ TRZCE
+ FLNTH	^ FLNTH	+ FLNTH	^ FLNTH	− FLNTH
^ CNCVE	^ CNCVE	^ CNCVE	^ CNCVE	^ CNCVE
− VSMTR	+ VSMTR	− VMSTR	+ VSMTR	+ VSMTR
− HSMTR	+ HSMTR	− HSMTR	+ HSMTR	+ HSMTR

[1]–[3]: [+ VSMTR, − HSMTR]

[1]–[5]: [+ VSMTR, − HSMTR]

a

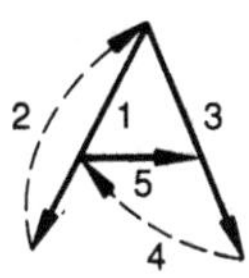

The handprinting or kinemic characterization of the letter 'A' is in terms of five strokes or kinemes, the second and fourth of which are off-the-page or invisible (traceless), as follows:

1	2	3	4	5
+ FLLG	− FLLG	+ FLLG	− FLLG	^ FLLG
− PROG	+ PROG	+ PROG	− PROG	+ PROG
+ TRCE	− TRCE	+ TRCE	− TRCE	+ TRCE
+ FULL	+ FULL	+ FULL	− FULL	− FULL
^ CLWS	^ CLWS	^ CLWS	^ CLWS	^ CLWS

b

Fig. 1 a, b. Phonemic (**a**) and kinemic (**b**) characterization of the letter A. Phanemes: *VRTCL,* vertical; *HRZTL,* horizontal; *FLNTH,* full length; *CNCVE,* concave; *VSMTR, vertically symmetrical; HSMTR,* horizontally symmetrical. Kinemes: *FLLG,* falling (downstroke); *PROG,* progressive (rightward stroke), *TRCE* trace; *FULL,* full length; *CLWS,* clockwise. Though drawn as curves here for clarity of illustration, kinemic concatenators are all $\wedge$CLWS by convention

ed. For instance, every visible or + TRCE stroke in the kinemic representation (unbroken lines) is matched by a visible or + TRACE line-segment in the phanemic one. (In the kinemic account, invisible or − TRCE strokes are strokes off the page, or concatenators of a sort; in the phanemic account invisible or − TRACE line-segments are concatenators per se. By convention, four-letter abbreviations are used for kinemes, five-letter ones for phanemes.[17]) However, despite the similarities between the kinemic and phanemic domains, their characterizations can be very different, so that inter-letter similarities vary greatly, depending on which of the two domains they are judged by.

Now, if under certain conditions letters tend to become more similar over time due to an evolutionary force, such mutual gravitations will assuredly look rather different in the two domains. The generalizations (suppressing differences; i.e., unlike features) will be very different, motivating any such evolutionary force. And of course two features in the phanemic characterization – the two symmetry attributes – have no counterpart in the kinemic characterization at all[18].

The evolutionary trend remarked on above, according to which letters become more alike, is one of four such trends, or "forces" as they may also be termed (if less securely). Here, as elsewhere, I have called this force by its obvious name, "homogenization." The other three are "facilitation," "heterogenization," and "inertia."[19] The latter two are relatively minor in effect, and, I think, in interest. Facilitation, however, is a force that over time leads to greater efficiency – i.e., a smaller expenditure of effort – in the processing of the signs of a semiotic system. Its chief effect is to ease the hand-execution of the letters. Most obviously, it reduces the amount of effort expended in manipulating the writing instrument with no visible result. For example, the A program of Fig. 2 is clearly more efficient in this sense than is that of Fig. 1a. [This aspect of facilitation, termed "efficiency" or ε, is computed as $\varepsilon = V/V + I$, where V is the sum of the lengths of all the letter's visible strokes and I is the sum of the lengths of all its invisible strokes. Length is measured in *vexils*, where for any alphabet in any given font the vexil is the altitude of the letter-space. For values of ε computed for the Greek majuscules, see Watt (1983). Another aspect of facilitation, akin to the linguistic notion of markedness, is a function of the relative difficulty of a letter's strokes.] If conservatively applied, facilitation results in no visible change (the traces left by the two very different programs of Figs. 1a and 2 are identical); however, if applied with less regard for maintaining the visible outcome – such as under the pressures of writing faster and with less effort – facilitation will begin to alter the resulting traces. This process is exemplified in Fig. 3a, where a more facile program has introduced a slight change (attested in many epichoric Greek alphabets), and even more in Fig. 3b, where minuscule α is a facilitated version of Fig. 3a. Just as this example suggests, both the Greek minuscules and the independent modern minuscules are in most respects simple facilitations of the corresponding

[17] This convention came into being in Watt (1980, 1981).

[18] Axial symmetry in the handprinting of letters could only be attained by using two writing instruments and executing two lines simultaneously, each the mirror image of the other and at every point equidistant from the chosen axis.

[19] These were first discussed, a little differently, in Watt (1979).

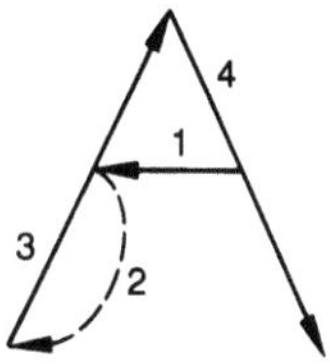

Fig. 2. A more facile program for executing A, as measured by ε, the ratio of the sum of the lengths of the visible strokes (*unbroken lines*) to the sum of the lengths of both visible and invisible strokes (*broken line*)

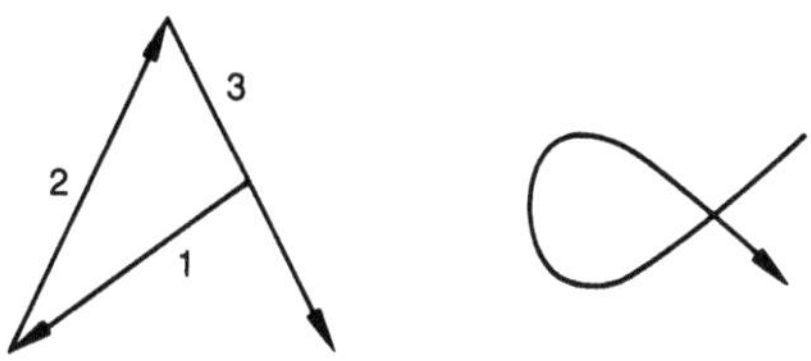

a **b**

Fig. 3. a A program for A that is more facile than that of Fig. 2; the resulting figure is found in several epichoric Greek alphabets [e.g., in that of Phokis (Jeffery, 1961, p. 99), of Aigina, where it was characteristic (Jeffery, 1961, p. 109), of Megara, where according to Jeffery it was normal during the fifth century B.C. (Jeffery, 1961, p. 132), and in the Ionian and Doric Islands (Jeffery 1961, p. 230, p. 308)]. It is also found, though "rarely," in Attica (Jeffery, 1961, p. 66). **b** Virtually the same program as for **a**, but with curvilinearization; the result is minuscule α. This is the most obvious (though not the only) source of α

majuscules, where the paramount consideration appears to have been minimization of invisible strokes and maximization of ε.

So far we have spoken exclusively of the facilitation of letter-production, ignoring that of letter-*recognition:* i.e. of facilitating not the programs but the patterns. We will return briefly to this topic below. But even now it should be evident that facilitation of the programs can easily lead to a *dis*facilitation of the patterns, in that the latter become harder to discriminate from each other, to recognize. Thus, intuitively at least, minuscule Greek γ and ν are harder to discriminate than the corresponding majuscules Γ and N; or compare υ and ν against Y and N. No further proof could be wanted, for the time being at least, of the hypothesis that the forces of alphabetic evolution can differentially affect the set of patterns and the set of programs: thus the thesis that the visual (phanemic) and the handprinting (kinemic) aspects of the alphabet must be separable as two distinct though related analyses is supported.

We will return to the force of facilitation, but at this point the key topic here – developing a model of alphabetic evolution and examining its cognitive implications – will be more quickly advanced by taking up another evolutionary force, one which [perhaps surprisingly given the claimed ubiquity of human indolence (Zipf, 1935)] is as powerful as facilitation, and in fact, at a certain evolutionary stage, more powerful still.

Homogenization and the Minor Forces

Shortly after 800 B.C. the Greeks inherited an alphabet that looked like this:

A, B, Γ, Δ, E, F[20], I[21], H, θ, ʃ, K, *⅃, M, N, Ξ, O, Π, *Ƴ, Q[22], P, ≷, T

But this was not the alphabet they adopted. At two points in this sequence (marked above by asterisks) they immediately made changes: ⅃ was changed to L (modern L) or even Λ (modern Greek Λ); and Ƴ they either dropped or changed to M (Jeffery, 1961; pp. 30–31, 32–33)[23].

Similarly, every fall millions of first-graders inherit an alphabet that looks like this:

A, B, C, D, E, F, G, H, I, *J, K, L, M, N, O, P, Q, R, S, T, U, V, W, X, Y, Z

Again, this is not the alphabet they adopt. At one point in this sequence – again marked by an asterisk – they typically at some stage in the learning process make a change: they substitute Ⴑ for J[24].

These two cases of letter-reversal seem similar; and so they are. The ancient Greeks and the modern first-graders both inherit an alphabet whose letters, to simplify for the moment, are overwhelmingly right-facing, or dextral, in that (if they are asymmetrical on the vertical axis) they consist mostly of a vertical staff, the "vexillum," plus an augmentation to the right. (Or, in the cases of G and Q, they consist of another letter plus an augmentation to the right.) In both cases there are a couple of letters which simply sidestep the question of dextrality – Greek ʃ and ≷; modern N, S, and Z – but even more conspicuously there are clear *exceptions* to dextrality – Greek ⅃ and Ƴ; modern J. Lastly, in both cases the changes introduced are those that remove precisely those exceptions, symmetrizing Ƴ to M (or dropping it altogether), and completely reversing ⅃ and J so that they are dextral[25]. In

[20] See footnote 8.

[21] The early form of Z, which developed from I to Z through a facilitation process as it could be made as one continuous visible stroke sequence. However, Z was dropped by the Romans and G was put in its place by Spurius Carvilius (Diringer, 1968, p. 420).

[22] At the time of its adoption by the Greeks, Q had its early form Ϙ.

[23] What form ⅃ took was largely determined by what form Γ (*gamma*) had already taken or would be altered to take, the point being to keep the two letters discriminable. For instance ⅃ could become Γ, close or even identical to the usual form of *gamma*, if *gamma* became Λ, as on the islands of Paros, Naxos, and Delos (Jeffery, 1961, p. 289). If *gamma* held to its original Γ form, as in Argos, then ⅃ could instead take the compromise form ⊦ or ⊢ (Jeffery, 1961, p. 152). In a few epichoric alphabets (e.g., Rhodian) *lambda* and *gamma* were briefly merged (Jeffery, 1961, pp. 345–349).

[24] The phenomenon is well known in the school room; see Frith (1971) for discussion.

[25] This is the place to confess that I have somewhat simplified the Phoenician-Greek transmission, for ease of presentation. The Phoenician writing system was of course sinistrograde (right-to-left) at the time of transmission, and so its letters (except for L and Ƴ) were sinistral (left-facing). This fact, and the Greek change to dextrograde via boustrophedon, are discussed below in the section Dextrality and Sinistrality.

both cases the alphabet after these alterations is more homogeneous (in this instance, more dextral) than before. In both cases it can plausibly be argued that this homogenization is undeliberated and unconscious (it surely cannot be that millions of first-graders decide to make this change calculatingly), and that it occurs at a point where facilitation does not seem to be an issue (none of the other letters are being facilitated, and in any case it is not obvious how ⌃ is more facile than ✓[26]. The point that this sort of homogenization is mostly unconscious is underscored by additional evidence. If it is true that there are really two alphabets, the patterns and the programs, then homogenization should turn up in the latter as well as the former. And it does. That is, users of the alphabet tend not only to make the letters *look* more alike, but also to execute them more similarly. What's more, they tend to do this even when the resulting program is in fact *less facile* than would otherwise have been possible. One can easily determine for oneself that T made with "→" first is slightly more facile than T made with "↓" first: yet typically it is made the latter way. Of course many people are taught in first grade to make T in this fashion, but this does not explain it. Many individuals who were taught to make E in the order vexillum first, then the three bars in order from the top, later facilitate it by making an "L" first, then the top and middle bars. Why then do they not also change to making T with "→" first? The reason for changing E cannot be simply that the new way imitates L, for similarly, making T with "→" first would imitate Z. "L"-first E is truly a change motivated by facilitation, which makes the preservation of "↓"-first T the more surprising. What seems to be the case is that as the overwhelming majority of letter programs begin with a downstroke, 'T' has been influenced to be begun in the same way. The A program of Fig. 1 b illustrates this point even more perspicuously. A could more easily be begun with an upstroke. In fact, this same upstroke is already present in the composition of other letters, such as the second stroke of V, the second and fourth strokes of W, and the third visible stroke of M. For that matter, it occurs as the second (invisible) stroke of A itself. As with T, the less facile A program is taught in school; and again it is not facilitated by most individuals, despite their facilitating the composition of other letters (often facilitating their own handwriting to the point of illegibility). The obvious – and I think the correct – explanation for this inconsistency is that a tendency to homogenize (or at least to retain the maximum homogeneity of) a very salient component of letter-composition, their beginning strokes, is stronger than the tendency to facilitate them. Thus, the same tendency to homogenize that we saw among the patterns is found also among the programs. Indeed, the failure in some

[26] Or ʅ more facile than J. If ✓ was made with the "↘" stroke first and then '↗', then of course ʅ avoids the awkward "↗" stroke, but the further change to ⌃ must either introduce the "↗" stroke or, if the letter is begun with "↙", introduce a discontinuity, since the writing instrument must then be lifted from the surface to execute "↘" and complete the letter.

dyslexic/dysgraphic children to achieve or retain maximum kinemic homogeneity may now be seen as a deficit in its own right, one we could aptly call "dyskinia."[27]

Does not this tendency to homogenize – seen also in language-change, though here perhaps more clearly – remind us of something else? That is, can we speak, as I have above, of an "evolutionary" tendency, of alphabetic or semiotic or even cognitive "evolution"?[28] If so, is this evolution of a Darwinian or neo-Darwinian type; is it analogous to that found in biological forms or to some other sort?

First and foremost, the tendency to homogenize letters has a particular goal: homogeneity. This is teleological; and the hallmark of both Darwinian and neo-Darwinian theories of evolution is that they posit a process that is *not* teleological, a process by which changes happen as the result of accidental mutations that serendipitously confer on a given life form a selectional advantage in its particular environment. Even if the means of alphabetic and biological mutation were similar, this factor would be outweighed by so glaring a difference. And of course (even allowing for the strained analogy in which an alphabet is somehow equated to a biological entity), the two means of mutation are anything but alike. Life forms evolve when their genetic codes are altered in such a way that their progeny, if any, are somatically different; while alphabets, to the best of our knowledge, don't have genetic codes, don't suffer alterations to them, and don't have progeny. Turning from mutation to the second major aspect of Darwinian evolution, selection, it is doubtful whether this applies to alphabetic change either. In both Darwinian and neo-Darwinian evolutionary accounts the inheritance of a mutation is not enough: the in-progress evolutionary event will probably die aborning unless the resultant difference (longer fangs, bigger brain, etc.) is beneficial to the progeny in the sense of giving them an advantage over their competitors within their ecological niche. In the case of an alphabet that has achieved, say, greater homogeneity, from the Darwinian standpoint its competitor would perhaps be another, less homogeneous alphabet, and its ecological niche would be the brain in which the two alphabets vied for dominance – a picture that is more than faintly ridiculous. As a metaphor, how-

[27] Inability to acquire normal writing skills, or dysgraphia, could thus be recognized as constituting two deficits: "dysphania," reduced ability to achieve the orthodox final product, and "dyskinia," reduced ability to execute the orthodox letter-composition sequences. Dysphania must always be accompanied by some dyskinia, but not vice versa. For a most illuminating discussion of normal vs dyskinic children's errors, especially "mirror-writing," see Vellutino (1979, pp. 135–141). Vellutino does not, of course, use the term "dyskinia."

[28] As the term is used here, "alphabetic evolution" is a special case of "semiotic evolution" (which includes the evolution of all systems of communication, including language) which in turn is a special case of "cognitive evolution," the change over time of what the mind knows, as a function of each individual mind. Apart from these senses I don't know what "cultural evolution" means; see below.

ever, it is perhaps slightly more productive than was the preceding reference to mutation.

The reader may be aware that, though the concept of *alphabetic* evolution is new, the notion that languages evolve is almost as old as the modern notion of evolution. Darwin himself, influenced here as so often elsewhere by Sir Charles Lyell, wrote that the "... formation of different languages and of distinct species, and the proofs that both have been developed through a gradual process ... [are] ... curiously parallel" (Darwin, 1882, p. 90). The Lyell work in question is *Antiquity of Man,* wherein Lyell averred that there are "fixed laws by which, in the general struggle for existence, some terms and dialects gain the victory over others" through the possession of some selectional advantage such as "brevity or euphony" (Lyell, 1863; p. 462). He also asserted, even more remarkably as we will see in a moment, that linguistic change is characterized by a "progressive improvement" reminiscent of biological change wherein "species of higher grade have special organs, such as eyes, lungs, and stomach ... [with functions] which in simpler organisms are all performed by one and the same organ" (Lyell, 1863, pp. 466–468). In the same year August Schleicher, the noted philologist, claimed that language, which had at one time followed a progressive evolutionary path comparable to organic development, was, since the advent of modern man and of fully-developed language (Greek?), probably degenerating. The title of his book was *Die Darwinsche Theorie und die Sprachwissenschaft* (Schleicher, 1863).

Remarks such as these, wherein biological change is metaphorically reinterpreted as linguistic or alphabetic (more broadly, semiotic) change, may seem odd when cast in their Darwinian framework, in which a language that has mutated in the direction of greater "brevity or euphony" vies (in some unspecified arena) with some less mutated language, and wins, the result being "progression," for example. Such views only seem odder as one strains the metaphor further. For surely it is of as much interest to look for the linguistic or semiotic analog of biological mutation as to dwell on the battles in which, *after* mutation, individuals compete. Resolved to take things to their logical extremes, then, we see that in the Darwinian view the environment in which linguistic or semiotic mutation takes place (the brain) must be functionally equivalent to that in which biological mutation takes place (the gonads); and in both mutational loci, if the comparison is to be maintained, the force(s) producing the mutation(s) must also be somehow equivalent. It is here that things get sticky, for Darwin, like his despised predecessor Lamarck (1809), held that an individual produces a mutated offspring by virtue of passing on characteristics that it has acquired during its lifetime (e.g., a longer neck developed by straining upward for foliage). Darwin even worked out a theory of "pangenesis" to explain how this happened: he resorted to an unseen substance, the biological equivalent of phlogiston one must suppose, that was produced by muscles on exercise and that headed unerringly for the gonads, where it altered the animal's genetic material and so destined it to have mutated offspring[29]. This is of course quite absurd as a biological theory, but as an account of the "mutation" of a language or an alphabet it is not so bad after all, pangenesis apart, since there is obviously a sense in which, in passing on a linguistic or alphabetic usage that one has altered, one *is* passing on an acquired trait. In fact the difficulty with this view of linguistic or semiotic mutation is not that it is too Lamarckian, but that it is not Lamarckian enough. For Lamarck, quite unlike Darwin, held that there is a fundamental source of evolutionary change that is quite independent of mutation: viz., the tendency of each species to evolve toward a condition of "perfection" or "complexity of organization," resulting in an increased capacity to dominate (or survive in) its original environment (the Garden of Eden?). Lamarckian evolution, then, is profoundly teleological, or goal-directed, and is aptly characterized as studying "la marche de la nature" (Schiller, 1971, p. 87). And what seems evident is that the sort of alphabetic evolution we have spoken of just above is rather Lamarckian in this sense: that the Greek alteration of  to  is indeed teleological in that it is the product of the mind's apparent tendency to homogenize; that, in short, homogeneity is analogous to Lamarck's much-ridiculed "perfection."[30]. Thus the brain in which linguistic or alphabetic mutation takes place differs from the gonads in which biological mutation takes place in this respect among others: the process that takes place within the brain (or mind) is determined by the organ itself: by the tendency of the brain (or mind) to homogenize elements of the systems it learns; and this determination is reminiscent of the purposefulness that Lamarck, in 1809, found in the world as a whole. (The world as the mind of God.)[31].

We will return shortly to take this theme further, but first we should conclude our survey of the four main forces of alphabetic evolution, by briefly taking up the two minor forces, heterogenization and inertia. To characterize all four forces together: (1) homogenization tends to make the letters more alike; (2) heterogenization tends to make them less alike; (3) facilitation tends to make them easier to execute or discriminate; and (4) inertia is a conservative force that resists change of any kind. The two major forces, (1) and (3), are "positive" in the sense that they induce change; the two minor forces, (2) and (4), are largely or entirely "negative" in the sense that they inhibit change (inertia) or both inhibit and occasionally undo it (heterogenization). As with the major forces, the minor forces have been discussed in

[29] Pangenesis is sketched by Hull (1984: *xli*, citing Darwin 1868).

[30] However, since according to Lamarck's general theory each species evolves in such a way as to adapt to its environment, given the diversity of environments this tendency should lead towards greater *hetero*geneity.

[31] The notion that kinds of evolution other than biological – e.g., so-called cultural or social evolutions – must have Lamarckian elements is not, of course, original. Herbert Spencer thought that these types of evolution were basically Lamarckian (Hull, 1984: *lvii-lviii*), a reasonable conclusion more recently echoed by B.F. Skinner (1972; p. 130).

some detail elsewhere (e.g., Watt, 1979); here, therefore, they will only be sketched.

Heterogenization is directly opposed to the force of homogenization, and is based on the need to retain or increase discriminability among the letters. Thus as homogenization makes the letters more and more alike, reducing discriminability, a counterforce will tend to exert itself to reverse this trend or at least to arrest it. Though not directly opposed to facilitation, heterogenization could also act so as to arrest a facilitative tendency whose result, albeit accidentally, was to make the letters it had facilitated too much alike. Typically, however (as we will see shortly), the phase of alphabetic history during which facilitation is strongest is also that during which heterogenization is weakest. In its primary role, that of opposing over-homogenization, the force of heterogenization acts merely to arrest the process; and since the force of inertia acts in precisely the same way, it might seem that the concept of heterogenization is hypothesized to no purpose. Not so, because occasionally heterogenization does act in a positive manner, to make the elements of a semiotic system *less* alike, in order to avoid loss of discriminability. The best modern example of this comes not from the letters but from the numerals. Continental Europeans normally make the numeral 1 as 1, which renders it practically indistinguishable from 7, necessitating the crossing of 7 to make 7. Another example, though more doubtful, would be the Etruscan ⊞, a version of the Ξ they inherited from the Greeks, perhaps augmented in this way to distinguish Ξ from the similar I. One could also regard as an instance of positive heterogenization the addition of a "tail" to P (for *rho*) in alphabets in which archaic Γ (*pi*) evolved to P instead of to π as in Greek; this is the source of modern R.[32] An even clearer example is the divergence of *gamma* and *lambda* after they had merged, in the few places where that happened (see footnote 23 above). Ordinarily, though, heterogenization acts in a purely negative manner, for example inhibiting the homogenization of L to Γ (for *lambda*) in alphabets in which the Γ pattern had already been preempted by inherited *gamma* (Jeffery, 1961, pp. 30–31)[33].

Assuming that a tendency to homogenize is a constant though (as we will see below) not one of unvarying power, the force of heterogenization, which must be invoked in any case to explain its rare positive instances, basically prevents any alphabet from homogenizing to the point where wholesale mergers commence, terminating presumably in the merger of all letters into one.

---

[32] So that the resemblance between P and R is accidental; see footnote 13 above.

[33] See also footnote 23 above. Note that reversing ⅃ yields L; inverting L yields Γ, where either simple reversal, or reversal + inversion, produces a letter more homogeneous with the others. Inversion alone, yielding ⅂, is (to my knowledge) never found, even though it would of course have resulted in an increase in homogeneity in its own right. If the tendency to homogenize is indeed behind these changes in *lambda* then the lack of attestation for the sequence ⅃ → L → Γ (McCarter, 1975, p. 84) may be only a lacuna in the epigraphic record, since the apparent failure of that record also to attest ⅂ can be taken as indicating that inversion necessarily followed reversal and was dependent on it in some way.

To turn to the fourth evolutionary tendency, without positing a force of inertia no "blocking" force would be available to prevent facilitation from taking matters to *its* extreme, at which point all letters would be maximally facile, having all become a mere point, perhaps. Actually, the inertial force is well motivated on general grounds, aside from the need to maintain alphabetic discriminability, since it is the expression, here applied to the alphabet and other semiotic systems, of society's general reluctance to change anything (fads aside). It is presumably to this force that we should turn for an explanation of why the Latin A (even though the figure Λ has not been preempted as in Greek), has kept its superfluous crossbar; why E has kept its middle bar; why Z did not revert to I when other letters were becoming symmetrical; and many other such examples.

The question of how these four evolutionary tendencies can be accommodated in the explicit characterization of the alphabet has been partially answered (details in Watt 1980, 1981, in press). Specifically, those attributes that these tendencies amplify or check have been explicitly incorporated. To take but one example, the fact that facilitation tends to attack all letters at once is accommodated by describing D and ▷ as only variants of each other, all such pairs either to be made curvilinear or left angular. Still, in future work the forces must themselves be explained as far as possible. Among the many facts to be explained is why they vary in their apparent power, one being paramount at one time, another at a different time. What has been described above is only one step toward the goal of providing a principled explanation for alphabetic (and other semiotic) evolution. Others are described below, but still others must await future research.

We pause again to remark the fact that all four of the forces discussed here are of a decidedly Lamarckian cast, in that each of them has a "goal" toward which it nudges the alphabet, as if toward "perfection." The only possible exception is inertia, whose goal must be defined, rather weakly, as maintenance of the status quo; the other three, however, heading for greater homogeneity, heterogeneity, or facility, fit the Lamarchian pattern to a T.

At this point it may be profitable to try to go beyond this characterization and the dubious revival of a long-dead doctrine. Let us explore, for at least one of the forces, a somewhat deeper understanding that might be termed "neo-Lamarckian." For homogenization this is not difficult. The schoolchildren who reverse J to L do not immediately hit on this "mistake"; typically, they progress in three stages:

1. They reverse any of the letters at random (except perhaps those forming their own names, which they may learn correcly early on), as if they were altogether indifferent to which way the letters face.
2. Suddenly they begin to make all of the letters right-facing, including J which therefore they make as L. N, S, and Z, which are neither right- nor left-facing, they may continue to get wrong, often making S like a curvilinear Z or Z like an angular S (i.e., reversing one or the other).

3. They learn that J is left-facing and they learn S, Z, and (if they are diligent) N.

The point is that they begin reversing J to ᒐ at just the stage at which they consistently get the other letters to face rightward; the mistaken ᒐ is part and parcel of the more general accomplishment. Indeed, it is misleading to term ᒐ a "mistake" at all, except in the sense that society (here represented by a teacher) refuses to condone this innovation. Rather, it is simply a by-product of the apparent generalization that "letters face rightward" – a generalization that is an expected part of the learning process and, plausibly, an essential part of it[34]. Put another way, part of the process of learning regularities is to forget or neglect the fact that there are exceptions. The result is homogenization, since an alphabet containing ᒐ is more homogeneous (more right-facing) than one with J. Even children who have previously made J more consistently than ᒐ may at this stage reverse themselves, in keeping with this generalization, as if they have decided that J, like ᙠ or ꓘ, say, has been a mistake all along. Even after J has been mastered it may be reversed on occasion, as if the right-facing regularity of the alphabet as a whole can still somehow suppress the one clear irregularity. This pattern of performance suggests that the child at stage 2 does indeed form the unconscious generalization sketched above. What form might such a generalization take?

Figure 4 presents a simple "alphabet" described by means of five primitive distinguishing features, such as AUG (has an augmentation) and AUG/TOP (the augmentation is in the top half of the letter-space). Thus, + AUG means "has an augmentation"; – AUG means "does not have an augmentation"; + AUG/TOP means "augmentation is in top half"; – AUG/TOP means "augmentation is elsewhere." As can be seen, the notation + AUG/TOP is almost redundant, since if a given letter is marked + AUG

	P	ᚳ	F	ᒪ	ᛘ	I
Contains curve	+	–	–	–	–	–
Contains Augmentation	+	+	+	+	+	–
Aug/Top	+	+	+	–	+	
Augmentation is complex	+	–	+	–	+	
Contains Diagonal	–	+	–	+	+	

Fig. 4. A simple "alphabet" characterized by five binary features. Except for ᒪ, the positive feature +AUG/TOP is redundant given the presence of the positive feature +CONTAINS AUGMENTATION

[34] The curious fact that overgeneralization (generalization beyond already familiar forms and/or beyond the set of canonical or socially acceptable forms) lags behind ordinary (canonical) generalization, is discussed in (Brown, 1973; pp. 289–291).

then it is marked + AUG/TOP as well, with only one exception marked − AUG/TOP. This degree of generality would be easily captured – and the individual specifications of all of these letters nicely caught – if the notation + AUG/TOP were omitted from all specifications, being then recovered by the following rule:

Rule A: [+ AUG] → [+ AUG/TOP]
Exception: [− AUG/TOP]

In ordinary English, if a letter is marked + AUG then it shall also be marked + AUG/TOP unless it is already marked − AUG/TOP. Now let us imagine a user of this "alphabet" who has a fallible memory. The user is new to this set of letters; has a somewhat shaky command of it; and sometimes forgets exactly how one of the letters should look. Suppose she or he forgets exactly where the augmentation of  should go: and is thus precisely in the position of having "omitted" the specification + AUG/TOP for that letter. Rule A immediately recovers this information. Now suppose the augmentation of  is forgotten: again it is as if the relevant specification for a letter had been omitted, but in this case Rule A, if actuated, will supply, not the correct , but the incorrect . + AUG/TOP is what the general part of Rule A specifies, and its exceptional part, governing cases where − AUG/TOP is present, fails because the presence of − AUG/TOP is precisely what our imaginary user has forgotten.

The fact that schoolchildren make J backward at just the point when they begin getting all the other letters right suggests that they have indeed formed an unconscious generalization along the lines of Rule A, where the feature in question is something like + AUG/RIGHT, "the augmentation is to the right." Once they have formed this generalization they need no longer worry about where a letter's augmentation goes. They will get them all correct except J, which unless or until they remember that it is − AUG/RIGHT they will inevitably get wrong. The hypothesis that the alphabet-learner – whether child or ancient Greek – forms such generalizations provides just that further step that was required for a neo-Lamarckian viewpoint[35].

Before leaving this topic it may be of service to generalize it. First of all, it is hard to believe that there could be four evolutionary forces, originating in properties of the human mind, that would attack alphabets only. This would be tantamount to holding either that human beings had evolved new writing-specific neural faculties since the invention of writing about 5000 years ago, or that for about his first 30000 years *Homo sapiens sapiens*, having cleverly anticipated the need for such faculties and evolved them at once, patiently kept them in abeyance. (Indeed, if the four forces literally attacked only alphabets, and not syllabaries or ideographic systems, one would

[35] A brief indication of some further steps toward this model of alphabetic evolution is to be found in the concluding paragraphs of this chapter.

have to hold beliefs still more limited and preposterous.) Rather, these forces must be more general in nature: in fact, they are known to be so, since they are all found in the developmental history of language. Homogenization is evident in the tendency to regularize declension, conjugation, and pronunciation: for example, in the contemporary tendency of "houses" to rhyme with "blouses," in the passage of "kneel/knelt/knelt" to "kneel/kneeled/kneeled," and, more locally, of "dive/dived/dived" to "dive/dove/dived," or even, among divers, to 'dive/dove/diven' (on the analogy of "drive/drove/driven" and so on), and in many other like regularizations. As in alphabetic homogenizations, those in language involve regularization either to the prevailing model (adding "-ed" for both past and past participle) or to a "strong" paradigm similar in sound (as with "dive/dove" after "drive/drove"). Heterogenization too is found, most conspicuously in cases of 'homonymic clash' (Williams 1944) where, if words become homophonous due to general sound changes, one may become entirely replaced by a nonhomophonous synonym or by a circumlocution. (cf. Rhodian 'gamma' and 'lambda,' which merged and then split, as recounted in footnote 23 above.) Facilitation is found in the dropping of consonants from clusters (/artik/ for "Arctic"), in epenthesis (Portuguese "estação" for Latin "stationis"), and in like phenomena. Finally, inertia is found in the tendency of languages, especially insular ones (such as Icelandic), to resist change of any kind.[36]

In another domain altogether, homogenization is apparent in the various mistakes that people make when reproducing simple figures (Fehrer, 1935). In general, then, at least three of the four forces appear to be general cognitive tendencies. They show up so clearly in alphabetic changes because the alphabet is after all a comparatively simple semiotic system, but they play other roles as well. Only heterogenization, if its origin is in the need to maintain discriminability among elements used to convey meaning, may be restricted to semiotic systems as such.

Dextrality and Sinistrality

Since so much of the foregoing discussion depends on letter reversals and more generally on dextrality and sinistrality, it would be impertinent to end the discussion without considering these factors at a little greater depth. First, some terms:

Dextral: a letter is "dextral" if it faces rightward.
Sinistral: a letter is "sinistral" if it faces leftward.

---

[36] See (Kiparsky, 1976, p. 98) for a comment on "small, homogeneous . . . populations."

Dextrograde: a writing system is "dextrograde" if its sequence of symbols runs from left to right.
Sinistrograde: a writing system is "sinistrograde" if its sequence of letters runs from right to left.
Ambigrade: a writing system is "ambigrade" if its sequence of letters runs both dextrograde and sinistrograde, the choice between the two directions being subject to some set of conditions. [37]
Boustrophedon: a special condition on an ambigrade system, dextrograde and sinistrograde alternating line by line.
Reversal: a letter is "reversed" if it faces backward, i.e., is the enantiomorph of its orthodox orientation.
Inversion: a letter is "inverted" if it is upside down.

About 85% of humans are right-handed (Geschwind & Behan 1984, p. 214); at least that proportion of writers will hold the writing instrument in the right hand. This has implications for preferred writing direction under certain conditions. If the writing instrument is held by some combination of thumb, forefinger, and middle finger, the natural inclination is to rest the side of the hand on the writing surface, to the right of where the writing instrument contacts that surface. With no further considerations, it would be easy to write from right to left as from left to right; however, there are three additional factors that militate against one direction or the other. The first is the fact that if one rests one's hand on the writing surface there is a risk of smudging what has just been written if that hand follows the writing instrument instead of preceding it. Another is the fact that if the hand follows the writing instrument it will obscure what has just been written, making it more effortful to check. These two considerations both militate against right-handers writing from right to left, thus favoring a dextrograde script. The third consideration is the fact that early scribes often propped up the writing surface with the left hand while writing with the right; this made the right edge of the writing surface more accessible to the instrument and militated in favor of a sinistrograde script. Thus each direction has points in its favor, and one should not be surprised to find examples of both, in the early stages of a writing system and even (given inertia), at later stages. By the same token one should not be surprised to find both directions in use at once, as in boustrophedon, which alternates line by line. [Neither should one be surprised to find systems in use, such as Chinese, in which writing is sinistrograde by (top-down) columns; in what follows we ignore this option, for simplicity of exposition.]

It is particularly to be noted that reading direction simply follows writing direction and that apparently nothing militates independently in favor of reading either dextrograde or sinistrograde, provided that the letters are

[37] "Ambigrade" is the only neologism among these terms.

turned in the direction of writing: dextral for dextrograde, sinistral for sinistrograde[38].

It should also be noted that the considerations applying to the right-handed majority apply in reverse to the left-handed minority. The net result is that the first two of the three special considerations cited above seem to motivate right-handers to prefer writing from left to right, left-handers from right to left. Except for boustrophedon, a writing system will almost invariably be either sinistrograde or dextrograde, thus forcing the right-handed or left-handed writer, respectively, to use a system that is less natural. Almost invariably, but not quite: there is one case, perhaps not so well known as it ought to be, where one writes in whichever direction is preferred. On the Philippine island of Mindoro two tribes, the Hununóo and the Buid, have a pure ambigrade system, with right-handers writing dextrograde and left-handers writing sinistrograde, readers of both persuasions being accustomed to read either (Conklin, 1949, p. 269, esp. footnote 5)[39]. Moreover, there are isolated cases in which left-handed persons have chosen to write sinistrograde when the writing is meant for their eyes only – Leonardo da Vinci is perhaps the most famous example – or perhaps when the dextrograde direction has not yet been firmly established (Jeffery, 1961, p. 47).

For a society adopting a new script to change its writing direction is far from uncommon; indeed, if the system being adopted is sinistrograde, a switch is almost to be expected, given the ubiquity of right-handers. The people usually credited with inventing true writing, the Sumerians (just after 3000 B.C.), wrote from left to right, as a result of initially writing from top to bottom in columns that proceeded leftward, and then progressing to a stage at which the writing surface was rotated 90° counterclockwise, making what had been top-to-bottom now left-to-right. The Semitic (Akkadian) peoples who inherited the Sumerian cuneiform system maintained this direction, but the Elamites, who also (though more indirectly) borrowed the cuneiform system, wrote mostly from right to left (Diringer, 1968, p. 24). The Egyptians, whose writing system seems to have developed independently of the Sumerian cuneiform (though it may have been influenced by it), ordinarily wrote from right to left [though there are also a few inscriptions in dextrograde and in boustrophedon (Diringer, 1968, p. 34)], as did the Semitic tribes who borrowed the Egyptian penalphabet of consonantal signs (drop-

---

[38] That a sinistrograde script is easier to read when its letters are sinistral was shown, with subjects given 3–5 hours to familiarize themselves with the reversed letters (minuscules), by Kolers (1968).

[39] Conklin attributes the ease with which both right- and left-handers read both dextrograde and sinistrograde to "the syllabic form of the individual characters" (Conklin, 1949; p. 269). It may indeed be true that a syllabary is easier to read when reversed than an alphabet is, ceteris paribus, but I know of no other indication that this is true, and I know no reason why it should be so.

ping the Egyptian two-consonant signs and hieroglyphs, thus creating the first true consonantal alphabet, or "betagamma"). Modern Semitic writing systems such as Hebrew and Arabic are sinistrograde to this day. Yet the Ugaritians, who borrowed this sinistrograde system but wrote it in clay, hence remaking the letters as cuneiforms (Stieglitz, 1971), reversed the direction and wrote, as did the other cuneiform systems (such as Sumerian), from left to right. The Greeks also borrowed the Semitic alphabet, but after a brief period during which they wrote boustrophedon, with the first line generally sinistrograde, they adopted the more natural dextrograde direction (Jeffery, 1961; pp. 46–47). The Etruscans borrowed the alphabet from the Greeks and at first wrote mostly from right to left, with some inscriptions in boustrophedon; later, however, under the influence of the Romans, who had borrowed their alphabet and reversed its direction to dextrograde, they began writing dextrograde themselves (Diringer, 1968; p. 387, p. 390). Throughout all of these changes in writing direction there is no reason to suppose that the proportion of right-handed individuals underwent any alteration [though at some periods the scribes may have held their styli differently; see, for instance, the fist-grip depicted in an Assyrian sculpture, probably from the palace of Tiglath-pileser III (Driver, 1976, p. 22)]. Thus, whatever problems were occasioned by these changes, or by the use of boustrophedon, must have been universal.

A special problem is of course posed by boustrophedon. In its sinistrograde lines the letters must be sinistral, in its dextrograde lines, dextral: thus all the letters (except for those that were completely symmetrical on the vertical axis) must face now one way, now the other. Some letters are easy to reverse (B, P in their modern forms), others, as one can immediately demonstrate for oneself, are more difficult (N, S). These difficult letters are the same (with Z) that modern schoolchildren have trouble with (setting to one side J, not a Greek letter), and for the same reason: they face neither rightward nor leftward (or face in both directions). Thus it would not be surprising to find the Greeks making numerous reversal errors when writing in boustrophedon, especially with the letters just named: and it is certainly tempting to speculate that this might have been how they came to reverse the sinistral 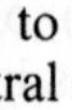 they had inherited, retaining the resulting  as a permanent fixture. Establishing this point would of course require the inspection of large numbers of boustrophedon inscriptions, which I have not been able to do; however it seems from the photographs available to me, if they are typical, that the Greeks were no more prone to making errors in boustrophedon than when writing in one direction only. [See, for example, the fine mistakenly-reversed N in a purely sinistrograde Messenian inscription tentatively ascribed to the period 500–475 B.C., reproduced by Jeffery (1961, Pl. 39). For two *inverted* As, see the earlier Eritrean aryballos, again in sinistrograde (possibly painted while in inverted position), said to be from ca. 650 B.C. and also reproduced by Jeffery (1961, Pl. 6).] Thus this convenient explanation for the famous *lambda* reversal may well prove to be erroneous. This indicates, perhaps, that, despite the strangeness (to our eyes at least) of boustrophedon or other ambigrade systems, the ease with which they are managed obviates any telling mistakes other than those to which all systems are prey[40].

Writing direction and covarying letter orientation are also implicated in what appear to be rather orderly stages of alphabetic evolution; stages distinguished by which of the four evolutionary forces, or combinations thereof, are dominant. These stages, which of course have fuzzy boundaries, seem to be approximately as follows:

Eographic. At this stage the letters are being mastered one by one, with no (or few) generalizations being formed. Historically this must have been a fleeting period indeed, given the apparent human tendency to generalize; but in children it can easily be observed early in the learning process, when they typically reverse letters at random, not yet having attained the generalization "letters face rightward." Writing direction too may not yet be stabilized.

Neographic. This is first stage at which stability is found, with writing direction and letter orientation becoming fixed (sinistrograde/sinistral; dextrograde/dextral; ambigrade/enantiomorphic). Phylogenetically, the neographic stage occurred in the Greek world during the first few years after the alphabet was received, before ease and speed became criterial and/or while writing was performed on difficult materials (e.g., by being incised). Thus the approximate Greek dates for the neographic stage would be from ca. 750 to ca. 650 B.C. (Jeffery, 1961, pp. 12–21); of course it varied from place to place. Ontogenetically a similar stage is observed when children grasp the dextrality generalization, at which juncture their writing becomes almost invariably dextrograde.

---

[40] McCarter has written that "... the option to write in either direction is the source of the confusion which leads to reversed letter-forms. Indeed this is the source of the occasional reversals instance in the Proto-Canaanite script itself" (McCarter, 1975, p. 119). No doubt some mistakes do come about in this way, but hardly all. The present-day storekeeper who puts out a boldly lettered **OPEИ** sign is not likely to have made this error because of an unfortunate habit of writing backwards. The reader/writer can readily determine how easy it is to write backward *if* one simply reverses the direction of one's strokes, maintaining their sequence. On the other hand, it is certainly worth noting that when the Greeks were writing boustrophedon, some scribes (or localities) preferred not to reverse certain letters. The early Phokian inscription on stone, with **A** used in both directions, is an example (Jeffery, 1961, Pl. 12, 1), although on a somewhat later Parian stone the same **A** can be found neatly reversed in sinistrograde, being dextral in dextrograde (Jeffery, 1961, Pl. 56, 28). In the same inscription, admittedly, Σ is to be found *un*reversed in a sinistrograde line, but it is not hard to find similar failures in inscriptions that are all sinistrograde, as witness the failure to reverse N in the second line of the Messenian inscription (in bronze) reproduced by Jeffery (1961, Pl. 39, 3). In the third line of that inscription, also sinistrograde, N is gotten right, as **И**.

Mesographic. During this stage the dominant force is that of facilitation, i.e., subordinating other criteria to that of best fitting the alphabet to the rapid writing of large quantities of material. (Inscriptional writing, as on stone, will lag behind quotidian writing on more casual materials in this respect, and "aulic" writing will lag behind "demotic" writing.) Older forms may be retained as living fossils for inscriptional or other formal purposes; for instance the majuscules are still used for capitalization conventions, while under the influence of facilitation the alphabet has otherwise gravitated toward the more fluid forms of the minuscules. The new, more fluid figures thus formed will be subject to homogenizing tendencies, mostly in the executive (kinemic) modality, making the programs more regular, while the forces of heterogenization and inertia will act as inhibitors[41].

Cenographic. This late stage really lies beyond, and somewhat peripherally to, the three preceding ones, for here an alphabet enters a more self-conscious design period during which such ancillary criteria as axial symmetry come into play. Some alphabets are far more subject to cenographic changes (Greek) than others (Arabic, Hebrew). The addition of serifs (Catich, 1968) is a cenographic phenomenon.

The present-day western European ("English" or "Roman") alphabet bears evidence of each of the last three stages. The majuscule L is neographic, formed at that point when the "letters face rightward" generalization was actuated. The forms of some of the other majuscules, for example Z and such curvilinears as C, are mesographs, formed under the influence of facilitation; as are the minuscules. The modern form of the letter M is a cenograph, having resulted from the symmetrization of archaic 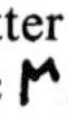. Lastly, since once an alphabet enters its mesographic stage it never really leaves it (its cenographic phase, if entered at all, is in a sense superimposed) the greatly facilitated forms of handwriting are best regarded as mesographs; indeed, they are the most mesographic of all.

It is perhaps an oddity of the foregoing view, in which historical events are to some extent paralleled by events in the schoolroom, that phylogeny recapitulates ontogeny, inasmuch as the identified stages of the former process are in a way arrested stages of the latter one.

Conclusion

In the foregoing glance at the alphabet and some of its implications for cognitive studies, in particular the study of cognitive evolution in its semi-

[41] An important step toward the understanding facilitation that is a prerequisite to having a full picture of mesographic events is the study of the comparative effortfulness or "markedness" of the various possible handprinting strokes. In fact, this study is well along (Mallon 1952; van Sommers, no date; Driver, 1976, p. 34).

otic (communicative) aspects, we have considered a particular approach to examining the alphabet, comprising two analyses each particular to a modality, and have then looked at the "neo-Lamarckian" evolution to which the alphabet, from this viewpoint, appears to be subject. In respect to homogenization, the most interesting (and during the neographic stage the most powerful) of the evolutionary forces that jointly determine the alphabet's history, we have considered a plausible first approximation to a mechanistic explanation for that force. Whenever it seemed apropos we have referred to issues of writing direction and letter orientation.

Obviously much more remains to be done in pursuit of a comprehensive theory of the cognitive representation of the alphabet and of the evolutionary forces that spring from such a representation and that change it. The comparative rates of alteration resulting from each of these forces are yet to be calculated, as are the details of how they vie with each other at various stages. Homogenization has properly been compared to gravitation, or mutual attraction; but whether anything like an "inverse square law" holds in this domain is entirely an open question. Another and far broader question left open, indeed not even discussed here, is that of the relationship between the sort of "cognitive evolution" discussed here and the sort of "cultural evolution" currently under active investigation by anthropologists and sociobiologists[42]. Perhaps these pages will provide a fresh beginning.

References

Albright, W.F. (1950). Some important recent discoveries: Alphabetic origins and the Idrimi statue. *Bulletin of the American Schools of Oriental Research, 118,* 11–20.

Alighieri, Dante. [ca. 1304] (1968). *De vulgari eloquentia.* A. Marigo (Ed.), Firenze: Felice Le Monnier.

Boyd, R., & Richerson, P.J. (1985). *Culture and the evolutionary process.* Chicago: University of Chicago Press.

Brown, R. (1973). *A first language.* Cambridge: Harvard University Press.

Brown, R., & Hildum, D.C. (1956). Expectancy and the perception of symbols. *Language, 32,* 411–419.

[42] I mean the well known current area of study variously called "cultural evolution," "social evolution," and "evolutionary epistemology." Or "cultural Darwinism" and so on (we might as well say 'cultural Lamarckism,' and so on). Particularly to be marked are: Dawkins (1976, pp. 203–215), in which the "meme" is proposed as the transmitter of cultural information, analogous to the gene; the papers collected in Plotkin (1982); Lumsden (1983); Boyd & Richerson (1985); and Chap. 1 of this book. Interest in the biological/cultural analogy seems to be a sporadic feature of the scholarly investigation of culture and society; for an early attempt to make this interest serious see the article by Gerard, Kluckhohn, & Rapoport (1956), who go so far as to propose experiments. A thumbnail sketch of the earlier history of such endeavors is provided by Honigmann (1976; pp. 273–283). More broadly, the search for "patterns," "parallels," or even "cycles" in cultural history is very old indeed, and still thrives in some quarters, sometimes with postulated "causes" that test one's credulity – one of the "causes" recently proposed is the intrusion into human cognitive affairs of the Evil One (Gowans, 1974, p. 98).

Catich, E.M. (1968). *The Origin of the serif.* Davenport, Iowa: St. Ambrose College (Catfish Press).
Conklin, H.C. (1949). Preliminary report on field work on the islands of Mindoro and Palawan, Philippines. *American Anthropologist, 51,* 268–273.
Cross, F.M. Jr., & Lambdin, T.O. (1960). A Ugaritic abecedary and the origins of the Proto-Canaanite alphabet. *Bulletin of the American Schools of Oriental Research, 160,* 21–26.
Darwin, C. (1868). *The variation of animals and plants under domestication.* London: Murray.
Darwin, C. (1882). *The descent of man.* London: Murray.
Dawkins, R. (1976). *The selfish gene.* New York: Oxford University Press.
Diringer, D. (1968). *The alphabet* (vol. 1). New York: Funk & Wagnalls.
Driver, G.R. (1976). *Semitic writing* (3rd ed.). London: Oxford University Press.
Eden, M., & Halle, M. (1961). The characterization of cursive handwriting. In Cherry, C. (Ed.), *Information theory: Fourth London symposium.* Washington: Butterworths.
Fehrer, E. (1935). An investigation of the learning of visually perceived forms. *American Journal of Psychology, 47,* 187–221.
Fischer-Jørgensen, E. (1952). The phonetic basis for identification of phonemic elements. *Journal of the Acoustical Society of America, 24,* 611–617.
Frith, U. (1971). Why do children reverse letters? *British Journal of Psychology, 62,* 459–468.
Gerard, R.W., Kluckhohn, C., & Rapoport, A. (1956). Biological and cultural evolution: Some analogies and explorations. *Behavioral Science, 1,* 6–34.
Geschwind, N., & Behan, P.O. (1984). Laterality, hormones, immunity. In Geschwind, N., & Galaburda, A.M. (Eds.), *Cerebral dominance: The biological foundations.* Cambridge: Harvard University Press.
Gibson, E.J., Osser, H., Schiff, W., & Smith, J. (1963). An analysis of critical features of letters, tested by a confusion matrix. In Cooperative Research Project #639: A basic research program on reading. Washington: U.S. Office of Education.
Gowans, A. (1974). *On parallels in universal history discoverable in arts & artifacts. Watkins Glen, NY: Institute for the Study of Universal History Through Arts and Artifacts.*
Hall, N. (1980). *The moon and the virgin.* New York: Harper and Row.
Harris, Z.S. (1961). *Methods in structural linguistics.* Chicago: University of Chicago Press.
Herrick, E.M. (1969). *The graphonomy of English,* preliminary edition (dittoed). Kalamazoo, MI.
Honigmann, J.J. (1976). *The development of anthropological ideas.* Homewood, IL: The Dorsey Press.
Hull, D.L. (1984). Lamarck among the Anglos. In Elliot, H. (Trans.), *Zoological philosophy,* by Lamarck. Chicago: University of Chicago Press.
Jeffery, L.H. (1961). *The local scripts of archaic Greece.* Oxford: The Clarendon Press.
Justeson, J.S., Norman, W.M., Campbell, L., & Kaufman, T. (1985). *Middle American Research Institute Publication 53: The foreign impact on lowland Mayan language and script.* New Orleans: Tulane University Press.
Kiparsky, P. (1976). Historical linguistics and the origin of language. *Annals of the N.Y. Academy of Sciences, 280,* 97–103.
Kolers, P.A. (1968). The recognition of geometrically transformed text. *Perception & Psychophysics, 3,* 57–64.
Kolers, P.A., & Perkins, D.N. (1969). Orientation of letters and their speed of recognition. *Perception & Psychophysics, 5,* 275–280.
Kolers, P.A., & Perkins, D.N. (1975). Spatial and ordinal components of form perception and literacy. *Cognitive Psychology, 7,* 228–267.
Kolers, P.A., Wrolstad, M.E., & Bouma, H. (Eds.) (1979). *Processing of visible language* (vol. 1). New York: Plenum.
Kolers, P.A., Wrolstad, M.E., & Bouma, H. (Eds.) (1980). *Processing of visible language* (Vol. 2). New York: Plenum.

Lamarck, J.B.P.A. (1809). *Philosophie zoologique* (Historiae Naturalis Classica, Tomus X). Weinheim: Engelmann and Wheldon & Wesley, 1960.

Lumsden, C.J. (1983). Cultural evolution and the devolution of tabula rasa. *Journal of Social and Biological Structures, 6,* 101–114.

Lyell, C. (1832). *Principles of geology* (12th rev. ed.) London: Murray, 1975.

Lyell, C. (1863). *The geological evidence of the antiquity of man, with remarks on the theories of origin of species by variation.*

Mallon, J. (1952). *Paléographie romaine.* Madrid: Instituto Antonio de Nebrija de Filologia.

Mendelson, J.M., Siger, L., Kubzansky, R.E., & Solomon, P. (1964). The language of signs and symbolic behavior of the deaf. In Rioch, D.M., & Weinskin, E.A. (Eds.), *Research publications of the association for research in nervous and mental diseases* (Vol. 42): *Disorders of communication.* Baltimore: Williams & Wilkins.

McCarter, P.K., Jr. (1975). *The antiquity of the Greek alphabet and the early Phoenician scripts* (Harvard Semitic Monographs No. 9). Missoula, Montana: Scholars Press, for the Harvard Semitic Museum.

Pirozzolo, F.J., & Wittrock, M.C. (Eds.) (1981). *Neuropsychological and cognitive processes in reading* (Perspectives in Neurolinguistics, Neuropsychology, and Psycholinguistics). New York: Academic.

Plotkin, H.C. (Ed.) (1982). *Learning, development, and culture: Essays in evolutionary epistemology.* New York: John Wiley.

Pritchard, R.M., Heron, W., & Hebb, D.O. (1960). Visual perception approached by the method of stabilized images. *Canadian Journal of Psychology, 14,* 67–77.

Rankin, B.K., III, Sillars, W.A., & Hsu, R.W. (1965). *On the pictorial structure of Chinese characters* (Technical Note 254). Washington: National Bureau of Standards.

Rankin, B.K., III, Siegel, S., McClelland, A., & Tan, J.L. (1966). *A grammar for component combination in Chinese characters* (Technical Note 296). Washington: National Bureau of Standards.

Reber, A.S., & Scarborough, D.L. (Eds.), (1977). *Toward a psychology of reading.* Hillsdale, NJ: Lawrence Erlbaum Associates.

Samuels, S.J., LaBerge, D., & Bremer, C. (1978). Units of word perception: Evidence for developmental change. *Journal of Verbal Learning and Verbal Behavior, 17,* 715–720.

Sandars, N.J. (1971). *Poems of heaven and hell from ancient Mesopotamia.* London: Penguin.

Schiller, J. (1971). L'Echelle des êtres et la série chez Lamarck. In Schiller, J. (Ed.), *Colloque International 'Lamarck'.* Paris: Librairie Scientifique et Technique A. Blanchard.

Schleicher, A. (1863). *Die Darwinsche Theorie und die Sprachwissenschaft.* Weimar: Böhlau.

Segert, S. (1983). The last sign of the Ugaritic alphabet. *Ugarit-Forschungen, 15,* Neukirchen-Vluyn: Neukirchener Verlag des Erziehungsvereins GmbH.

Skinner, B.F. (1972). *Beyond freedom and dignity.* New York: Knopf.

Stephens, L.D., & Justeson, J.S. (1978). Reconstructing "Minoan" phonology: the approach from universals of language and universals of writing systems. *Transactions of the American Philological Association, 108,* 271–284.

Stieglitz, R.R. (1971). The Ugaritic cuneiform and Canaanite linear alphabets. *Journal of the Near Eastern Society, 30,* 135–139.

Taylor, I. (1980). The Korean writing system: An alphabet? A syllabary? A logography? In Kolers, P.A., Wrolstad, M.E., & Bouma, H., (Eds.), *Processing of visible language* (Vol. 2). New York: Plenum.

Terry, P., Samuels, S.J., & LaBerge, D. (1976). The effects of letter-degradation and letter spacing on word recognition. *Journal of Verbal Learning and Verbal Behavior, 15,* 577–585.

Townsend, J.T. (1971). Theoretical analysis of an alphabetic confusion matrix. *Perception & Psychophysics, 9,* 40–50.

Townsend, J.T., Hu, G.G., & Evans, R.J. (1984). Modeling feature perception in brief displays with evidence for positive interdependencies. *Perception & Psychophysics, 36,* 35–49.

Tversky, A. (1977). Features of similarity. *Psychological Review, 84,* 327–352.

Tzeng, O.J.L., & Hung, D.L. (1981). Linguistic determinism: A written language perspective. In Tzeng, O.S.L., & Singer, H. (Eds.), *Perception of print: reading research in experimental psychology.* Hillsdale, NJ: Lawrence Erlbaum.

Tzeng, O.J.L., & Singer, H. (Eds.) (1981). *Perception of print: Reading research in experimental psychology.* Hillsdale, NJ: Lawrence Erlbaum.

van Sommers, P. (no date) Drawing and copying: Constraints in execution. North Ryde, NSW, Australia: School of Behavioural Sciences, Macquarie University (mimeo).

Vellutino, F.R. (1979). *Dyslexia: Theory and research.* Cambridge: MIT Press.

Watt, W.C. (1975). What is the proper characterization of the alphabet? I: Desiderata. *Visible Language, 9,* 293–327.

Watt, W.C. (1980). What is the proper characterization of the Alphabet? II: Composition. *Ars Semeiotica, 3,* 3–46.

Watt, W.C. (1981). What is the proper characterization of the alphabet? III: Appearance. *Ars Semeiotica, 4,* 269–313.

Watt, W.C. (1983). Mode, modality, and iconic evolution. In Borbé, T. (Ed.), *Approaches to semiotics,* Vol. 68: *Semiotics unfolding.* Amsterdam: Mouton.

Watt, W.C. (1984). Signs of the times. *Semiotica, 50,* 97–155.

Watt, W.C. (1987). The Byblos matrix. *Journal of Near Eastern Studies, 46,* 1–14.

Watt, W.C. (in press). What is the proper characterization of the alphabet? IV: Union. *Semiotica.*

Williams, E.R. (1944). *The conflict of homonyms in English.* (Yale Studies in English, Vol. 100). New Haven: Yale University Press.

Zipf, G.K. (1935). *The psycho-biology of language.* Boston: Houghton Mifflin.

CHAPTER 9

Logical Principles Underlying the Layout of Greek Orthography

DERRICK DE KERCKHOVE[1]

Types of Orthographies

There are two principal categories of writing systems; those representing words, images, or ideas contained in language, and those representing the sound of the languages. Among the latter, there are two main types, those representing so-called "concrete" sounds (Havelock, 1976, 1982), called syllabaries because each sign represents a fully pronounceable syllable, and those that represent phonemes, i.e., parts or segments of sounds, namely the alphabets. There are two kinds of alphabets: those like Hebrew and Arabic which use consonants alone to indicate the radicals of the words, and those like Latin or Greek which use a combination of consonants and vowels to represent fully the sequence of syllables that form the words. (Some syllabaries, such as the Indic Nagari and its derivates, and also the Ethiopian script and the Korean Hangul, are in fact very sophisticated systems of phonemic articulation that include the phonemic analysis within the syllabic character. They present a different situation which must be investigated on its own terms.)

The occurrence of one or the other of these various possibilities seems to be somewhat dependent upon the phonological structure of the language being represented. Thus, pictograms suited the largely monosyllabic language of the Sumerians (Driver, 1954; Higounet, 1976), just as ideograms are still considered adequate today to represent the mainly monosyllabic vocabulary of several Chinese languages (Hagège, this volume). It seems logical indeed to adopt a writing system based on pictorial representation when the language contains a great number of homophones that require unequivocal discrimination (Etiemble, 1973). However, when they were adopted to serve a complex polysyllabic language such as the North Semitic Akkadian, the Sumerian pictograms, already well on the way to cuneiform stylization, were altered to represent sounds rather than images or ideas, following the so-called "acrophonic principle" (Diringer, 1948; Driver, 1954; Gelb, 1963; Kristeva, 1981). This was the earliest stage of full, or almost full, *phonetization* of pictograms in the Mediterranean.

---

[1] McLuhan Program in Culture and Technology, and Department of French, University of Toronto, 39A Queen's Park Cr., Toronto, Ontario, M5S 1A1, Canada.

Table 1. Types of orthographies and their most common layout

I. Symbols for *Things: Pictography* (generally in *vertical* columns read to the *left*)
Main examples: Proto-Sumerian, Proto-Elamite, Proto-Sinaitic, Mayan, Egyptian Hieroglyphics

II. Symbols for *Thought: Ideography* (originally in *vertical* columns read to the *left*)
Main examples: Chinese, Japanese Kanji

III. Symbols for *Sounds: Phonography*

Two main categories

1. Symbols for *Concrete* Sounds: *Syllabaries* (90% *horizontal* to the *right*: $n = 209$)

Two subcategories

a) *Plain* Syllabaries
Examples: Japanese Kana, Minoan Linear B, Old Persic, Ugaritic
b) *Phonemic* Syllabaries
Examples: Early Indio scripts, Nagari, Ethiopic, Hangul

2. Symbols for *Abstract* Elements of Sounds: *Alphabets*

Two subcategories

a) *Consonantal* Alphabets (all *horizontal* to the *left*)
Main examples: Phoenician, Hebrew, Arabic
b) *Vocalic* Alphabets (all *horizontal* to the *right*)
Main examples: Greek, Latin, Cyrillic

Principal structural differences between consonantal and vocalic alphabets

- Presence of symbols for the sound of *vowels*
- *Rightward* direction of vocalic scripts
- Initial *indifference* to word parsing

The other strain, that of *phonemic differentiation*, was developed by the Canaanites in the turquoise mining areas where they labored for the Egyptians (Innis, 1950). Whereas the Egyptians could not seem to do without their complex ideographs, the Canaanites, otherwise known as the Phoenicians, managed with only 22 diacritical signs (which the Egyptians used acrophonically to help distinguish among similar hieroglyphic characters that had different meanings). The inspiration for such an economical system may have come from the much reduced set of cuneiforms used in the contemporary Ugaritic syllabary, but the issue of which came first is not fully resolved (Cohen, 1958; Irigoin, 1982; Sznycer, 1977).

However, whereas the use of consonants alone was perfectly suited to the nature of the Phoenician language, it was of little use to the Greeks (see discussion in Lafont, this volume; also Sampson, 1985). There is an important structural difference between the linguistic properties of Semitic and Indo-European languages. The major part of the basic vocabulary of any Semitic language consists of radicals that can be represented by two, three, or less frequently four or more consonants. The role of the vocalic sounds that highlight the consonantal ones is flexional, that is, *grammatical,* not *lexical.* The differing physiological requirements to produce the sound of consonants and vowels must be stressed here: vocalic sounds are produced by modulating the air flowing through the throat and mouth, whereas consonants are produced by presenting various obstacles to stop the flow. Thus, although it is impossible to speak without a combination of both, different languages make use of consistently different combination patterns.

Lexical morphemes, otherwise known as lexemes, radicals, or word roots, are the basic elements that carry meaning in any language. Because the radicals of their words make use of consonants exclusively, Semitic languages use the vocalic airflow not to evoke a lexeme, but to structure the relationships between the consonantal radicals during speech production. For example, in Semitic languages, the consonantal sounds represented by /k/, /t/, and /b/ evoke the basic idea of "writing"; the grammatical flexions are indicated by vocalic sounds in the intervals between the consonantal sounds. These sounds are not usually represented in the orthography because their values can be guessed at by reference to the context. Thus, in standard Hebrew /k/ + /a/ + /t/ + /b/ + /a/, i.e., "KaTaBa", means "he has written"; /k/ + /i/ + /t/ + /a/ + /b/, i.e., "KiTaB", means "book." But when both words are written, they appear simply as "KTB" and the context alone will sort them out (Cohen, 1958; Jurdant, 1984).

In contrast, Indo-European languages such as Sanskrit, Greek, or Latin, use a combination of vocalic airflow and consonantal obstacles to indicate the differences between basic words, relying on other recurrent patterns of combinations such as prefixes, suffixes, desinences, and other strategies to mark the grammatical relationships. For example, the Greek infinitive for "writing" is γϱαφειν. The lexeme of the "idea of writing" is not γϱφ but γϱαφ. The presence of the vowel α is essential to identify the word. Contrary to Semitic words, there are many Greek lexemes that can only be distinguished by their vocalic component: such is the case for οϱοσ ("limit," hence the English word "hour"), εϱοσ ("love"), ιεϱοσ ("sacred," hence the word "*hiero*glyph"), and Οϱυσ (the Egyptian god Horus).

The Dutch philosopher Spinoza used a metaphor to describe the relationships of the written characters to the vocalic sounds during the reading process of Hebrew: "As for this difference between the letters and the vowels, it can be very simply illustrated by the example of the flute: the vowels are the sound of the music; the letters are the holes as they are covered by the fingers" (Spinoza, 1968). Although this image is useful to describe Semitic writing, it is inadequate to properly differentiate Semitic and Indo-European languages. Each linguistic system is based on completely different types of sonorities. Because they are based on modulated consonants, Semitic languages function like stringed instruments, which have to be plucked and bowed to produce a sound. By comparison, it is Greek and Latin which are more like organ pipes, because the sounds of these languages are produced by modulating the vowels.

This difference deeply affects and marks both writing systems in spite of many similarities between them. The fact that the Greeks retained the alphabetic appearance of their model is largely due to the convenience of keeping most of the consonantal characters with their original values (Naveh; Lafont, this volume). It was the Greeks, not the Phoenicians, who discovered the principle of *syllabic synthesis,* by which I mean the combination of discrete and usually unpronounceable elements to form the representation of a pronounceable sound. We can deduce that the Greeks were aware of their accomplishment, in the way Æschylus described the alphabet in *Prometheus Bound:* γραμματων τε συνθεσεισ, i.e., the "assembling of letters". No such invention can be detected or even deemed necessary for the Phoenicians. While providing evidence for the schooling methods of Ancient Greeks, Henri-Irénée Marrou points out the fact that consonants and vowels were taught separately in early grammar schools in Greece (Marrou, 1964).

In fact, the principle of the Greek system is much closer to Korean Hangul or even Indian scripts than to its original model. James Fevrier (1959) and later Sznycer (1977) tried to resolve the controversy as to whether Phoenician and Hebrew scripts were alphabets or syllabaries by stating that they must be considered as syllabaries with unmarked vocalic sounds (see also Lafont, this volume). This, however, is not very helpful. It obscures the fact that the Phoenicians invented a truly original system, but one only suited to Semitic languages. In fact, the Phoenician system escapes the definition of both syllabic and alphabetic categories. For want of a better definition, it must be called a *consonantal alphabet* (Havelock, 1976).

There are three main features that distinguish the Greek from the Phoenician alphabet (see Table 1): the presence of characters for the sound of vowels, the orientation of the written line to the right, and a relative indifference to word and sentence segmentation. This last difference, subject however to great variations over time and space, comes from the fact that Semitic scripts must depend upon word parsing, whereas in vocalic alphabets and syllabaries such segmentations are largely optional. These distinctive features are not arbitrary insofar as the succession of consonantal letters without proper separation would burden the reader's task with the need to parse the words before benefiting from the information given by context.

However, one thing that should have alerted us long ago to take the matter seriously is the fact that among all Semitic – that is consonantal – scripts and their derivates, which number about 50, there is not a single one that at any time has been consistently written in any direction other than to the left. Conversely, there is hardly a single vocalic script, among several hundred, that has not been either written rightward from the start, or been reverted to the rightward direction within 100 years of its initial appearance. For example, the first Greek epigraphs were often, though not exclusively, written right to left, as were their Phoenician models, but eventually they gave way to patterns that were oriented rightward, the way ours and those of all other full alphabets are. Besides Greek, there are several other cases of initially leftward orthographies such as the Etruscan and Latin alphabets and the Ethiopic syllabary, which were reoriented rightward after they began to include fixed signs for vowels (Cohen & Sainte Fare Garnot, 1963).

Almost all syllabaries, which by definition must include a vowel value in their signs, are written rightward. The main exceptions are Japanese Kana and Korean Hangul, which were written leftward until recently, when both forms adopted the rightward direction in the wake of westernization. These exceptions are interesting for different reasons. If my theory is correct, both should have been written rightward in the first place. However, Kana was invented long after the Japanese had adopted the Chinese characters (which constitute the body of Kanji signs) and consequently would follow the direction of the established writing system, which was laid out in vertical columns read to the left. The case of Hangul is even more interesting, in that these characters can be considered as syllabic structures that present a square shape guided by the model of ideograms. In this case, there would be little resistance to following the pattern of the Chinese script, which is known to have strongly influenced the Korean system. This is particularly plausible as the script was written vertically until the recent westernization (Taylor & Taylor, 1983).

There are two other lesser-known cases of syllabaries that were written leftward for only a limited period of time: the early stage of Etruscan writing (Cohen, & Sainte Fare Garnot, 1963) and the Indian Kharostri (Cohen, 1958). Both exceptions can be explained by the fact that these scripts were adapted directly from Semitic models, Phoenician for Etruscan and Aramaean for Kharostri. The first was eventually reversed and became our Latin alphabet, while the second lasted no more than 200 years, from the seventh to the fifth century B.C. (Cohen, 1958).

The Redirection of the Greek Alphabet

As we have suggested, to write their own language with the 19 consonants and the three semivowels – the glottal occlusives *alef*, *yod*, and *waw* – of the Phoenicians, the early Greek writers had to accomodate some of those letters to represent vowel sounds needed for their own vowel-based morphemes. If, as the statistical evidence indicates, there is indeed a rule that somehow requires vocalized scripts to be written rightward, the presence of vowels in Greek would suffice to explain why it and other vowel-based orthographies are written rightward. In fact, the shift from left to right in Greek epigraphy was neither smooth nor immediate, nor uniformly adopted over all the Greek territories (Jeffery, 1961).

One of the earliest scholars to hint at the possible cultural relevance of the change of direction of Greek writing was the Canadian cultural historian Harold Innis (1950) whose succinct description can serve as the basis for the evidence: "Following the Semites the Greeks began by writing from right to left, but they continued in a boustrophedon style and finally, by the end of the fifth century, wrote from left to right and generally reversed individual Semitic letters" (p. 81).

According to A. G. Woodhead (1981), it is not possible, for lack of precisely datable evidence, to claim that all the earliest Greek epigraphs were written from right to left. Even in the eighth century B.C., acknowledged as the earliest period of *extant* Greek epigraphs (although not necessarily acknowledged as the time when the system was invented; see Naveh, 1982, this volume; for discussion of early alphabetization of Greece, see Burns, 1981), there may already have been some isolated instances of writing from left to right, just as there are later samples of writing in random directions that follow unspecified designs of up, down, and sideways. It is assumed that scriptforms were subjected from the beginning to a certain amount of experimentation, which followed various trends in various regions (Jeffery, 1961). However, the second period of Greek alphabetization is well documented. It is possible to date the appearance of what is known as the *boustrophedon* style to the beginning of the sixth century B.C. (Guarducci, 1967; Woodhead, 1981).

The boustrophedon was a stunning departure from previous models of scriptforms, as it was written continuously in both directions, alternately to the right and to the left in uninterrupted horizontal lines starting from the top of the writing surface. There was no separation provided for individual words, no segmentation of any kind, and the orientation of individual letters faithfully followed the alternate directions of the lines. Several other styles, though less common because they seem to have been developed exclusively for writing on pottery and other uneven surfaces, have been recorded (Jeffery, 1961; Dain, 1963; Guarducci, 1967), as *plinthedon* (writing on the sides of a brick), *speiredon* (spiralling script), and *kionedon* (writing in columns).

About 50 years after the appearance of boustrophedon, another style arose in Attica, the region of Athens, and later in other Greek communities. This is the *stoichedon*, a word derived from the military term κατα στοιχουσ which means "by orderly ranks." Woodhead says that this new genre, which is evidenced on Athenian walls and stelas from the middle of the sixth century B.C., required that all the letters be written in the same direction to achieve an effect of orderly regularity according to the equidistant positioning of each individual letter.

R. P. Austin (1938) suggests that the guiding principle of the creation of the new style was the need to align the vertical traces of each individual letter to insure a vertical continuity from the top to the bottom of the writing surface. The search for a geometrically precise equidistance is evidenced in some epigraphs by faint traces of a grid pattern on the stones that bear the inscriptions. The stoichedon style was incompatible with the bidirectional practice of the boustrophedon, because one of the criteria of its regularity and lapidary beauty was the alignment of the vertical segments of all the same letters that reappeared in the text. Since the vertical segments in several letters, notably B, Γ, E, K, and P, are not centered but are to the left of the character, such letters when presented in the retrograde direction of the boustrophedon would appear to the right and disturb the alignment. This theory is plausible although there is no direct evidence to support it.

The choice of an initial orientation was not stabilized one way or the other at the inception of the style. There are several vestiges of stoichedon from the middle of the sixth to the end of the fifth century B.C. that are written leftward, although the majority read rightward. The third and last phase of the incipient alphabetization of Greece came out of a bureaucratic decision that was taken in Athens in 403/2 B.C., by Archinos, then the archon of Athens, to homogenize the writing of official documents (Dain, 1963; Guarducci, 1967). This decree, according to Austin and Woodhead merely reinforced and institutionalized the predominant trend of writing from left to right that had existed from the earliest beginnings of the stoichedon style.

Reviewing the developments of Greek writing, one can tentatively conclude that the first period of alphabetization demonstrated a condition of lateral indeterminacy biased by a pronounced tendency to follow the pattern inherited from the Semitic models (Jeffery, 1961; Woodhead, 1981). The second period was marked by an ambilateral practice that extended to the orientation of individual letters. At this point, in spite of the well-established practice of word breaks in the Phoenician models, there was no attempt at parsing words, sentences, or even paragraphs: this pattern has been dubbed *scriptio continua* by classical epigraphists (Jeffery, 1961; Dain, 1963; Guarducci, 1967) as letters follow each other in a single continuous sequence snaking from the top to the bottom of the writing surface (Cohen, 1958; Dain, 1963). The third period introduced the final stage of rightward direction which we still observe today after 2500 years. It is worth remarking here that the conventional separations between words that we observe today, was not regularized in Greek manuscripts until the Byzantine period. Punctuation was practiced only by the grammarians of Alexandria, and only for a period of time ranging possibly from the second century B.C. to the fourth century A.D. (Cohen, 1958; for a similar suppression of separations between words in Roman cursive scripts in the third century A.D., see Marichal, 1963, Saenger, 1982).

The characteristic development of Greek orthography, from an initial leftward direction, including word separation, followed by a boustrophedon stage, and ending in a uniformly rightward direction without word separation, is also observed in many different adaptations of Phoenician, Greek, or Etruscan scripts on coastal regions of the Mediterranean sea and on the Italian Peninsula. This was the case, notably, for Illyrian, Venetian, Osc, Umbrian, Picenian, Messapian, and other local orthographies that would eventually be absorbed by Latin (Cohen, 1958, pp. 188 ff.). A reorientation to the right, but without going through the boustrophedon stage, also occurred with Ethiopic, at the time when vowels were included to create a syllabary out of the Semitic consonantal system that the ancient Ethiopians had been using (Lafont, this volume).

The explanations proposed for these mysterious, although relatively orderly developments of Greek and related orthographies range from considerations of esthetics (Austin, 1938; Dain, 1963; Guarducci, 1967) to possible changes in the posture and the writing materials of the scribes. To be sure, there may be material and cultural reasons for these shifts of direction, and they have been convincingly propounded by several authors (Jackson, 1981; Sirat, 1976, this volume). Other scholars such as Etiemble (1973) throw up their arms and claim that the orientation results from a combination of chance and the inertia of custom.

Chance occurrences, custom, and material causes would be satisfactory explanations for the changes if there were sufficient reason to believe that serendipitous processes with serendipitous results were always at work. The examplary strength of the original model and the force of custom may be sufficient to account for some constancies in individual cases, but, precisely because the correlations between the presence of vocalic signs and rightward direction are so generalized over such wide spans of time and space, other, more logically grounded arguments must be invoked. There is at least one epigraphist, Jean Filliozat (1963), who suggests that material conditions have little to do with the practice of complex orthographies and that explanations for their layout must be sought in psychological constraints:

> There are many examples (in Indic scriptforms, on many different materials ranging from granite to different types of palm leaves) which prove that writing materials do not impose their conditions to the calligraphist, nor even to the cursive writer; it is man, in India at least, who imposes his will to the material. What is the nature of this willful activity? This can only be revealed by the kinds of psychological investigations which will show why some cultures adopt simple, and others, complicated systems of writing (Filliozat, 1963).

Contextual Vs Sequential Orthographies

The main fact to consider here is that by adding vowels to the Phoenician consonants, the Greeks created a writing system based on a fully sequential phonemic representation (de Kerckhove, 1984). This feature is shared with the most ancient syllabaries; indeed, the reading of syllabaries depends on deciphering the continuous sequence of syllabic characters. By comparison, the most sophisticated Semitic scripts are approximations, or, as Lafont suggests in this volume, "shorthand" systems, which require the reader to supply the oral context to decipher the text. What is implied here is that the reader of Arabic or Hebrew, for instance, must supply the values of the vocalic intervals between the consonants.

The problem of reading syllabaries is different in that, although the vocalic values are included (either within the syllabic sign as in Japanese Kana, or as stand-alone vowels that constitute a full syllabic sign, as in the Akkadian, the Ugaritic, and the Old Persic syllabaries), the syllabic combination is not always a straightforward process. The "neutral" vocalic component that is included in most open syllabic characters must be eliminated or contracted to enable the reader to achieve a consonantal grouping such as CCV. For example, the name of one of the main Persian cities was Ekbatan. It was rendered in the cuneiform syllabaries as /EK/+/KA/+/BA/+TA/+/NA/; to read the word as it was supposed to sound, it was necessary to omit the vocalic components of the syllabic characters for /KA/ and /NA/. For syllabaries, the problem is not to supply missing vocalic values, but to eliminate them where they are superfluous.

The reason why vowel signs are dispensable in Semitic scripts is that it is almost always possible to recognize the meaning and the structure of the line of writing by the order of the consonants, provided that the reader can supply the vocalic sound. The way the vocalic interval is achieved depends on contextualizing the consonantal signs. Admittedly, there is a well-known case in Biblical history that shows that this was not always an easy or unambiguous process – I am referring here to the story of the prophet Daniel who had to be brought in before Nebuchadnezzar, among a vast assembly of learned people who were admitting defeat, to decipher the famous "writing on the wall," *mene tekel pharsin,* which must have appeared in Aramaean as MNTKLFRSN. However, this kind of difficulty must have been the exception rather than the rule, because it did not encourage Semitic scribes, of the past or the present for that matter, to systematically employ the use of diacritical signs, the *matres lectionis,* which are used mainly for teaching reading.

This economy of signs is not available to Indo-European languages such as Greek or Latin, because, as we have seen, the morphemic structure of their vocabularies includes vocalic sounds to distinguish one word from another. From the standpoint of logical structures, the Greek alphabet and its derivatives share two important features with both the consonantal systems and the syllabaries. They are based, like Semitic writing, on phonemic differentiation *and* on a sequence of separate phonemes. However, like the syllabaries, they need to be read as a combination of contiguous characters, the difference being, of course, that, in syllabaries, each character constitutes a syllable, while it constitutes a phoneme in vocalic alphabets.

The conclusion I would like to draw from this differentiation is that it is not specifically to the presence of vowel characters that the vocalic alphabets or the syllabaries owe their rightward direction, but, as a result of the inclusion of these vocalic characters, to a fully linear sequence of characters. It is the priority given to the sequence over the iconic features of the characters that is responsible for the orientation of the orthography to the right.

Logical Principles and Hypotheses

Other logical principles can be suggested from reliable correlations between types of writing systems and the way they are spatially presented on the surface of writing. From a survey of the principal types of writing, following Gelb's classification (1963) and from the accurate descriptions and datings found in Cohen (1958) and Février (1959), I have derived six principles of relationships between the structure of orthographies and their layouts (Table 2).

Table 2. Logical principles of the layout of orthographies

I. Axis

1. *Iconic* structures favor *vertical* reading
2. *Linear* structures favor *horizontal* reading
3. The conversion of pictography to phonography will *horizontalize* vertical scripts

II. Direction

1. *Contextual* relationships favor the *left* direction
2. *Contiguous* relationships favor the *right* direction
3. The conversion of consonantal alphabets to vocalization will *redirect* scripts to the right

III. Linearity

1. Decipherment depending on *contextualized* symbols requires *parsing*
2. Decipherment depending on *contiguous* symbols favors continuous *linearity*
3. Vocalized alphabets and syllabaries have tended to adopt patterns of *scriptio continua* and/or *boustrophedon* styles

IV. Economy

1. *Syllabic* symbols are *dedicated* and explicit
2. *Phonetic* symbols are *modular* and implicit
3. *Phonetic* systems can represent any linguistic structure through a limited number of symbols combined through the *syllabic synthesis*

V. Uniformity

1. *Iconic* systems emphasize the *shape* of the symbols
2. *Modular* systems emphasize the *sequence* of the symbols
3. *Modular* systems will favor abstract *uniformity* while *iconic* ones favor concrete *stylization*

VI. Abstraction

1. *Syllabic* symbols present an *analogical* relationship to spoken language
2. *Phonemic* symbols present a *digital* relationship to spoken language
3. Consonantal alphabets and syllabaries *converge* with oral speech while vocalic alphabets are *independent* of it

Axis (*Iconic* vs *Linear* Processing)

Rule 1. Iconic writing systems, whether pictographic, hieroglyphic, or ideographic, although occasionally found to be written horizontally, are predominantly written vertically (Proto-Sumerian, Sumerian, Proto-Elamite, Aztec/Maya pictographic, Egyptian hieroglyphic, Byblos pictographic, Proto-Indic pictographic, Chinese ideographic, Korean, Japanese Kanji). It is possible that this is due to the fact that it is easier to read individual characters vertically than horizontally (Ghent, 1961; Kolers, 1983; Kolers & Perkins, 1975; Salapatek, 1968; Taylor & Taylor, 1983).

Rule 2. Not counting all their variations, Semitic, Indic, Greco-Roman, Cyrillic, all the cuneiforms, and all the phonological systems generally, whether syllabic, consonantal, or vocalic alphabetic, numbering over 200 different styles, are written horizontally. It is suggested that it is easier to read *linear* sequences of connected signs horizontally than vertically (for developments on this observation, see de Kerckhove, this volume; for a discussion of vertical vs horizontal spatial cognition, see Olson & Bialystok, 1983).

Prediction. Consequently, the shift from the vertical to the horizontal line that seems almost always to follow a critical stage of phonetization of an originally pictographic or ideographic system (such as Sumerian, Egyptian, and Chinese via the Japanese and Korean adaptations; see Driver, 1954; Sirat, this volume, for material causes), could be credited to the facilitation of the reading process: when each sign ceases to stand on its own as a pictographic or ideographic representation and has to be linked with other signs to form a phonological sequence, the prediction is that the axis of the strings of characters will become horizontal.

Observations

1 There is one exception to rule 1: Mongol, though a phonologically based script, the creation of which was ordered by Genghis Khan to simplify the Chinese system, can be written vertically, following the Chinese model, or more often horizontally (Cohen, 1958).
2 It is worth remarking that, in many instances when pictograms or ideograms were included in later developments of phonologically based orthographies, such as Akkadian and Elamite, they were subjected to the rule of horizontalization. The cases of Japanese Kana and Korean Hangul are different in that the force of custom may have been instrumental in keeping the new phonological systems vertical; however, they are in the process of being generally horizontalized today (see I. Taylor, this volume).

Direction (*Contextual* vs *Contiguous* Sequencing)

Rule 1. One factor that may guide the direction of reading and writing is the nature of the relationships between the individual characters; if this relationship is *contextual,* as is the case for consonantal alphabets, reading will depend primarily on pattern recognition.

Rule 2. If this relationship is *contiguous,* as is the case for vocalic alphabets and syllabaries, reading and writing will depend primarily on sequential recognition. Experimental evidence (Bentin & Carmon, 1983; Tzeng & Hung, 1984) suggests that, for reasons that are not yet quite clear, sequential features are generally faster and more accurately identified in the right than in the left half of the visual field. Conversely, feature recognition is better and faster effected in the left half of the visual field (Bersted, 1983; Hellige & Webster, 1981).

Prediction. Consequently, the types of scripts that require speed of feature recognition to establish appropriate contextual relationships (such as consonantal alphabets) will be read to the left of the visual field better than to the right; while those that require speed of connecting the contiguous relationships of the sequence of characters (such as the Greek or Latin vocalic alphabets, and most but not all vocalized syllabaries) will be read better to the right than to the left of the visual field.

For example, take a sign A: to decipher it calls upon recognition of pattern; but together, the signs K and A also require the recognition of sequence. If a phonological orthography contains signs for vocalic sounds, the most urgent thing to do may be to recognize the sequence of the letters to effect their combination. However, if the orthography does not specify the vowels, then the most urgent task is not to combine the characters, since they are not supposed to form syllables by combination, but to recognize the patterns or the configurations they present, because that is how the missing elements, the vowels, can be guessed at.

Observations

Depending upon whether or not the classification includes minor variations in style, there are anywhere between 20 and 50 different consonantal alphabets, all of them derived more or less directly from the Phoenician original. The main ones are Phoenician itself, three variations of Hebrew (Old, Square, and Modern), Aramaic, Punic and Neo-Punic, South-Arabic, Sogdian, Syriac, Nabatean, Arabic (with a dozen major styles), Mandean, and Safaitic. These are all written *leftward.* The force of custom and especially religion, with respect to the conservative influence of Islam, may be invoked. However, I propose a more systematic approach in suggesting that all these orthographies share a common feature; i.e., they all depend on a *contextual* deciphering process (see above). This feature is also shared with the iconic writing systems, and, appropriately, all these orthographies are written in vertical columns that read from right to left. The suggestion, which will be expanded in my other chapter in this volume, is that it may be easier to contextualize the distinctive features of individual visual forms to the left of the visual field than to the right.

Conversely, all the cuneiforms, all the aforementioned Indic, Greco-Roman, and Cyrillic variations, and some other isolated scripts such as Cretan Linear B (though not A; see Ventris and Chadwick, 1956), Ethiopic, Vai, and Inuit, are written horizontally to the right. Although there are several interesting exceptions to the general rule (see observations below), it may be easier to connect the *contiguous* signs in their appropriate sequence to the right of the visual field than to the left (for discussion, see de Kerckhove, this volume).

The exceptions to rule 2 are not statistically significant and thus do not invalidate the hypothesis, but they deserve to be identified and, with a closer analysis, possibly explained. The main exceptions, as mentioned above, are the Kharostri and early Etruscan syllabaries, which were written leftward, probably under the direct influence of their Semitic models.

This hypothesis does not preclude the possibility or deny the evidence that even vocalized alphabets can be and are indeed read contextually and ideographically, especially with respect to high-frequency words and word orders (Seidenberg, 1985). It merely asserts that the initial training processes involved in learning to read require that the proper sequencing of contiguous characters be mastered before the growing competence of the reader can enable him or her to skip some analytical procedures in a large spectrum of proficiencies, that range from the laborious first-grader's letter-by-letter spelling exercises to high-speed reading (Read, 1985).

Linearity (*Discriminative* Vs *Determinative* Processing)

Rule 1. Scripts that require contextual evidence for the decipherment of each character or each group of letters (such as consonantal alphabets) will also require some form of parsing to avoid delays and confusion in processing. The process of decoding such scripts is based on recognizing the features of the characters and on *discriminating* among several potential combinations, retaining the one that is the most appropriate to fit the graphemic and the semantic contexts.

Rule 2. Scripts that depend primarily on syllabic and/or phonemic synthesis predicted on contiguous sequential relationships (such as most syllabaries and all vocalized alphabets) will not systematically require word separations and will tend to favor a layout of characters in uninterrupted sequences. The reason for this may be that the process of decoding such scripts is based on recognizing the features of the characters and combining them to determine largely unambiguous sequences requiring little or no help from the surrounding graphemic context.

Prediction. Consequently, many syllabaries and vocalized alphabets (Greek, Latin, Indic scripts, Ethiopic, but not the cuneiforms) seem to have favored a high incidence of *scriptio continua* and even boustrophedon (Cypriot, Greek, Latin) styles, leaving it up to the recoding processes to effect appropriate segmentation wherever applicable (Saenger, 1982). Such scripts did not systematically practice full or even partial segmentation for centuries, although at a later stage, probably at the time when silent reading became the norm and when the benefits of a given literacy were extended to the majority of the population, most vocalized alphabets began to adopt word separation for the convenience of faster word recognition.

Economy (*Dedicated* Vs *Modular* Symbols)

Rule 1. Syllabaries are often classified as "concrete" (or "explicit") writing systems because each syllabic symbol is usually dedicated to a single pronounceable sound (Havelock, 1982).

Rule 2. Alphabets are classified as "discrete" (or "implicit") writing systems because the phonemic symbols must be combined with one another before they can constitute a pronounceable sound.

Prediction. The discrete units presented by the abstract letters of vocalic alphabets, by providing a "double articulation" of phonemic signs into syllabic structures, and of syllabic structures into groups of sign-to-sound correlations, overcome the difficulty inherent in syllabaries and reduce the need for a great quantity of signs, thus achieving both optimal economy and reliability for the visual strings to adequately represent the phonological strings. Alphabetic systems can thus express great quantities of different sounds with a modest quantity of signs. Vocalic alphabets can serve as translating devices or algorithms for any linguistic structure.

Observation

There are many syllabaries such as the Indic scripts, Korean Hangul, Inuit, and Ojibway/Cree that are modular to a degree in that they have achieved the idea of syllabic synthesis, namely combining the same basic consonantal element with vocalic variables. In spite of such flexibility, these syllabaries do not achieve full modularity with respect to polysyllabic structures, but still have to contend with syllabism rather than with pure phonetism. The fact that the syllabic principle adheres closely to a concrete representation of fully pronounceable sounds may be the fundamental reason why these syllabaries, in spite of benefiting from all the resources of a complete phonemic analysis (Nagari, Hangul, Cree/Ojibway, Inuit) have remained faithful to it, overriding the implied phonological analysis. In such cases, the syllabic character is akin to an "ideogram for sound", and the vocalic marker is no more than an added precision to avoid possible confusion.

Uniformity (*Explicit* Vs *Discrete* Coding)

Rule 1. Iconic scripts (ideograms and pictograms) as well as consonantal alphabets and some syllabaries are explicit because they must emphasize the features and the contours, or the *shape* of the symbols, to avoid confusion and facilitate the deciphering process.

Rule 2. Modular systems that rely first and foremost on the combination of discrete contiguous symbols to effect the syllabic synthesis will tend to emphasize the connection and the linear *succession* of signs.

Prediction. Consequently, the explicitness of the symbols will find expression in calligraphy, monumental epigraphy, and systematic stylization, which both reflect and give to the characters a sacred quality (for an examination of the relationships between Chinese and Japanese calligraphy and their writing systems, see Jones & Aoki, this volume; see also Safadi, 1979, for a similar study of Semitic calligraphy). Conversely, the technical dimension of information processing implied in the discrete symbols of the alphabets promote uniformity and homogenization.

Observations

1 It is only due to the highly iconic qualities of photoengraving and mechanical reproduction that the western alphabets have recovered some of the sensory qualities that prevailed in the medieval manuscript culture, and which have never been lost by other formal writing systems (McLuhan, 1969).

2 The case of the "abstract" quality of consonantal alphabets is not settled, as experts do not agree on whether each consonant actually represents an unpronounceable phonemic unit (Cohen, 1958; Sznycer, 1977) or merely a syllabic indicator whose vocalic content is not specified and cannot be specified until the reading process is completed (Février, 1959; Gelb, 1963; Havelock 1982; Lafont, this volume). The fact is that all consonantal alphabets have produced some of the world's finest examples of calligraphy and monumental inscriptions, and I believe this to be an indicator that they are on the concrete end of the spectrum, more so than most rightward written syllabaries (Safadi, 1979).

Abstraction (*Analog* Vs *Digital* Coding and Recoding)

Rule 1. Syllabic systems present analogical features between the visual and the auditory stimuli of speech production; syllabic symbols are dubbed concrete not only because each one represents a naturally pronounceable sound, but also because the sequence of syllables presents a chain of signs that is approximately analogous to the chain of sounds of human speech.

Rule 2. Vocalized alphabets are usually classified as abstract (or discrete) systems because most of their individual characters (except for vowel symbols) are single digits that bear no relationship by themselves to a pronounceable sound of human speech; it is only by first effecting the visual synthesis of a combination of letters that the subsequently phonological recoding of a full syllabic sound can be achieved.

Prediction. Consequently, only the fully vocalized alphabets are intimately and, for the purpose of learning to master them, exclusively bound to the timing and sequencing processes affecting visual stimuli prior to their recoding into concrete syllabic sound patterns. This must be emphasized, because one might be led to conclude from the above hypotheses that because syllabaries always imply vocalic components and because most are in fact written to the right, they require the application of the same rules as vocalic alphabets. In fact, all they have in common with such alphabets is that they are processed contiguously; the resemblance stops there. Indeed, although syllabaries such as the Indic scripts and Korean Hangul also require a measure of phonemic analysis for individual syllabic signs, they tend to be closely bound to the oral dimension of the text. As Lafont describes them in this volume, such scripts are bound in a phonological syntax rather than a morphological one. It is necessary to refer to the full oral context of each sentence, not only to understand its meaning, but to make sure that it has been properly read. That feature is shared with consonantal alphabets, and for this reason, syllabaries can never reach the level of abstraction that is characteristic of vocalic alphabets.

Observations

1 This hypothesis does not preclude or deny the evidence that dual (visual and/or phonetic) recoding processing can and often does occur in reading vocalized alphabets (for a review of theories of recoding, see Humphreys and Evett, 1985). It merely asserts that the proper sequencing of the visual stimuli must be mastered before any other convenient recoding strategies can be brought to bear on the reading process.

2 A rather confusing, though entertaining distinction between "Chinese" and "Phoenician" readers was introduced by Baron and Strawson (1976) to distinguish those readers of our alphabet who demonstrably and reliably prefer visual over phonetic recoding strategies. While the designation of phonetic recoders as "Phoenicians" could be relevant (in spite of the fact that Bentin, Bargai & Katz [1984] provide evidence that readers of Semitic scriptforms do not rely much on phonemic mediation), the other category is probably unsuited to visual recoders. It is difficult to imagine that visual recoding of the alphabetic sequence could lead to an idea or an image directly, as may be the case for readers of ideograms. Rather, word recognition must depend on the evocation of a concept that, in turn, must be combined with other concepts to form the image or idea. Even readers of Chinese characters appear to depend partly on phonological indicators in the total character to distinguish isomorphic signs from one another (Jones & Aoki, this volume; Taylor & Taylor, 1983).

Conclusions

Keeping in mind that the form of its orthography reflects the specific nature of an Indo-European language, the Greek alphabet and its derivates have followed several principles according to a logical development of languages:

1. Phonography favors a horizontal layout.
2. Contiguous decipherment, for reasons that will be examined in the concluding section of this book, favors a rightward orientation of the script.
3. The linear sequencing of characters emphasizes the continuity of the script, which is the reason why the Greek, like the Etruscan and Latin alphabets, was once amenable to the boustrophedon layout.
4. The modular structure of the phonemic differentiation allows for a manageable and relatively constant number of characters applicable to any combination of speechforms.
5. The modular structure also emphasizes the need for uniformity, allowing for only limited stylization of the individual characters.
6. By taking the analysis of sound structures one step further away from direct sign-to-sound correlations, the Greek system introduced a level of abstraction that would all but remove the script from the context of its production in oral forms.

Indeed, the most important conclusion to draw from the above principles is that vocalized alphabets are the only systems of linguistic representation that are neither convergent with (as syllabaries or consonantal alphabets) nor foreign to (as Chinese or Egyptian orthographies) the specific properties of human speech, but *parallel* to them. The visual sequence of any vocalized alphabetic script can be completely removed from the conditions of its creation during the recoding and interpreting process. The text, as it were, "can speak for itself." Therefore the Greek alphabet was the first writing system to provide humans with a technique of information processing bearing all the qualities and complexities of speech without being burdened by the need to be reinserted in a binding oral context. Its basic process was the atomization of speech: its parsing into its smallest indivisible elements, the phonemes, also known as στοιχεια by the Greeks. This parsing, or fragmenting effect would eventually lead to the fragmenting and the classification of much of human experience and knowledge through the full externalization of language in visual classifiable form (de Kerckhove, 1981, 1982).

What remains to be examined is whether the rightward orientation of the Greek script was not predicated upon neurophysiological rather than merely technical criteria, and furthermore, whether such neurophysiological constraints did not entail other determining changes at higher levels of information processing. This is the subject of Chap. 20 in this volume.

References

Austin, R.P. (1938). *The stoichedon style in Greek inscriptions.* Oxford University Press.

Baron, J., & Strawson, C. (1976). Use of orthographic and word-specific knowledge in reading words aloud. *Journal of Experimental Psychology: Human Perceptual Performance, 2,* 386–393.

Bentin, S., & Carmon, A. (1983). The relative involvement of the left and right cerebral hemispheres in sequential vs. holistic reading: electrophysiological evidence. Paper presented at the annual BABBLE meetings, Niagara Falls, Canada, March.

Bentin, S., Bargai, N., & Katz, L. (1984). Orthographic and phonemic coding for lexical access: evidence from Hebrew. *Journal of Experimental Psychology: Learning Memory and Cognition, 10,* (3), 353–368.

Bersted, C.T. (1983). Memory scanning of described images and undescribed images: hemispheric differences. *Memory and Cognition, 11,* 127–136.

Burns, A. (1981). Athenian literacy in the fifth century B.C. *Journal of the History of Ideas, 42,* (3), 371–387.

Cohen, M. (1958). *La grande invention de l'écriture et son évolution.* Paris, Imprimerie Nationale.

Cohen, M., & Sainte Fare Garnot, J. (Eds.), (1963). *L'écriture et la psychologie des peuples.* Paris: Armand Colin.

Dain, A. (1963). L'écriture grecque du VIIIe siècle avant notre ère à la fin de la civilisation byzantine. In Cohen, M., & Sainte Fare Garnot, J. (Eds.), *L'écriture et la psychologie des peuples.* Paris: Armand Colin.

de Kerckhove, D. (1981). A theory of Greek tragedy. *Sub-Stance, 29,* 23–36.

de Kerckhove, D. (1982). Ecriture, théâtre et neurologie. *Etudes françaises, Avril,* 109–128.

de Kerckhove, D. (1984). Effets cognitifs de l'alphabet. D. de Kerckhove, & D. Jutras (Eds.) *Pour comprendre 1984.* Occasional Paper No 49, pp. 112–129. Ottawa: UNESCO.

Diringer, D. (1948). *The alphabet: a key to the history of mankind.* London: Hutchinson.

Driver, G.R. (1954). *Semitic writing: from pictograph to alphabet.* London: Oxford University Press.

Etiemble (1973). *L'écriture.* Paris: Gallimard.

Février, J.G. (1959). *Histoire de l'écriture* (2nd ed.). Paris: Payot.

Filliozat, J. (1963). Les écritures indiennes: le monde indien et son système graphique. In Cohen, M., & Sainte Fare Garnot, J. (Eds.), *L'écriture et la psychologie des peuples.* Paris: Armand Colin.

Gelb, I.G. (1963). *A study of writing: the foundations of grammatology* (2nd ed.). Chicago: Chicago University Press.

Ghent, L. (1961). Form and its orientation: a child's eye view. *American Journal of Psychology, 74,* 177–190.

Guarducci, M. (1967). *Epigrafia greca,* VI. *Caratteri e storia della disciplina, la scrittura greca dalle origini all'età imperiale.* Roma: Libreria dello Stato.

Hagège, C. (1985). *L'homme de paroles.* Paris: Fayard.

Havelock, E.A. (1976). *Origins of western literacy.* Toronto: Publications of the Ontario Institute for Studies in Education.

Havelock, E.A. (1982). *The literate revolution in Greece and its cultural consequences.* Princeton University Press.

Hellige, J.B., & Webster, R. (1981). Case effects in letter-name matching: a qualitative visual field difference. *Bulletin of the Psychonomic Society, 17,* (4), 179–182.

Higounet, C. (1976). *L'écriture.* Paris: PUF.

Humphreys, G.W., & Evett, L.J. (1985). Are there independent lexical and non-lexical routes in word processing? An evaluation of the dual-route theory of reading. *The Behavioral and Brain Sciences, 8,* 689–740.

Innis, H.A. (1950). *Empire and communications.* Oxford: Oxford University Press.

Irigoin, J. (1982). Les Grecs et l'écriture, quelques jalons historiques. *Corps écrit.* VI. *L'écriture,* Paris: Presses universitaires de France.

Jackson, D. (1981). *The story of writing.* New York: Taplinger.

Jeffery, L.H. (1961). *The local scripts of archaic Greece.* Oxford: Clarendon.

Jurdant, B. (1984). *Ecriture, monnaie et connaissance.* Unpublished doctoral thesis. Strasbourgh, GERSULP.

Kavanagh, J.F., & Mattingly, I.G. (Eds.), (1972). *Language by ear and by eye: the relationships between speech and reading.* Cambridge, MA: MIT Press.

Kolers, P.A. (1983). Polarization of reading performance. In: Coulmas, F., & Ehlich, K. (Eds.), *Writing in focus.* Berlin: Mouton.

Kolers, P.A., & Perkins, D.N. (1975). Spatial and ordinal components of form perception and literacy. *Cognitive Psychology, 7,* 228–267.

Kristeva, J. (1981). *Le language, cet inconnu.* Paris: Le Seuil.

Lafont, R. (Ed.), (1984). *Anthropologie de l'écriture.* Paris: Centre Georges Pompidou.

Marichal, R. (1963). L'écriture latine et la civilisation occidentale du 1er au XVIe siècle. In Cohen, M., & Sainte Fare Garnot, J. (Eds.), *L'écriture et la psychologie des peuples.* Paris: Armand Colin.

Marrou, H.I. (1964). *Histoire de l'éducation dans l'Antiquité.* VI, *Le monde grec* (2nd rev. ed.)., Paris: Le Seuil.

McLuhan, M. (1969). *The Gutenberg galaxy: the making of typographic man.* Toronto: University of Toronto Press.

Naveh, J. (1982). *Early history of the alphabet: an introduction to west semitic epigraphy and palaeography.* Jerusalem: Magnes.

Olson, D.R., & Bialystok, E. (1983). *Spatial cognition: the structure and development of mental representations of spatial relation.* Hillsdale, NJ: Lawrence Erlbaum.

Olson, D.R., Torrance, N., & Hildyard, A. (Eds.), (1985). *Literacy, language and learning.* Cambridge: Cambridge University Press.

Read, C. (1985). Effects of phonology on beginning spelling: some cross-linguistic evidence. In Olson, D.R., Torrance, N., & Hildyard, A. (Eds.), *Literacy, language and learning,* Cambridge: Cambridge University Press.

Saenger, P. (1982). Silent reading: Its impact on late medieval script and society, *Viator, 13,* 369–414.

Safadi, Y.H. (1979). *Islamic calligraphy.* Boulder: Shambala.

Salapatek, P. (1968). Visual scanning of geometric figures by the human newborn. *Journal of Comparative & Physiological Psychology, 66,* 247–258.

Sampson, J. (1985). *Writing systems.* Oxford: Oxford University Press.

Seidenberg, M.S. (1985). The time course of phonological code activation in two writing systems. *Cognition, 19,* 1–30.

Sirat, C. (1976). *Ecriture et civilisations.* Paris: Editions du CNRS.

Spinoza, B. de (1968). *Abrégé de grammaire hébraïque.* Paris: Vrin.
Sznycer, M. (1977). L'origine de l'alphabet sémitique. *L'espace et la lettre,* Cahiers Jussieu V3, Paris, Union Générale d'Editions coll. 10/18.
Taylor, I., & Taylor, M. (1983). *The psychology of reading* (pp. 233–267). New York: Academic.
Tzeng, O. J. L., & Hung, D. L. (1984). Orthography, reading, and cerebral lateralization. In Quing, C., & Stephensen, H. (Eds.), Current issues in cognition (pp. 179–200). National Academy of Sciences Publications, USA, and Amerian Psychological Association (joint publication).
Ventris, M., & Chadwick, J. (1956). *Documents in Mycenian Greek.* Cambridge: Cambridge University Press.
Woodhead, A. G. (1981). *The study of Greek inscriptions* (2nd ed.). Cambridge: Cambridge University Press.

CHAPTER 10

The Material Conditions of the Lateralization of the Ductus *

COLETTE SIRAT [1]

The modern alphabet had its beginnings in about 1700–1300 B.C. Today, two principal branches are in use: the Greek writing system and its derivatives (Latin, Russian, etc.), written from left to right, and Hebrew and Arabic, which derive from Aramaic and are written from right to left.

Chronologically, writing in vertical lines preceded horizontal writing. The top-to-bottom direction was characteristic of three scripts that lay at the foundation of literate civilizations. One, the Chinese system, retained this direction, while Sumerian cuneiform eventually came to be written horizontally from left to right, and Egyptian hieroglyphs came to be written horizontally from right to left. None of these three systems was alphabetic (Cohen, 1958). The earliest true alphabet, which historians variously call Phoenician, Canaanite, Proto-Canaanite or Proto-Hebrew, arose at the crossroads of cultures that wrote horizontally. Each of its branches settled on a direction after two or three centuries of trials, with the Aramaic script being written from right to left, and the Greek script from left to right (Diringer, 1968; Naveh, 1982). Once these directions were firmly established they were conserved without change.

It should be noted, however, that societies have often changed the alphabets that transcribe their languages. When a people adopt a writing system for the first time, or switch to a new one, they incorporate its direction – which of course is only a concomitant circumstance of a much larger phenomenon, the incorporation of ideologies, aesthetics, techniques, and often language. It was originally political conquests, then cultural conquests, that led to cultures adopting new alphabets. The conquests of Alexander the Great (333 B.C.) and the hellenization of the Orient transformed the writing systems of many peoples of the Middle East. During the fourth century A.D., when they accepted the Christian religion whose doctrine was written in Greek, the Egyptians abandoned their right-to-left script for the Coptic alphabet written from left to right (in contrast to the Jews who, faithful to their religion, were also faithful to their alphabet with its right-to-left direction). The subsequent spread of Islam during the seventh century A.D. im-

* This chapter gives the main ideas from a paper read before the French Académie des Inscriptions et Belles-Lettres: Comptes rendus des séances de l'année 1987, pp. 6–56.

[1] Ecole Pratique des Hautes, Etudes, Institut de Recherche et d'Histoire des Textes, 40, Avenue d'Iéna, 75116 Paris, France.

posed the right-to-left Arabic alphabet on its converts[2], although in Turkey (in 1928) the influence of occidental civilization led to the abandonment of the Arabic alphabet for the Latin one. Conversely, in Algeria, the revival of Islam after the independence led to the officializing of Arabic, and children there learn to write from right to left. In the Hebrew state, Israelis write from right to left. Not only may whole societies switch alphabets, but the individuals who can read and write in more than one direction are in the tens of millions. Immigrants, of course, may have to adopt new systems of writing whatever their original culture and writing may be. Chinese, Japanese, Arabs, and Israelis can often easily write not only in their familiar directions, but also in English or other occidental languages written in Latin characters going from left to right. Fluency in two or three alphabets is not at all unusual: it is prevalent in the history of the Jews from at least two or three centuries before the Christian era until our time (Sirat, 1986), as well as for many other peoples.

When the practice of writing was still sporadic, its direction might vary depending on the role it played. For instance, to the Attic ceramists in the sixth to fifth centuries B.C. writing was an iconic element: letters went in varied directions, but not randomly. They drew lines in relation with the images represented in the design, underlining movements and gestures to weave a network orienting the comprehension of the intended meaning (Lissarague, 1985). It must be emphasized that concepts of backward and forward, mirror reading, or shifting from one direction to the other do not apply here. The direction of writing was simply taking into account the center of interest: generally the person in the scene who was speaking the words or the person for whom they were destined. Similarly, in Egyptian writing when a text and an image were present together, the horizontal graphic lines would be oriented toward the depicted human figures. If a figure were located in the center of the line, the two halves of the line would lead toward it; if it were located to the right, the line would read from left to right (Fisher, 1977). This is because hieroglyphic writing never completely lost its iconic character: writing served to complete images, but did not replace them (Sainte Fare Garnot, 1963; Vernus, 1982).

In the two examples cited here, writing and writing direction cannot be discussed as entities in themselves. Rather, they are intertwined with particular cultural conceptions that differ from modern occidental conceptions. In other words, the role that writing plays in various civilizations may vary since its functions may vary: one cannot detach a writing system and its direction from a whole set of intellectual, social, and technical conceptions. This is not without consequence for our study. Many modern analyses of social and neurological phenomena are based on occidental societies, which

[2] Not all the Egyptians converted to Islam. Coptic language and script were in use in the important Christian community until the eighth century at least. Today, they are the liturgical language and writing of the Coptic Church.

are highly alphabetized in particular ways. The validity of these studies is not questioned, but precautions are required before extending their findings to ancient or different societies.

The dates at which the directions of the two main branches of the alphabet were fixed – right to left for some scripts, left to right for others – are roughly known: the tenth to eighth century B.C. for Aramaic, the seventh or sixth century B.C. for Greek. The reasons for these choices must be sought in those times, for from then on they were supported and maintained by tradition as integral parts of culture; the eventual appearance of printing, which enabled the mechanical multiplication of written material, bringing no changes.

One might assume that in these Mediterranean civilizations, symbolisms of left and right had an influence on the choice of writing direction. However, this was not the case: to the Semites, as well as for the Greeks and Romans, the left was generally considered inauspicious (viz. Latin *sinister*) whereas the right was considered a good omen. There were exceptions to this, but they appeared equally in both cultures (Needham, 1973; Guillaumont, 1985).

Neither can the explanation of why writing systems took the directions they did be sought in reading processes, for reading and writing are essentially different activities, historically and physiologically (Sirat, 1987). From a historical point of view these two activities did not always go together; there were a great number of readers who could not write, and some scribes read very badly (Petrucci, 1984). More importantly, in societies where orality predominates, writing comes first. It is a reminder for memory, it gives words a lasting existence. Of course, it could be read, but it was generally read on a one-to-one ratio: one writer, one reader. Publication was oral, the reader reading aloud. There were exceptions. Some books were not meant to be read at all: the Egyptian Books of the Dead are a good example. Cuneiform school texts were meant to be read by more than one pupil. Physiologically, reading is equally easy going from top to bottom, from left to right, or from right to left, since the head and eyes, as opposed to the hand, do not have a preferred direction: the eye can follow, and does follow, writing in any direction that it takes as we experience it every day[3]. When books were scarce, the Jews were accustomed to reading from any angle. That is, a book would be placed in the middle of a group of up to ten people and each individual would read the text from a different position. Attestations of this come from Spain in the Middle Ages and Yemen until the twentieth century (Sirat, 1986).

[3] In our societies, children are influenced by the writing direction of the culture they live in a long time before they learn to read and write. To cite only one example: when the same television game is produced in the US and Israel, the figures enter from the right in the US, from the left in Israel.

The act of reading a given work can be repeated indefinitely, and may vary greatly from reader to reader. When we in the 1980s read an inscription or a manuscript that was written 2000 years ago, the process in some ways resembles, yet also differs from that followed by those who read it 1000 years ago. Similarly, the innumerable people who preceded or will follow us would read it with both similarities and differences. There are differences not only in comprehension, but also in physiological rhythm, which varies according to the individual, the text, and the epoch. As long as the inscription or the manuscript lasts, reading may be repeated: it is thus *multiple.* It is a selective and discontinuous process, occurring at very different speeds. In oral reading, the pronounciation of words requires a linear rhythm, slower and more homogeneous. However, reading aloud, even quietly, is today a rare occurrence, limited to children, poor readers, or theatrical work. Silent reading has gradually imposed itself since the thirteenth century (Saenger, 1982), and has sharpened many of the differences between reading and writing – between the intellectual grasp of the meaning of a text by means of visually perceived cues, and an activity that involves the whole of the body.

In contrast to reading, the action of the writer is a unique event (if writing on an ordinary surface it is performed in one act; if inscribing on stone, two acts are necessary: writing out the inscription and then engraving it). It is *dynamic,* taking place in three-dimensional space, in a linear time span. The movements of the hand and body, guided by the eye and the brain, determine the trajectory of the instrument as it leaves traces on the writing material, whether that be papyrus, paper, or parchment. In order to reconstitute the act of the scribe, one must reconstitute the unseen motions that lead to the creation of the visible traces. This gesture of the scribe, this dynamic movement, has been called *ductus* by Jean Mallon. In reconstituting the ductus one must take the human body into account, for the entire body participates in the act of writing.

The graphic space in which the writer moves is limited by constraints of body mechanisms, of the material and instrument used, of the form of the letters that one wants to trace. These constraints form a whole, because it is their conjunction that determines the limits within which the writer's activity is carried out. These limits are plastic and flexible as long as the activity is that of an isolated individual; one can write in any position, with any instrument, and in any direction for a short time – if experienced enough, one can even write and paint using the foot or mouth. However, when a society or part of a society writes – once writing has become a regular community activity, even if it be limited to a caste of scribes – it is necessary that it be carried out with a modicum of comfort.

A successful system, one able to impose itself on millions of people, has to be moderately easy to perform and "rational." In order for it to be adopted by different cultures, as was the case with cuneiform, Greek, Roman, and Arabic, its motions have to be performed in a relatively comfortable manner[4], taking both the writing material and the instrument into account. Writing systems that did not succeed are many, and those that we know of through archeological findings are probably only a modest sample. The reasons for success or failure cannot always be known with certitude, but without doubt the harmony of conjunction between the scribe's movements, the material, and the instrument was an important factor.

Humans of almost every era have used a wide variety of writing materials: hard substances like stone, ivory, or wood; flexible ones like clay or wax; smooth ones like parchment, papyrus, or paper. The characters have been engraved with a chisel or stylus, or applied using colored inks. In ancient societies abundance was not the rule, and one wrote with whatever usable materials were at hand: the number of writings found on *ostraca* (potsherds) prove this. However, within each system it was the most commonly employed materials and instruments that gave the writing its characteristic aspects; the standard practice gave to the letters the tone, form, ductus, and direction that were maintained should the writing come to be performed on other materials with different instruments (Sirat, 1976). If we look at the ostracon in Fig. 13, we can see the characteristic style of writing that indicates the common use of papyrus and rush.

At the adoption of a writing system, the common practice would not yet be fixed. Often no one material would immediately achieve unanimous acceptance, and techniques and direction would remain variable. It was only as writing became institutionalized in the school system that one direction of writing would become standardized. Thus, this study will deal only with those societies where writing was known and taught by relatively large numbers of people. The motions described in this paper will refer only to right-handers, for until relatively recently, left-handers were reeducated to write using the right hand[5].

We will here examine aspects of society, rather than just aspects of writing. That is, we will look at writing as an activity occurring within a particular cultural setting, in a given historical period. A number of points must thus be taken into consideration:

- The form of the written letters.
- The writing materials, instruments, and book forms.
- The movements of the scribes and their representation in works of art.
- The scribes' posture, in the context of their societies' customs.
- The scribes' social status.
- The existence of schools and of a formalized teaching of writing.

[4] We do know of long-lasting customs that were not easy to perform, but generally they were carried out by the slave classes of society, and were beneficial for the master class. Here, we are speaking only of the societies in which mostly free individuals were performing the action.

[5] Many left-handers achieved fame but they did not exert any influence on the history of writing.

Thus, all preserved sources of information must be investigated to address these questions: the writings themselves, the texts about writing, and iconographic representations of persons writing. All of these, however, reveal only very fragmentary images of reality. In order to sketch out this reality we need a leitmotif, and I propose the gestures of the scribe – the physiology of writing.

The Physiology of Writing

The physiological instruments for writing, the human body, eye, arm, and hand, have certain characteristic mechanisms of configuration, orientation, and musculature. Analyses of these mechanisms can begin with modern observations, for human physiology has not changed since the beginnings of writing – a very short period of time in evolutionary terms, 5000–6000 years. Body uses can vary, and years of apprenticeship may have allowed certain movements that now appear impossible to our eyes. However, a number of basics still exist: making an effective movement without muscular force is an impossibility; the relative length of the arms and legs has not changed; nor has the orientation of the muscles of the arm, wrist, and hand; nor has the amplitude of the cubital slope of the wrist in relation to the radial slope. These are the physiological foundations of the movements of the scribe.

Let us first look at the position of the hands as they are commonly placed when writing with pen and paper, as noted by Callewaert (Fig. 1). The "combined" mode allows the moving up and down of the two upper phalanges of the forefinger and middle finger, while the solid, broken, and fixed modes do not allow this. In the combined mode the pen is oblique in relation to the writing surface, while in the other positions it is more vertical. Looking at the hands is not enough, for two sorts of movements are combined: the tracing of each letter, and the horizontal progression along the line, right to left or left to right. Horizontal progression is accomplished by rotation of the forearm around the elbow (Fig. 2). It is this movement of shifting the radius around the cubitus that gives horizontal writing its rapidity and lesser degree of fatigue, compared with vertical writing which keeps the elbow in a fixed position and uses the movements of the shoulder.

Let us broaden our analysis of the writer's position. We note how in Fig. 3:

- The hand is contracted and rests only lightly on the support.
- The pen is almost vertical (85° angle).
- The elbow does not rest on the table.
- The body is contracted and leaning forward.
- The paper is placed parallel to the body.

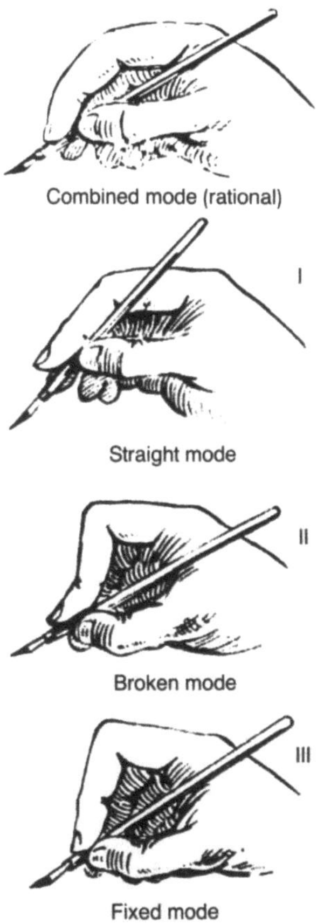

◀ **Fig. 1.** Positions of the hands. (Callewaert, 1962, plate XVI)

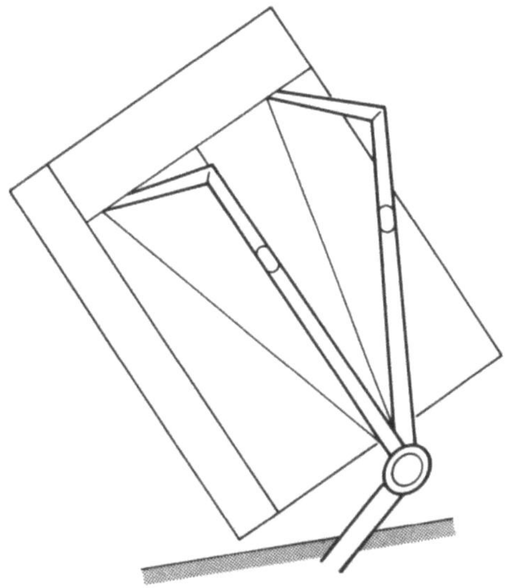

Fig. 2. Movement of translation from the elbow, toward the left or the right. (Callewaert, 1962, plate IX)

In Fig. 4:

- The hand, relaxed, rests completely on the table.
- The pen is held almost horizontally, at an angle of 20° – 25°.
- The elbow rests on the table.
- The torso is upright.
- The paper is placed diagonally in relation to the body.

Let us deal first with the elbow, since its position is a consequence of the position of the hand. The rotating movement of the forearm can be easily accomplished if the arm muscles are relaxed and the elbow rests on a support. In Fig. 3, the support is provided but the contraction of the muscles prevents the elbow from resting on it. The horizontal progression gives the direction of the line. If it were the only movement, it would result in a straight line: from left to right or right to left. When we speak of a right-to-left or a left-to-right alphabet, we are speaking of that horizontal progression. However, the formation of individual letters is accomplished by shorter

Fig. 3. Writing using the shoulder and arm muscles. (Callewaert, 1962, plate VIII)

movements in other directions which depend on the shape of the letters: in every alphabet, the vertical strokes are drawn from top to bottom, but oblique strokes can be written in either direction, and circles are necessarily traced in part in the opposite direction from that of the lines. Depending on the script, there may be a predominance of straight, rigid, or curved strokes. In Fig. 3 there is a predominance of straight strokes, and in Fig. 4 there is a predominance of curved ones.

The writing position shown in Fig. 3 is close to that employed when drawing on a vertical surface, for whole-arm movements are predominant. The muscles of the upper arm and forearm are being used, while those of the hand are contracted and hold the writing instrument in a tight grip. The entire arm is moving in angular segments, with the motion of the humerus controlled mostly by the deltoid muscle of the shoulder. The writing instrument is held at an angle of almost 90° to the writing surface.

The position shown in Fig. 4 is characterized by relaxation of the arm muscles and the use of the muscles of the fingers to guide the pen. The muscular effort of the arm, the forearm, and the fingers associated with a vertical or oblique instrument is more conducive to writing controlled, angular, separated strokes, whereas a relaxed posture, in conjunction with an almost horizontal instrument, leads to a more spontaneous circular, connected writing.

In scripts where stick-like forms predominate, the characters can show more contrast. There can be both thick and thin strokes, since one can use

Fig. 4. Writing in a relaxed position. (Callewaert, 1962, plate IX)

not only the tip of the pen but its edges. Let us disregard the ballpoint pen, and recall those pens that were in use until recently. When such a pen is held vertically, it is controlled by the extremity of the top phalanx of the index finger. One can turn it very easily, so not only is the grip applied by the first three fingers as strong as possible, but the tip of the index finger allows one to turn the nib and use its edge, permitting the formation of both thick and filiform strokes. It is obvious that the mobility of the pen depends on this part of the index finger in conjunction with the tips of the thumb and the middle finger, as can be seen when one tries to turn a brush. This is the reason why Chinese scribes hold their brushes by joining the tips of the first three fingers.

When the pen is oblique, almost horizontal in relation to the surface of the paper, the top joint of the index finger is flat and grips the instrument firmly, but the movement cannot be as accurate as when the pen is vertical. In order to turn the pen one would have to use the thumb, which would make the movement much looser and longer. In reality, when the pen is oblique it stays in the same position: the thickness or fineness of the downstrokes and upstrokes depends on the size of the nib and the degree of pressure exerted by the hand, so that writing will be quicker; it will also not contain very thick or very thin strokes.

Thus, muscle posture and the form of writing are intimately linked. To produce flowing and connected writing, it is necessary to relax the brachial muscles and use the digital muscles. In this case, whether the direction of the line be from left to right or from right to left, the curves will form of themselves and the characters will be linked one to the other, as in Naskhi Arabic writing (Fig. 5). When a rigid, contrasted, perfectly calligraphic and controlled writing style is desired, one has to use all the muscles of the arm, the forearm, and the hand. The wrist is locked in the position for maximum efficiency and accuracy of the motor muscles: a slight dorsal extension of $40° - 45°$ and a slight cubital slope (adduction) of $15°$. This means that the raised wrist is interposed between the directing eye and the stroke being written. With this position of the hand, horizontal strokes are traced more easily and straighter from left to right than from right to left.

A calligraphic Hebrew script, as in Fig. 6, has a ductus whereby all the horizontal strokes are drawn from left to right although the overall direction of the script is from right to left. If we reconstitute the ductus, the gestures that link the strokes to each other, it is apparent that the leftward movements are as numerous as the visible rightward ones. These backward steps multiply the number and the slowness of the movements. In the Latin B, whose ductus has been reconstituted by J. Mallon (1982; Fig. 7), one also sees a backward step. These backward steps do not occur as frequently in the stick-line Latin alphabet as in the Hebrew alphabet, but they do exist. In a writing with curved characters, backward motions opposite to the direction of the lines are reduced to a minimum, which explains the rapidity of cursive writing.

Fig. 5. Arabic Naskhi writing. Letter of the eighth century. Louvre, *Département des antiquités islamiques,* inv. 7030

Fig. 6. Ductus of a Rashi type writing. (According to Greenspan, 1982, p. 93)

Fig. 7. B latin. (From Mallon, 1982; detail of plate, pp. 22–23)

From the aspect of the ductus, the differences are thus greater between two scripts that have the same line direction but differ in form (stick-like or curved), than between scripts that go in different directions but share a form. This is one of the reasons why, from a global point of view, the visual resemblance between writings of the same form is so strong, even though one may be Latin and the other Hebrew (Sirat, 1981 a). Let us further define this in terms of positions.

In Fig. 3 the hand holds the instrument almost at its tip, lying directly over the line being written so that the muscular effort is more efficient and accurate. We have already said that in the case where the muscles of the shoulder, arm, and hand must be used in the position of the greatest accuracy and efficiency, the hand (of the right-hander) screens the characters, and that only horizontal strokes being traced from left to right can be controlled by sight. However, even when writing from right to left, the control is not that easy, and it is in order to facilitate visual control that the writer in Fig. 3 is leaning toward the left; this deviation is added to the general tension of the muscles provoked by the gesture itself.

In Fig. 4, the length of the writing instrument between the written letter and the hand is much greater. Here, the hand is placed *under* the line of writing and does not interfere with eye control. Thus, for a curved, not so controlled script, when the hand is placed under the line of writing, by itself the direction of the horizontal strokes of the letter does not influence speed, except that it is faster to write them in the same direction as the line of writing.

The figures reproduced here show children writing on flat tables. In Fig. 4, the paper is placed diagonally, a position that optimizes visual control. With flat tables, it is easy to move the paper around, change the forearm orientation, and, more generally, find an easy and comfortable position. These writing tables first made their appearance in the fifteenth and sixteenth centuries, contemporaneous with the beginnings of the flowing script invented by Italian writing masters and the spreading of printing; they characterize a particular stage of occidental civilization. In the past, dining tables were trays that were brought out at meal time and placed on tripods. Fixed tables were rare, and were not used for writing but for the placement of objects. Inclined writing desks date from the Middle Ages, and they too were unknown at the time that interests us. On the other hand, scribes were able to cut the calamus at any angle they wanted, an advantage not offered by our modern writing instruments.

The Direction of the Alphabet

When the linear alphabet was born, between 1700 and 1300 B.C., other writing systems had already been in use for twenty centuries in the Middle East. The alphabet needed twenty more centuries in order to supplant the two systems that preceded it, Egyptian and cuneiform. The concept of a simplified graphic notation was "in the wind," and we know of a number of alphabet-like systems that did not succeed. Very probably there were many others we do not know of, whose material traces have been entirely lost. The history of the alphabet is thus the history of technical and cultural choices, in a region where solidly entrenched writing traditions crossed each other, were superimposed upon each other, and fought each other.

The Egyptian Tradition

Egyptian writing was characterized by the use of a roll of papyrus as the writing material and rushes as the instrument – both smooth materials – and of ink (Czerny, 1957). In the fifteenth century B.C., common writing was hieratic (Fig. 8), a system that coexisted with hieroglyphs, whose formal beauty and iconic function were more suited to inscriptions on stone and religious texts like the Roll of the Dead. In hieratic texts, columns followed each other from right to left, and within each column the horizontal lines were also directed from right to left.

Let us compare the position of an Egyptian scribe (Fig. 9) to the position shown in Fig. 4:

– Both the elbow and the hand, relaxed, rest on the surface of a loincloth stretched over the knees.

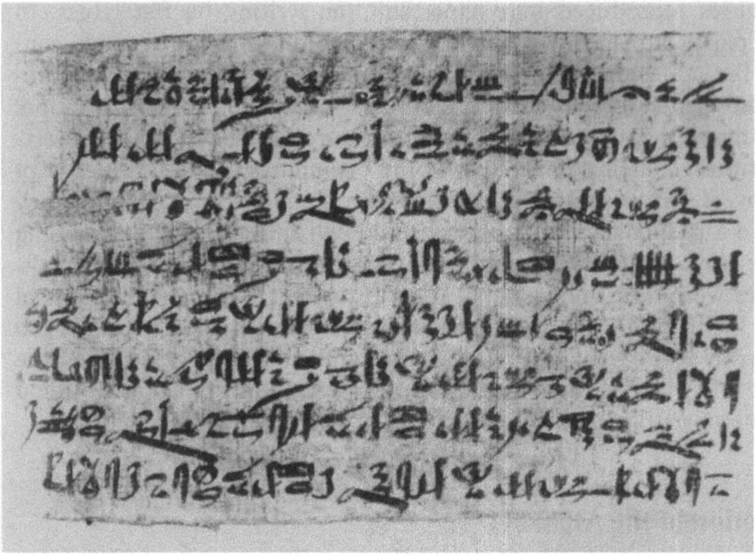

Fig. 8. Hieratic papyrus. Louvre AE/E 3230. (André-Leicknam and Ziegler, 1982, *103*, 159)

- The rush used has disappeared, but the forms of the hieratic characters are proof of a quickness linked to the oblique position of the writing instrument.
- The torso is upright; in this position the eyes of a person 156 cm in height would be 46 – 47 cm away from the papyrus.
- The roll of papyrus is slightly diagonal in relation to the scribe's body.

The posture depicted here not only provided a large area of support for the elbow, forearm, and wrist, but also allowed control of that portion of the papyrus already written on. The scribe held the unused portion of the roll in the left hand, while writing with the right. The written portion fell by the right side of the body, and as writing progressed, the papyrus could be rolled up again if it began to unravel. This control using the right hand was very simple when sitting on the floor, because the elbow of the scribe would be at waist-height, and the length of the forearm was greater than the height of the lap. If the roll were suddenly moved by some accident, it could be easily reached and put back in place.

In the absence of writing tables, it would not be possible to write on a roll[6] in any position other than this. This is made clear by numerous statuettes of the god Imhotep, a famous scribe and architect who lived during the

[6] We deal here only with rolls written in perpendicular columns. Rolls could be written, and were written during the Middle Ages, in one lengthy vertical column, in a posture different from the one described in this paper.

Fig. 9. Egyptian scribe of the New Empire. Berlin, Ägyptisches Museum, inv. 22621, ca. 1370 B.C.

Fig. 10. Imhotep. British Museum, inv. 64495, ca. 650 B.C. ▶

Ancient Empire in about 2600 B.C. and was deified in roughly the seventh century B.C. These statuettes (Fig. 10) date from the Saitian era (approximately the eighth century B.C.). They show Imhotep seated on a chair and writing. His arms are not resting on anything. The surface provided by his knees, which are close together, is just large enough for a column of the roll, which thus cannot be rolled forward or backward or placed slightly at an angle when he arrives at the bottom of the column. Let us imagine that Imhotep, despite his very uncomfortable position, has completed a few metres of this roll. The written portion will fall to the ground from the height of the seat – actually, a slightly greater height since the thickness of his thighs must be added. Imhotep would have to constantly bend to the ground, at risk of losing his balance, in order to put the already written roll back into place. Instead of rolling itself up, the roll would unwind as would a ball of wool let fall by a knitter!

In fact, Imhotep's upper body is in the position of a scribe, and his lower body is in the seated hieratic position of a god. Being a god, he could not be portrayed sitting on the ground; being a scribe he had to be shown writing.

The act of writing required a seated posture on the ground, but this posture was not worthy of a god, nor (with a few exceptions) of a man highly placed in society. Gods, pharaohs, and important persons were always represented as standing or seated in a chair (Baker, 1966). The further on in time, the greater is the number of chairs and armchairs seen in representations and found in archeological excavations; they became very common in the Hellenic period. Shoemakers, sculptors, and blacksmiths practiced their arts sitting on stools, but commoners generally sat on the ground, squatting on their heels. To sit on the ground was a posture of humility, characteristic of common people, of a worshipper before a god, of a child before a parent. However, a large number of statues of scribes have been preserved, and all are shown seated on the ground (Vandier, 1958; Schäffer, 1974). Èven pharaohs had themselves represented sitting in the scribe's position. Writing had a power and a dignity that surpassed and transcended the framework of proper behavior generally admitted in Egyptian society.

One further detail that is important is clothing. Under the Ancient Empire all men, except gods and pharaohs, wore essentially the same garment: a loincloth wrapped around the middle and held in by a belt. (Differences in rank were indicated by jewellery, hair style, or by a sort of apron worn over the loincloth made of cloth of varying richness.) This loincloth is present in all representations of scribes, stretched over the knees and serving as a support for the papyrus roll. Under the Eighteenth Dynasty in the New Empire the fashion was transformed, with the loincloth remaining the clothing only of common people and that worn during religious ceremonies. In the first case it was obviously the least expensive type of garment; in the second, its ancient nature conferred a sacred character upon it. For more distinguished persons, the everyday garb became an over-garment covered with robes in a transparent pleated material. The top of the body was covered by a type of shirt, fairly large, which came down to the feet and was often draped in a wide belt forming a knot in the front (Houston, 1920). From the New Empire onward, the sitting scribes were depicted as wearing a light shirt covering the torso. However, from the waist down they still wore only the old loincloth. After all, how would they have been able to spread out a roll on a robe pleated like a kilt?

In fact, two contradictory characteristics are associated here: the humility of the body posture, and the sacred nature of writing. In Egyptian society the scribes held power, since written words – representations of living things – were believed to seal the fate of human beings in this world as in the next. Thot, the god of writing, weighed the soul during the Last Judgement, and the Books of the Dead, written by the scribes, were indispensable in order to undergo successfully the trials that would allow one to reach "Paradise" (Sainte Fare Garnot, 1963). Priests initially, the scribes were also from very early on administrators, ministers, astrologers, and doctors. Writing opened the doors to all social and religious functions. It is thus not surprising that the position of the scribe was the one most often chosen to be represented: it was the most honorable and coveted position, and the fact that the scribe is seated on the ground like other commoners had little importance in such a context (Erman & Ranke, 1976).

The existence of schools for scribes is revealed by numerous schoolbooks (Erman, 1966; Lichteim, 1973). In the maxims that served as models for the teaching of writing, the scribes glorified themselves in a manner that seem to us to be excessive, but probably corresponded to reality (Drioton, 1949). They often signed their names, and thus we know the names of many. In the Middle Empire, the prophecies of Nepherti were often copied out. In this text, the pharaoh Snefru (of the Fourth Dynasty) is depicted speaking with the wise man and acting as the scribe, "stretching out his hand towards the box that contains the writing instruments, taking out a roll and a palette, and putting in writing the words of the reader-priest Nepherti ..." (Drioton, 1949).

The Cuneiform Tradition

Cuneiform writing was characterized by the use of tablets made of clay, which was a firm material support, and of a stylus made of reed, wood, or metal as the instrument (Fig. 11). Common cuneiform writing was very angular, made by the impression of the stylus in the clay. Chronologically it came after pictographic writing, linear drawings engraved by a vertically held point (Oppenheim, 1964; Driver, 1976).

Even though the strokes were not linked to one another, cuneiform writing was in fact "cursive": that is, written very rapidly. The instrument was held obliquely and not vertically, as can be seen by the deepness of the written sign. Characters were traced in horizontal lines directed from left to right. The form and weight of the tablets varied greatly; some were the size of a postage stamp, most could be held in the palm of the hand, some were very large and heavy. On the large tablets, the impression of the instrument was no deeper at the top of the "page" than at the bottom, implying that when the tablets were not held in the hand at a good distance from the eye, they were placed on a horizontal support so that the movement of the hand could be regular, in the same way that the choice of a small number of directions of strokes had allowed a quicker and more flowing writing (Powell, 1981). The columns followed each other from left to right on the front of the tablet and from right to left on the back. Despite the sometimes considerable weight of the tablets the traces are not blurred, indicative of the firmness of the clay used for them; thus, the scribes were not afraid of erasing the already written signs (Driver, 1976).

The gestures of the cuneiform scribe are not as easy to reconstitute as those of the Egyptian scribe, for whom representations were very numerous throughout the period of Egyptian writing. Representations of cuneiform scribes do not go further back than the ninth–eighth centuries B.C. (Gadd,

Fig. 11. Cuneiform tablet, roughly 1500 – 1300 B.C. Louvre AO 7095 (cf. André-Leicknam and Ziegler, 1982, n° *157*, 215)

1936) and these are always depictions of royal scribes standing up on fields of battle. With only two exceptions, these scribes are not writing on tablets of clay but on tablets of waxed wood, which had the same firm consistency and were always ready for use (Galling, 1971).

As our previous discussion would indicate, the representations of the Sumerian scribe (Fig. 12) show that cuneiform writing was written "with a raised hand," mobilizing the brachial muscles and keeping the wrist locked. When not standing, how did this scribe sit – on the ground like Egyptian scribes, or on a seat? In support of the use of a seat, no representations from the Sumerian epoch, such as the Akkadian and Kassite eras, depict persons sitting on the ground. The fact that this is seen consistently in both the bas-reliefs and the thousands of seals that have been found (Baker, 1966) is unlikely to be due to chance. Like other persons of their time, the scribes were very probably seated on chairs.

Clothing is of no importance here, because even if the scribe's skirt could be stretched between the knees, as were those of the Egyptian scribes sitting

Fig. 12. Assyrian scribes. British Museum, Assyrian Basement, 124953–124960, Palace of Sanherib, 604–681 B.C.

on the floor, it would not have served as the support for a tablet made of clay or waxed wood. One would have needed a piece of wood to form a support, and in any case, the relaxed posture of the Egyptian scribe was not appropriate for the degree of muscle tension required in cuneiform writing. Here, physiology and iconography work together to make the use of seats more probable.

The importance of scribes in the societies that used cuneiform writing (Oppenheim, 1965; Rainey, 1969; Sack, 1981) is well described in a few lines by R. Labat (1976), speaking of the scribes in Sumeria:

The documents of this era (the third millenium B.C.) allow us to count the existence of a few thousand scribes already. There were among them subaltern scribes, scribes of high rank, scribes appointed to the service of the palace or of the temples, scribes specialized in a particular bureaucratic category. At the bottom of the tablets they wrote, a number of them placed after their signatures the names and professions of their fathers: these were governors, ambassadors, administrators of temples, officers, high-ranking functionaries, priests, managers, and foremen. We can thus see that the scribes belonged to the richest families, or at least to the families that were the highest placed in the cities. We can also observe that among all their names there is not a single woman. Being a scribe was a man's profession. It was also one of the most respected professions of Sumerian society. It was so well respected that a number of scribes acceded to the highest positions of the government, and some even became kings. Tel Anam, king of Uruk, who was at first an archivist or Lugal-usum-gal, prince of Lagas, to whom we owe a personal copy of an ancient document. The manner in which he signed this tablet reveals the value that he attached to the knowledge of writing. After his name he wrote a sort of hybrid neologism in which he put together the words for scribe and prince (dub-pa-te-si) just as if a modern sovereign were to add to his name "writer-perator". In Lagas, there is among the scribes the son of Prince Gudea. (pp. 82–83).

A number of schools for scribes played a predominant role in the conservation and diffusion of culture and writing. The prevalent atmosphere of these schools is known through preserved texts, and it is clear that long years might be passed there, encouraging the development of a caste spirit (Kramer, 1949; Gadd, 1956). Over the centuries, cuneiform characters served to write numerous languages with different structures: Sumerian, Akkadian, Eblaitic, Elamitic, Hussite, Canaanite, Aramaic, etc. The system of schools of scribes spread to Persia, Armenia, Asia Minor, Syria, Palestine, and, for a period of time, to Egypt (Hallo, 1980; Durand, 1977).

Not all the users of the clay tablet and the reed adopted cuneiform writing. Cypro-Minoan writing is linear (Ventris and Chadwick, 1955). Linear A has not been deciphered, but Linear B is Greek and dates from the thirteenth and twelfth centuries B.C. The form of the signs, as well as the materials used, show that they were written with a raised hand, and, as we would predict, the horizontal lines ran from left to right. The tablets were palm sized, and were probably held in the left hand. We know nothing of the posture of the scribes, their clothing, their rank in society, or the existence or nonexistence of schools (Bennett, 1960). However, J. Chadwick (1967) has noted that:

Contrary to the Akkadian cuneiform tablets, which frequently bore the name of the scribe who wrote them, none of the Mycenaean tablets has a signature of this type. It seems that the writing of a tablet did not offer matter for glorification; we do not have a parallel with the Ugarit scribe, who wrote after his name the title: master-scribe. Apparently it was not judged necessary to note the name of the scribe who took responsibility for the correctness of the notation. (p. 127).

The Alphabet

At the same time as the Cypro-Minoan tablet, the alphabet, still in its infancy, sought its path through various materials and instruments of the two traditions of writing. It is found written in cuneiform characters not only in Ugarit, but also in Palestine: in Beit Shemesh, Ta'anach, and Nahal Tavor. In scripts whose letters were close to the forms we use today, the writing direction might be vertical or horizontal, going from the left to the right, or in boustrophedon (Naveh, 1982).

By the ninth–eighth centuries B.C. the choice was made: the alphabet came to follow the Egyptian tradition, commonly written in ink on a smooth surface with a rush. We have not actually found alphabetic texts written on a papyrus with ink and a rush that date from this period, but inscriptions on other less fragile and more easily preserved surfaces evidence the writing style that was typical for papyrus; the thick and thin strokes and the curves that would result from the relaxed movement of the rush gliding across a smooth surface from right to left (Fig. 13). The Aramaic scribes, who accompanied their cuneiform-writing colleagues on the bas-reliefs of the Assyrian palace, are shown using rolls (Fig. 12). There are wooden tablets in the hand of the scribe of Barrakab, but they are of plain wood, not waxed, and the tool box of the Egyptian scribe indicates to us that the rush and the ink were the work instruments of the Aramaic scribe (Galling, 1971).

The ordinary body posture of the scribes in alphabetic writing is not known to us – they are represented as standing on the fields of battle – but if they wrote on rolls made of leather or papyrus, they must have sat on the ground, for as we have seen it would not be possible to write on a roll in any other position. The humility of the posture was largely compensated for by the honorable nature of writing. As in Egyptian society, writing was considered worthy of divinity: in the Bible, it is God himself who writes the Tablets of the Commandments given to Moses after the departure from Egypt. In Israel, the schools of scribes were close to the kings and courts, and, as in Egypt and the Akkadian and Babylonian kingdoms, there were numer-

Fig. 13. Ostracon in alphabetic writing, eighth century B.C. (Israel Department of Antiquities) Bode Museum

ous classes of scribes with diverse functions, religious as well as bureaucratic. The priests were scribes, as were the prophets (Demsky, 1976). The schools of scribes probably played a role both in the codification and the interpretation of written laws (Lemaire, 1981). According to Judges VIII, 14–15, it also seems that at least some of the common people knew how to write.

A great revolution seems to have occurred in the eighth and seventh centuries B.C., when the alphabet spread among merchants and craftsmen over the whole of the Middle East. This movement of true alphabetization was seen amongst the Phoenicians, the Canaanites, the Israelites, and the Greeks. Reading and writing ceased to be the privilege of scribes and became accessible to all people, whatever their rank might be. The Phoenician alphabet, like other consonantal alphabets, had already determined its direction: right to left. In Greece, when the direction was determined in the sixth to fifth centuries B.C., it went from left to right. Before then, Greek writing had been written either vertically, from left to right, from right to left, or in boustrophedon (Jeffery, 1961, 1976; Naveh, 1982), and, as mentioned earlier, ceramists arranged writing direction around images.

What influenced the Greeks ultimately to choose a script going from left to right in a horizontal direction? Let us first look at the forms of the letters reproduced in Fig. 14. In this papyrus – the Persians of Timothy of Milet, which is said to date from the third century B.C. and would be the oldest Greek papyrus – we recognize the characteristic traits of archaic Greek writing. The lines are rigid; this is a stick-like writing with no traces of thick or thin strokes, and no curves. The scribe has used all muscular resources, writing with a "raised hand." What could have been the writing surface, the instrument, and the book forms that were most common? Seeing the stick-like forms of this writing, one immediately thinks of a firm support and an equally firm instrument.

In fact, all the representations of the Greeks in the act of writing depict them as writing on tablets of wax with a stylus. The most ancient known representation is a small statue of baked clay, part of a series of everyday scenes from Boiotia that has been dated from 520–480 B.C. (Fig. 15). The scribe, seated on a stool, holds on his knees a codex of waxed wood, a *diptych*, on which he writes lengthwise, in contrast to the Assyrian scribes who wrote widthwise and sometimes in two columns. In his hand he has a stylus whose other end is in the form of a spatula, in order to erase. This spatula is reminiscent of what is in the hand of the little scribe Tarhuypiyas on a Neo-Hittite stele from Marash, dating from the end of the eighth century B.C. (André-Leicknam and Ziegler, 1982, p. 331), which also depicts the hinged codex used by the Assyrian scribes of Fig. 12.

Between this statuette of roughly 500 B.C. and the third century A.D., the Greeks and Romans have left us quite a large number of representations of people writing. Without exception, all are writing on wax tablets, and all are depicted either standing or sitting on seats. The very large majority are chil-

Fig. 14. The Persians of Timothy of Milet. Berlin, Staatliche Museen *Papyrussammlung* P 9875. Second half of fourth century or third century B.C.

dren, young people, and women. Adult men are extremely few in number: with the exception of the male scribe in Fig. 16, I know of only one other, a scene of a census dating from roughly 100 B.C. (Louvre, MA 975). Thus, for eight centuries, the Greeks and the Romans were represented as writing on tablets; never before the second-third centuries A.D. do we see someone writing on a scroll (Parassoglou, 1979).

The wax tablet was evidently not the only material for writing. Among other materials cited by Greek authors are the leaves of trees, olive leaves in particular; the bark of trees, especially lime trees, from which tablets were also made (it is the living part of the tree, *liber,* which has led to the French word for book, *livre*); cloth material, clay and pottery, specifically the ostraca, pottery fragments that were used for ostracisms and of which large numbers have recently been discovered covered with inscriptions; the walls on which one wrote graffiti; and metals such as lead and bronze. Laws were posted on wooden tablets covered with stucco, and according to Herodotus, the ancient Ionians wrote on skins (Schubart, 1960).

Fig. 15. Greek scribe. Louvre CA 684, between 520 and 480 B.C.

Fig. 16. School scene. Louvre G 457, second half of fifth century B.C.

However, it is not very probable that the most common material for books in Greece of the fifth century was either leather or any of the other materials mentioned above. Papyrus is a very serious candidate as the most commonly used material in Greece, at least for preserving texts. The multitude of papyri found in Egypt (Turner, 1971, 1980) prove that from the time of the conquests of Alexander, the Greeks in Egypt very frequently used papyrus and ink, with the calamus as the instrument: a hard reed, more rigid than the rush used in Egyptian writing, and able to trace the stick-like forms of Greek letters.

In Greece itself, much earlier, paintings on vases represented persons reading rolls. These representations, most often of men, began in approximately 500–490 B.C. (Immerwahr, 1964, 1973). They became more numerous at the beginning of the Christian era, and persons reading rolls are very numerous in Greek and Roman statuary. Even though this situation may seem paradoxical to our eyes, writing on tablets and reading rolls existed at the very time when the direction of writing became fixed. This is often represented in school scene (Fig. 16). On a cup with red figures by the Painter of Eritrea, the poet Linos is seen instructing the young Mousaios to read a roll while the young man is writing on tablets. This cup dates from the second half of the fifth century B.C., but schools existed well before then, and the direction of the lines was fixed in the schools where the Greeks were taught to write on tablets and read on rolls.

The existence of schools and the teaching of writing is well attested (Marrou, 1948; Girard, 1891; Beck, 1975, 1978). In Chios in 496 B.C. the roof of a school fell in, burying 119 children (Herodotus, VI, 27); in Astypalaia, in 492 B.C. the pugilist Cleomedes massacred 60 children in a school (Pausanias, *Description of Greece,* VI, 9, 6, quoted in Marrou, 1948, p. 522). In 480 B.C. the Troizenians received the women and the children forced away from Athens, and hired, with their own city paying the cost, schoolmasters to teach them to read (Plutarch, *Themistocles,* 10, quoted by Marrou, 1948, p. 83). At the time of Plato, the teaching of reading and writing was taken for granted [*Protagoras,* 325e and 326c–e (Turner 1965) and *Charmides,* 159c]. Many other findings indicate that Greek citizens read and wrote (Harvey, 1966); in 508–507 B.C., Clistenes instituted ostracism (Vanderpool, 1969), implying that practically all citizens knew how to write.

A.E. Havelock (1976, 1982) has claimed that the Greeks were not "literate." This is a difficult view to uphold, for an important characteristic of the teaching of the "letters" was its democratic nature: it was given to all, boys and girls, citizens and slaves. On the other hand, what is striking (and here Havelock is right) is the contempt in which writing was held, in comparison to speech. It is known that children of both sexes went to school and that citizens could read the texts of the laws and write letters. However, nowhere is writing cited as an accomplishment worthy of praise. Whereas sculptors and potters signed their works, and the names of great orators, politicians, philosophers, and winners of Olympiads are known to us through inscriptions and literary quotations, we do not know of any scribe signing his name when finishing his work, as was the case for Minoan tablets. Even more surprising, no author is said to have actually written his own works: for Aristotle, we know we only have the course notes taken by his students. The question of the beauty or usefulness of writing is never discussed in Greek literature. The only passage in Plato that alludes to it is quite negative: reading and writing, as well as all literature, should be taught in 3 years (between 10 and 13 years of age): "thus as far as reading and writing are concerned, they should be practiced up to the point that [youth] should be able to do both; and at the same time not being concerned with the perfection of quickness and beauty for those children who, during the prescribed period, have been poorly served by nature" (*The Laws,* 810).

The reason for this was that writing was considered to be a servile activity. The free citizens – those who made history, contributed to culture and science, and incidently fixed the direction of the lines of the alphabet – wrote on tablets, while the rolls were written by slaves. Very few servile activities were the subject of representations; one hardly ever finds cooks or laborers depicted, and never the scribes who wrote on rolls. All these professions were reserved for slaves, and did not have importance in the framework of Greek, and later Roman, life.

The almost absolute dichotomy that we see in Greece between writing as a profession and writing as a means of communication is based on two principal reasons. The first is intellectual, affirming the superiority of speech over writing. Writing was never the "sacred character" or the "divine writing," but an inadequate transcription of discourse. This definition of writing has been picked up by linguistics and has remained the base of western conceptions. Different conceptions exist in Semitic and oriental civilizations, but they are very seldom taken into account by westerners. The other reason is related to uses and customs, which prohibited a free Greek from sitting on the ground. In fact the Greeks, like the Assyrians before them, never sat on the ground but always on seats. The only persons represented sitting on the ground are young girls playing dice. Can one see Athena in her beautiful pleated dress seated on the ground in the scribe's position writing on a roll? Turner (1977) describes the Greek scribe as sitting in the position of the Egyptian scribe. And this was very probably the way that the Greek rolls were written by slaves, since physiologically (and in the absence of flat tables) it was impossible to write them in any other posture.

When the free Greek wrote, it was on tablets. In Greek and Latin societies, the upgrading of the position of the scribe, the promotion of this profession to the rank of a socially honorable activity, took place very slowly under the influence of the Christian religion, reflecting the Jewish influence. The evangelists, when they write, are always depicted sitting on seats and writing on a codex of papyrus or parchment, which took the place of the wooden codex. However, and this held right up to the Middle Ages, never were calligraphers or scribes at the top of the social scale as they were in the Egyptian, Assyrian, Babylonian, Jewish and later Arabic societies.

The alphabets derived from Greek, brought by Hellenistic culture, have conquered Europe and America. We continue to write from left to right, to write while sitting down on chairs, and to underrate the typists who have taken the place of the Greek scribes.

References

André-Leicknam, B., & Ziegler, C.L. (1982). *Naissance de l'écriture.* Paris: Editions de la Réunion des Musées Nationaux.

Anonymous (1958). *City invincible. A Symposium on Urbanization and Cultural Development in the Ancient Near East.* Chicago: Chicago University Press.

Baker, H.S. (1966). *Furniture in the ancient world.* London: The Connoisseur.

Beck, F.A.G. (1975). *Album of Greek education. The Greeks at school and at play,* Sydney: Cheiron Press.

Beck, F.A.G. (1978). *The schooling of girls in ancient Greece. Classicum IX,* 1–9.

Bennett, E.L. (1960). *Anonymous writers in Mycenean palaces. Archaeology, 13,* 26–32.

Callewaert, H. (1962). *Graphologie et physiologie de l'écriture.* Louvain-Paris: E. Nauwelaerts.

Černy, J. (1952). *Paper and books in ancient Egypt.* London: H.K. Lewis.

Chadwick, J. (1967). *The decipherment of linear B* (2nd ed.). Cambridge: Cambridge University Press.

Chiera, E. (1938). *They wrote on clay.* Chicago: Chicago University Press.

Christin, A.M. (Ed.) (1982). *Ecritures. Colloque international de l'Université Paris-VII,* 22–24 April 1980. Paris: Le Sycomore.

Christin, A.M. (Ed.) (1985). *Ecritures II.* Paris: Le Sycomore.

Cohen, M. (1958). *La grande invention de l'écriture.* Paris: Imprimerie Nationale.

Cohen, M., & Sainte Fare Garnot, J. (1963). *L'écriture et la psychologie des peuples.* Paris: Colin.

Demsky, A. (1976). *Literacy in Israel and among neighboring peoples in the biblical period* (in Hebrew). Unpublished doctoral dissertation, University of Jerusalem.

Diringer, D. (1968). *The alphabet: a key to the history of mankind.* London: Hutchinson.

Drioton, E. (1949). *La pédagogie aux temps des Pharaons.* In Revue des Conferences françaises en Orient, May 1949.

Driver, G.R. (1976). *Semitic writing* (3rd ed.). Oxford: Oxford University Press.

Durand, J.M. (1977). *Diffusion et pratiques des écritures cunéiformes au Proche-Orient ancien.* In *L'espace et la lettre* (pp. 13–59). Paris: Union Générale d'Editions. Cahiers Jussieu, 3. University of Paris VII.

Erman, A. (1966). *The ancient Egyptians:* a source-book of their writings. London: Methuen.

Erman, A., & Ranke, H. (1976). *La civilisation égyptienne.* Paris: Payot.

L'espace et la lettre, écritures, typographies. (1977). Paris: Union Générale d'Editions. Cahiers Jussieu, *3,* University of Paris VII.

Fisher, H.G. (1980). The orientation of hieroglyphs. Paris: Conference publique du Collège de France.

Gadd, C.J. (1936). *The stones of Assyria, the surviving remains of Assyrian sculpture, their recovery and their original position.* London: Chatto and Windus.

Gadd, C.J. (1956). *Teachers and students in the oldest schools.* London: University of London inaugural lectures.

Galling, K. (1971). *Tafel, Buch and Blatt.* In Goedike, H. (Ed.), *Near-Eastern studies in honor of W.F. Albright* (pp. 207–223). Goedike, Baltimore: Johns Hopkins Press.

Girard, P. (1891). *L'éducation athénienne aux Ve et IVe siècles avant J.-C.* (pp. 65–69). Paris: Hachette.

Greenspan, J.S. (1982). *Hebrew calligraphy, a step-by-step guide.* New York: Schocken.

Guillaumont, F. (1985). *Laeva prospera:* remarques sur la droite et la gauche dans la divination romaine. In *D'Heraklès à Poseidon, mythologie et protohistoire* (pp. 159–177). Geneva: Droz; Paris: Champion.

Hallo, W.W. (1980). *The expansion of cuneiform literature.* In Proceedings of the American Academy for Jewish Research Jubilee volume, Jerusalem, pp. 307–322.

Harvey, F.D. (1966). Literacy in the Athenian democracy, *Revue des Études Grecques, 79,* 585–633.

Havelock, E.A. (1976). *Origins of western literacy.* Toronto: Ontario Institute for Studies in Education.

Havelock, E.A. (1982). *The literate revolution in Greece and its cultural consequences.* Princeton: Princeton University Press.

Houston, M.G. (1972). *Ancient Egyptian, Mesopotamian and Persian costume.* London: Adam and Charles Black.

Immerwahr, H.R. (1964). *Book rolls on Attic vases studies.* In Henderson, C. (Ed.) Classical, mediaeval and renaissance studies in honor of B.L. Ullman (vol. 1, pp. 17–48). Rome: Edizione de Storia e Letteratura.

Immerwahr, H.R. (1973). *More book rolls on Attic vases, Antike Kunst, 16,* 143–147.

Jeffery, L.H. (1961). *The local scripts of Ancient Greece. A study of the origin of the Greek alphabet and its development from the eighth to the fifth centuries.* Oxford: Clarendon Press.

Jeffery, L.H. (1976). *Archaic Greece, the city-states c. 700–500* B.C. London: Methuen.

Kramer, S.N. (1949). Schooldays: A sumerian composition relating to the education of a scribe. *Journal of the American Oriental Society, 69,* 199–215.

Labat, R. (1963). *L'écriture cunéiforme et la litterature mésopotamienne.* In Cohen, M. Sainte Fare Garnot J (Eds.), *L'écriture et la psychologie des peuples* (pp. 73–92). Paris: Colin

Labat, R. (1976). *Manuel d'épigraphie akkadienne* (new ed.). Paris: Imprimerie Nationale.

Lemaire, A. (1981). *Les écoles et la formation de la Bible dans l'Ancien Israël,* Fribourg: Editions Universitaires

Lichteim, M. (1973). *Ancient Egyptian Literature. A book of readings.* Berkeley: University of California Press.

Lissarrague, F. (1985). *Paroles d'images: Remarques sur le fonctionnement de l'écriture dans l'imagerie attique. In Christin, A.M. (ed.) Ecritures II.* Paris: Le Sycomore.

Mallon, J. (1982). *De l'écriture.* Paris: Editions du Centre National de Recherche Scientifique.

Marrou, H.I. (1948). *Histoire de l'éducation dans l'antiquité.* Paris: Le Seuil.

Naveh, J. (1982). *Early history of the alphabet.* Jerusalem: Magnes Press.

Needham, R. (Ed.) (1973). *Right and left, essays on dual symbolic classification.* Chicago: Chicago University Press.

Oppenheim, A.L. (1964). *Ancient Mesopotamia, portrait of a dead civilisation.* Chicago: Chicago University Press.

Oppenheim, L.A. (1965). *A note on the scribes in Mesopotamia.* In *Studies in Honor of Bruno Landsberger* (pp. 253–256). Chicago: Chicago University Press.

Parassoglou, G.M. (1979). Some thoughts on the postures of the Ancient Greeks and Romans when writing on papyrus rolls. *Scrittura e Civiltà* III, 5–21.

Petrucci, A. (1984). Lire au Moyen Age. In *Mélanges de l'Ecole française de Rome, 96,* 603–616.

Powell, M.A. (1981). Three problems in the history of cuneiform writing: origins, direction of script, literacy. *Visible Language, 15,* 419–440.

Rainey, A.F. (1969). *The scribe at Ugarit, his position and influence.* In *Proceedings of the Israel Academy of sciences and humanities, section of humanities* (vol. 3, pp. 126–147).

Sack, R.H. (1981). *The temple scribe in Chaldean Uruk. Visible Language, 15,* 407–418.

Saenger, P. (1982). Silent reading: its impact on late medieval script and society. In *Viator, 13,* 369–414.
Sainte Fare Garnot, J. (1963). Les hiéroglyphes, l'évolution des écritures égyptiennes. In *L'écriture et la psychologie des peuples.* (pp. 51–71). Paris: Colin
Schäffer, H. (1974). *Principles of Egyptian art.* Oxford: Clarendon Press.
Schubart, W. (1960). Das Buch bei den Griechen und Römern (3rd ed.). Leipzig: Kochler und Amelung.
Sirat, C. (with the collaboration of M. Dukan) (1976). *Ecriture et civilisations.* Paris: Centre National de la Recherche Scientifique.
Sirat, C. (1981 a). *L'examen des écritures.* Paris: Centre National de la Recherche Scientifique.
Sirat, C. (1981 b) (with L. Avrin: *Micrography as art). La lettre hébraïque et sa signification.* Paris: Centre National de la Recherche Scientifique.
Sirat, C. (1986a). Les moyens d'investigation scientifique et les manuscrits hébreux du Moyen Age. *Scriptorium, 40,* 278–296.
Sirat, C. (1986b). Les manuscrits en caractères hébraïques, réalité d'hier et histoire d'aujourd'hui. *Scrittura e Civiltà, 10,* 239–288.
Thompson, E.M. (1966). *A handbook of Greek and Latin palaeography (new ed.).* Chicago: Chicago University Press.
Turner, E.G. (1965). *Athenians learn to write. Platon, Protagoras 326 d. Bulletin of the Institute of Classical Studies, 12,* 67–69.
Turner, E.G. (1971). *Greek manuscripts of the ancient world.* Princeton: Princeton University Press.
Turner, E.G. (1977). *Athenian Books in the fifth century* B.C. (rev. ed.). London: University College Publications.
Turner, E.G. (1980). *Greek papyri, an introduction* (2nd ed.) Princeton: Princeton University Press.
Vanderpool, E. (1969). *Ostracism at Athens* (pp. 3–16) Cincinnati: University of Cincinnati Press.
Vandier, J. (1958). *Manuel d'archéologie égyptienne III. Les grandes époques, la statuaire.* Paris: Picard.
Ventris, M., Chadwick, J. (1955). *Documents in Mycenean Greek.* Cambridge: Cambridge University Press.
Vernus, P. (1982). Les jeux d'écriture. In André-Leicknam, B., & Ziegler, C. (Eds.) *Naissance de l'écriture* (pp. 130–133). Paris: Editions de la Réunion des Musées Nationaux.
Williams, R.J. (1972). Scribal training in ancient Egypt. *Journal of the American Oriental Society, 92,* 214–221.

CHAPTER 11

Psychology of Literacy: East and West

INSUP TAYLOR[1]

> Reading is the magic key that unlocks the door
> to the wonderland of stories and information.
> (Taylor & Taylor, *The Psychology of Reading*, 1983: 397)

Warning and Introduction

In this chapter, I discuss largely, though by no means exclusively, Far Eastern literacy. (The Far East will be henceforth referred to simply as "the East.") And I tend to minimize the relations among literacy, brain, cognition, and culture. Literacy is simply obtaining sound and meaning from printed language; as such, literacy by itself may not have a great influence on cognition and culture. Literacy may have greater influence on cognition and culture, if it serves as "the magic key that unlocks the door to the wonderland of stories and information." And, as I will argue, cortical activities are similar whether one reads in an Eastern or Western script.

In his informative and entertaining book entitled *Chinese Looking Glass*, Bloodworth (1967) observes:

> The ideograms are like a series of pictures, absorbed passively by the eye with all its Taoist spontaneity, whereas western alphabetical script draws the reader forward along rails of writing as a man is taken along by a closely-reasoned logical argument. Chinese, therefore, is not a language of ideas or of rational debate, some say (p. 193).

He goes on to say "But this argument can be taken too far." I will try to show that this argument does not hold much water. A logography is not as pictorial as has been claimed, nor is an alphabet completely logical.

The warning out of the way, I now get on with my chapter.

Before discussing literacy, we must learn about the writing systems and scripts of the world, some used in the East, some in the West, and some elsewhere. A variety of writing systems used today differ in appearance, number of symbols, complexity, and above all, the linguistic units coded. In other works I have described four writing systems and ten of their extant varieties or scripts (see Table 1), and then discussed how they are learned and used (Taylor, 1986; Taylor & Taylor, 1983). In this chapter, I will give only an overview of these scripts and their uses.

---

[1] McLuhan Program in Culture and Technology (and Scarborough Campus), University of Toronto, 39A Queens Park Cres., Toronto, Ontario, M5S 1A1, Canada.

Table 1. Variety of writing systems and scripts. (From Taylor, 1986; with the editors permission)

System	Script	Ling. Unit	No. of Symbols
Logography	Colored shape	Word	130?
	Blissymbol	Word	500?
	Chinese character	Morpheme	50000?
Syllabary	Cree-Eskimo	Syllable	44–48 (+ ?)
	Vai	Syllable	210 (+ ?)
	Japanese Kana	Syllable	46 (+ 60)
Alphabet	English	Phoneme	26
	Finnish	Phoneme	37 (+ 6)
	Hebrew	Consonant	22 (+ 9)
Alphabetic syllabary	Korean Hangul	Phoneme	24
		Syllable	2000?

Under "No. of Symbols," a question mark indicates that the number is unspecifiable. A logography codes meaning unit, while a syllabary, an alphabet, and an alphabetic syllabary code phonetic units. Historically, writing systems devleoped in this order.

I will consider eight themes in this chapter:

1. The characteristics of language dictate, to some degree, the type of script preferred by that language.
2. The way a script represents meaning and/or sound has a great influence on how word recognition in that script is learned.
3. However, a script is not the sole influence on reading achievement and literacy attainment; socioeconomic factors also play a role.
4. The process of comprehending prose passages is, largely but not entirely, similar across different language/scripts.
5. The effects of literacy on cognitive and linguistic functions (other than reading and writing) are limited to specific tasks.
6. Reading, whether in Western or Eastern scripts, involves a sequential and phonetic activity. As such, it is preferentially, but not exclusively, controlled by the left hemisphere of the cortex.
7. The East and the West differ in culture and also in writing systems, but the differences in the two domains are not necessarily mutually causative.
8. In modern times, the East has adopted much of Western culture, narrowing, though not eliminating, the differences between the two.

Logography: Chinese Characters

Each logograph represents a small meaningful unit, a word or morpheme. It may not be associated with any sound, or it may be associated with more than one sound. Consider one kind of logograph, the Arabic numeral 4. Its meaning remains constant, even to a deaf-mute who cannot read it aloud, or to speakers of diverse languages who sound it out variously as *four* (English), *quatre* (French), *yotsu* or *shi* (Japanese), and *net* or *sa* (Korean). In this sense, a logograph represents meaning directly and sound indirectly through its meaning.

For rudimentary "reading and writing," a handful of logographs are adequate. Two such examples are colored shapes for chimpanzees (Premack & Premack, 1972) and Blissymbols for cerebral-palsied children (Bliss, 1965; Kates, MacNaughton, & Silverman, 1978). For proper reading and writing, a full logography, a system of logographs, is necessary. The Chinese system is a prime example of a logography, and furthermore it appears to be the only full logography used in the modern world. It has been used continuously for over 4,000 years. Today it is the sole writing system in the People's Republic of China and Taiwan, the major system in Japan, and a supplementary one in South Korea.

Chinese Characters in Chinese, Japanese, and Korean

By way of preview, Fig. 1 shows the same sentence in Chinese, Japanese, and Korean. The word *Chinese* is identical in the three languages, though it is pronounced somewhat differently: /zhong guo ren/ in Mandarin, /chu goku jin/ in Japanese, and /chung guk in/ in Korean. Only Chinese has tone variations, between four and nine, depending on dialect. Note the tiny circles that mark the end of the sentences. These and other punctuation marks have been introduced in modern times, under the influence of Western writing.

As for the direction of writing and reading in these Eastern languages, by tradition sentences are written vertically, but in modern times, in order to accommodate Arabic numerals and alphabetic letters, they tend to be written horizontally. The horizontal direction is used also for word processing on a computer (see "Learning characters in China" below). Or, ordinary texts

a 我是中国人。

b 私は中国人だ/です。

c 나는 中国人이다/입니다.

d 나는 중국인이다/입니다.

Fig. 1. "I am a Chinese" in Chinese (**a**), Japanese (**b**), and Korean (**c, d**)

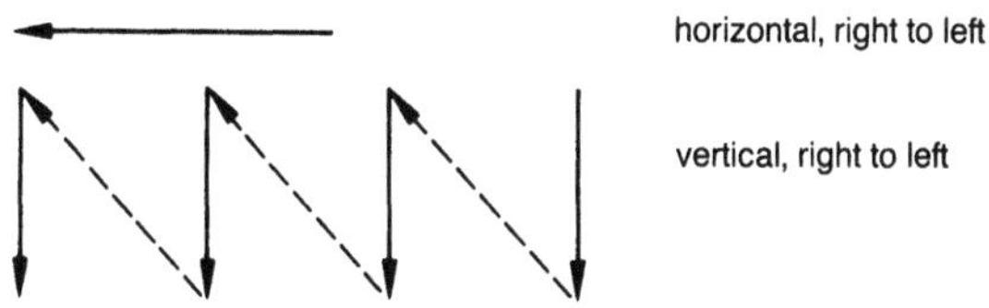

Fig. 2. Traditional writing/reading directions in the East

tend to be written in a vertical direction, and technical texts in a horizontal direction.

In the East, from early on horizontal writing was occasionally used in visual art and displays, dictated by the shape of available space. In doing so, the direction was right-to-left, in harmony with the vertical writing that moved up-to-down and then right-to-left, as shown in Fig. 2. However, in modern times, under the influence of Western writing, the direction in horizontal (even vertical) writing is left-to-right (occasionally, one finds shop signs in the reverse direction). In short, writing direction is not overly rigid.

In preparing Fig. 1, I first wrote the sentences vertically, but the slash (/) in the Japanese and Korean sentences looked awkward. Besides, in this horizontal text, vertical sentences would have taken up more space than horizontal ones.

In the logography and syllabary used in the three languages, both directions are practical because a content morpheme tends to require only one to three symbols, which is certainly within a horizontal span of vision, and possibly within a vertical one as well (a visual span is longer horizontally than vertically; Salapatek, 1968). By contrast, in alphabetic orthography, a content word tends to have a long array of letters and is more suitable for horizontal than vertical writing. Schoolchildren in the United States could read more accurately a horizontal than vertical typography, whether reading text presented normally or in a tachistoscope (that minimizes the effect of a reading habit) (Coleman & Hahn, 1966).

Speaking of the Western writing direction, Etiemble (1973) claimed that it results from a combination of chance and the inertia of custom. Speaking of the Eastern writing direction, I will add the factors of convenience and imitation.

Returning to Fig. 1, the Japanese and Korean sentences require postpositions that indicate case roles of nouns, and verb endings that inflect for tense, levels of politeness, and so on. The sentences end in present tense verbs in two alternate levels of politeness, a neutral level before the slash, and a polite level after it. The "grammatical morphemes" (inflectional endings, postpositions, function words, etc.) in Japanese and Korean tend to be written in phonetic scripts. Indeed, it is the necessity of expressing them that compelled Japanese to invent phonetic scripts. The Chinese sentence requires neither postpositions nor verb endings that inflect.

Here we see how three languages use the same script, similarly for some purposes, and dissimilarly for others. The different uses are necessiated by differences in languages: the Chinese language belongs to the Sino-Tibetan family, whereas Japanese and Korean belong to the Ural-Altaic family. Japanese and Korean are similar only in syntax, but not in speech sounds, and hence not in vocabularies. Speakers of the three languages cannot communicate to each other by speaking; they can communicate to a limited degree through writing in characters.

Learning Chinese Characters in China

Recognition of individual characters can be learned by the Three-Phased Learning method: whole → analysis into components → mature whole (see M.M. Taylor, this volume). Initially, a character as a whole visual pattern is associated with its sound (syllable) and meaning (morpheme). In the analysis phase, ideographic compounds and semantic-phonetic compounds are analyzed into their components. Each character can be further analyzed into an ordered set of strokes in writing. In the mature whole phase, the components come together again in wholistic perception, based on securer knowledge of the components and strokes. Whereas the initial global shape perception can differentiate only a small number of classes of characters, the mature global perception based on components can make much finer distinctions, permitting a large vocabulary to be recognized.

Because characters do not code sounds directly, some kinds of phonetic scripts are required for teaching the sounds of characters. Two kinds of auxiliary phonetic scripts are used today: in the People's Republic of China, *pinyin* (the letters of the Roman alphabet), and in Taiwan *zhuyin fuhao* [a set of 37 symbols that represent the initial consonant (C) and the final vowel (V) or VC]. These auxiliary phonetic scripts are learned before the characters, and then are used to annotate the characters to be learned. They can be used also in word processing (Becker, 1984).

Because character writing involves analysis of a pattern into an ordered sequence of strokes, as well as full recall of patterns and fine motor coordination, it requires much practice. Accordingly, character writing is regularly assigned as homework wherever characters are learned. In not too distant a future, schoolchildren may learn to write on a word processor, thus lightening their heavy homework load. For some Chinese children in North America, the use of word processors to learn to write is in fact taking place now. All the same, learning how to write characters should never be abandoned, for it helps in learning to recognize characters. It is part of the analysis phase in the Three-Phased Learning. In Liu's (1978) experiment in Taiwan, fourth-graders who learned characters in three aspects (sound, meaning, and writing) scored higher on tests of sound and meaning than those children who learned, in the same interval of time, only sound and meaning.

Learning to read in Chinese involves a heavy dose of rote memorization. This emphasis on memorization is a vestige of the long (over 1000 years) tradition in China, even up to the beginning of the twentieth century, where prospective mandarins studied for written examinations on Confucian classics. In Liu's (1984) survey in Taiwan, schoolchildren have to memorize every lesson in Chinese language textbooks. In junior high schools, pupils memorize every lesson written in the classic literary style as well. Such lessons comprise about 30% in grade 7, and as much as 60% in grade 9. In the People's Republic of China, too, a vast quantity of rote learning remains part of primary school syllabus, despite Mao Zedong's opposition to "stuffing students with memorized passages like Beijing ducks" (Unger, 1977). In spite of all these efforts that go into memorizing and homework, the reading achievements of schoolchildren in Taiwan are not necessarily higher than those in Japan and the United States (Stevenson, Stigler, Lucker, Lee, Hsu, & Kitamara, 1982).

Complex as is the Chinese system, it does not prevent a nation from attaining full literacy. In Taiwan, the illiteracy rate is a low 0.43%; it was 20% in 1950 (Ministry of Education, cited by Liu, 1979, personal communication). In the People's Republic of China, the illiteracy rate was 23.5% at the last mammoth census of 1982 (reported in *The Globe and Mail,* 1982, October 23); it is estimated to have been 80% before the birth of the Republic of China. In the 1930s only 3000000 newspapers were sold daily, to be read by perhaps one person in 25 (Bloodworth, 1967, p. 192). Obviously, raising a literacy rate depends a great deal on improvement of political, social, and economic conditions.

Kanji in Japan and Korea

Japan has true phonetic scripts, called Kana (see below), which are used to annotate the sounds of Kanji to be learned. The sounds of Kanji are harder to learn in Japan than in either China or Korea, because the majority of Kanji have at least two different sounds, one Chinese-derived and one Japanese-native. Chinese-derived sounds approximate the Chinese sounds of Kanji, whereas Japanese-native sounds are none other than Japanese words for the concepts expressed in Kanji. And many Kanji have a few variants of Chinese sounds as well as of Japanese sounds.

Take the three Kanji that together make up the word "Chinese" in Fig. 1. They are pronounced in the Chinese way as /chu goku jin/. In variants of the Chinese sounds, /jin/ is pronounced as /nin/, and /goku/ as /kok/ in words such as /nin pu/ ("man worker" = "coolie") and /kok ka/ ("national anthem"). Then, each of the three Kanji for "middle kingdom man" has Japanese sounds, /naka, kuni, hito/. /Hito/ itself becomes /bito/ in /hito bito/ ("man man" = people). To complicate the picture further, a reader is not always sure when a Kanji is to be read in a Chinese or a Japanese way. When a word is made up of more than one Kanji, each may sometimes by read by a different method.

Learning 2000–3000 Kanji, each of which has multiple sounds, takes time and effort. And mastery is seldom complete: even highly educated people may not know unusual sounds of some Kanji, or cannot fully recall infrequent Kanji for writing. Word processing on a computer is similar to that described for Chinese, except that grammatical morphemes in Japanese may constrain the number of content morpheme alternatives (Becker, 1984; see above, "Learning Chinese Characters in China").

However, learning by whole-pattern a few hundreds of Kanji for simple concepts is easy, especially if the sounds are bypassed. Even language-handicapped toddlers and preschoolers can learn them (Steinberg, Harada, Tashiro, & Yamada, 1982). In 10 months, three normal toddlers tutored by their mothers under Steinberg's guidance, learned 300–500 Kanji and Kana words (Steinberg, Yoshida, & Yagi, 1985). Some preschoolers "pick up" up to 170 Kanji before entering school (Muraishi & Amano, 1972; see "Kana" below).

In South Korea, Kanji learning is relatively painless, because the official 1800 Kanji, are sufficient, and because each Kanji has only one Chinese-derived sound. The official Kanji are taught in middle and high schools, but not in primary schools. Even official Kanji are not used in government publications and books for children; they appear in scholarly publications and the political and economics sections of newspapers.

In Japan and Korea, there have been sporadic movements to reduce drastically, or even abolish, Kanji. Kanji are in fact abolished in North Korea. Actually abolishing Kanji may not be a good idea. The limited number of Kanji is not too difficult to learn; they also serve useful purposes, such as disambiguating homophonic Chinese loan words or making key content words stand out in texts (see Fig. 1, and also "Comprehension subskills" below). They also provide "the magic key" to the written treasures of the past, and serve as a means of communication among speakers of three different languages.

Syllabaries

In a syllabary, one symbol represents one syllable. Languages differ enormously in the number of syllables used, from as few as 100 (e.g., Japanese) to as many as a few thousand (e.g., English and Korean). A pure syllabary is feasible only for a language with a limited number of syllables such as Japanese. Since the number of symbols required is relatively small, the shapes of the symbols need not be complex.

Of the several varieties of syllabaries used today, I will discuss five: Cree and Eskimo briefly, and Vai and two Japanese Kana at some length. A sample of CV symbols in these syllabaries is shown in Table 2 in Vowel (row) x Consonant (column) matrix or chart.

Table 2. CV symbols in five syllabaries

Script	C	Vowel /a/	/e/	/i/	/o/
Cree		b	q	p	d
Eskimo		q	p	b	d
Vai	/k/	[illegible]	[illegible]	[illegible]	[illegible]
Katakana		カ	ケ	キ	コ
Hiraqana		か	け	き	こ

For the Vai syllabary, see Scribner & Cole (1981, Fig. 3.2); for Cree and Eskimo, see Jensen (1970, Figs. 203, 206); for Kana, see Taylor & Taylor (1983, Chap. 4).

Cree and Eskimo Syllabaries

The Cree and Eskimo syllabaries were invented in the 19th century by Christian missionaries. Each CV symbol can be analyzed into C (coded by the shape) and V (by the orientation of the shape). Learning simple shapes distinguished by left-right or up-down orientation may appear easy, but it actually confuses young children (Gibson, Gibson, Pick, & Osser, 1962; Tanaka & Yasufuku, 1975). In reproducing 12 Cree CV symbols (Taylor & Taylor, 1983, Table 2-2), I myself made an orientation confusion error by listing for /ka, ke, ki, ko/ "p q p d," which should have been (I believe) "p q p d." Oops, wrong again! They should be "b q p d;" I checked and rechecked Jensen's (1970) Fig. 203. It is even more confusing when the symbols for /na ne ni no/ are the same "b q p d" but lying on their sides!

Vai Syllabary

The Vai people in Liberia, Africa, have a syllabary that evolved from a picture-word writing. The invention of the syllabary in the early part of the 19th century is credited to a native Vai (Gelb, 1963; Jensen, 1970). As can be seen in Table 2, each symbol as a whole pattern codes a CV syllable; that is, each symbol is not analyzable into C and V components, as in the Cree-Eskimo syllabary. With a few exceptions, symbols for related sounds are not similarly shaped.

Apparently Vai speakers require 2–3 months of lessons to achieve some functional literacy in the script (Scribner & Cole, 1981). In spite of the simplicity of the syllabary, literacy in the Vai script is low, and is confined largely to two-thirds of male adults. The Vai script is not the tool of formal education; nor is it the tool of obtaining knowledge from printed texts. In other words, Vai-script literacy is not "the magic key that unlocks the door to the wonderland of stories and information." Instead, it is used for such mundane activities as letter writing and record keeping. Vai-script literates tend to be farmers who also do crafts or grow cash crops. Literacy is perhaps a unnecessary luxury for most Vai people who are subsistence farmers. (See also "Literacy, Culture, and Cognition" below).

Japanese Kana

The Japanese created two syllabaries, Katakana and Hiragana, between the eighth and twelfth century A.D. Each Kana set has about 100 symbols. The two differ slightly in the shapes of symbols (see Table 2), and also in function (Katakana is used to write European loan words and Hiragana, grammatical morphemes). Otherwise, there is one-to-one correspondence between the symbols of the two Kana sets.

Kana is so easy that many children learn it, or rather "pick it up," before entering school. In a large survey of preschoolers' reading activities, a month before entering school 88% of the children could read 60 or more Hiragana (as well as 8–168 Kanji) (Muraishi & Amano, 1972). A few factors other than the writing system also mattered: gender (in favor of the girls over the boys); the fathers' education (in favor of the higher over lower education); the years spent at a nursery school (in favor of 2–3 years over 1 year). Concerning the last factor, the children are not taught reading at the nursery schools; rather, they are exposed to printed words and sentences that identify the owners of objects and give simple instructions. The children's questions about the printed materials are also answered. The moral: When printed materials are used to convey vital information, preschoolers are motivated to learn them.

Japan boasts one of the highest literacy rates in the world, 99%, with illiteracy confined largely to the mentally retarded (Marshall, 1986; Sakamoto & Makita, 1973). One factor responsible for this salutary situation could be initiation into reading with the simple Kana; another situation could be the favorable socioeconomic conditions.

Alphabet

Alphabet: Its Nature and Variety

In an alphabet, each letter codes one phoneme, in principle but not necessarily in practice. Since most languages of the world use between 20 and 37 phonemes (Maddieson, 1984), an alphabet needs only a small number of letters. These letters can therefore be shaped simply.

Today the alphabet has all but conquered the world: It is used in almost all Indo-European languages as well as in such non-Indo-European languages as Hebrew, Finnish, Turkish, and Vietnamese. The widespread adoption of an alphabet does not necessarily attest to its superiority over other writing systems; it simply attests to the fact that other systems are impractical for most languages. A logography is impractical because of its large quantity of symbols and because of its unsuitability in representing grammatical morphemes that inflect. A syllabary is impractical to most languages as they have over 1000 syllable types. Thus most languages have no choice but adopt an alphabet.

Learning to Read in an Alphabet

Learning an alphabet itself is easy, because it has a small number of simply shaped letters. Many preschoolers learn it before entering school. Learning to read in an alphabet is not necessarily easy, for several reasons.

In some alphabets, such as Hebrew and Arabic, Vs are indicated not by proper letters but by optional marks over, under, or inside Cs. Because of the "invisibility" of the tiny marks "tucked under" C letters, beginning Hebrew readers make many errors in sounding out Vs (Feitelson, 1980). (For adults, however, the presence or absence of the V marks does not affect the latency for pronunciation; Navon & Shimron, 1981.)

The phonetic unit coded in an alphabet, the phoneme, is abstract, small, and unstable. Stop Cs by themselves are unpronounceable, and most phonemes change their sound values in different phonetic contexts. As part of the phonics method of reading instruction, phonemes must be put together into syllables and words by a process called phonic blending. Preschoolers find it easier to blend syllables into words (e.g., *can.dy; tea.cher*) than phonemes (e.g., *p.a.y; u.p*) (Brown, 1971; Coleman, 1970). Finally, several meaningless phonemes must be strung together before the meaning of a word emerges.

Understandably, the phoneme is not the unit people most easily become aware of. Historically, the alphabet was the last writing system to emerge, after logography and then syllabary (Jensen, 1970). Gelb (1963, p. 203) points put: "Almost all the writing introduced among primitive societies in modern times stopped at the syllabic stage." Children find it more difficult to segment a word into phonemes than into syllables (Liberman, Shankweiler, Fischer, & Carter, 1974).

A single word in an alphabet tends to require a long array of letters, longer than a word in a syllabary or logography. For example, *gentleman* requires nine letters in English, six Kana in Japanese, three syllable blocks in Korean Hangul (see below), and two Kanji (in translation). The longer an array, the more material must be processed visually, posing a problem in sequencing and grouping letters and sounds for word identification. People often have trouble seeing and remembering the order of items in a list, especially in the middle. In English, transposition errors involving letters in medial positions, such as"-er-" and "-re-" in *there* and *three*, are intractable among grade 2 poor readers (Park, 1978–1979).

For an alphabet to be a useful representation of speech, its letters have to code sounds consistently and accurately. But preserving close letter-sound correspondence is difficult in the face of sound changes that can occur over time, over geographic regions, and in word derivation. For example, for *heal* and *health*, if the latter is spelled "helth" to represent its sound more closely, the relatedness of the two words becomes less transparent. The same observation applies to *train*, which is pronounced as "trine" (rhymes with *wine*) in Australia, and as "trane" (rhymes with *rain*) in most other English-speaking countries.

The ideal is to have perfect letter-sound correspondence in a language whose sounds do not change unless root meaning also changes.

Learning English Orthography

Partly because of the irregularity and complexity of English orthography, "early readers" (children who read before going to school) are few, on the one hand, and "non-readers" (schoolschildren who cannot decode words) are many, on the other. Only 1%–3.5% of children can read at a grade 2 level when they enter school at age 6 (Durkin, 1966 a, b). (In Finland, where orthography is regular, 15% of children can read at a grade 2 level when they enter school at age 7; Kyostio, 1980.) Once in school, up to grade 3–4 most children are learning to decode words (Calfee, Venezky, & Chapman, 1969). Even at grade 5, children cannot correctly pronounce some less frequent irregular words that are within their listening vocabulary (Adams & Huggins, 1985).

As for nonreaders, one study reports that one-fifth of all grade 3 children tested in three schools in a racially mixed, urban inner city were "total nonreaders" (Gottesman, Croen, & Rotkin, 1982). Another study reports high rates of nonreaders among black working-class children in Philadelphia (Baron & Treiman, 1980). The nonreaders could not decode any of the test words, which were familiar "exception" or irregular words (*was, said, come*), regular words (*has, maid, dome*), and pseudowords (*mas, haid, gome*).

Irregular words should be difficult wherever English is learned. But a high rate of nonreaders perhaps tends to occur among severely disadvantaged children. All the studies on English orthography cited above have been carried out in the United States.

According to the U.S. Department of Education, 13% of 3400 Americans age 20 or over flunked a basic literacy test given by the Bureau of the Census. An additional 20% of those offered the test refused to take it presumably for fear of revealing their illiteracy. The educators fault the American school system with its dull word-drill work-books and force-feeding of new words by the "look-and-say" method (reported in *Time*, May 5, 1986). The illiteracy rate may differ from one English-speaking nation to another, since it is affected not only by writing system but also by socioeconomic factors and teaching methods.

Alphabetic Syllabary

Hangul Syllable Blocks

In an alphabetic syllabary, each symbol codes a phoneme, as in an alphabet, but between two and four alphabetic symbols are packaged into a syllable block, which is the unit of printing and reading. The Korean script called Hangul appears to be the only alphabetic syllabary. (To be precise, the second vowel in "Hangul" is 'u,' which is a high central vowel with lips spread; it sounds like a vowel between /i/ and /u/). Unlike in the five syllabaries shown in Table 2, in the Korean alphabetic syllabary, the two margins of the chart can show Hangul alphabetic symbols, Cs vertically and Vs horizontally. (Some of the V and C symbols are themselves combinations of two or more simple symbols.) Each of the 19 C symbols can combine with each of the 21 V symbols to produce 399 (19 × 21) CV syllable blocks. Then, underneath each of these CV blocks the final C can be placed to produce 7581 (399 ×19) CVC blocks, not all of which are actually used. Underneath a handful of CVs, a second C can be placed to produce CVCC blocks.

Table 3 shows four alphabetic symbols that are packaged into syllable blocks of three levels of phonetic complexity (CV, CVC, CVCC), with com-

Table 3. Hangul Syllable-Blocks in Three Complex Levels (I. Taylor, 1980, p. 170)

Level	Alphabet Symbol	Syllable Block	Syllable structure	Morpheme
I	ㅏ	아	V /a/	suffix; ah
I	ㄷ ㅏ	다	CV /da/	all
II	ㅏ ㄹ	알	VC /al/	eqq
II	ㄷ ㅏ ㄹ	달	CVC /dal/	moon
III	ㄷ ㅏ ㄹ ㄱ	닭	CVCC /dalg/	hen

mensurate levels of visual complexity. With increasing levels of complexity, the likelihood of one syllable block by itself unambiguously representing a morpheme increases.

The three levels of complexity helps word recognition. In an experiment, a target syllable block was recognized faster against the background of other blocks in all three complexity levels than in only the complexity level of the target (I. Taylor, 1980).

Learning and Using Hangul

Hangul is easy to learn for reasons such as:

- Symbol-sound correspondence is high;
- The reading unit is a concrete and stable syllable;
- Although the number of syllable blocks is large, each syllable block need not be learned by rote; it can be deduced from the CV chart.

Because of the perfectly systematic placement of two to four alphabetic symbols in a syllable block, word processing on a computer is easy and fast (Chong, Han, & Kang, 1983). A writer types words in alphabetic symbols, and a computer packages them into syllable blocks.

Korea enjoys a high rate of literacy, with illiteracy confined to the mentally retarded or to the very old who have not benefited from compulsory education. The simplicity of Hangul, as well as favorable socioeconomic conditions, must be responsible for this welcome state of affairs.

Comprehension in Different Writing Systems

Whatever the writing system the reader's goal is the same: to comprehend (obtain the meaning conveyed in the text), and to retain at least the gist of what has been comprehended. Comprehension skill can be analyzed into such subskills as recognizing words and organizing them into larger syntactic units. Intuitively, comprehension subskills should be similar across different language/scripts, though some subskills may receive more (or less) emphasis in different language/scripts. Since the subskills that are similar across different language/scripts have been discussed in Taylor and Taylor (1983), here I will single out one subskill that may not be the same in different scripts: "Construct a message based on content words, with the help of function words if necessary."

Processing Content and Function Words: English

The English language has content words and function words. A portion, about 60, of the function words can be considered "prototypical" in that they possess all the following seven characteristics (Taylor & Taylor, 1984, unpublished paper):

- Have syntactic roles
- Form a closed set
- Occur frequently
- Are short
- Have little semantic content
- Are not used alone as a complete utterance
- Are unstressed in normal speech

The prototypical function words are: the articles (*the, a*); ten (out of over 60) short and frequent prepositions (*of, . . . at*); the copula *be* and its inflected forms; the auxiliary verbs *do* and *have* and their inflected forms; possesive and objective pronouns (*his, . . . me*); conjunctions (*and, . . . but*); *to (come)* before a verb; the filler *there (is)* and demonstratives (*this, . . . those*).

In reading a passage silently, a skilled reader's eyes tend to skip over prototypical function words and fixate on content words. When they fixate on function words, they do so only for short durations. In the following sentence, which was part of a paragraph, the numbers over the words are the total time the word was fixated in milliseconds (Just & Carpenter, 1980, p. 330).

1566 267 400 80 267 617 767 450 450
Flywheels are one of the oldest mechanical devices known to man.

Read the following "telegraphic sentence" without function words, and see whether you have any trouble comprehending the sentence.

One carrying corn allowed go, one with load gold held.

In this sentence the 11 original content words have been retained while the 12 original function words have been deleted, preserving the word order. The original sentence is shown below; underneath each word is the number of subjects (out of 17) who deleted it in an experiment in which they were asked to delete words that were not essential for comprehension and that could be restored by other readers (Taylor & Taylor, 1984).

The	one	that	was	carrying	the	corn	was	allowed	to	go,
14	1	16	17	1	17	0	13	0	9	0

but	the	one	with	the	load	of	gold	was	held.
9	16	1	3	17	8	13	0	12	0

Processing Grammatical Morphemes in Nonalphabetic Scripts

Differential processing of grammatical morphemes and content words is facilitated in languages such as Japanese and Korean in which grammatical morphemes tend to be written in phonetic scripts while content words are written in Kanji (see Fig. 1). In all-Hangul texts, postpositions and verb endings come at the end of a phrase or sentence just before a space, and hence are distinguishable from content words.

Differential processing of content words and grammatical morphemes does not appear easy with texts in Vai scripts (Scribner & Cole, 1981, Fig. 9.1). The boundaries of words and phrases are not marked with extra spaces, and two types of words are written in the same script.

Differential processing is difficult with Chinese text too. In the Chinese language a handful of particles and affixes indicate completed action, question, and so on, but they do not inflect and are written in one character, just as are some content morphemes (Chao, 1968; Kratochvil, 1968). In other words, in text grammatical morphemes do not look different from content morphemes, and hence the strategy of differential processing cannot be easily applied. Perhaps partly for this reason, and partly for the high density of content morphemes, in an eye-movement study Chinese readers averaged ten saccades per line (frequent fixations), compared with English readers who averaged four saccades (Stern, 1978). Differential processing of the two kinds of morphemes may be possible if the characters for common grammatical morphemes are simplified more drastically than are content morphemes.

Types of scripts matter in text comprehension, though perhaps not as much as in word recognition.

Literacy and Cortex

The human cortex contains a left hemisphere (LH) and a right hemisphere (RH) connected by the corpus callosum. The two have different but complementary processing modes.

Processing Models of the Two Hemispheres

The modes and linguistic stimuli preferentially processed by the LH and the RH are summarized in Table 4. The linguistic materials are dichotomized because they require the specialized processing modes given in the upper panel. Table 4 is prepared by reviewing an extensive literature on physio-

Table 4. Specialization of the LH and the RH

LH	RH
Processing mode	
Sequential	Simultaneous
Analytic	Wholistic
Verbal	Imagistic
Logical	Intuitive
Linguistic material	
Syntax	Receptive vocabulary
Speech output	Prosody and gesture
Phonetic symbol	Single logograph
Literal comprehension	Pragmatic, contextual

logical and behavioral measures of people with intact brains, people with damage in either LH or RH, people with either LH or RH removed, and people with split brains (the connection between the LH and RH is severed).

When a single word is presented briefly in a tachistoscope, it tends to be perceived more accurately and/or speedily by one or the other hemisphere (see Jones & Aoki, this volume, for details of experimental procedures). Are a logograph and a word in phonetic script perceived differently by the two hemispheres? To answer the question, numerous experiments have been carried out on symbols and words of English, Kanji, Kana, and Hangul.

Which Hemisphere for Eastern Scripts?

A single character (for Chinese subjects) or Kanji (Japanese subjects) is perceived more accurately and/or speedily in the left half-visual field (LVF), which projects to the RH, than in the right half-visual field (RVF), which projects to the LH (Hatta, 1978; Tzeng, Hung, & Cotton, 1979). In the same two studies, a pair of vertically arranged characters or Kanji forming a word was processed better by the LH than by the RH.

Jones and Aoki (this volume), noting that "phonological elements" or phonetic components of semantic-phonetic compounds tend to be on the left side of Kanji, suggest that the processing of phonetic components is easier in the RVF (LH) while the reverse holds for the semantic components. Contrary to their suggestion, single characters or Kanji, whether they contain phonetic components or not, are processed better by the RH (Tzeng, et al., 1979).

Single characters or Kanji, regardless of their inner structure and complexity, appear to be processed as wholistic visual patterns by the RH. But two or more characters or Kanji making up a word involves sequential processing of the LH. Even single Kanji are processed by the LH, however, if the subject's task involves purely phonetic processing or "deep" semantic (and perhaps relational) processing (Hatta, 1980; Hayashi & Hatta, 1982; Sasanuma, Itoh, Kobayashi, & Mori, 1980).

A single word in phonetic script tends to be recognized more accurately and/or speedily by the left hemisphere, whether in English (Mishkin & Forgays, 1952), Kana (Hatta, 1978) or Hangul (Endo, Shimizu, & Nakamura, 1981 a, b). Most of Hatta's noun stimuli contained two Kana, and being transliteration of Kanji words, were in an uncustomary script. But Endo et al.'s noun stimuli were single Hangul syllable blocks and in a customary script, and hence were more directly comparable to single Kanji stimuli.

In Japan, some brain-damaged patients lose the ability to process phonetic Kana while retaining the ability to process logographic Kanji; other patients show the reversed pattern. Impaired Kana processing, both reading and writing, seems to be associated with lesions in the left temporal area, and impaired Kanji processing with lesions in the left parietooccipital area (Paradis, Hagiwara, & Hildebrandt, 1985; Sasanuma, 1980). This selective impairment of Kanji and Kana should be seen as an example of more general dissociation of semantic processing from phonetic processing, for similar dissociation patterns can be found in patients who learned to read in alphabetic scripts (see below).

A word often consists of two to three Kanji, and a phrase in Japanese and Korean almost always contains grammatical morphemes written in a phonetic script. And a sentence or text is a sequence of these words and phrases. In the process of comprehending a sentence, readers store its words in working memory and integrate the words into larger units unitl they obtain the meaning of the sentence as a whole. Such storing and integrating seem to be done in working memory in a phonetic code whether in English alphabetic orthography (Baddeley & Lewis, 1981) or in Chinese logographic characters (Tzeng, Hung, & Wang, 1977). In both writing systems, a sentence consisting of words with similar sounds is more difficult to process than is a sentence consisting of words with dissimilar sounds.

To conclude, reading proper (as opposed to recognizing single Kanji) involves a sequential and phonetic process, which has been shown to be the province of the left hemisphere. Kanji writing involves analysis of individual Kanji into its components as well as into a sequence of strokes. It also involves a sequence of kinesthetic movements. Kanji writing, if studied, may show involvement of LH processing. For all these reasons, processing of both Kana and Kanji is impaired when damage occurs in the LH. As we shall see below, reading, whether in the East or West, may involve also RH processing.

Which Hemisphere for Western Scripts?

The RH was once believed to be "word-deaf" and "word-blind." There is mounting evidence that the RH participates in language understanding, if not in production.

In order better to study the linguistic capability of the RH, Zaidel (1978a, b), devised a special contact lens to confine the stimuli to either the right or the left visual field for prolonged periods. He studied two split-brain patients (L.B. and N.G.) and one patient who underwent LH hemispherectomy. The RH of these patients has neither phonetic coding nor letter-to-sound conversion rules, and hence cannot compose speech. The RH has poor short-term memory and has trouble relating as many as four content words, such as *happy little girl jumping;* it also has only rudimentary syntax. What the RH does have is a visual lexicon whose words are recognized as visual gestalts, or an auditory lexicon whose words are recognized as auditory gestalts. The RH recognizes an object through template matching. The RH of one patient could "evoke the sound image" of a word to the extent of matching two pictures with homonymous names (e.g., *male-mail*) without being able to name either one. He read "ideographically," that is, recognized words directly as visual gestalts without intermediate phonetic coding (Zaidel & Peters, 1981).

One might wonder whether the patients with split brains or hemispherectomy are prone to bilateral language representation because of their long-standing abnormality. According to Gazzaniga (1983, p. 532), "[RH language] . . . in almost every case can be attributed to brain pathology occurring prior to commissural section." He asserts that the normal RH is nonlinguistic. But Zaidel (1983) disagrees, pointing out that there is little reason to believe that the patients L.B. and N.G. have the kind of LH lesion that would result in RH takeover of language function.

Let us look at the language functions of RH-damaged patients without long-standing abnormalities. They do score poorer than normal controls in tasks on auditory and visual semantic, though not on phonemic, discrimination (Gainotti, Caltagirone, Micelli, & Masullo, 1981). One right-handed, RH-damaged patient cannot address his lexicon directly from the global form of a word but has to use a phonetic route. The necessity to sound out words before identifying them results in abnormally slow word naming that increases with word length (Ogden, 1984). RH damage may also impair the affective component of productive language that is modulated through prosody and gesturing (Ross, 1983).

While preserving basic linguistic abilities, RH-damaged patients show inadequacies in subtle linguistic-pragmatic functions. They have problems recognizing what parts of stories go together, determining whether items in a story are plausible in context, and understanding jokes. They tend to accept literal readings of metaphoric statements without finding them funny. In telling a story, they may fail to recognize it as fiction, and may inject themselves into the plot or argue with the story's premises (e.g., Brownell, Michel, Powelson, & Gardner, 1983). In short, the RH provides a pragmatic framework within which the literal understanding of the LH is placed.

Those who lost phonetic ability called "phonemic dyslexics" cannot pronounce pseudowords such as DAKE, and read concrete and picturable words better than abstract ones (e.g., Coltheart, Patterson, & Marshall, 1980). Japanese phonemic dyslexics have trouble reading Kana aloud, while retaining their ability to obtain meaning from Kanji. By contrast, surface dyslexics can decode pseudowords such as DAKE but have trouble decoding irregularly spelled common words such as *yacht*. Japanese surface dyslexics would have trouble with Kanji, while retaining Kana processing ability.

According to Marin (1980), phonemic dyslexia is caused by an extensive lesion in the left frontal, temporal, and parietal lobes, as well as the angular gyrus and the subcortical structure (the areas served by the left middle cerebral artery). Surface dyslexia may be caused by a diffuse loss of tissue outside the language areas (Deloche, Andreewsky, & Desi, 1982; Sasanuma, 1975; Warrington, 1975). Note that the two contrasting types of acquired dyslexia are both associated with lesions in the LH, though perhaps in different areas or paths within it.

Now let us consider normal people with intact brains. English words are processed by the LH, as demonstrated in the seminal experiment by Mishkin and Forgays (1952). Since then, researchers have identified several conditions that can alter, even reverse, LH processing. Consider the following two stimuli, FISH vs fIsH: the same four-letter English-monosyllabic word is written in all capital in the former, and in case alternation in the latter. Evidence of LH processing was found stronger in the case-alternating word, in which wholistic processing is made difficult (Tzeng & Hung, 1980). In a lexical decision task, even at a brief exposure (20 ms) during which identification of three-letter English words was impossible, subjects' decisions were better than chance, and better in the LVF (RH). As exposure duration increased, subjects were better at identifying the words in the RVF (LH) (Bradshaw, Hicks, & Ross, 1979). The subjects may have used fast template matching in short exposure, and slower analytic processing in longer exposure.

The cortical activities can be monitored on-line while a normal person is reading text. In reading English text, EEG signals were stronger from the LH if the text was technical, and from the RH if it was a story high on imagery (Ornstein, Herron, Johnstone, & Swencionis, 1979). In subjects reading sentences that might end in a semantically anomalous word, EEG signals were stronger in the LH for the first few words that determine the syntactic structure of the sentence, but stronger from the RH for the semantically anomalous last word (Kutas & Hillyard, 1982). The LH tends to deal with functional relationships, the RH with pictorial and semantic ones.

Measurements of blood flow to the brain, using radioactive xenon, show that silent reading activates four areas in the cortex in addition to the primary visual area (which is not reached by this technique): the visual association area, the frontal eye field, the supplementary motor area, and Broca's speech center (Lassen, Ingvar, & Skinhoj, 1978). Reading aloud activates two more centers: the mouth area and the auditory area. Presumably, the auditory area is activated by hearing the speech, and the mouth area by producing it. Both hemispheres show the same pattern of activity.

Corresponding regions of the two hemispheres tend to change flow rates simultaneously during rest state and also during simple sensory and motor activities. When tasks become more complex, involving higher level mental activity, flow rates increase less symmetrically; there is a greater increase in the LH with linguistic tasks, and in the RH with visuospatial tasks (Prohovnik, Hakansson, & Risberg, 1980).

To conclude, reading, whether in Western or Eastern scripts, involves a sequential and phonetic activity. As such, it is preferentially controlled by the LH. At the same time, some reading materials, whether in Eastern or Western scripts, call for wholistic, imagistic, and pragmatic processing; as such, they involve also RH processing. It is possible that the relative contribution of LH and RH processing might differ between Eastern and Western reading/writing. This possibility cannot be clarified conclusively with the available evidence. We need studies that monitor cortical activities during reading and writing in Eastern scripts.

Literacy, Culture, and Cognition

Since learned colleagues in this volume argue eloquently for all kinds of differences – cultural, cognitive, and cortical functions – attributable to the differences in writing systems, I will be the devil's advocate and argue for the opposite. The truth, as usual, may lie in between. Far be it from me to question cultural differences – their origin, nature, and function. Rather, I will ask more tractable questions, such as: How do cultural differences affect cognitive functions? How does literacy in a particular script affect cognition?

Cultural Differences in Cognition

High verbal skills are prized in Jewish but not so much in Chinese culture. Confucius, whose codes of conduct permeate all Chinese and Eastern countries, warned: "Talking easily leads one into trouble because when you talk, you use so many words, and it is easy to let them out of your mouth, but difficult to take them back" (Lin, 1938: 181).

Such cultural differences show up as cognitive differences in psychological tests. One study tested the mental abilities of first-graders in the United States from four different cultural backgrounds: Chinese, Jewish, black, and Puerto Rican (Lesser, Fifer, & Clark, 1965). The four mental abilities tested were: verbal, number, reasoning, and space conceptualization. Each cultural group showed a distinctive pattern of relative abilities regardless of social class, with Chinese and Jews scoring much higher than blacks and Puerto Ricans on reasoning, number facility, and space conceptualization. However, the Chinese group scored much lower than the Jews and lower than the blacks, on verbal ability. In a follow-up study of the same subjects 6 years later, Lesser (1976) found these cultural differences to be stable.

In Chinese education, learning of past cultural heritage, such as classic poetry and history, are valued more than originality. Reflecting this tradition, Chinese subjects in Taiwan, from grade 2 to college age, scored lower than American subjects of similar age on the Torrance test of creative thinking: idea fluency, originality, and flexibility (Liu & Hsu, 1974; Torrance, 1966). The scores for creative thinking of Taiwanese college students were lower, not only than those of the American students, but also than those of the Taiwanese senior high school students. Even on a nonverbal, figural Torrance test of creative thinking, Chinese college students scored lower than American students except on the fluency subtest (Paschal, Kuo, & Schurr, 1980).

Memorizing is a major part of language art syllabus in some cultures, such as Chinese, but not in others, such as American. Accordingly, verbal memory is indeed found to be higher than that reported in the American literature. For example, for paired-associate learning. English-speakers require four trials to attain 6–11 items correct, whereas Chinese speakers require only one trial for comparable results (Huang & Liu, 1978; Paivio, 1965).

In a cross-cultural comparison of the mathematical abilities of children (Kindergarten, grades 1 and 5), Taiwanese children in Taipei scored higher than Japanese children in Sendai, who in turn scored higher than American children in Minneapolis (Stevenson, Lee, & Stigler, 1986). The discrepancies in scores among the three national groups widened from Kindergarten to grade 5. The differences in performance may be attributable partly to the amount of practice, which decreases from Taiwan to Japan and is lower again in the United States, as shown in Table 5.

The pace of learning may be faster in the Eastern countries than in the United States in elementary and secondary schools, but not necessarily in universities, where original thinking and understanding count. In fact, the number of Japanese students who come to the United States for post-secondary education rose dramatically between the 1960s and 1983 from 2168 to 13610, a 527% increase! (Marshall, 1986).

In mathematics, Oriental children even in the United States score higher than whites, blacks, and Mexicans (Flaugher, 1971). The Confucian codes of

Table 5. Time spent in study. (Based on data given by Stevenson, et al., 1986)

Country	Academic activity (%)		Homework (min)		Days in school
	Grade 1	Grade 5	Grade 1	Grade 5	
Taiwan	85.1	91.5	77	114	240
Japan	79.2	87.4	37	57	240
US	69.8	64.5	14	46	178

conduct – respect for learning and the superior – may be practised in the homes of the Orientals even in the West.

In addition, dare one speculate genetic differences among different ethnic groups? According to Lynn (1982), over the course of one generation, the mean performance IQ (PIQ) in Japan has risen by 7 points. He doubts whether a rise of this magnitude could be accounted for by a change in the genetic structure of the population. He attributes the rise to such environmental improvements as health and nutrition. Today, the mean PIQ of children in Japan is 111, 10 points higher than that of children in the United States and other developed countries, such as Britain, France, and Germany (Lynn, 1982). It is hard to believe that the health and nutrition of children in Japan are so much better than those in other developed countries. Since invoking racial differences in genetics is a taboo, we may invoke cultural differences in the quantity and quality of intellectual stimulation given to children. One factor that is ruled out as a major influence is the writing system, since the language of mathematics is more or less universal.

Effects of Literacy on Cognition

Written language is thought to promote abstract concepts, analytic reasoning, new ways of categorizing, a logical approach to language, and so on (e.g., Goody & Watt, 1968; Havelock, 1978). As Scribner and Cole (1981) observed: "It is striking that the scholars who offer these claims for specific changes in psychological processes present no direct evidence that individuals in literate societies do, in fact, process information about the world differently from those in societies without literacy" (p. 7).

To resolve the controversy let us turn to research findings. A few lucky (?) chimps have been taught to "read and write" with colored plastic shapes. A large red shape (a silhouette of a chimp?) stands for Sarah the star pupil; a smaller yellow irregular shape stands for *give;* a still smaller blue rectangle stands for *apricot;* and the sequence of the three symbols reads "Sarah give apricot" (Premack & Premack, 1972). Once the chimpanzee has attained "literacy," it can solve certain kinds of problems that it cannot solve otherwise (Premack, 1984). Specifically, it can solve problems on a conceptual

Table 6. Types of literacy and cognition. (From Scribner & Cole, 1981, Fig. 14.1; with the publisher's permission)

Cognitive Task	Test	English-schooled	Vai	Qur'anic	Arabic
Categorizing	Form/number sort	×	×		×
Memory, recall	Incremental			×	×
	Free	×			
Logical reasoning	Syllogism	×			
Encoding/decoding	Rebus reading	×	×		
	Rebus writing	×	×		×
Semantic integration	Word	×	×	×	×
	Syllable		×		
Verbal explanation	Game instruction	×	×		
	Grammatical rule	×	×		
	Sort geometric figure	×			
	Logical syllogism	×			
	Name switch	×			

The type of literacy has an effect on cognitive/linguistic tasks in the cells marked with x's. For "semantic integration," see "Literacy and Linguistic Functioning". The last item "name switch" tests whether people agree that the sun could arbitrarily be renamed "moon," and vice versa.

rather than a sensory basis. For example, while the normal chimp can match, say, half an apple with half an apple, or three-fourths of a cylinder of water with three-fourths of a cylinder of water, it is only after the chimp has been language-trained that it can match, say three-fourths of an apple with three-fourths of a cylinder of water, that is, match equivalent proportions of objects that do not look alike. Through language training, the chimp learns that a symbol can stand for an object. This symbolizing function of language, rather than "literacy," must be responsible for the chimp's improved problem-solving ability.

Scribner and Cole (1981) have carried out an extensive study on the effect of literacy on cognition among the Vai of Liberia, Africa (see Table 2 for letter samples). They administered a variety of cognitive and linguistic tasks to four types of literates among Vai speakers: English-schooled, Vai script, Qur'anic (for reading the Koran without much understanding), and Arabic language/alphabet. Table 6 summarizes their results.

In interpreting the data, remember that literacy activities in three types of literacy – Vai, Qur'an, Arab – are restricted; they are associated with neither formal schooling nor text reading. Only English is associated with years of formal schooling; it is also the official language, the language of commerce and government. In this language, reading materials such as newspapers, magazines and books are available.

First, formal schooling with instruction in English increased the Vai people's ability to provide a verbal explanation of the principles involved in performing the various tasks. Schooled individuals, more than did nonschooled ones, gave task-oriented and informative justifications; they more often made use of class and attribute names. Speaking English never substituted for school variables, and on verbalization measures, school, not English reading scores, was the best predictor (Scribner & Cole, 1981, pp. 130–131).

Second, the effect of nonschooled literacy is found on those tasks that the researchers adapted to mirror specific literacy practices. For example, Vai-script literates consistently outperformed other types of literates when talking about good or bad sentences. It turns out that letter writing and instruction in Vai puts much emphasis on "correct Vai." Qur'an literacy has an effect on memory, in particular incremental recall. Thirdly, on no tasks did all nonliterates perform at lower levels than all literates.

To conclude, literacy per se, unaccompanied by formal education or text reading, has only a minor influence on cognition. Any effect of a particular script is confined to specific tasks emphasized in learning and using that script.

Literacy and Linguistic Functioning

Literacy may enable readers to handle text or speech in units coded by that script. Such segmenting ability is considered part of "metalinguistic knowledge" essential for reading. Liberman and Mann (1981) assert:

> First, he [the beginning reader] must realize that spoken words consist of a series of separate phonemes. Second, he must understand how many phonemes the words in his lexicon contain and the order in which these phonemes occur. Without this awareness, he will find it hard to see what reading is all about. (p. 126)

Is segmenting ability a prerequisite to learning to read, as claimed by Liberman and Mann, or is it a by-product of learning to read, as Taylor and Taylor (1983; Chap. 14) and Valtin (1984) conclude? Several studies show correlation between segmenting ability and reading achievement in early grades (e.g., Fox & Routh, 1976; Liberman, Shankweiler, Liberman, Fowler, and Fischer, 1977). The children's interest in, and facility with, linguistic materials may underlie both segmenting ability and reading ability.

Let us turn once more to Vai literacy (Scribner & Cole, 1981). Vai script is written without extra spaces between words, phrases, and sentences. Vai-script literates were asked to segment a written letter into *koali kule,* which literally means 'a piece of speech' or 'an utterance' but actually corresponds to word, phrase, or sentence, depending on context. The Vai literates could analyze the text into units, which, however, tended to be meaning-based units, phrases such as "my big brother." In an auditory integration task that required subjects to integrate words or syllables into a sentence, schooling and Vai-script literacy were similar in performance when the unit was a word, but expert Vai-script literacy was better when the unit was a syllable. Recall that a symbol in the Vai script codes a syllable (Table 2). Even among Vai literates, "advanced Vai" (mean 16 years of use, with Vai teaching experience) were better than beginning Vai (8 years of use) of the same age in their ability to handle syllables as the unit of communication. Such differences between advanced and beginning Vai literates led Scribner and Cole (1981) to observe: "We are in a better position to implicate prior experience when we find systematic and consistent differences among individuals of different levels of proficiency within a single literacy group ..." (p. 185).

Syllable segmenting is particularly easy in Japanese because its reading unit is the syllable, whose structure is simple (V, CV), as it is in the Vai script. A high percentage (60%) of Japanese preschoolers can do the task, even when they cannot read any Hiragana. But the percentage shoots up to 90% as soon as the preschoolers can read one to five Hiragana, and close to 100% when they read 60 or more Hiragana (Amano, 1970). The task of locating a designated syllable in a word, say /ko/ in *kotori*, is more difficult, and only 20% of the preschoolers can do the task before they can read their first Hiragana; the percentage jumps to 42% when the children can read one to five Hiragana, and to 95% when they can read all 71 basic and secondary Hiragana.

A person can learn to read without awareness of words, and if word boundaries are not marked in text, the reader does not necessarily becomes aware of a word as a unit. Literacy in a syllabary does lead to implicit use of syllable as a unit.

Segmenting words and syllables into phonemes is difficult, but can develop when people learn to read in an alphabet, and by phonics. In Portugal, illiterate adults scored far poorer than controls who became literate in adulthood in phoneme deletion and addition tasks (*farm* → arm; *ant* → pant). In fact, 50% of the illiterates failed every single test, whereas none of the literates did (Morais, Cary, Alegria, & Bertelson, 1979). In the United States, the proportion of children who could do phoneme segmenting at ages 4, 5, and 6 was 0%, 17%, and 70% respectively (Liberman et al., 1974). Once again, the segmenting ability developed dramatically when children started reading and writing at school.

In Belgium, one group of grade 1 children had been taught for 4 months by a whole-word method and the other by a phonics method. The phonics group did somewhat better on syllable segmenting, but spectacularly better on phoneme segmenting, than did the whole-word group (Alegria, Pignot, & Morais, 1982). In Canada, preschool readers who had a grade 2 reading level did poorly on phoneme segmenting and blending tasks. These preschoolers had learned to read largely on their own by the whole-word method (Patel & Patterson, 1982).

Even in laboratories, a training method that simulates aspects of the phonics method of teaching a letter-sound relation in alphabetic orthography leads to phoneme segmenting ability (e.g., Ehri, 1984). In one study, two experimental groups of preschoolers (aged 4 and 5) who were given an intensive training in analysis of words by initial, medial, and final phonemes scored higher in reading and spelling in schools than two control groups, one that received training in categorizing pictures of words based on concepts and the other in no task (Bradley & Bryant, 1983). Of the two experimental groups, the group that learned word analysis with plastic letters read and spelled better than the group who learned word analysis with colored pictures only. Such results are expected: in this study, too, the experimental groups were in fact trained in some aspects of the phonics. If the intensive training (40 individual sessions spread over 2 years!) given to the preschoolers had been in reading proper, the children might have become even better readers and spellers. In addition, if they had used their reading ability as "the magic key," they may have obtained useful and interesting information by reading stories and labels on objects.

Segmenting ability, if it is of any great importance, will develop in learning to read, to an extent that depends on the method of learning and the script in which reading is learned.

East and West: Mark, the Twain Shall Meet

That East and West differ in cultures is not disputed here; that the twain differ in cognitive styles has some observational support (see "Cultural differences in cognition" above).

What I will question is: Are the differences between the East and the West in cognitive styles attributable to the differences in writing systems? First of all, in terms of writing systems, it is not possible to divide the world cleanly between East and West. The People's Republic of China uses alphabetic *pinyin,* and Taiwan a sort of syllabic *zhuyin fuhao,* as auxiliaries to a logography, while both Japan and South Korea use bona fide phonetic scripts along with many logographs, and North Korea uses only a phonetic script. All these Eastern countries have adopted whole-heartedly the Western punctuation system, and partly the Western horizontal writing direction. In the West, all may use phonetic scripts, a variety of alphabets, but they also use Arabic numerals and other logographs (e.g., &, +). More importantly, in processing textual materials, the East and the West do not differ greatly.

In modern times (since the nineteenth century), the East has adopted from the West important political and economic institutions and systems – education, government, health care, transportation, banking, sport, and so on. Even art is not sacrosanct: Western art – music, theater, visual art – is taught at school, and is appreciated by a select public (as it is in the West). On a more mundane level, the majority of people in the East wear Western clothing and sport Western hair styles. They also avidly learn English, which is a major subject in schools, sometimes from grade 3 on. They learn English the better to absorb Western technology and culture, which they believe will lead to a high living standard and just society. The British historian John Roberts titled his TV Ontario series (March to April, 1986) appropriately "Triumph of the West." He asserts that Eastern influence on Western culture and civilization is marginal, compared with the overpowering Western influence on the East.

Why has the East lagged behind the West in science and technology? The reason cannot be the writing system, for ancient and medieval China, while using the same writing system used today, was far ahead of the West in observation of comets, eclipses, sun spots, and other heavenly phenomena, as well as in invention of silk, porcelain, gunpowder, printing, paper, the magnetic compass, the mechanical clock, the seismograph, etc. It was the Chinese who first recorded Halley's Comet in 240 B.C. It was the Chinese who left us the oldest specimen of packing paper made in the second century B.C. and writing paper in A.D. 100. To sneak in Korean achievement, Koreans used the first cast bronze type in A.D. 1234; there are extant pages printed from cast bronze type in A.D. 1403, 50 years before the famous Gutenberg type.

Joseph Needham, the author and editor of the multivolume opus *Science and Civilization in China* (1954–1985), pleads with the West for better appreciation of the Chinese achievements. He also observes:

> There is a commonly received idea that the ideographic language was a powerful inhibiting factor to the development of modern science in China ... This is grossly over-rated. It has proved possible ... to draw up large glossaries of definable technical terms used in ancient and medieval times for all kinds of things and ideas in science and its applications. The Chinese language at present day is found to be no impediment by the scientists of the contemporary culture (Needham, 1963, p. 137).

We can ponder a few reasons why the early achievements in science and technology were not sustained. In all these achievements the Chinese tended to be practical and technical rather than theoretical and conceptual. Thus, extremely accurate astronomical observations did not lead to the theory of cosmology (e.g., Nakayama, 1973). Other culprits must be sought in Taoism, which teaches, "Do not seek to probe the workings of nature, and all things will then flourish of themselves," and in the long-established civil service examination that emphasized memorization of Confucian classics at the expense of original thinking and scientific facts.

> Confucianism maimed the spirit of progress by teaching the Chinese that only what was ancient was good; Taoism then smilingly lured them away from the solid path of scientific method, and left them a mighty heritage of mumbo-jumbo, an aversion for the rational and contempt for Q.E.D., a distrust of dogma and deep-frozen truths, and a marvellous talent for spurious argument (Bloodworth, 1967, p. 180).

Originally, Confucianism, with its teachings of moderation and social order, loyalty to one's family and superiors, had salutary effects on Chinese, and Korean, peoples. Over the years, however, it has become anachronistic.

In a Confucian society, a woman existed solely to serve men: as a child, her father; as a wife, her husband; as a mother, her sons. Confucianism has become also hollow: people follow it outwardly without much inner conviction. Taoism too has had good and bad effects. The salutary influence of Taoism on Eastern art is seen in familiar landscape paintings – a canvas filled with misty mountains and space, against which a man or a few men are solitary, and tiny. It is seen also in Eastern calligraphy, which prizes spontaneity of brush works – reworking of a brush stroke is simply not done. Taoism does not encourage active and aggressive probing of the nature.

Japan in the nineteenth century was the first Eastern nation to successfully modernize in the Western mode. She was the first to adopt Western science and culture. In fact, the two-Kanji word for "science" was first coined in Japan and then exported to China. Japan has now already caught up with, even surpassed, the West in technology, if not in science. Korea in the nineteenth century was not interested in the West; she was called "the hermit kingdom." Today, South Korea is rapidly catching up with Japan in technology and export industry. The People's Republic of China, too, under Deng Xiaoping, is moving toward modernization, which in the East is almost synonymous with Westernization. Better communication and traffic between the East and the West can only hasten this process of East-West rapprochement and dialogue.

One hopes that the East will always preserve the core of its culture, including language, writing systems, art, and custom, because they are worth preserving. The world without the differences between the East and the West will be one hell of a dull place!

References

Adams, M.J., & Huggins, A.W.E. (1985). The growth of children's sight vocabulary: a quick test with educational and theoretical implications. *Reading Research Quarterly, 20,* 262–281.

Alegria, J., Pignot, E., & Morais, J. (1982). Phonetic analysis of speech and memory codes in beginning readers. *Memory & Cognition, 10,* 451–456.

Amano, K. (1970). Formation of the act of analyzing phonemic structure of words and its relation to learning Japanese syllabic characters (Kanamoji). *Japanese Journal of Educational Psychology, 18,* 76–89 (in Japanese with English abstract).

Baddeley, A.D., & Lewis, V.L. (1981). Inner active processes in reading: the inner voice, the inner ear, and the inner eye. In A.M. Lesgold, & C.A. Perfetti (Eds.), *Interactive processes in reading.* Hillsdale, NJ: Lawrence Erlbaum.

Baddeley, A.D., Eldridge, M., & Lewis, V. (1981). The role of subvocalization in reading. *Quarterly Journal of Psychology, 33 a,* 439–454.

Baron, J., & Treiman, T. (1980). Use of orthography in reading and learning to read. In Kavanaugh, J.F., & Venezky, R.L. (Eds.), *Orthography, reading and dyslexia.* Baltimore: University Park Press.

Becker, J.D. (1984). Computerized typing and editing can now be extended to all living languages of the world. *Scientific American, 251,* 96–107.

Bliss, C.K. (1965). *Semantography* (2nd Ed.). Sydney: Semantography.

Bloodworth, D. (1967). *The Chinese looking glass.* London: Secker & Warburg.
Bradley, L., & Bryant, P.E. (1983). Categorizing sounds and learning to read: a causal connection. *Nature, 301,* 419–421.
Bradshaw, G.J., Hicks, R.E., & Ross, B. (1979). Lexical discrimination and letter-string identification in the two visual fields. *Brain and Language, 8,* 10–18.
Brown, D.L. (1971). Some linguistic dimension in auditory blending. In *Reading: the right to participate.* (20th Yearbook of the National Reading Conference), Clemson, S.C.: National Reading Conference.
Brownell, H.H., Michel, D., Powelson, J., & Gardner, H. (1983). Surprise but coherence: sensitivity to verbal humor in right-hemisphere patients. *Brain and Language, 18,* 20–27.
Calfee, R.C., Venezky, R.L., & Chapman, R.S. (1969). *Pronunciation of synthetic words with predictable and unpredictable letter-sound correspondences.* Technical Report 71, Wisconsin Research and Developmental Center for Cognitive Learning.
Chao, Y.R. (1968). *A grammar of spoken Chinese.* Berkeley: University of California Press.
Chong, J.Y., Han, S.J., & Kang, T.J. (1983). *Hangul Word Processor III.*, JAE Consultants, 1983.
Coleman, E.B. (1970). Collecting a data base for a reading technology. *Journal of Educational Monograph, 61 (4),* 1–23.
Coleman, E.B., & Hahn, S.C. (1966). Failure to improve the readability with a vertical typography. *Journal of Applied Psychology, 50,* 434–436.
Coltheart, M., Patterson, K., & Marshall, J.C. (Eds.) (1980). *Deep dyslexia: a review of the syndrome.* London: Routledge & Kegan Paul.
Deloche, G., Andreewsky, E., & Desi, M. (1982). Surface dyslexia: a case report and some theoretical implications to reading models. *Brain and Language, 15,* 12–31.
Durkin, D. (1966a). *Children who read early.* New York: Teachers College, Columbia University.
Durkin, D. (1966b). The achievement of pre-school readiness: two longitudinal studies. *Reading Research Quarterly, 1,* 5–36.
Ehri, L.C. (1984). How orthography alters spoken language competencies in children learning to read and spell. In Downing, J., & Valtin, R. (Eds.), *Language awareness and learning to read.* New York: Springer.
Endo, M., Shimizu, A., & Nakamura, I. (1981a). Laterality differences in recognition of Japanese and Hangul words by monolinguals and bilinguals. *Cortex, 17,* 1–9.
Endo, M., Shimizu A., & Nakamura, I. (1981b). The influence of Hangul learning upon laterality difference in Hangul word recognition by native Japanese subjects. *Brain and Language, 14,* 114–119.
Etiemble, M. (1973). *L'ecriture.* Paris: Gallimand.
Feitelson, D.I. (1980). Relating instructional strategies to language idiosyncracies in Hebrew. In Kavanagh, J.F., & Venezky, R.L. (Eds.), *Orthography, reading and dyslexia.* Baltimore: University of Park Press.
Flaugher, R.L. (1971). *Patterns of test performance by high school students of four ethnic identities.* Princeton, NJ: Educational Testing Service.
Fox, B., & Routh, D.K. (1976). Phonetic analysis and synthesis as word attack skills. *Journal of Educational Psychology, 68,* 70–74.
Gainotti, G., Caltagirone, C., Micelli, G., & Masullo, C. (1981). Selective semantic-lexical impairment of language comprehension in right-brain damaged patients. *Brain and Language, 13,* 201–211.
Gazzaniga, M.S. (1983). Right hemisphere language following brain bisection (a 20-year perspective). *American Psychologist, 38,* 525–549.
Gelb, I.J. (1963). *A study of writing.* Chicago: University of Chicago Press, 1963.
Gibson, E.J., Gibson, J.J., Pick, A.D., & Osser, H.A. (1962). A developmental study of the discrimination of letter-like forms. *Journal of Comparative and Physiological Psychology, 55,* 897–906.

Goody, J., & Watt, I. (1968). The consequences of literacy. In *Literacy in traditional societies.* Cambridge: Cambridge University Press.

Gottesman, R. L., Croen, L., & Rotkin L. (1982). Urban second grade children: a profile of good and poor readers. *Journal of Learning Disabilities, 15,* 268–272.

Hatta, T. (1978). Recognition of Japanese Kanji and Hirakana in the left and right visual fields. *Japanese Journal of Psychology, 20,* 51–59 (in Japanese with English abstract).

Hatta, T. (1980). *Kanji processing levels and cerebral hemisphere differences.* Part IV 29, Osaka Educational University.

Havelock, E. A. (1978). *The Greek concept of justice: From its shadow in Homer to its substance in Plato.* Cambridge, Mass: Harvard University Press.

Hayashi, R., & Hatta, T. (1982). Visual field differences in a deeper semantic processing task with Kanji stimuli. *Japanese Psychological Research, 24,* 111–117.

Huang, J. T., & Liu, I. M. (1978) Paired-associate learning proficiency as a function of frequency count, meaningfulness, and imagery value in Chinese two-character ideograms. *Chinese Psychological Journal,* 1978, *20,* 5–17 (in Chinese with English abstract).

Jensen, H. (1970). *Sign, symbol and script.* London: George Allen & Unwin.

Just, M. A., & Carpenter, P. A. (1980). A theory of reading: from eye fixations to comprehension. *Psychological Review, 87,* 329–354.

Kates, B., MacNaughton, S., & Silverman, H. (1978). *Handbook of Blissymbolics.* Toronto: Blissymbol Communications Institute.

Kratochvil, P. (1968). *The Chinese language today.* London: Hutchinson.

Kutas, M., & Hillyard, S. A. (1982). The lateral distribution of event-related potentials during sentence processing. *Neuropsychologia, 20,* 579–590.

Kyostio, O. K. (1980). Is learning to read easy in a language in which the grapheme-phoneme correspondences are regular? In Kavanagh, J. F., & Venezky R. L. (Eds.), *Orthography, reading and dyslexia.* Baltimore: University Park Press.

Lassen, N. A., Ingvar, D. H., & Skinhoj, E. (1978). Brain function and blood flow. *Scientific American, 239,* 62–71.

Lesser, G. S. (1976). Cultural differences in learning and thinking styles. In Messick, S. (Ed.), *Individuality in learning and thinking.* San Francisco: Jossey-Bass.

Lesser, G. S., Fifer, G., & Clark, D. H. (1965). Mental abilities of children from different social-class and cultural groups. *Monographs of the Society for Research in Child Development, 30* (4).

Liberman, I. Y., & Mann, V. A. (1981). *Should reading instruction and remediation vary with the sex of the child?* Status Report on Speech Research SR-65, Haskins Laboratories.

Liberman, I. Y., Shankweiler, D., Fischer, F. W., & Carter, B. (1974). Explicit syllable and phoneme segmentation in the young child. *Journal of Experimental Child Psychology, 18,* 201–212.

Liberman, I. Y., Shankweiler, D., Liberman, A. M., Fowler, C., & Fischer, F. W. (1977). Phonetic segmentation and recoding in the beginning reader. In Reber, A. S., & Scarborough D. L. (Eds.), *Toward a psychology of reading.* Hillsdale, N.J.: Lawrence Erlbaum.

Lin, Y. T. (1938). *The wisdom of Confucius.* New York: The Modern Library.

Liu, I. M. (1978). Methods of acquiring Chinese vocabulary. *Bulletin of the Sun Yat-sen Cultural Foundation, 22,* 159–187.

Liu, I. M. (1979). Personal communication, May 17.

Liu, I. M. (1984). A survey of memorization requirement in Taipei primary and secondary schools. Unpublished manuscript.

Liu, I. M., & Hsu, M. (1974). Measuring creative thinking in Taiwan by the Torrance Test. *Testing and Guidence, 2,* 108–109.

Lynn, R. (1983). IQ in Japan and the United States shows a growing disparity. *Nature, 297,* 222–223.

Maddieson, I. (1984). *Patterns of sounds.* New York: Cambridge University Press.

Marin, O. S. M. (1980). CAT scans of five deep dyslexic patients. In Coltheart, M., Patterson, K., & Marshall, J. (Eds.), *Deep dyslexia.* London: Routledge & Kegan Paul.

Marshall, E. (1986). School reforms aim at creativity. *Science, 233,* 267–270.
Mishkin, M., & Forgays, D.G. (1952). Word recognition as a function of retinal locus. *Journal of Experimental Psychology, 43,* 43–48.
Morais, J., Cary, L., Alegria, J., & Bertelson, P. (1979). Does awareness of speech as a sequence of phones arise spontaneously? *Cognition, 7,* 323–331.
Muraishi, S., & Amano, K. (1972). *Reading and writing abilities of preschoolers: a summary.* Tokyo: The National Language Research Institute (in Japanese).
Nakayama, S. (1973). The empirical tradition: science and technology in China. In Toynbee, A. (Ed.), *Half the world: the history and culture of China and Japan.* New York: Holt, Rinehart and Winston.
Navon, D., & Shimron, J. (1981). Does word naming involve grapheme-to-phoneme translation?: Evidence from Hebrew. *Journal of Verbal Learning and Verbal Behavior, 20,* 97–109.
Needham, J. (1954–1985). *Science and civilization in China* (Vol. 1–5). New York: Cambridge University Press.
Needham, J. (1963). Poverties and triumphs of the Chinese scientific tradition. In Crombie, A.C. (Ed.), *Scientific change.* London: Heinemann.
Ogden, J.A. (1984). Dyslexia in a right-handed patient with a posterior lesion of the right cerebral hemisphere. *Neuropsychologia, 22,* 265–280.
Ornstein, R., Herron, J., Johnstone, J., & Swencionis, C. (1979). Differential right hemisphere involvement in two reading tasks. *Psychophysiology, 16,* 398–404.
Paivio, A. (1965). Abstractness, imagery, and meaningfulness in paired-associate learning. *Journal of Verbal Learning and Verbal Behavior, 4,* 32–38.
Paradis, M., Hagiwara, H., & Hildebrandt, N. (1985). *Neurolinguistic aspects of the Japanese writing system.* New York: Academic.
Park, R. (1978–1979). Performance on geometric figure copying tests as predictors of types of errors in decoding. *Reading Research Quarterly, 14,* 100–118.
Paschal, B.J., Kuo, Y.-Y., & Schurr, K.T. (1980). *Creative thinking in Indiana and Taiwan college students.* Paper read at the 5th Conference of the International Association of Cross-cultural Psychology, 1980.
Patel, P.G., & Patterson, P. (1983). Precocious reading acquisition: psycholinguistic development, IQ, and home background. *First Language, 3,* 139–153.
Premack, A.J., & Premack, D. (1972). Sarah, a young chimpanzee learns 130 words and a bit of grammar. *Scientific American, 227,* 92–99.
Premack, D. (1984). Upgrading a mind. In Bever, T.G., Carroll, J.M., & Miller, L.A. (Eds.), *Talking minds: the study of language in cognitive science.* Cambridge, MIT Press.
Prohovnik, I., Hakansson, K., & Risberg, H. (1980). Observation on the functional significance of regional cerebral flow in 'resting' normal subjects. *Neuropsychologia, 18,* 203–217.
Ross, E.D. (1983). Right-hemisphere lesions in disorders of affective language. In Kertesz, A. (Ed.), *Localization in neuropsychology.* New York: Academic.
Sakamoto, T., & Makita, K. (1973). Japan. In Downing, J. (Ed.), *Comparative reading.* New York: Macmillan.
Salapatek, P. (1968). Visual scanning of geometric figures by the human newborn. *Journal of Comparative & Physiological Psychology, 66,* 247–258.
Sasanuma, S. (1975). Kana and Kanji processing in Japanese aphasics. *Brain and Language, 2,* 369–383.
Sasanuma, S. (1980). Acquired dyslexia in Japanese: clinical features and underlying mechanisms. In Coltheart, M., Patterson, K., & Marshall, J.C. (Eds.), *Deep dyslexia.* London: Routledge & Kegan Paul.
Sasanuma, S., Itoh, M., Kobayashi, Y., & Mori, K. (1980). The nature of the task-stimulus interaction in the tachistoscopic recognition of Kana and Kanji words. *Brain and Language, 9,* 298–306.
Scribner, S., & Cole, M. (1981). *The psychology of literacy.* Cambridge, Mass.: Harvard University Press.

Steinberg, D.D., Harada, M., Tashiro, M., & Yamada, A. (1982). One congenitally deaf 1-year-old learn to read. *Hearing Impaired, 376,* 22–30 and 46 (in Japanese).
Steinberg, D., Yoshida, K., & Yagi, R. (1985). Teaching reading to one and two year olds at home. *The Science of Reading, 29,* 1–17 (in Japanese with English summary).
Stern, J.A. (1978). Eye movements, reading, and cognition. In Senders, J.W., Fisher, D.F., & Monty, R.A. (Eds.), *Eye movements and higher psychological functions* (Vol. 2). Hillsdale, N.J.: Lawrence Erlbaum.
Stevenson, H.W., Stigler, J.W., Lucker, G.W., Lee, S.-Y., Hsu, C.-C., & Kitamara, S. (1982). Reading disabilities: the case of Chinese, Japanese and English. *Child Development, 53,* 1164–1181.
Stevenson, H.W., Lee, S.-Y., & Stigler, J.W. (1986). Mathematics achievement of Chinese, Japanese, and American children. *Science, 231,* 693–699.
Tanaka, T., & Yasufuku, J. (1975). Development of the cognition of letters (2). *Osaka University of Education Report, 24,* 85–99 (in Japanese with English abstract).
Taylor, I. (1980). The Korean writing system: An alphabet? A syllabary? A logography? In Kolers, P.A., Wrolstad, M.E., & Bouma, H. (Eds.) *Processing of visible language.* v. 2. New York: Plenum.
Taylor, I. (1986). The variety of scripts and reading. *Proceedings of the 12th annual symposium of the Deseret Language and Linguistic Society.* Provo, Utah; Brigham Young University.
Taylor, I., & Taylor, M.M. (1983). *The psychology of reading.* New York: Academic.
Taylor, I., & Taylor, M.M. (1984). English function words. Unpublished paper.
Torrance, E.P. (1966). *Torrance tests of creative thinking: norms-technical manual.* Princeton: Personnel Press.
Tzeng, O.J.L., & Hung, D. (1980). Reading in a nonalphabetic writing system. In Kavanagh, J.F., & Venezky, R.L. (Eds.), *Orthography, reading and dyslexia.* Baltimore: University Park Press.
Tzeng, O.J.L., Hung, D.L., & Wang, S-Y. (1977) Speech recoding in reading Chinese characters. *Journal of Experimental Psychology: Human Learning and Memory, 9,* 621–630.
Tzeng, O.J.L., Hung, D.L., Cotton, B., & Wang, S.-Y. (1979). Visual lateralization effects in reading Chinese characters. *Nature, 282,* 499–501.
Unger, J. (1977). Introduction: primary school reading texts and teaching methods in the wake of the cultural revolution. Chinese Education, 10, 4–29.
Valtin, R. (1984). The development of metalinguistic abilities in children learning to read and write. In Downing, J., & Valtin, R. (Eds.), *Language awareness and learning to read.* New York: Springer.
Warrington, E.K. (1975). The selective impairment of semantic memory. *Quarterly Journal of Experimental Psychology, 27,* 635–657.
Zaidel, E. (1978a). Lexical organization in the right hemisphere. In Buser, P.A., & Rougeul-Buser, A. (Eds.), *Cerebral correlates of conscious experience.* (INSERM Symposium No. 6), Amsterdam: Elsevier/North Holland Biomedical.
Zaidel, E. (1978b). Concepts of cerebral dominance in the split brain. In Buser, P.A., & Rougeul-Buser, A. (Eds.), *Cerebral correlates of conscious experience* (INSERM Symposium No. 6), Amsterdam: Elsevier/North Holland Biomedical.
Zaidel, E. (1983). A response to Gazzaniga: language in the right hemisphere. Convergent perspectives. *American Psychologist, 38,* 342–346.
Zaidel, E., & Peters, A.M. (1981). Phonological encoding and ideographic reading by the disconnected right hemisphere: Two case studies. *Brain and Language, 14,* 205–234.

Part 4 Neuropsychological Considerations

Introductory Remarks

Is the biology of mind relevant to the puzzle of writing right and writing left? The answer seems to be yes, with a principal clue emerging from the division of cognitive labor shared between the brain's cerebral hemispheres. Lecours and Nespoulos begin the discussion of laterality and its implications for writing, surveying the field from the perspective of its historical trends. Grant discusses the neuroanatomy of the brain's language systems, setting the neuropsychological studies in structural perspective. The pivotal role of lateralization is treated in turn by Lecours, Mehler, Parente, and Vadeboncoeur, reporting studies in literate and nonliterate subjects, by Jones and Aoki for the neurological handling of the Kanji and Kana characters in Japanese subjects, and by Tzeng and Hung, who provide a broad synthesis of the work relating script variations to cerebral lateralization. The amassed findings suggest the presence of a complex interplay between biological and cultural factors in literacy, with reading and writing of different orthographies affecting the likely options in neurophysiological development and ontogenetic constraints influencing the way different orthographies are handled by the brain.

CHAPTER 12

The Biology of Writing

ANDRÉ ROCH LECOURS[1] and JEAN-LUC NESPOULOUS[1]

In a very beautiful and gripping book written in the form of the memoirs of a doctor living in Ancient Egypt "in approximately the year 1350 before Jesus Christ," Mika Waltari (1977) describes several neurosurgical techniques in practice at that time. Amongst these, he details an experimental operation carried out by the "Royal Trepanner", Ptahor, on the person of a "robust slave who had lost the ability to speak and who could no longer move his limbs after being injured by hitting his head against a rock during a brawl." The hero, Sinouhe, tells us how Ptahor has the slave bound, cleans his head, observes the presence of a fracture and a puncture wound, uses a drill, a saw and a pair of pliers to open and lift the skull, and instructs the neophytes of the "House of Life" how to take out the blood clots that are compressing the "white folds of the brain . . . so that each student has the time to observe and fix in his mind the external aspect of a living brain." He then "closes the hole with a silver plate cleansed in the fire," after which the slave is released, sits up and begins to swear.

Combining imagination and erudition, Waltari has without doubt drawn the essential elements of his description from some ancient document: it would not be possible to invent this story out of nothing, and it appears that the doctors of Ancient Egypt knew that certain types of damage to the brain could interfere with the use of speech and language, disorders now classified under the term *aphasia.* This conclusion seems warranted from the stories of a patient described in Edwin Smith's papyrus (McHenry 1969), a document that leads us to believe that as early as 3500 B.C., Egyptian doctors had gained an understanding of the localization of the cerebral lesions responsible for unilateral limb paralysis (hemiplegia), and had distinguished between a paralysis that interferes with the ability to write and a paralysis that leaves this capacity intact. That is, they seem to have realized that the brain does not necessarily function in a unified and global fashion, but is made up of functionally specialized structures whose activities are linked to the determination of specific behaviors.

---

[1] Centre Hospitalier Côte-des-Neiges, 4565 Queen Mary, Montreal, H3W 1W5, Prov. Quebec, Canada.

It is possible that this deduction was first formulated on the banks of the Nile, but in any case it was lost for thousands of years, as was so much other knowledge – at least, it apparently did not spread to the West. Until the end of the eighteenth century, Western medicine taught that the brain was a homogeneous organ acting under the influence of the "will" and distributing a "vital energy" to all parts of the body, with the nerves serving to channel and distribute this energy according to a principle of "sovereign indifference". At the beginning of the nineteenth century, Franz Joseph Gall, a doctor and anatomist from Baden, conceived of a new doctrine: that the brain was composed of a mosaic of juxtaposed organs, each governing a well-defined "moral or intellectual faculty." Gall based his teachings on the comparative study of irregularities on the outer surface of the skull, defined on the one hand through meticulous inspection and palpation, and on the other hand through observing the aptitudes and behaviors of his subjects (Gall & Spurzheim, 1810–1818). Thus, having noted while still a schoolboy in the Black Forest that his friends who were skilled at reciting verses had "eyes like those of calves," he inferred that the "cerebral site for language is situated immediately above the eye sockets": enlarged in these individuals, the specific organ of language pushes the eyes downward and forward, leading to exophthalmos (Marie 1906). Following this reasoning, Gall "localized" the seats of parental love, knowledge of wine, magnanimity, sexual prowess, and many more. This new discipline carried the name of *phrenology*. The project of course miscarried, but popular speech still shows some traces of it; for instance, modern French incorporates the phrase *la bosse des mathematiques* (the mathematician's bump). More importantly, Gall's fundamental intuition has remained: under the bumps, there is the brain.

It is without doubt the study of phrenology as instituted by Gall that gave birth to the anatomoclinical approach used by neurologists today. This approach consists of correlating specific clinical perturbations with the cerebral lesions that cause them, by defining the precise localization of these lesions in the brain. The first perturbations to be thus studied were those of language and speech, the aphasias. Jean-Baptiste Bouillaud (1825), a devoted admirer of Gall; Paul Broca (1861 a, 1861 b, 1865); and Carl Wernicke (1874, 1881) were the most famous of those who undertook the "phrenology" of the convolutions of the brain as opposed to the bumps of the skull. Another important contributor was Siegmund Exner (1881), one of the first to be interested in the cerebral representation of writing.

Our objective here is to discuss the biology of literacy. The acquisition of this skill is almost always dependent on the previous acquisition of oral language – that is, one first learns to speak, and subsequently learns to read and write with reference to words that can already be pronounced, rather than with reference to the tangible objects that these words represent. Hence, this theme is only partially dissociable from a discussion of the biology of language in general.

The Language Zone

If one cites only published works, the primordial discovery of the biological foundations of language is attributable to the French surgeon and anthropologist Paul Broca (1865). His discovery was that once a certain point in maturation has been attained, linguistic behaviors are governed by the left cerebral hemisphere (which also governs the activities of the right hand, normally the hand that writes). Broca's crucial observation was that, in general, diseases of the left hemisphere but not of the right may cause severe disorders of language. The representation of language in the brain thus appears to be *lateralized:* i.e., there exists a functional asymmetry, analogous to that existing for handedness.

First brought to light by the anatomoclinical method, lateralization for language was later confirmed by more sophisticated approaches: electrical stimulation of the cerebral cortex (Penfield & Roberts, 1959); selective deactivation of a hemisphere by unilateral injection of a barbituric substance into the artery that feeds it (Wada & Rasmussen, 1960); dichotic listening tasks, which simultaneously present different linguistic messages to each ear (Kimura, 1961); and tracing of radioactive substances incorporated into cerebral blood flow (Ingvar & Schwartz, 1974). All the data are in agreement: as a rule, language is situated on the left side.

Not all the convolutions and structures of the left cerebral hemisphere are equally implicated in the regulation of linguistic abilities. Those that play a particularly important role are collectively known as the *language zone.* Each hemisphere is composed of a number of lobes, named for the parts of the skull that cover them, and each lobe contains a number of convolutions, or folds, on its surface. The most important components of the language zone – its most specialized components – are the posterior half of the third frontal convolution of the left frontal lobe, or *Broca's area* (1865); the posterior half of the first convolution of the left temporal lobe, or *Wernicke's area* (1874); and the supramarginal convolution and the angular convolution in the parietal lobe (Marie 1906). This last region is one of the functional units that play a particularly important role in governing reading and writing (Dejerine, 1892).

Recognizing the existence of a language zone, and attributing particular functions of encoding and decoding oral or written language to its various components, does not mean that we conceive of the brain, as did Franz Joseph Gall, as a mosaic of juxtaposed organs constituting the "centers" of autonomous psychological functions. The following analogy may serve to clarify the modern conception of brain centers. The fact that a traffic accident occurring during rush hour on the Jacques Cartier Bridge perturbs the flow of traffic in the city of Montreal much more than a comparable accident in most other cities does not imply that the bridge is the "seat" of traffic function in Montreal, but one could certainly conclude that it plays an essential role in this regard. In the same way, many cerebral structures participate in the regulation of linguistic behaviors, including certain regions within the right hemisphere (Joanette, 1980), but the structures on the left that constitute the language zone play so essential a role that any disease that attacks them will perturb language abilities in a particularly dramatic fashion.

Anatomical Asymmetries

If one looks closely at a human brain, one first has the impression that the hemispheres are symmetrical to the point of being mirror images of each other. This impression is in part corroborated by studies (macroscopic and microscopic) showing that each of the structures of one hemisphere has its counterpart in the other hemisphere and can normally exchange information with it through direct or indirect anatomical links. It is, however, still true that certain structures are more developed in one hemisphere than in the other. In the majority of human brains, there are thus parts of Broca's and Wernicke's areas which are anatomically more important than their homologues in the right hemisphere (anatomical asymmetry). One is thus naturally tempted to establish a link between these two sets of facts, but such a link is not intrinsically necessary and one must still prove that these individuals show both an anatomical and a functional asymmetry. This has in fact been demonstrated, if one has no methodological objections, by the researchers at the Montreal Neurogical Institute: the subjects who represent the rule in the human species, that is to say those who speak with the left hemisphere, as shown by the fact that they suffer from transitory aphasia following the injection of a barbituric substance into the arteries feeding the left hemisphere of their brains, are also those whose show a greater development of the Wernicke's area in the left hemisphere than in the right, as shown by special radiographical slides (Ratcliff, Dila, Taylor & Milner, 1980). If one then knows that this anatomical asymmetry can be observed in the fetal brain as early as the 28th or 29th week of gestation (Tezner, 1977), that is to say before birth and, a fortiori, well before any realization of any sort of language capacity, one can believe that the dominance of the left brain for language rests, in the human species, on an innate biological predisposition. The question is settled.

Realization of the Biological Predisposition

Innate Language Capacities of the Newborn Infant

Present at the end of the seventh (lunar) month of gestation, anatomical asymmetries affecting the language zone can still be observed at birth (Witelson & Pallie, 1973) and persist throughout life. However it is well known that newborn babies do not speak, a fact which presents a certain number of advantages. It seems, however, even though absolute agreement has not yet been reached on this subject (Vargha-Khadem & Corbalis, 1979), that at birth, or shortly after, the left cerebral hemisphere of the newborn infant is better able to integrate specifically linguistic auditory information than the right hemisphere (Entus, 1977; Mehler, 1980, personal communication; Molfese & Molfese, 1979; Segalowitz & Chapman, 1980). If this is the case then one should immediately ask whether there exists a significant relationship between this left functional specialization and two other factors: on the one hand, although the brain of the human fetus is deprived of visual information, it is, however, exposed to all types of auditory information well before birth, especially auditory information generated by the mother's speech; on the other hand, the biological maturation of certain components of the auditory pathways of the human nervous system, as opposed to the maturation of the corresponding components of the visual pathways, is largely prenatal (Yakovlev & Lecours, 1967; Lecours, 1975).

Of all those proponents of innateness, Frederick II of Prussia was probably the most enthusiastic and the most determined. Certainly, his constitutional privileges protected him from reprisals! Unable to decide whether our species' "original language" was Greek, Hebrew, or Latin, he entrusted a group of newborn babies to nurses who were under orders to take great care of the infants and to provide them for some years with all that was necessary, without however ever speaking to them or speaking aloud in their presence. Frederick's prediction was that these children would spontaneously begin to speak, at the appropriate moment, in the original language. It appears that the nurses followed their orders, because these children grew up and died without ever having spoken Greek, Hebrew, or Latin, and also without ever having spoken German (Dudson, 1970). Feral children show a similar lack of spontaneous speech (Singh & Zingg, 1966). If he had known that his subjects were probably endowed from the start with a robust Wernicke's area, the experimentalist king might have been able to conclude that although we are furnished with an innate predisposition that favors dominance of the left brain for language, the human organism must also be exposed to the complementary effects of the linguistic environment in order for this predisposition to be realized. The coin thus has two sides.

The Cerebral Representation of Language in Children

As with adults, children may show linguistic perturbations following unilateral brain lesions. However, acquired aphasias in children are distinct in a number of ways. Up to a certain age they obviously affect the acquisition as well as the use of speech, and they have, moreover, to a certain extent on the level of language use, their own semiology (and it is a truism to say that this would not affect written language before schooling if this should occur). Two other differences are particularly revealing. First, in children up to 5 or 6 years of age, aphasia can result from either a left-sided or a right-sided lesion (Basser, 1962); after this age, right-sided lesions no longer cause aphasia, while damage to the left results in it more systematically (Wechsler, 1976). Second, from the age of 5 or 6 years up to adolescence, the child can regain linguistic function completely or almost completely following a severe aphasia caused by a left cerebral lesion (Alajouanine & Lhermitte, 1965; Basser, 1962; Branco-Lefèvre, 1950; Guttman, 1942).

One is thus led to conclude that, although it is indicated by an anatomical asymmetry well before birth, the biological predisposition for left functional lateralization for language is only fully realized through the influence of external factors: i.e., linguistic immersion. Initially, cerebral representation is relatively diffuse and bihemispheric; it later becomes lateralized to the left, and eventually limits itself to certain predestined convolutions (the language zone). For a few years, then, the right hemisphere retains the ability to take over following serious damage to the left. This functional reserve is quite large, yet the developing child will normally abandon it in order to continue the process of linguistic lateralization up until adulthood, at which point the potential of the left brain is maximal and that of the right brain is greatly reduced or nonexistent. As the child of 5 or 6 is more or less the linguistic equal of the adult (we are not speaking of encyclopedic knowledge), one should ask what advantage is gained by continuing this process of functional lateralization. We will return to this question shortly.

Manual Dominance and the Cerebral Representation of Language in the Adult

True right-handedness, which can be more or less enforced according to the individual, is not as ubiquitous as is generally believed. Approximately 65% of all humans show complete preference for the right hand, about 25% more are ambidextrous (but usually write with the right hand if they are literate, which leads to the generally accepted notion that the overall population is 90% right-handed), and the remaining 10% show complete left-hand prefence (Subirana, 1969). Of the populations that have been studied to date, at least 99% of true right-handers use the left hemisphere for speech (Milner, Branch & Rasmussen, 1964), retaining in adulthood a limited or virtually nonexistent right hemisphere linguistic capacity. It is on the basis of this majority population that, for a century now, aphasiological research directed at anatomical identification of the language zone and delineation of brain-language relationships has been carried out. Almost 70% of ambidextrous and left-handed subjects also speak with the left hemisphere, indicating that the dominance of the left brain for language is even more universal than its dominance for manual dexterity. The remaining 30% of this group speak either with the right hemisphere or both sides of the brain (Milner, Branch & Rasmussen, 1964).

The study of aphasia in ambidextrous and left-handed adults reveals that, whatever the side and degree of functional lateralization for language, the prognosis for recovery following damage is on the whole much better than for right-handed individuals (Courtois, Lecours & Lhermitte, 1979), particularly if there is a family history of other ambidextrous or left-handed members. Non-right-handed adults thus constitute a group that, like children, has not pursued to the maximum the process of lateralization for language. One might think, given the frequency of cerebral lesions after a certain age, that this type of cerebral organization would be preferable to that of the right-handed majority. This is possibly true from one perspective, but it also appears to pose certain disadvantages. The most apparent of these is that families with ambidextrous or left-handed members are also those where lexic disorders are the most frequent and long-lasting (Critchley, 1964; Hallgren, 1950), which suggests that a strong left lateralization may be biologically necessary for the optimal acquisition of reading and writing.

The Japanese

The Parisian newspaper *Le Monde,* on October 19th 1980, carried a banner headline (as if it were a new discovery) stating that "the Japanese speak on the left." Following some incorrect generalizations (". . . the Japanese have a more developed left hemisphere than other humans") the author, Gabriel Racle, reported certain interesting findings. According to Tsunoda (1978), left cerebral lateralization is even more complete in native Japanese speakers than in anyone else. The reason apparently lies in certain particularities of the language that relate to an exceptionally forceful semantization of the vowel sounds. Racle does not mention another curiosity regarding Japanese. This language can be written in two ways: one script, Kana, is almost completely syllabic, while the other, Kanji, is related to ideographic Chinese writing. Recent aphasia research (Sasanuma, 1975) has shown that cerebral lesions do not perturb the use of these two codes in the same way, demonstrating that the particularities of one's linguistic environment (if one thinks for example of agglutinating languages or even of tone languages) can affect the biological organization of the brain. If the language zone had first been defined in Japan rather than in Europe, its official limits might have been somewhat different.

Cerebral Representation in Illiterates: Interaction of Biological and Social Factors

It is now accepted that exposure to an oral linguistic environment is necessary for the realization of the biological predisposition favoring the lateralization of language in the left brain. But is this lateralization predetermined to the point of being exclusive and unchangeable? To answer this, one must take into account that almost all the studies leading to modern theories of brain-language relations have been carried out in Europe and North America, hence on largely literate populations. It is conceivable that their conclusions are in part the reflection of a cultural bias.

Ernst Weber (1904) was apparently the first to propose that another environmental factor, exposure to *written* language, plays a determining role in the process of strong left lateralization. This issue by far transcends the domain of biology, since by definition such exposure must take place within an educational context, and thus is strongly affected by cultural factors. One could however approach the problem from a biological angle in the following way. Children are generally preliterate up to about age 6, and left lateralization for language (in right-handers) is usually established by this age. However, this lateralization is not yet so established as to exclude a later dominance by the right hemisphere in the event of injury to the left. One could not discount the possibility a priori that a right-handed illiterate adult exhibits a degree of functional lateralization similar to that of the preliterate child. One could easily generate experimental paradigms to test this. The one most closely conforming to ethical standards is the prediction that an illiterate adult, even with left lateralization for speech, would retain the potential for take-over by the right hemisphere, and thus in the event of left cerebral damage has a better prognosis for recovery of speech than someone literate. It might be assumed that aphasia research had long since put this hypothesis to the test, but this is not the case. A few systematic studies of aphasia in illiterates have been carried out, but results are contradictory: two of these studies conclude that the acquisition of written language does play a role in the functional lateralization of language (Cameron, Currier & Haerer, 1971; Gorlitzer von Mundy, 1957), while a third concludes that it does not (Damasio, Castro-Caldas, Grosso & Ferro, 1976). However, none of these studies have taken into account the parameter of prognosis, which is essential (Lecours, 1980). This research remains to be carried out, and hopefully will shed more light on the question.

We have asked what advantage the immature brain might be seeking to gain, from the age of 6 years (the usual beginning of education) in continuing the process of left lateralization for language. Might not this advantage lie in the mastery of written communication? Today, writing can to a certain extent be replaced by machines, but we would certainly not have evolved very far culturally without it (imagine the number of words written before the dictaphone was invented – and after all, what is the purpose of even this machine, but to ultimately increase written output?). Was not the invention of writing the best way for humans to communicate, lie, influence, and exert power beyond the barriers of space and time? How many wealthy illiterates are there, in either the industrialized or nonindustrialized nations (with the possible exception of the wives of rich men in cultures that do not encourage education for women)? The biological price that must be paid in literate societies is the loss of linguistic potential of the right cerebral hemisphere. Our species, however, is not one to sacrifice its evolutionary momentum for a consideration such as that. We may not have conquered death by inventing writing, but we view as immortal those who practice this art with success.

References

Alajouanine, T., & Lhermitte, F. (1965). Acquired aphasia in children. *Brain, 88,* 653–662.

Basser, L.S. (1962). Hemiplegia of early onset and the faculty of speech with special reference to the effects of hemispherectomy. *Brain, 85,* 427–460.

Bouillaud, J.-B. (1825). Recherches cliniques propres à démontrer que la perte de la parole correspond à la lésion des lobules antérieurs du cerveau et à confirmer l'opinion de M. Gall, sur le siège de l'organe du langage articulé. *Archives Générales de Médecine, 8,* 25.

Branco-Lefèvre, A.F. (1959). Contribuiçao para o estudo da psicopatologia da afasia em criancas. *Archivos Neuro-Psiquiatria, 8,* 345–393.

Broca, P. (1861 a). Remarques sur le siège de la faculté du langage articulé suivies d'une observation d'aphémie (perte de la parole). *Bulletin de la Société d'Anatomie, 6,* 330–357.

Broca, P. (1861 b). Nouvelle observation d'aphémie produite par une lésion de la moitié postérieure des deuxième et troisième circonvolutions frontales. *Bulletin de la Société d'anatomie, 6,* 398–407.

Broca, P. (1865). Sur le siège de la faculté du langage articulé. *Bulletin de la Société d'Anthropologie, 6,* 337–393.

Cameron, R.F., Currier, R.D., & Haerer, A.F. (1971). Aphasia and literacy. *British Journal of Disorders of Communication, 6,* 161–163.

Courtois, G., Lecours, A.R., & Lhermitte, F. (1979). Dominance cérébrale et langage, In Lecours, A.R., & Lhermitte, F. et al. (Eds.) *L'Aphasie,* (pp. 371–400). Paris: Flammarion.

Critchley, M. (1964). *Developmental Dyslexia,* London: Heinemann.

Damasio, A.R., Castro-Caldas, A., Grosso, J.T., & Ferro, J.M. (1976). Brain specialization for language does not depend on literacy. *Archives of Neurology, 33,* 300–301.

Dejerine, J. (1892). Contribution à l'étude anatomo-pathologique et clinique des différentes variétés de cécité verbale: I et II. *Compte Rendu de la Société de Biologie de Paris, 44–61.*

Dudson, F. (1970). *Tout se joue avant six ans,* Paris: Nelson.

Entus, A.K. (1977). Hemispheric asymmetry in processing of dichotically presented speech and nonspeech stimuli by infants. In Segalowitz, S.J., & Gruber, F.A. (Eds.) *Language Development and Neurological Theory* (pp. 63–73). New York: Academic Press.

Exner, S. (1881) *Untersuchungen über die Lokalisation der Funktionen in der Großhirnrinde des Menschen.* Vienna: Braumuller.

Galaburda, A.M. (1980). La région de Broca: observations anatomiques faites un siècle après la mort de son découvreur. *Revue Neurologique, 136,* 609–616.

Galaburda, A.M., Sanides, F., & Geschwind, N. (1978). Human brain: cytoarchitectonic left-right asymmetries in the temporal speech region. *Archives of Neurology, 35,* 812–817.

Gall, F.G., & Spurzheim, H. (1810–1818). *Anatomie et physiologie du système nerveux en général et du cerveau en particulier, vol. 5, pp. 1810–1818. Paris: Shoell.*

Gorlitzer von Mundy, V. (1957). Zur Frage der paarig veranlagten Sprachzentren. *Nervenarzt, 28,* 212–216.

Guttman, E. (1942). Aphasia in children, *Brain, 65,* 205–219.

Hallgren, B. (1950). Specific dyslexia (congenital word-blindness): a clinical and genetic study. *Acta Psychiatrica et Neurologica Scandinavica, Suppl. 65.*

Ingvar, D.H., & Schwartz, M.S. (1974). Blood flow patterns induced in the dominant hemisphere by speech and reading. *Brain, 97,* 273–277.

Joanette, Y. (1980). *Contribution à l'étude anatomo-clinique des troubles du langage dans les légions cérébrales droites chez le droitier.* Thesis, faculty of méd. Montréal: University of Montreal.

Kimura, D. (1961). Cerebral dominance and the perception of verbal stimuli. *Canadian Journal of Psychology, 15,* 166.

Lecours, A.R. (1975). Myelogenetic correlates of the development of speech and language. In Lenneberg, E., Lenneberg, E., (Eds.), *Foundations of Language Development,* vol. I., pp. 121–135. New York: Academic.

Lecours, A.R. (1980). Asymétries anatomiques et asymétries fonctionnelles: l'aphasie des illettrés, *Cahiers de Psychologie, 23,* 283–304.

Marie, P. (1906). Que faut-il penser des aphasies sous-corticales (aphasies pures)? *Semaine Médicale, 26,* 565–571.

McHenry, L.C. (1969). *Garrison's History of Neurology.* Springfield: Thomas.

Milner, B., Branch, C., & Rasmussen, T. (1964). Observations on cerebral dominance. In de Reucke. A.V.S., & O'Connor, M. (Eds). *Disorders of Language,* pp. 200–222. London: Churchill.

Molfese, D.L., & Molfese, V.J. (1979). Hemisphere and stimulus differences as reflected in the cortical response of newborn infants to speech stimuli. *Developmental Psychology, 15,* 505–511.

Penfield, W., & Roberts, L. (1959). *Speech and Brain Mechanisms.* Princeton: Princeton University Press.

Penfield, W., & Roberts, L. (1963). *Langage et mécanismes cérébraux.* Paris, Presses Universitaires de France.

Ratcliff, G., Dila, C., Taylor, L., & Milner, B. (1980). The morphological asymmetry of the hemispheres and cerebral dominance for speech: a possible relationship. *Brain and Language, 11,* 87–98.

Sasanuma, S. (1975). Kana and kanji processing in Japanese aphasics, *Brain and Language, 2,* 369–383.

Segalowitz, S.J., & Chapman, J.S. (1980). Cerebral asymmetry for speech in neonates: a behavioral measure. *Brain and Language, 9,* 281–288.

Singh, J.A.L., & Zingg, R.M. (1966). *Wolf-children and feral man.* London: Archon.

Subirana, A. (1969). Handedness and cerebral dominance. In Vinken, P.J., & Bruyn, G.W. (eds.) *Handbook of Clinical Neurology,* vol. 4, pp. 248–272. Amsterdam: North Holland.

Tezner, D. (1977). *Étude anatomique de l'asymétrie droite-gauche du planum temporal sur 100 cerveaux d'adultes.* Thesis, faculty of médicine, University of Paris.

Tsunoda, T. (1978). *The Japanese brain: working mechanisms of brain and East-West culture.* Tokyo: Taishu Kan.

Vargha-Khadem, F., & Corbalis, M. (1979). Cerebral asymmetry in infants. *Brain and Language, 8,* 1–9.

Wada, J., & Rasmussen, T. (1960). Intracarotid injection of sodium amytal for lateralization of speech dominance. *Journal of Neurosurgery, 17,* 266–282.

Waltari, M. (1972). *Sinouhé l'Égyptien.* Paris: Orban.

Weber, E. (1909). Das Schreiben als Ursache der einseitigen Lage des Sprachzentrums. *Zentralblatt für Physiologie, 18,* 341–347.

Wechsler, A.F. (1976). Crossed aphasia in an illiterate dextral. *Brain and Language, 3,* 164–172.

Wernicke, C. (1874). *Der aphasische Symptomenkomplex.* Breslau: Cohn and Weigert.

Wernicke, C. (1881). *Lehrbuch der Gehirnkrankheiten.* Kassel: Fischer.

Witelson, S.F., Pallie, W. (1973). Left hemisphere specialization for language in the newborn. Brain, 96, 641–646.

Yakovlev, P.I., Lecours, A.R. (1967). The myelogenetic cycles of regional maturation of the brain. In Minkowski, A. (ed.) *Regional Development of the Brain in Early Life,* pp. 3–70. Oxford: Blackwell.

CHAPTER 13

Language Processing: A Neuroanatomical Primer

PATRICIA ELLEN GRANT[1]

Introduction

In attempting to resolve the relationship between writing and the brain, one must first examine the underlying neuroanatomical structures. Numerous studies have been carried out to determine the specifics of functional localization in the brain, ranging from animal experimentation to cognitive testing. This introductory chapter discusses cortical and subcortical areas believed to be important in language processing, and describes the major cortico-cortical connections involved. These anatomical and functional findings are then drawn together to describe entire circuits participating in language processes, and a neurological model for reading and writing is proposed. In the final section the possibility of a neuroanatomical basis for the directionality of various script types will be discussed. This chapter will therefore assist readers, especially those unfamiliar with neuroanatomy, to put into context the material presented in this section. Throughout this chapter it will be assumed that the brain processes of a right-handed individual with left hemispheric dominance for sequential analysis are being discussed.

The present state of knowledge concerning neuroanatomy and functional localization has been extensively reviewed in whole or in part by a number of authors (Kandel & Schwartz, 1985; Shepherd, 1983; Nieuwenhuys, Voogd, & van Huijzen, 1981; Barr & Kiernan, 1983; Lassen, Ingvar & Skinhøj, 1978; Bear, 1983; Ross, 1981; Crossen, 1985; Taylor & Taylor, 1983; Damasio & Geshwind, 1984). The data felt to be most relevant to language processes are summarized here.

Cortical Areas Involved in Language

The various cortical areas discussed below are illustrated in Fig. 1. The exact location and boundaries of all the regions are not known, but experimental evidence to date favors the organization given here.

---

[1] Room 7313, Medical Sciences Building, University of Toronto, Toronto, Ontario, M5S 1A8 Canada.

Primary Sensory Areas

Visual

The primary visual cortex (V 1) is located in the occipital lobe, around the calcarine fissure, covering Brodmann's area 17. In each hemisphere, the primary visual cortex receives information concerning the contralateral visual hemifield. All impulses resulting from light impinging on photoreceptors pass through at least two synapses within the retina before ending on ganglion cells. These cells, also located within the retina, are morphologically classed as X, Y, or W cells. W cells, making up only 10% of the total number of ganglion cells, project to the superior colliculus. This pathway is involved in reflexive coordination of head and eye movement. Y cells, also 10% of the total, project to both the superior colliculus and the magnocellular regions of the lateral geniculate nucleus (LGN), although the majority terminate in the LGN. The fast-conducting axons of these cells participate in directing visual attention and carry information concerning general features and movement. The X cells, constituting 80% of the total, propagate information regarding fine detail, form, and color to the parvocellular and magnocellular regions of the LGN at a slower rate than the Y cells. Within the six-layered LGN, projections from the ipsilateral eye end in layers 2, 3, and 5, whereas those from the contralateral eye end in layers 1, 4, and 6. These projections create a point-to-point mapping of the retina on the LGN. Efferents from the LGN continue on to the ipsilateral primary visual cortex. In the retina there is a region called the *fovea* which contains the highest density of the X-type neurons, i.e., those responsible for high-acuity vision. The fovea also has the highest density of innervation in the retina. This is reflected in the proportionately large area of visual cortex devoted to processing foveal information, an area that begins in the region of the occipital pole and extends approximately one-third of the way into the calcarine fissure (Fig. 2). The resulting high resolution of images falling on the foveal area allows the detection of fine details required for activities such as reading. When devoting one's visual attention to an object, the eyes are always positioned such that the object's image falls on the fovea.

The primary visual cortex itself is organized into columns and layers. According to Hubel and Wiesel (1959), one type of column is concerned with particular axes of orientation in the visual world, meaning that cells in one of these columns respond maximally to a stimulus whose edges are oriented in one direction, while cells in another column respond maximally to a stimulus whose edges are oriented in another direction. Adjacent columns, 30 – 100 μm apart, show a progressive shift in axis orientation of about 10°. Interspersed with the orientation columns are peg-shaped columnar regions that serve as wavelength discriminators. Livingstone and Hubel (1984) have described these cortical pegs as being involved in color vision. A third type, ocular dominance columns, are involved in binocular vision and depth per-

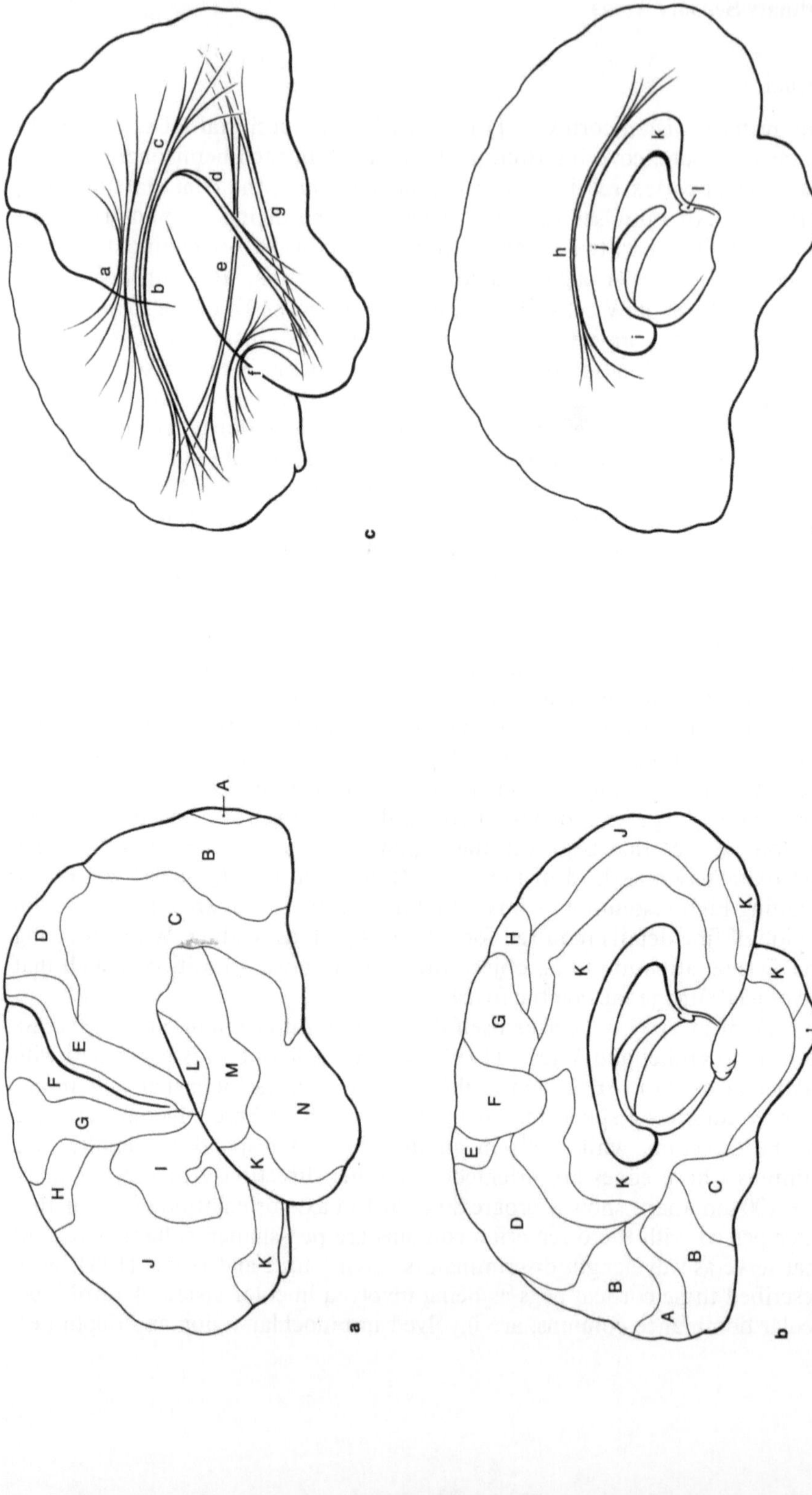

Fig. 1

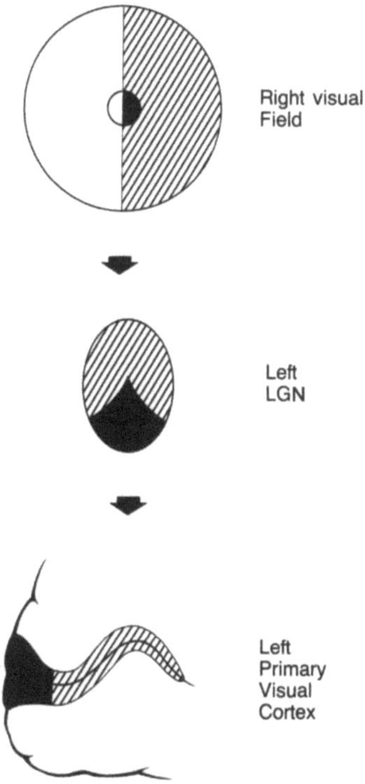

Fig. 2. Projection of the visual field onto the LGN and calcarine sulcus of the primary visual cortex, showing the relative areas devoted to processing central vs peripheral visual field information

Fig. 1. a, b Functional subdivisions of the cerebral hemisphere and the corresponding Brodmann's area

	Functional Subdivision	*Brodmann's area*
A	Primary visual (V 1)	17
B	Secondary visual (V 2) and V 3, V 3 a, V 4	18, 19
C	Parietal-temporal-occipital association cortex	39, 40
D	Higher order somatic and visual	5, 7
E	Somatosensory	1, 2, 3
F	Primary motor	4
G	Premotor, supplementary motor	6
H	Frontal eye field	8
I	Expressive motor	44, 45
J	Prefrontal association cortex	9, 10, 46, 47
K	Limbic	23, 24, 38, 28, 11
L	Auditory	41, 42
M	Secondary auditory	22
N	Visual temporal area	20, 21

c, d Location of major fiber tracts. *a* Superior occipitofrontal fasciculus; *b* Superior longitudinal fasciculus; *c* Posterior branch of *b*; *d* Anterior branch of *b*; *e* Inferior occipitofrontal fasciculus; *f* Uncinate fasciculus; *g* Inferior longitudinal fasciculus; *h* Cingulum; *i* Splenium of corpus callosum; *j* Truncus of corpus callosum; *k* Genu of corpus callosum; *l* Rostrum of corpus callosum

ception. Most ocular dominance columns, orientation columns, and the cortical pegs receive input from one eye only and are thus concerned with either the contralateral nasal or the contralateral temporal visual field. All of these columns combine to constitute a single "hypercolumn", described by Hubel and Wiesel (1974) as a unit of cortex, approximately 2 mm^2 in surface area, that contains the entire set of individual columns required to completely analyze one small region of the visual field (Fig. 3). These findings suggest that the role of the primary visual cortex is feature detection, in terms of perception of straight line segments, color, and depth.

Each column in turn consists of six layers. Layer I receives input from layer III. Layers II and III project to higher order cortical areas, including the extrastriate cortex, the medial temporal cortex (Brodmann's areas 20, 21), the posterior parietal cortex (Brodmann's area 7), and the frontal eye field (Brodmann's area 8). Layer III also sends projections to other ocular dominance columns. Layer IV receives input from the lateral geniculate nucleus (LGN), as does layer VI, but layer VI also sends efferents back to the LGN. Layer V projects to the superior colliculus for reflex actions. This layering allows some functional separation analogous to compartmentalization within the cell body.

In the rhesus monkey, areas involved in processing macular vision have been shown to be the only areas of the primary visual cortex to have in-

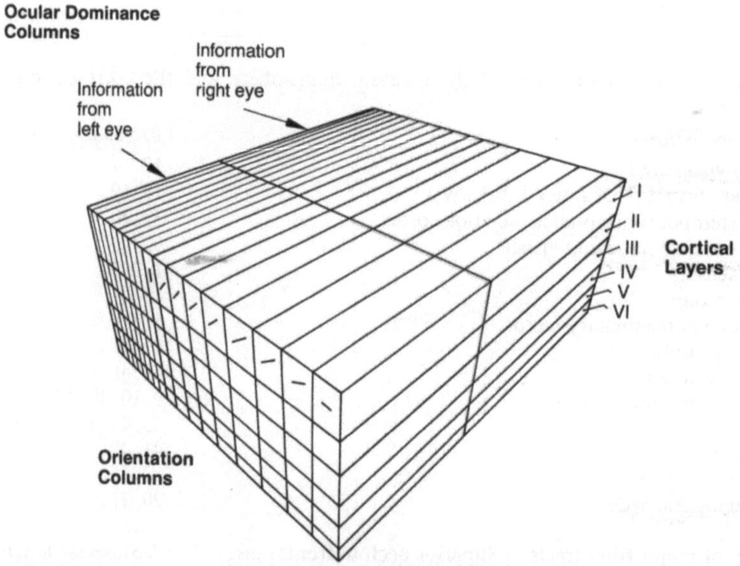

Fig. 3. The structure of a hypercolumn in the primary visual cortex. Cortical layers, ocular dominance columns and orientation columns required to process all the visual information in the area of space served by the hypercolumn are shown. The cortical pegs are not illustrated but are oriented parallel to the other columns and are interspersed with them

terhemispheric connections, with information travelling in the splenium of the corpus callosum (Myers, 1965). This gives each cerebral hemisphere access to primary visual information from the entire macular región, a region of the retina that encompasses the fovea. Although bilateral projections from the *macula* have not been found in humans, cortical lesions above and below the calcarine fissure or in the geniculocalcarine tract result in loss of vision in the entire contralateral visual field except the macular region. However, this phenomenon, known as *macular sparing,* may simply be due either to slight shifting of gaze during testing; to the fact that the cortical representation of this area is relatively large, so that lesions are unlikely to destroy the entire field; or, when the lesion is the result of an infarction, to anastomoses with patent arteries that supply the macular area (Barr & Kiernan, 1983).

Thus, the primary visual cortex would be essential for any type of reading. Foveal vision allows for high-resolution analysis, and interhemispheric connections, if present in humans, would make the results of primary visual cortex analysis of the middle visual field directly available to both hemispheres.

Auditory

The primary auditory cortex is located in the temporal lobe in the two transverse convolutions of Herschl (Brodmann's areas 41 and 42). Sound waves entering the ear are first converted to mechanical waves by the tympanic membrane and the three ossicles. In the cochlea, these mechanical waves are transmitted as changes in fluid pressure, and converted to nerve impulses by the *hair cells,* which respond to different sound frequencies in a tonotopic manner. The impulses are then relayed to the cochlear nucleus by the eighth cranial nerve, whose branches terminate in such a way that the cochlear nucleus is also tonotopically organized with multiple representations. This nucleus contains neurons of various types that break the input down into different functional properties. This information is then transmitted and processed by parallel pathways that subserve different functions. All projections from the dorsal cochlear nucleus go directly to the contralateral inferior colliculus, while the ventral cochlear nucleus sends contralateral information both directly and via a multisynaptic route, as well as connecting ipsilaterally via the superior olivary complex. There may also be interhemispheric projections that contribute to the ipsilateral information reaching the inferior colliculus. Due to the bilateral nature of these pathways, lesions on one side do not result in deafness in either ear. Within the inferior colliculus, the bilateral information is tonotopically organized. Efferents project mainly to the ipsilateral medial geniculate nucleus (MGN), but a few are sent contralaterally to both the other inferior colliculus and MGN. The MGN is divided into ventral and magnocellular portions. In animals, fibers to the ventral division have been shown to terminate in a tonotopical spiral pattern similar to that of the cochlea. Studies in the cat have shown that this ventral subdivision projects only to the primary auditory cortex, whereas the magnocellular subdivision has diffuse projections to surrounding cortical areas as well (Shepherd, 1983).

In most mammals, the primary auditory cortex itself contains a large region that is tonotopically organized. In both the macaque monkey and the cat, separate representations of the frequency range have been shown to exist, as well as two types of frequency columns, which are organized in an alternating pattern (Brugge & Merzenich, 1973; Merzenich & Brugge, 1973; Kandel & Schwartz, 1985). The frequency to which these columns are tuned gradually increases posteriomedially. In *summation* columns, binaural cellular response is greater than the response to either ear individually, while in *suppression* columns, the greatest response occurs when stimuli reach them from a single ear. In the cat, callosal connections have been analyzed by observing the deposition of radioactive amino acids in the primary auditory cortex opposite that of injection, and by observing degeneration of the primary auditory cortex after complete callosal sectioning. These studies show alternating bands of cortex that may branch and connect and receive commissural connections similar to those observed in the ocular dominance columns of the primary visual cortex. Others have also found connections via the anterior commissure (Pandya, Hallet & Mijherjee, 1969). Unilateral lesions of the primary cortex do not affect frequency detection, but if large enough, they may affect the ability to *localize* sound, through detecting differences in amplitude and time of arrival of inputs originating in the two cochleae. It appears that it is the primary auditory cortex opposite the origin of the sound that is most important in its localization (Kandel & Schwartz, 1985, pp. 907–908). On the basis of lesion studies and sodium amytal tests, it has also been suggested that the hemisphere controlling speech is most directly connected to the contralateral ear. On dichotic listening tests, individuals with left hemispheric speech showed a right ear advantage for speech stimuli, while those with right hemispheric speech showed a left ear advantage (Springer & Deutsch, 1981, pp. 66–70).

The primary auditory cortex is arranged in layers similar to those of the primary visual cortex. Efferents are sent from layer VI to the MGN, and from layer V to the inferior colliculus. Layers II and III establish connections with higher order cortical areas.

These findings show that the primary auditory area plays a role in frequency analysis and sound localization. The resolving power of the frequency analysis is crucial for the ability to distinguish different phonetic sounds in human speech.

Somatosensory

The somatosensory cortex is located in the postcentral gyrus of the parietal lobe, which is also known as Brodmann's areas 1, 2, and 3. The somatosensory cortex is an integral part of fine motor control, essential in actions such as writing, providing a route whereby muscle movements can be monitored and tuned.

Proprioceptive information concerns itself with location and movement, which is important in the control of the fine motor movements of the wrist, hand, fingers, and thumb when writing. Impulses from these body regions travel in the fasciculus cuneatus to the nucleus cuneatus, cross a synapse, and continue in the contralateral medial lemniscus to the ventral posterior (VP) nucleus of the thalamus. From there, they project to an area covering the middle third of the postcentral gyrus (Shepherd, 1983, p. 285; Nieuwenhuys, Voogd, & van Huijzen, 1981, pp. 316–330). Projections from the primary somatosensory area reach the contralateral primary somatosensory area, the ipsilateral secondary somatosensory area, the posterior parietal cortex, and the motor cortex, as well as the thalamus, brain stem and spinal cord. These connections allow integrated right and left muscle movement and feedback to higher order cortical areas.

Primary Motor Cortex

The primary motor cortex is located in the precentral gyrus of the frontal lobe, also known as Brodmann's area 4. It receives information from a number of areas. Direct projections are sent from the primary somatosensory cortex, while the ventral anterior (VA) and ventral lateral (VL) nuclei of the thalamus provide information from the cerebellum and basal ganglia that is required for directing finely tuned movements. The premotor cortex also projects to this area and is believed to be involved in providing planned motor programs. The organization of the motor cortex is similar to that of primary somatosensory cortex with respect to the representation of body parts. Those areas controlling the contralateral wrist, hand, fingers, and thumb movements are located in the middle region of the precentral gyrus. The motor cortex, being responsible for the execution of voluntary muscle movement, sends projections directly to the skeletal motor neurons of the opposite side of the body. There is no ipsilateral control of the distal voluntary musculature, including those involved in writing. Thus, if the fibers from the motor cortex are severed, paresis of the contralateral voluntary musculature results. Efferents from the motor cortex to the musculature of the face and vocal chords, however, are more complicated. The motor nucleus of the trigeminal nerve, which supplies ipsilateral muscles of mastication and some smaller surrounding muscles, receives bilateral innervation. The facial motor nucleus, sending efferents to the muscles of facial expression, receives bilateral afferents from the motor cortex for the motor neurons supplying the upper, but only contralateral afferents for those controlling the lower facial musculature. The nucleus ambiguous, which innervates muscles of the soft palate, pharynx and larynx, also receives bilateral afferents, whereas the hypoglossal nucleus, supplying the intrinsic muscles of the tongue as well as a few extrinsic muscles, receives mainly contralateral input from the motor cortex. Thus the cortical area just above the lateral sulcus controls voluntary movement of all the contralateral muscles involved in speech and facial expression, with the exception of the upper muscles of the face, the soft palate, the larynx and the pharynx, all of which are bilaterally controlled. The fine movements of the hand and the facial musculature enable written and spoken language, respectively, to occur. Thus, this cortical region is intrinsically involved in the normal development of language skills.

Higher Order Sensory Cortices

Secondary Visual Cortex (V 2), V 3, V 3 a, V 4, and Visual Temporal Area

The secondary visual cortex is located in the occipital gyri of the occipital lobe (Brodmann's area 18). Visual III, III a, IV, and V are also located in this area as well as in the superior temporal sulcus (part of Brodmann's area 19). The visual temporal area is found in the anterior and inferior regions of the temporal lobe (Brodmann's areas 20 and 21). Areas 18 and 19 receive inputs mainly from the ipsilateral primary visual cortex but also from the ipsilateral pulvinar of the thalamus. Connections to the ventral inferior temporal lobule, the inferior and superior parietal lobules, and the premotor area of the frontal cortex have been shown, as well as reciprocal connections with the contralateral higher order visual areas via the splenium of the corpus callosum. Areas 20 and 21 receive afferents from areas 18 and 19 and project to areas of the frontal cortex.

Little is known about the function of these areas, but there are indications that in addition to a hierarchical organization resulting in feature abstractions of increasing levels of complexity, there exist multiple parallel pathways, each processing one aspect of the stimulus at an increasing level of complexity. In this view, no one pathway is responsible for the detection of an entire object, but only for some particular aspect of it. Thus, simultaneous activity of multiple pathways is required for visual recognition. Support for this hypothesis comes from two areas (for review see Kandel & Schwartz, 1985, pp. 381–382). One is the existence of various visual agnosias such as color, object, or movement agnosia. The other is the existence of two major largely bidirectional pathways proposed on the basis of functionally separate projections from the primary visual cortex to higher visual areas (Van Essen & Maunsell, 1983). The first of these two major pathways is as follows:

V 3
V 1 → V 2 ⇄ ↓ Inferotemporal cortex
V 4 ↗

This pathway, thought to be a continuation of the X pathway, is suggested to be concerned with color and form. This is supported by behavioral studies that show the inferotemporal cortex to be involved in visual discrimination learning. The second pathway,

$$V1 \rightarrow V2 \rightarrow V3 \rightarrow \begin{matrix}\text{Visual Temporal}\\ \text{Area}\end{matrix} \rightarrow \begin{matrix}\text{Posterior Parietal}\\ \text{Cortex}\end{matrix}$$

is thought to be a continuation of the Y pathway, and to be concerned with movement and attentional aspects. Many interconnections between these and other parallel pathways exist to allow for encoding and analyzing complex visual stimuli.

These higher order visual cortices allow integration of information between hemispheres and perform complex analyses resulting in the association of visual stimuli, such as words, to meanings derived through past experiences.

Posterior Parietal Cortex

This area consists of the superior parietal lobule (Brodmann's areas 5 and 7), with Brodmann's area 5 involved in processing somatic information and area 7 in processing visual information. The superior parietal lobule receives inputs from the primary somesthetic area (Shepherd, 1983). It has reciprocal connections with the pulvinar, lateral posterior (LP), and lateral dorsal (LD) nucleus of the thalamus and sends projections to the inferior parietal lobule. Lesions of this area cause deficits in understanding the significance of information received, and large lesions can cause tactile agnosia and loss of awareness of the relative spatial location of different body parts. When the damage occurs in the dominant hemisphere aphasia often results, whereas lesions in the opposite hemisphere often result in an inability to perceive information relayed by nonverbal aspects of language such as emotional tone (Kandel & Schwartz, 1985, p. 680). Thus, this region is required to attach literal and metaphorical significance to the written word as well as to coordinate movements required for writing.

Secondary Auditory Cortex

The secondary auditory cortex is located in the superior temporal gyrus, including the planum temporale, of the temporal lobe (Brodmann's area 22). Afferents to this area arise from the dorsal MGN of the thalamus and primary auditory cortex. There are reciprocal connections between frontal premotor areas, VP nucleus and pulvinar of thalamus, and parietal-temporal-occipital association cortex. Other efferents are sent to limbic areas such as the cingulate gyrus and parahippocampal gyrus, which are the gyri that receive or carry much of the input from other regions.

On the left side, the planum temporale contains Wernicke's area, which draws on experience to recognize and interpret auditory input. The cor-

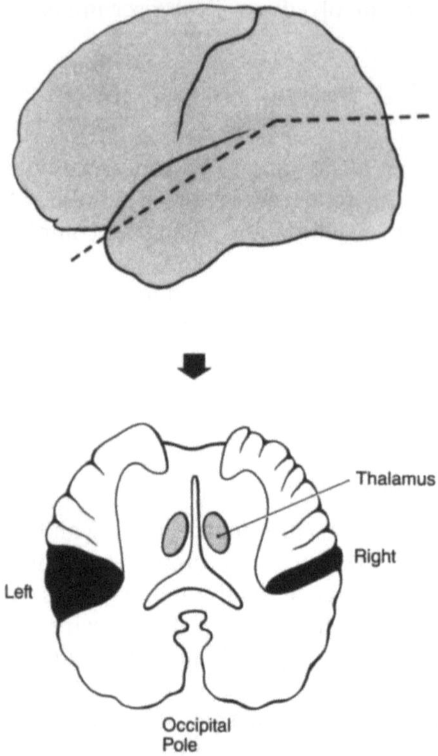

Fig. 4. Anatomical asymmetry of the planum temporale (*shaded regions*) shown by a horizontal section in the plane of the lateral fissure

responding region on the right has been shown to be involved in the interpretation of the emotional intonations of speech and the emotional content of facial expressions (Ley & Bryden, 1979). This area is therefore extremely important in language comprehension; the left side in understanding the literal content of language, and the right side, the emotional content (Rizzolatti, Umilta, & Berlucchi, 1971). Anatomically, the left planum temporale has been shown to be much larger than the right (Fig. 4), possibly reflecting an emphasis on its style of processing when interpreting language.

Higher Order Motor Cortices

Premotor and Supplementary Motor Cortex

These cortical areas are found in the frontal lobe, rostral to the postcentral sulcus (Brodmann's area 6). Both of these areas have crude somatotopic representations. They receive inputs from VA and VL thalamus, areas 5 and 7 of the posterior parietal lobe, and the frontal eye field (FEF; area 8). The supplementary motor area (SMA) sends projections to the premotor area, and both project in a somatotopic manner to the primary motor cortex. The SMA is thought to be responsible for the planning of sequential motor tasks and, along with the premotor area, the modification of precise motor movements. Speaking, reading, and writing all require such skills.

Frontal Eye Field

The FEF is located in the frontal cortex, rostral to area 6 (Brodmann's area 8). This area is responsible for voluntary conjugate movements of the eyes, through polysynaptic projections to the motor nuclei of the extrinsic musculature of the eye. Afferents from VA thalamus, SMA, and parietal cortex influence activity. This area is active in both hemispheres during reading.

Expressive Motor Areas

The expressive motor areas are located in the inferior frontal gyrus of the frontal lobes (Brodmann's areas 44 and 45). In the left hemisphere this area is called Broca's motor speech area, but the corresponding area in the right hemisphere has no specific name. In both hemispheres this area receives input from the VA thalamus, which is known to be involved in planning motor activities. Lesion studies show that Broca's motor speech area is involved in both the smooth production of speech and the comprehension of complex grammar (Blumstein, 1981; Mohr et al., 1978), and PET scans show it to be activated during both silent and vocalized speech (Lassen, Ingvar, & Skinhøj, 1978). Reciprocal connections with higher order auditory areas responsible for language reception are made via the uncinate fasciculus, giving Broca's area access to language input. The corresponding area in the right hemisphere also becomes activated. This leads to the speculation that it is responsible for programming motor movements underlying emotional expression. Lesion data have supported this theory (Ross & Mesulam, 1979). The connections here are similar to those on the left, but here the uncinate fasciculus reciprocally connects this area with regions involved in the reception of the emotional content of language and gestures. Thus, both the left and right inferior frontal gyri are required to decipher the full meaning of spoken and written language.

Association Cortices

Parietal-Temporal-Occipital

The parietal-temporal-occipital association cortex is located at the junction between the parietal, temporal, and occipital lobes (Brodmann's areas 19, 21, 22, 37, 39, and 40). This association cortex is polymodal, reciving auditory, visual, somesthetic, and kinesthetic information from the higher order areas. It also receives afferents from prefrontal cortex, limbic areas and from VA, VL, and LP thalamus. Efferents are sent to the FEF, temporal cortex, the reticular activating nucleus, and limbic areas. Visual, auditory, and somatosensory information is sent to the frontal lobe. In the left hemisphere, this area has been implicated in associating language symbols with other modalities (Blumstein 1981; Luria 1973), while on the right it has been shown to be involved in the nonsyntactical processing of language (Kaczmarck, 1984). The parietal-occipital-temporal association cortex is therefore crucial in the development and recollection of polymodal associations that give language its meaning.

Prefrontal

The prefrontal association cortex is located in the rostral part of the dorsolateral surface of the frontal cortex (Brodmann's areas 9, 10, 46, and 47). This region is well developed only in primates and is thought to play a role in selecting the appropriate behavioral responses based on foresight, judgement, sensory inputs, and past experience. Reciprocal connections exist with the medial dorsal (MD) nucleus of the thalamus, which is known to be involved in memory and emotions. Inputs to the prefrontal area arise from VA, VL, and pulvinar nuclei of the thalamus and from frontal, parietal, occipital, and limbic cortical regions. These connections provide the frontal cortex with access to input from all modalities. In monkeys, the central superior frontal gyrus, shown to be involved in preservation of response patterns, has major projections to temporal lobe regions involved in the perception of complex visual and auditory information. The implication is that the prefrontal area is involved in selecting the appropriate language responses and influences the perception of both auditory and visual language.

Limbic

The limbic area consists of parts of the temporal and parietal cortices as well as the cingulate and parahippocampal gyri, the temporal pole, and the orbital frontal cortex (Brodmann's areas 11, 23, 24, 28, and 38). These areas play an intricate role in both memory and emotions. The orbitofrontal cortex is involved in relating emotions to social and environmental experiences. Inputs arrive from MD thalamus and the amgydala, as well as from VA, VL, pulvinar, and intralaminar thalamic nuclei. Both the angulate and enterorhinal gyri receive widespread neocortical input and send efferents on to the hippocampus. The posterior angulate gyrus, receiving visual and auditory association fibers, is thought to be involved in visuospatial integration. The limbic cortex participates in language processing through its involvement in memory and emotions, and both written and spoken language rely on memories and emotional responses for successful communication.

Subcortical Regions Involved in Language

Thalamus

The thalamus is believed to play an important role in language function. All lesions in this center are associated with similar syndromes, characterized by flowing speech with varying degrees of comprehensibility and relatively preserved comprehension and repetition of other people's speech. Perseveration and lack of spontaneous speech may also occur (Crosson, 1985). It is not surprising that the thalamus has a role in language function, for it serves as a relay for incoming sensory information en route to cortical areas; provides sensory feedback to motor cortices; and allows for integration of input from association cortices.

The thalamus is divided into a number of functionally distinct nuclei which are described below (Fig. 5).

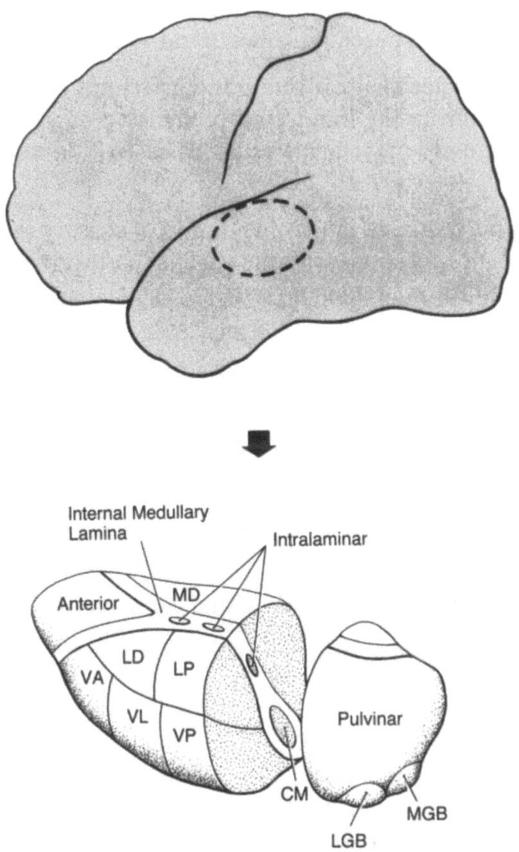

Fig. 5. Location and structure of the left thalamus

Medial Dorsal Nucleus. The MD nucleus receives afferents from the amygdala and hippocampus, and sends efferents to prefrontal and orbitofrontal cortex and the striatum. This nucleus is involved in both emotional and memory circuitry, with bilateral lesions resulting in retrograde amnesia. Access to emotional information and facts from memory is required by language centers in order to communicate thoughts.

Intralaminar Nuclei. These nuclei may play an indirect role in language functions. They receive inputs from motor and premotor cortices, VA thalamus, the reticular formation, the globus pallidus, and the spinal cord. Efferents project to the striatum. These nuclei are thought to play a role in arousal and hence may be involved in the activation of muscles used in language production.

Lateral Dorsal Nucleus. The LD nucleus has reciprocal connections with the angulate gyrus. Through these connections it is thought to play a role in emotional expression.

Ventral Anterior Nucleus. This nucleus relays information from the globus pallidus to the premotor cortex and thus plays a role in planning motor actions. Stimulation results in involuntary speech (Schaltenbrand 1975).

Ventral Lateral Nucleus. The VL nucleus is involved in motor actions relaying information from the cerebellum to the motor and premotor cortex. Its role is therefore in the precise muscle movements required in writing and speech production.

Lateral Posterior Nucleus. This nucleus aids in the integration of sensory information by setting up reciprocal connections with the parietal cortex. It assists in the development and recollection of memory patterns associated with written and spoken language.

Pulvinar. This nucleus is thought to provide an important link between parietal, temporal, and occipital association areas, and hence is an important element in the integration of sensory information. It receives inputs from the superior colliculus and primary visual cortex, and has reciprocal connections with the parietal, temporal, and occipital cortices. The pulvinar also sends projections to the VA thalamus and is therefore indirectly connected with frontal areas, which may allow it to influence anterior language areas.

Lateral Geniculate Nucleus. See discussion of primary visual cortex above.

Medial Geniculate Nucleus. See discussion of primary auditory cortex above.

Striatum (Caudate and Putamen)

The caudate and putamen are believed to be involved in planning motor and higher mental functions (Springer & Deutsch, 1981). The caudate's involvement in language functions is considered to be the more direct, since stimulation of the head of the left caudate may arrest speech or induce fluent but inappropriate speech (Van Buren, 1966). Lesions in the left putamen have varying effects. Both nuclei receive afferents from most regions of cortex, but the caudate receives more from the prefrontal cortex and the putamen more from the motor cortex. The caudate also receives input from VA, DM and intralaminar nuclei of the thalamus and the substantia nigra. Both nuclei send the majority of their efferents to the globus pallidus and virtually none to the cerebral cortex. It has been suggested that the caudate influences language through an indirect route to the anterior language zones via the globus pallidus and VA thalamus (Crossen, 1985).

Globus Pallidus

The globus pallidus (GP) is thought to be involved in programming motor and higher mental functions. The lateral medial portion is implicated in language functions, as lesions here result in word-finding difficulties, paraphasia, and subjective feelings of uncoordinated thought and speech patterns (Svennilson, Torvik, Lowe, & Beksell, 1960). Stimulation results in arrest of ongoing speech (Hermann, Turner, Gillingham, & Graze, 1966), probably through inhibition of VA thalamus, which activates anterior language areas. Projections from the GP also reach VL thalamus to modify activation of the motor cortex.

Cortico-Cortical Fiber Tracts

Intracortical (Association) Fibers

Association fasciculi join regions of cortex within the same hemisphere (Fig. 6), while short-range, or *arcuate* fibers join adjacent gyri. Longer range tracts joining different regions of cortex are illustrated in Fig. 1. These tracts may carry information concerning different modalities.

The *arcuate fasciculus,* joining the inferior frontal gyrus and the higher order auditory cortex, is known to be important in language processes. On the left side it is thought to carry phonological information, and on the right, to mediate information concerning the emotional content of language (Ross, 1981). It also provides connections between frontal motor cortex and higher order visual cortex and between premotor cortex, visual temporal area, and inferior and superior parietal lobules.

The *cingulum* is a major component of the limbic system. It provides a pathway for communication between a number of neocortical areas and the limbic system. It feeds information into the cingulate gyrus, the parahippocampal gyrus, and the hippocampus.

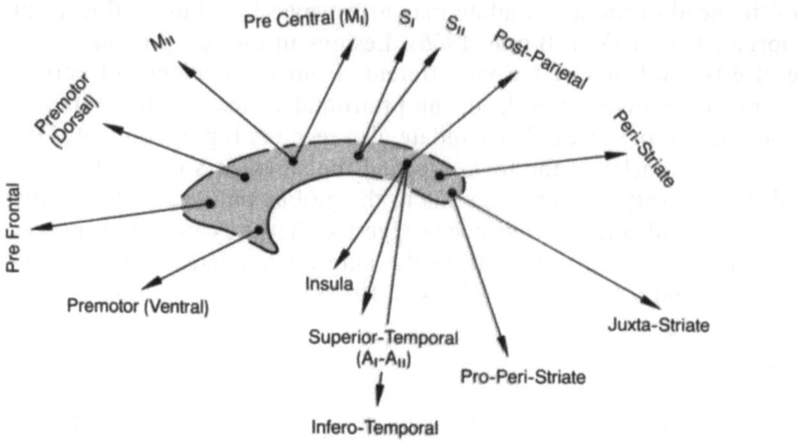

Fig. 6. Topography of commissural fibers in the corpus callosum. (Adapted from Pandya & Karol, 1971)

Intercortical (Commissural) Fibers

Commissural fibers provide a means of communication between the two hemispheres. On the basis of animal studies, they are thought to join either corresponding areas of cortex or those that are closely related in function. The *corpus callosum* is the major commissure, carrying most of the neocortical connections. It is subdivided into four anatomical subsections which are, from anterior to posterior, the rostrum, genu, truncus, and splenium. The rostrum connects orbitofrontal regions and the genu connects prefrontal regions. The pericentral regions, parietal cortex, and temporal cortex send efferents to the opposite hemisphere via the truncus and the higher order visual cortex via the splenium (Fig. 6). Although in humans, no information from the primary visual, primary auditory, primary motor and primary sensory cortices has been shown to cross over to their corresponding areas in the opposite hemisphere, in animals transfer of information regarding the foveal region, parts of the auditory region, and sensorimotor regions have been reported (Myers, 1965). In a recent study, the anterior four-fifths of this commissure was found to be 11% larger in non-right-handers (Witelson, 1985). The corpus callosum is important in language functions, allowing integration of the left and right processes and in particular allowing communication between the two inferior frontal gyri. The *anterior commissure* carries the remaining few neocortical connecting fibers including, in monkeys, some connecting primary auditory cortices.

Circuitry in Language Processing

Traditionally, both language input through reading or listening and language output through writing or speaking have been considered to be functions of the left hemisphere. However, these initial conclusions were drawn on the basis of cognitive testing of the literal denotative aspects of language carried out on lesioned patients originally literate in alphabetic scripts (Luria, 1973). Thus only the semantic deficits were illustrated. Recent studies (e.g., Ross, 1981) have shown that the right hemisphere also plays an essential role in language functions.

The left hemisphere has long been known to be specialized for the recognition and interpretation of complex auditory input, as reflected in the anatomic asymmetry in the region of the planum temporale (Fig. 4). Its mode of processing, which is influenced by conscious effort, excels in sequential, analytical, and syntactical tasks requiring precise responses. It seems that input into this hemisphere is translated into sequential information and analyzed predominantly by logical orderly means. This is supported by the observation that left temporal lobe epileptics characteristically show "ideative" traits due to increased limbic input into sensory association areas of the left hemisphere, resulting in emotions being expressed as reflections on abstract topics such as religion and philosophy (Bear, 1983). Epilepsy, being an increase in cortical activity, is assumed to enhance or overemphasize the normal functions of the affected area.

The right hemisphere has been shown to be superior in recognizing patterns, colors, and pictures, as well as in making associations and doing spatial manipulations. The style of processing utilized by this hemisphere seems to be more of the gestalt type, once the appropriate associations have been set up through experiences, and less a result of conscious effort. It excels at parallel, spatial tasks and is required for the comprehension of metaphorical and emotional speech. These parallel tasks involve processing of several aspects of a problem simultaneously instead of sequentially. It seems likely that all activities in the right hemisphere are carried out on this level. This is supported by the observation that right temporal lobe epilepsy causes emotions to be expressed more directly, resulting in an altered affect or mood and impulsive behavior (Bear, 1983).

Language in humans seems to be a predisposed trait. This is supported by evidence such as the asymmetry in size of the planum temporale apparent even in the fetus. From birth through the first few years of life, children show the ability to distinguish between the acoustic sounds of all phonemes, the basic units of sound. As they mature, however, they retain the ability to distinguish slight variations in only those phonemes used in the language to which they are exposed (Eimas, 1985). This ability to recognize frequently used phonemes develops at the expense of the others and is thought to occur through synaptic facilitation, the alteration of the strength of neuronal connections as a result of experience. At the same time, general rules about grammar and syntax are formed, merely through language exposure. The fact that the rules are extracted in such a manner suggests that there is a genetically programmed predisposition to recognize these general rules.

Ross (1982) proposed that the right hemispheric areas controlling language are organized in a similar manner to those in the left, based on his observation that increased blood flow occurred in homologous regions during language tasks in both hemispheres. If this is so, then it would seem likely that these regions would show a similar development. It is possible that, as with phonetic discrimination on the left, the right auditory cortex develops, facilitating pathways in such a way that the ability to recognize frequently encountered tones of intonation and melody are developed at the expense of others. This would imply a genetically based development of tonal discrimination and emotional detection.

As well as building up these structural neurological relations, synaptic facilitation has been proposed as a means of building up long-term memory stores of meanings associated with auditory and visual cues, as well as their associated informational and emotional contents. The mammillary body and MD thalamus seem to be involved in the encoding at the time that learning is occurring. If this is true, the hippocampus would then mediate the transition from short-term to long-term memory, with the right hippocampus possibly transferring patterns of sensory input, such as faces, and the left transferring sequences of sensory input, such as verbal input. The hippocampus is thought to play a role in the development of *declarative memory.* Declarative memory, expressible in a declarative verbal statement and often acquired in one exposure, requires conscious effort to both create and recall. Such memory processes would be employed when learning the correspondence between objects and written words. The amygdala, located beside the hippocampus in the temporal lobe, is thought to be more active in the acquisition of *reflexive memory,* which develops gradually through repeated exposure to a stimulus or through repeated performance of a task. No conscious effort is required in either learning or recalling, and its acquisition is reflected by improved performance. This type of memory may result in the development of motor patterns required for speech (Kandel & Schwartz, 1985).

Other structures that may also be involved in memory circuitry are the posterior cingulate gyrus, the inferior temporal gyrus, the prefrontal cortex, the parahippocampal gyrus, and the mammillary bodies. The exact circuitry is not known, but areas such as the cingulate gyrus and enterorhinal area (area 28) receive inputs from much of the neocortex and send information on to the hippocampus. Recent memory and the ability to visually discriminate sequential but not simultaneous stimuli have been linked to the hippocampus. The two hippocampi can also communicate via the hippocampal commissure.

Many of the centers involved in memory are intricately linked to those involved in emotional circuitry. Four major areas have been linked in this way to emotions: the cingulate gyrus, the orbitofrontal cortex, the amygdala, and the hypothalamus. Higher cortical centers feed into this system mainly through the cingulate gyrus, while output to the body is regulated by the hypothalamus via hormones. As suggested earlier, the two hemispheres are thought to express emotion in different ways; the left ideatively and the right emotively.

In our earliest language exposure, which is auditory, it seems plausible that reflexive memory processes would allow common words, colloquial phrases, and certain commonly used grammatical forms to acquire meanings that include both emotional and semantic significance. Both hemispheres would be essential in the formation of such links: the left hemisphere may lead in deciphering the grammatical structures and determining literal meaning, while the right may lead in deciphering metaphors and determining the emotional tone. The hemispheric localization of these processes is supported by lesion data (e.g., Wapner, Hamby, & Gardner, 1981; Winner & Gardner, 1977). It seems possible, therefore, to think of the hemispheres as two interdependent parallel systems, processing various aspects of language in a complementary manner. Many opportunities for integration and comparison of information are made possible through callosal connections. It would seem that a complete understanding and effective expression of all aspects of language, including allegories, metaphors, and emotional connotations embedded in grammatical structure, would require a cooperative effort from the two hemispheres. For example, an innocent defendant appealing to a jury has to choose the correct words and grammatical structures to most powerfully convey his or her emotional feelings of innocence. The actual expression of emotion has been shown to be a right brain function, involving the right inferior frontal gyrus, whereas the grammatical form in which the sentences are expressed and the actual choice of words are functions of the left inferior frontal gyrus. The goal is not only to arouse emotional feelings in the jury (relying on right brain reception), but to logically portray innocence (a left brain function). Thus, when considering the implications of dichotic tests showing left brain preference for spoken language, the use of sentences or word sequences lacking metaphorical content and emotional impact must be taken into account. Assumptions based on quantitative analysis of positron emission topography (PET) scans should also be made with care, taking into account both the grammatical and metaphorical complexity of the text involved when reading tasks are performed.

Language that is written and read can achieve the same impact as oral language only if it is able to gain access to the same lexicon. For ease of transition between the oral and the written, a script must have some type of phonetic association, whether it be through the association of each symbol with a word, a syllable, or a phoneme. In order to write the script, kinesthetic patterns must be stored. Thus, when learning to read and write a particular script, it seems likely that parallel channels of access to the lexicon are established. One pathway would access the lexicon via auditory memory pathways that were established due to exposure to conversational language. Another pathway would access the lexicon via visual memory pathways, either directly or indirectly. The indirect route would first require finding the associated pronunciation of the script and then the lexicon could be accessed by the auditory route. The direct pathway would be utilized when there existed a direct correspondence between visual input and meaning. Such direct pathways may develop from the indirect pathway after repeated exposure to the same visual input. When learning how to write, kinesthetic pathways to the lexicon may also be set up inadvertently as a result of repeated writing of the same unit. The end result would be a diffuse storage of language information.

Cerebral blood flow studies using PET (Lassen, Ingvar, & Skinhøj, 1978) have shown which centers are most active during reading. Silent reading activates the inferior frontal gyrus (Broca's area on the left), the SMA, the FEF, higher order auditory cortex (Wernicke's area on the left), and visual association cortex. Primary visual cortex is presumably active as well, but this area was not measured in this study. Similar regions in both hemispheres were found to be active. The relative activity of the various regions in the left and right hemispheres would probably be a function of the content of the text, but no such studies have been documented to date. Based on the cerebral blood flow studies and on the present knowledge about function of the cortical areas, the circuitry for reading shown in Fig. 7 is proposed. The premotor cortex feeds information to the FEF which in turn controls the horizontal eye movements. The retinal images are sent to the primary visual cortex, via the LGN, where feature abstraction takes place. From there visual information is passed on to higher order visual cortices where interhemispheric communication begins and associations are recalled. This information is fed into the parietal-temporal-occipital association cortex, where associated memories in other modalities are accessed. Area 22 is then called upon, along with the inferior frontal cortex in complex cases, to combine and decipher the grammatical and emotional content of the text. Information is passed between the two hemispheres by callosal connections (not shown), allowing integration and complementation of left and right hemispheric processes. Feedback to the cortex is mediated mainly by the thalamus. Efferents from VA thalamus to the FEF and inferior frontal gyrus provide information about the motor movements being executed. The pulvinar, since it provides links between parietal, occipital, and temporal cortex for sensory integration, is likely to be involved in deciphering the script by assisting in drawing associations between visual input and semantic meaning.

To determine what areas are active while writing, regions active during the four tasks of speaking, reading silently, reading aloud, and moving the fingers were analyzed. Finger motions activated the premotor cortex bilaterally, especially in the SMA, as well as the primary motor and somatosensory regions for the hand and fingers. Reading aloud activated the mouth area of the somatosensory and motor cortices and the primary auditory cortex, and those regions activated when reading silently. Speaking activated the SMA, the mouth area of the somatosensory and motor cortices, the primary and higher order auditory cortex, and the inferior frontal gyrus (Lassen et al., 1978).

The difference between speaking aloud and thinking of what one wanted to say was assumed to be the same as the difference between reading aloud and silently: the vocalization activated motor and somatosensory areas of the cerebral cortex. Hand movements required to write were assumed to result from cortical activity similar to the finger movements studied (Lassen et al., 1978). Using this information as well as information concerning the function of the various cortical areas, a circuitry for writing has been proposed (Fig. 7). Auditory association cortex is assumed to be the area of cortex mediating associations between auditory sounds and meanings since speaking does not activate any other association cortices. When writing in a language such as English, the idea must first be translated into internalized speech. The recollection of the pronunciation of the word corresponding to the desired meaning activates area 22, the auditory association cortex. This information is passed on to the inferior frontal gyrus where complex grammatical structures are formed to best convey the desired emotional and informational content. This requires interaction and information exchange between at least the two inferior frontal gyri (expressive motor areas). The pre-

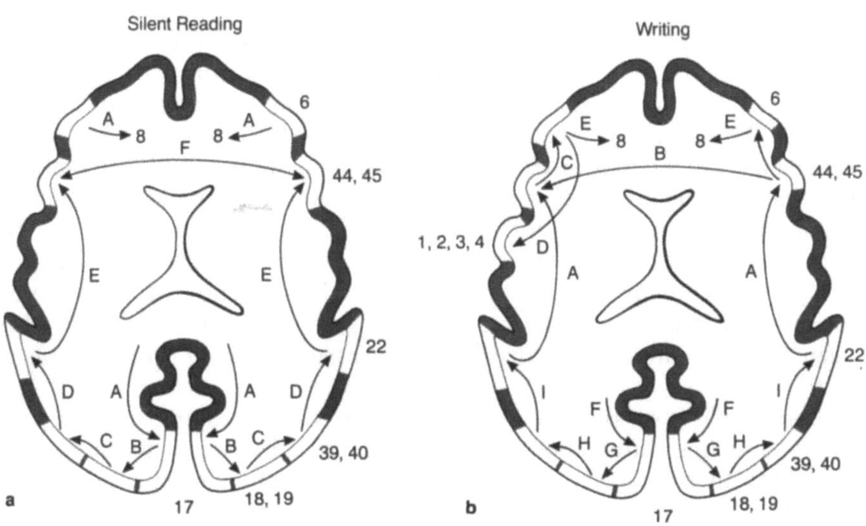

Fig. 7 a, b. General outline of information flow (indicated by *arrows*) for the processes of **a** silent reading and **b** writing. *Numbers* denote Brodmann's areas, *letters* indicate the sequence of information flow. For detailed description see text

motor area then receives the appropriate information to construct motor programs that can direct the motor cortex to write the desired words and the FEFs to keep what is written in view. The visual input regarding what was just written is processed, eventually reaching area 22 where comparisons between written and desired meaning can be performed. The visual pathway must be involved to confirm that the word produced does in fact correspond to the word that was intended: this comparison probably occurs in the parietal-occipital-temporal association cortex. Integration of left and right language functions occurs via the callosal connections. Similar subcortical structures from the right and left hemisphere are presumably involved in such integrations, and one would expect the feedback to motor and premotor cortices from VA and VL thalamus to increase due to the activation of peripheral musculature.

The diagrams in Fig. 7 are not intended to be proclamations of the exact circuitry involved in reading and writing, but rather a simplified outline of the general direction of information flow. The actual circuitry is undoubtedly much more complex, including many feedback loops, modulating inputs, and interhemispheric connections.

When analyzing reports on various PET studies, it becomes apparent that not only do the modality of the stimulus and the strategy used by the subject alter the pattern of blood flow, but the quality of each modality affects the activation and relative participation of the centers involved in language (Phelps & Mazziotta, 1985). Simple spoken words activate primary and higher order auditory cortex, whereas more complex verbal input activates the inferior frontal cortex as well. In one study, simple verbal stimuli caused asymmetrical increases in the left hemispheric auditory areas (Phelps & Mazziotta, 1985) whereas in a more complex task (involving listening to a Sherlock Holmes tale in order to memorize phrases from it) right hemispheric regions appeared to be equally if not more activated than the left (Phelps, Mazziotta, & Huang, 1982). Different distributions in the basal ganglia and frontal lobes were also observed. This seems to indicate that the right and left hemispheres are equally important in deciphering language when presented in complex form, and that the relative contributions of each hemisphere to this decoding process are very much dependent on the modality and quality of the stimulus presented. The strategies used to perform the task at hand, such as image creation or sequence recollection, have also been shown to influence the distribution of cortical activity (Phelps & Mazziotta, 1985).

In summary, we require a complex interaction of complementary processes occurring throughout the brain to extract the richest meaning of a text. Even then, however, this meaning is quite individualistic, relying on past experiences, present situations, and individual characteristics.

Language Processing and Script Type

On the basis of information presented above, we can begin to look at different script types that require different methods of analysis, and their different distributions of cortical activity.

Alphabetic scripts are the most common in Western societies. In these scripts, consonant and vowel segments from spoken language are represented by letters, although the correspondence between the sounds and the symbols is rarely perfect. Alphabetic scripts require sequential analysis into syllables, the natural units of speech, and then morphemes, the units of meaning, and words before the semantic lexicon can be accessed. Individual phonemes, units of sound that combine to form syllables, and syllables are not associated with any particular meaning. This sequential analysis is thought to be best executed by the left hemisphere, and processing is probably dominated by this side until the semantic lexicon is accessed, at which point the right hemisphere can begin to excel and interact as metaphorical and emotional aspects come into play. That is, such interactions may only be possible once the lexicon is known. This would mean that the left hemisphere would lead the right in processing literary comprehension, except for common words that can be recognized through a gestalt-type process in the right hemisphere (Sinatra & Stahl-Gemake, 1983). Assuming that letters could be recognized in the peripheral visual field (an ability that has not been proven), horizontal scanning would increase the speed of reading since more letters can be contained in the visual field at one time in the horizontal than in the vertical direction (Skoyles, 1986). Rightward horizontal scanning would also allow leading of the right visual field. Information would then reach the left hemisphere slightly ahead of the right, and perhaps allow the lexicon to be accessed through sequential analysis before the right hemisphere begins to call on associations with the word form. This may allow the image of the word to reach the right visual association cortices at the same time that associations with the lexicon presented by the left hemisphere are made. This could aid in the establishment of direct associations to the lexicon, via the right hemispheric gestalt processes that occur as a result of repeated exposure. Once these associations are available, the speed of reading increases. This is supported by the preservation of speed reading despite left hemispheric lesions (Springer & Deutsch, 1981). However, once this occurs, the advantage of a leading right visual field is lost. Normal reading involves a combination of the sequential and gestalt-type processes. Thus, it is possible that rightward scanning increases the rate of sequential analysis, the much slower process, relative to the gestalt-type processes, increasing the overall speed of reading.

When the symbols correspond to syllables, the script is called a syllabary. Like alphabets, syllabaries attempt to provide a visual representation of auditory sequences. Thus, a sequential analysis may also be performed to obtain meaningful units of sound. These units may then be made available to both hemispheres for an integrated analysis. As with alphabetic scripts, a rightward direction of reading may maximize the speed at which the text is deciphered.

Semitic consonantal alphabets, through their exclusion of vowel symbols, may shift the cortical regions of emphasis. Their consistent right-to-left nature suggests a possible advantage to a right hemispheric lead in analysis. These scripts, being more ambiguous than alphabetic and syllabic ones, require some knowledge of the language before the text can be deciphered. The requirement of increased familiarity makes it possible that the right hemisphere may lead in the analysis by making "best guesses" at the word meaning, with the subsequent complementary processing in each hemisphere fine-tuning it (Taylor & Taylor, 1983). This right hemispheric ability to recognize well-known words or phrases is supported by the observation that the right hemisphere is capable of producing often-used phrases.

In contrast to phonographic scripts are logographic scripts. Where phonographic writing systems correspond to spoken language, which in turn are related to units of meaning called morphemes, logographic writing systems have a direct relationship to morphemes. Use of logographic scripts may result in yet another cerebral emphasis of activity. The foveal analysis required of the more complex individual characters may predispose the scripts to a vertical arrangement, since vertical scanning is the easiest direction of scanning when the fovea must receive the information (Skoyles, 1986). Recall that for the analysis of alphabetic scripts, the peripheral field was assumed to be able to participate in the recognition process. It seems likely that higher order visual areas receive information about the entire symbol as a result of interhemispheric connections that may begin at the primary visual cortex as it performs the feature analysis. Since both the phonetic pronounciation and the meaning must be memorized, both hemispheres are probably involved in the analysis from the onset. The right hemisphere may recognize the individual symbols more effectively, but their attachment into meaningful sequences and the associations of the phonetic pronunciations are probably best handled through interaction with the left hemisphere. It is possible, therefore, that in this case the need for foveal analysis is the largest force influencing the direction.

Although there exist different languages and different types of scripts, it seems unlikely that we are predisposed to best understand one particular type. Rather, through learning and exposure to specific languages, particular methods emphasizing different cortical regions are developed due to the nature of the script. Just as mathematics is best suited to describe some phenomena and art others, different languages, because of their inherent natures, are better at expressing or developing different cognitive tasks or methods.

However, the extent to which a language shapes the mind, as opposed to the mind influencing the language, is difficult to determine. Their interdependence results in a mutual development. The hypothesis that the method of language processing has influenced script directions is plausible given the above scenarios, but things are not that simple. We must not neglect the backlash of a given script type on mental development, and the continual interaction of these phenomena as a society develops.

References

Barr, M. L., & Kiernan, J. A. (1983). *The human nervous system: an anatomical viewpoint*, 4th ed. New York: Harper & Row.

Bear, D. M. (1983). Hemispheric specialization and the neurology of emotion. *Archives of Neurology, 40*, 195–202.

Blumstein, S. E. (1981). Neurolinguistic disorders: language-brain relationships. In: Filskov, S. B., & Boll, T. J. (Eds.), *Handbook of Clinical Neuropsychology*, pp 227–256. New York: Wiley.

Brugge, J. F., & Merzenich, M. M. (1973). Responses of neurons in the auditory cortex of the macaque monkey to monaural and binaural stimulation. *Journal of Neurophysiology, 36*, 1138–1158.

Crossen, B. (1985). Subcortical functions in language: a working model. *Brain and Language, 25*, 257–292.

Damasio, A. R., & Geshwind, N. (1984). The neural basis of language. *Annual Review of Neuroscience, 7*, 127–147.

Eimas, P. D. (1985). The perception of speech in early infancy. *Scientific American, 252*, 46–52.

Herman, K., Turner, J. W., Gillingham, F. J., & Graze, R. M. (1966). The effects of destructive lesions and stimulation of the basal ganglia on speech mechanisms. *Confina Neurologica, 27*, 197–207.

Hubel, D. H., & Wiesel, T. N. (1959). Receptive fields of single neurones in the cat's striate cortex. *Journal of Physiology (London), 148*, 574–591.

Hubel, D. H., & Weisel, T. N. (1974). Uniformity of the monkey striate cortex and a parallel relationship between field size, scatter and magnification factor. *Journal of Comparative Neurology, 158*, 295–306.

Kaczmarck, B. L. J. (1984). Neurolinguistic analysis of verbal utterances in patients with focal lesions of frontal lobes. *Brain and Language, 21*, 52–58.

Kandel, E. R., & Schwartz, J. H. (1985). *Principles of neural science*, 2nd ed. New York: Elsevier.

Lassen, N. A., Ingvar, D. H., & Skinhøj, E. (1978). Brain function and blood flow. *Scientific American, 239*, 62–71.

Ley, R. G., & Bryden, M. P. (1979). Hemispheric differences in processing emotions and faces. *Brain and Language, 7*, 127–138.

Livingstone, M. S., & Hubel, D. H. (1984). Anatomy and physiology of a color system in the primate visual cortex. *Journal of Neuroscience, 4*, 309–356.

Luria, A. R. (1973). *The working brain*. New York: Basic.

Merzenich, M. M., & Brugge, J. F. (1973). Cochlear partition on superior temporal plane of monkey. *Brain Research, 50*, 275–296.

Mohr, J. P., Pessin, M. S., Finkelstein, S., Finkelstein, H. H., Duncan, G. W., & Davis, K. R. (1978). Broca's aphasia: pathological and clinical. *Neurology, 28*. 311–324.

Myers, R. E. (1965). The neocortical commissures and interhemispheric transmission of information. In: Hinger G. (Ed.), *Functions of the corpus callosum*, pp. 1–17. London: Churchill.

Nieuwenhuys, R., Voogd, J., & van Huijzen, C. (1981). *The human central nervous system. A synopsis and atlas*, 2nd ed. Berlin Heidelberg New York: Springer.

Pandya, D.N., & Karol, E.A. (1971). The topological distribution of interhemispheric projections in the corpus callosum of the rhesus monkey. *Brain Research, 32,* 31–43.
Pandya, D.N., Hallet, M., & Mijherjee, S.K. (1969). Connections of the neocortical auditory system in the rhesus monkey. *Brain Research, 14,* 49–65.
Phelps, M.E., Mazziotta, J.C. (1985). Positron emission tomography: human brain function and biochemistry. *Science 228,* 799–809.
Phelps, M.E., Mazziotta, J.C., & Huang, S.-C. (1982). Study of cerebral blood function with positron computed tomograhy. *Journal of Cerebral Blood Flow Metabolism, 2,* 113–162.
Rizzolatti, G., Umilta, C., & Berlucchi, G. (1971). Opposite superiorities of the right and left cerebral hemispheres in discriminative reaction time to physiognomical and alphabetical material. *Brain, 74,* 431–442.
Ross, E.D. (1981). The aprosodias: functional-anatomical organization of the affective components of language in the right hemisphere. *Archives of Neurology, 38,* 561–569.
Ross, E.D. (1982). The divided self. *The Sciences,* February, 8–12.
Ross, E.D., & Mesulam, M.M. (1979). Dormant language functions of the right hemisphere? Prosody and emotional gesturing. *Archives of Neurology, 36,* 144–148.
Schaltenbrand, G. (1975). The effects on speech and language of stereotactical stimulation in thalamus and corpus callosum. *Brain and Language 2,* 70–77.
Shepherd, G.M. (1983). *Neurobiology.* New York: Oxford University Press.
Springer, S.P., & Deutsch, G. (1981). *Left brain, right brain.* San Francisco: Freeman.
Sinatra, R., Stahl-Gemake, J. (1983). *Using the right brain in language arts.* Illinois: Thomas.
Skoyles, J.R. (1986). Did ancient people read with their right hemispheres? A study in neuropalaeographology. *New Ideas in Psychology,* 3(3), 243–252.
Svennilson, E., Torvik, A., Lowe, R., & Beksell, L. (1960). Treatment of Parkinsonism by stereotactic thermolesions in the pallidal ragion. *Acta Psychiatrica et Neurologica Scandinavia, 35,* 358–377.
Taylor, I., & Taylor, M. (1983). *The psychology of reading.* New York: Academic.
Van Buren, J.M. (1966). Evidence regarding a more precise localization of the posterior frontal-caudal arrest response in man. *Journal of Neurosurgery, 24,* 416–417.
Van Essen, D.C., & Maunsell, J. (1983). Hierarchical organization and functional streams in the visual cortex. *Trends Neuro Sci., 6 (9),* 370–375.
Wapner, W., Hamby, S., & Gardner, H. (1981). The role of the right hemisphere in the apprehension of complex linguistic materials. *Brain and Language, 14,* 15–33.
Winner, E., Gardner, H. (1977). The comprehension of metaphor in brain-damaged patients. *Brain, 100,* 717–729.
Witelson, S.F. (1985). The brain connection: the corpus callosum is larger in left-handers. *Science, 229,* 665–667.

CHAPTER 14

Orthography, Reading, and Cerebral Functions*

OVID J. L. TZENG[1] and DAISY L. HUNG[2]

Introduction

Research on reading has regained the attention of experimental psychologists in the last two decades (Tzeng, 1981). The new field is markedly interdisciplinary, and addresses questions of practical application as readily as questions of pure theory or knowledge. These questions have been attacked from a number of different perspectives, using various experimental paradigms. Some studies record patterns of eye movements and relate them to various phases of information pickup during reading comprehension (Thibadeau, Just, & Carpenter, 1982). Others look into specific characteristics of word perception, e.g., the word-superiority effect, and attempt to build interactive models of human information processing (Rumelhart, 1985). A third type of study examines the consequences of becoming literate in more than one writing system. For example, differences as subtle as phonemic awareness and as general as problem-solving strategies have been observed in such comparative reading studies (Scribner & Cole, 1978; Liberman, Liberman, Mattingly, & Shankweiler, 1980; Tzeng & Hung, 1981).

Finally, an emerging set of studies has focused on patterns of deficits manifested in the reading performance of various types of aphasic patients. This line of research cuts across the previous three, which consist of models and theories derived from testing normal readers, looking instead at the performance of patients known to have different types of neurological damage, and correlated linguistic and behavioral disorders. The merging proves to be very fruitful on both empirical and theoretical grounds. Since the publication of Marshall and Newcombe's seminal work (1973), hundreds of journal articles and book chapters have been devoted to further specification of the unique characteristics of various types of "acquired dyslexia" (Coltheart, Patterson, & Marshall, 1980) and how they may relate to different types of aphasic syndromes. An important, and not less interesting, observation is

---

* This work was supported in part by National Institutes of Health grant to the Salk Institute for Biological Studies (HD 13249), in part by a research grant from the Academic Senate of the University of California, Riverside, and in part by a Research Fellowship from the Wang Institute of Graduate Studies.
[1] Department of Psychology, University of California, Riverside, CA 92521, USA.
[2] The Salk Institute for Biological Studies, La Jolla, CA, USA.

that patterns of reading disorders depend not only on the location of brain damage but also on the typology of the writing system. Such observations have led many investigators to conclude that lexical access of words written in different scripts (e.g., alphabetic vs logographic) is accomplished by different neurolinguistic pathways, which are probably represented by different special functions of the two cerebral hemispheres.

In this chapter, we would like to examine the research issues related to the experimental study of reading, focusing particularly on the last two issues.

Orthographic Variations

Writing systems have been qualified as logographic, syllabic, or alphabetic according to the morphemic, syllabic, or phonemic representation level of the speech (Hung & Tzeng, 1981). Among the many writing systems existing in the world today, Chinese logographs are unique in that their relationship with the spoken language they transcribe is rather opaque. This relationship can be described as morphosyllabic in nature. However, the logographs and syllables do not have a one-to-one correspondence: the same syllable may be represented by different logographs with different meanings. The number of Chinese logographs has expanded to thousands, and they are complex in configuration (Hung & Tzeng, 1981, Tzeng & Wang, 1983; Wang, 1981).

There is another unique aspect of Chinese logographs that needs to be mentioned. Centuries ago, these logographs were adopted by the Koreans, the Japanese, and the Vietnamese to become their respective national writing systems. The sound systems of these languages are quite different from spoken Chinese, and there were major problems in adopting the Chinese writing system to transcribe them. Today, North Korea and Vietnam have dropped the use of Chinese logographs altogether and opted for an alphabetic system. However, South Korea and Japan maintained them, and created sound-based systems (the Hangul alphabet for Korean and Kana syllabaries for Japanese) to overcome the problem of mismatch between the writing system and the sound system. Let us take a closer look at the Japanese case.

The origin of the Japanese spoken language is quite different from that of Chinese. The former evolved from the Altaic family of languages, which includes Turkish and Mongolian (Miller, 1980). The latter, however, is not part of the Altaic group, and there are substantial differences in phonology between the languages. As a result of borrowing an orthography from a different spoken language, the Japanese have evolved two different pronunciations of the Kanji (the borrowed Chinese logographs) characters – a Japanese pronunciation and an approximation of the Chinese pronunciation. In addition, due to syntactic requirements, they have developed two syllable-based scripts in order to be able to represent function terms and loan words. These are called Kana script in general, and the Hiragana and Katakana syllabaries specifically. Nowadays an ordinary Japanese text contains all three scripts in their distinctive styles.

For most Indo-European languages, the writing system, patterned after that of the Greeks, evolved to an alphabetic script, with the number of written symbols extensively reduced. A full alphabet, marking vowel as well as consonant phonemes, developed over a period of about 200 years during the first millenium B.C. in Greece (Kroeber, 1948). The transition from the syllabic to the alphabetic system marked a gigantic jump with respect to the script/speech relationship. In fact, the development of vowel letters, which form the basis of the analytical principle of an alphabetic system, has been characterized as something of an accident rather than a conscious insight (Gleitman & Rozin, 1977). As a sound-writing script, an alphabetic system maps onto speech at the level of the *phoneme*, a linguistic unit smaller than the syllable but larger than an articulatory feature.

As we look back at these historical changes, we see that the evolution of writing seems to have taken a single direction: at every advance, the number of symbols in the script decreases, and as a direct consequence the abstractness of the relation between script and meaning increases and the link between graphemes and phonemes becomes clearer. This pattern of development seems to parallel the general trend of cognitive development in children and thus may have important implications for beginning readers of different orthographies. One of the major activities in learning to read is exploring the correspondence between the written script and the spoken language (Tzeng & Singer, 1981). Since the script/speech relations in different orthographies tap into different levels of speech perception, and since the size of the minimal character set required for transcribing the entire speech segments in a language depends on such mapping relations, these unique historical developments provide ample opportunity to study the effects of orthographic variations on visual information processing within and across languages, and with respect to both skilled and beginning readers. A question of psychological interest concerns the extent to which different orthographies undergo similar (or different) processing. The present chapter addresses these issues with respect to current experimental work on normal and aphasic reading.

Differing Patterns of Lexical Access

Fluent readers can read faster than they can talk, but for a child just learning to read, the opposite is usually true because every word has to be sounded out in order to get at the meaning. At some point during the process of acquiring reading skills, the transformation of visual code into speech code becomes automatic via some nonlexical symbol-sound correspondence rules, or becomes unneccessary altogether (the latter view has generally been referred to as the direct access hypothesis). In recent years, studies of word recognition in an alphabetic script like English have been dominated by concern over the nature of the code that allows the reader to go from print to meaning, a process called *lexical access.*

Almost 20 years ago, when experimental psychologists started to launch their first attacks on reading from the perspective of information processing, using reaction time as the dependent measure, a number of investigators held the view that phonological recoding was a necessary preliminary to lexical access (Gough, 1972; Gough & Cosky 1977; Rubenstein, Lewis, & Rubenstein, 1971). A considerable amount of evidence was collected to support the phonological recoding hypothesis. However, other investigators were cumulating abundant evidence to support the direct access hypothesis. It is now clear from both the experimental and neuropsychological literature that, for a large number of words, phonological recoding for the purpose of lexical access is not necessary: in fact, some form of orthographic or visual code is sufficient for the purpose of getting meaning from print (Henderson, 1982; Hung & Tzeng, 1981; McCusker, Hillenger, & Bias, 1981; Saffran & Marin, 1977; Seidenberg, 1985).

Adding Chinese logographs into the picture seems to complicate, rather than clarify, the issue. Early supporters of the direct access hypothesis always used the example of reading Chinese to reinforce their argument. The argument goes like this: Because Chinese logographs do not contain information about pronunciation, people must be able to read without speech recoding. However, this statement is not exactly correct. First, Chinese logographs consist of a majority of phonograms that at times do give clues to pronunciation. Thus, with the ability to pronounce a limited number of basic logographs, and knowledge of orthographical rules in the construction of logographs, readers of Chinese can in fact make reasonably successful guesses about how to pronounce logographs that share the same phonetic component, even those that they have never encountered before (Zhou, 1978). The procedure involved in this type of grapheme-sound conversion is of course very different from that involved in the GPC (grapheme-phoneme conversion) rules advocated by Coltheart (1980). But it is similar to Glushko's (1979) activation-synthesis model of the generation of phonological codes. Indeed, such a procedure of generating phonological codes by analogy was proposed by Tzeng (1981) as one of two mechanisms in speech recoding, and was recently thought to be used by fluent readers of English for most words (Kay & Marcel, 1981; Seidenberg, 1985). Empirical evidence for the operation of this type of speech recoding in reading Chinese has been provided by Fang, Horng, and Tzeng (1986) and Lien (1985). Second, the Chinese writing system also makes it very clear that we cannot assume a one-to-one correspondence with respect to semantics between a word in print and a meaning in the mental lexicon. Single logographs are often recombined to make up new words; hence, there is nothing in the lexicon to be accessed. Meanings of words become available through the reference back to phonology and contexts. In this sense, it is rather difficult, if not impossible, to conceive of access to the lexicon via some orthographic or visual configurational cues. To a lesser degree this may also be true with respect to English orthography.

Reading should not be equated with the lexical access of a single word; rather, it should be regarded as a more general linguistic activity such as iconic scanning and storage, lexical retrieval, short-term retention, syntactic parsing at both macro- and microlevels (Kintsch & Van Dijk, 1978), and semantic integration over the entire discourse. This kind of conceptualization immediately questions the validity of the view that reading logographs involves no grapheme-phonology translation. Thus, despite the bias towards direct grapheme-to-semantic processing, logographs may also activate phonological recoding processes. Erickson, Mattingly, and Turvey (1977) found increased errors in an immediate memory task when Kanji characters were phonologically related. Tzeng, Hung, and Wang (1977) found similar effects in Chinese readers when phonetically similar logographs were used in an immediate memory task and in a sentence judgment task in which subjects decided whether sentences were meaningful and grammatically correct.

One implication to be drawn from all of these findings is that phonological recoding is just one of the strategies for obtaining access to meaning, rather than an obligatory stage. There are at least two major ways in which such a recoding process is important. First, in blending the individual letters (or logographs) of words, the phonological recoding of the individual letter (or logograph) sound can plausibly be argued to be an important intervening stage, at least for children learning to read. A second way in which phonological recoding may be involved in reading is concerned with the question of whether fluent readers need to phonologically recode printed materials or are assisted by doing so. In this latter view the phonological recoding is regarded as a general strategy of human information processing, and thus, the orthographic difference in the printed materials becomes less important (Tzeng et al., 1977).

Differing Patterns of Code Activation

Because logographs represent units of meaning rather than units of sound, it has been suggested that logographic orthographies allow more rapid access of meaning than phonetic orthographies (Biederman & Tsao, 1979; Hatano, Kuhara, & Akiyama, 1981). Phonetic orthographies rely at least in part on phonological recoding processes; that is, the written symbols arouse names which then access meaning. Based on this view, logographic orthographies may allow more rapid access of meaning, although phonetic orthographies may allow more rapid access of names. Thus, reading Chinese may involve different cognitive processes than reading English. To obtain empirical evidence about differential script processing with logographic and alphabetic writing systems, most investigators have employed facilitation-interference paradigms like the Stroop color-word test (Stroop, 1935) or its variations (Besner & Coltheart, 1979). In the Stroop color-word test, a disruption and delay in naming the color of the ink occurs when the ink spells on incongruent color name (e.g., the word "red" written in green ink). The slowing of naming in the presence of conflicting words has been termed the "Stroop interference effect."

Biederman and Tsao (1979) carried out a study to see whether varying the type of orthography would produce different amounts of Stroop interference, by comparing Chinese (graduate students from Taiwan) and English readers in the Stroop task for their respective orthographies. They found a greater Stroop interference for Chinese readers, and suggested that the direct associations between symbol and meaning produced greater interference in the Chinese version of the task. Results of this study have provoked a great deal of interest among cognitive psychologists, and the following years saw a lot of studies attempting to replicate and/or extend the original findings on both sides of the Pacific Ocean. Smith and Kirsner (1982) failed to replicate Biederman and Tsao's finding in a bilingual study that compared French-English and Chinese-English bilinguals with a modified Stroop task. They asked their subjects to name objects represented by line drawings in the presence of conflicting words. They found that Chinese-English bilinguals experienced greater interference in the presence of conflicting Chinese logographs than in the presence of conflicting English words. However, this study suffered from the same problem encountered by most bilingual studies: the bilinguals are unlikely to have equal proficiency in both languages, especially with respect to reading. In Smith and Kirsner's study, this was particularly true since all their Chinese-English bilingual subjects were from Hong Kong, where the reading ability in Chinese has been shown to be lower than that of subjects from Taiwan. Another difficulty in comparing this study to that of Biederman and Tsao is that it used Chinese words containing two logographs, whereas the latter used words that could be represented by a single logograph.

To avoid these difficulties, several studies examined the Stroop interference effect with the Japanese writing system, in which both first-language logographic and phonetic systems map onto the same spoken language. With this experimental paradigm, Shimamura (1987) found that words written in Kanji indeed produced more interference than words written in Kana, despite the fact that the Kanji words were named faster. Related findings were obtained by Fang, Tzeng, and Alva (1981) and Hatta (1981) in a visual hemifield paradigm via visual hemifield presentation.

An interesting question arises at this point: Does the magnitude of interference – i.e., the time taken to name a color with a distracting word minus the time taken to name a plain color – differ among various scripts? According to Fang et al. (1981) the answer is a decisive yes. For a long time, researchers have realized that this interference is reduced for a bilingual reader if the printed color words and the response color names are from two different languages. Fang et al. replicated this finding, and further showed that there is a systematic relation between the amount of interference and the degree of similarity between the two scripts used. The systematic variation observed in the degree of interference across different scripts suggests that the linguistic code used in reading cannot be simply semantic or phonetic. Rather, the findings indicate that the reader spontaneously and automatically explores all sorts of cues, semantic, phonetic, as well as orthographic, in the printed symbols.

Can the above results be an artifact of naming the colors aloud? To eliminate such a possibility, several investigators conducted another type of experiment that was in essence a variant of the Stroop task, but required no oral response. Shimamura (1987) asked his English-speaking subjects to indicate the spatial location of stimuli in the presence of conflicting arrows or words (e.g., the word "left" in the right position). The result showed that conflicting arrows (i.e., logographs) produced more interference than conflicting words, yet words were named faster than arrows in another experiment which required oral response.

These studies provide convincing evidence to support the hypothesis that logographic symbols are comprehended faster than phonetic symbols. Thus, logographic symbols appear to allow more rapid access to meaning, whereas phonetic symbols allow more rapid access to verbal names. Such an effect has also been extended to other logographic symbols such as Arabic numerals and mathematical signs. In an interesting experiment carried out by Besner and Coltheart (1979), subjects were asked to point out manually which of two numbers, written logographically or alphabetically, was the larger numerically. In addition to their numerical variation, the members of each pair could also differ in physical size. In incongruent trials, the numerically larger number was physically the smaller, while in congruent trials it was the larger. Results of the experiment showed a "size congruity effect" – a longer mean response time for incongruent than for congruent trials – and not surprisingly this effect was obtained for logographic but not for alphabetic symbols. In other words, the Stroop-like interference disappeared when the logographic symbols, e.g., 2 and 6, were replaced by spelled-out alphabetic symbols, e.g., "two" and "six."

Once again, Besner and Coltheart's (1979) demonstration of a size congruity effect with logographic symbols led to further examination on various other scripts. Takahashi and Green (1983) and Tzeng and Wang (1983) have reported supporting results with Chinese logographic numerals. Vaid (1985) contrasted Hindi and English scripts and observed that the degree of phonological transparency in a script might play an determining role in the magnitude of the size congruity effect. In contrast, recently Fotz, Poltrock and Potts (1984), Peereman and Holender (1984), and Besner, Davelaar, Alcott, and Parry (1984) obtained a size congruity effect with both logographic and alphabetic numbers, although the effect was larger for the former. Investigators from the latter group tend to dispute the reality of the logographic factor in the demonstration of the size congruity effect. They mistakenly treat the concept "logographic" as an all-or-none feature. In fact, both logographic and phonetic concepts should be regarded as aspects of how a printed symbol comes to represent a segment of speech. For example, in the Japanese writing system, the pronunciation of a Chinese logograph (i.e., Kanji) is determined by the way it is used in a sentence. Thus, the phonological cues that are associated with many Chinese logographs do not facilitate sounding by a Japanese reader as much as they would by a Chinese reader. The Japanese Kanji, therefore, has been considered to be more strictly logographic than its Chinese equivalent (Hatano et al. 1981). From this perspective, the seemingly conflicting data reviewed above do show a consistent pattern: a large size congruity effect with logographic numbers and a small size congruity effect (or none at all) with alphabetic ones.

These findings have interesting implications for bilingual processing. In our laboratory we presented a group of Chinese native speakers who had learned English as young adults with all three types of stimuli: Arabic numerals, English spelled-out number words, and Chinese logographic number words. As before, we observed the interference with both Arabic and Chinese stimuli. Unexpectedly, however, these readers also showed an interference with the English words. How could we account for this last finding? Could it simply be due to the fact that these subjects learned English later in life? Or was it because they were transferring the processing strategy for logographs to alphabetic spelling? To choose between these two hypotheses, we next worked with a group of native Spanish speakers who had also learned English as young adults. They were asked to perform the same task with Arabic numerals, English number words, and Spanish number words. The results were unequivocal: significant interference occurred only with the Arabic numerals, not with Spanish or English number words. Therefore, the interference observed with English words for the Chinese-English bilingual readers was not due to their later learning of that language, since we did not observe a similar effect with the Spanish-English readers. The remaining hypothesis, then, is that the Chinese-English subjects had transferred their reading processes from logographs to English words.

The evidence we have reviewed so far, from both the color and number Stroop experiments and with both oral and manual responses, supports the contention that the relationship between script and speech underlying all types of writing system plays an important role in reading behavior. A reader of a particular script must assimilate the orthographic characteristics of that system. That is, if the configuration of a logograph is important in deciphering it, then the reader has to pay special attention to the position of every element it contains. As a consequence, we should expect the processing of logographs to involve more visual memory than the processing of alphabetic scripts. There is plenty of such evidence in the literature. Let us now turn to this issue.

Differing Patterns of Memorial Processes

Studies of the cognitive processes involved in reading have provided further evidence that logographic and phonetic orthographies are not processed in exactly the same way. Park and Arbuckle (1977) compared long-term memory performance when words were presented using either Chinese logographs or the Korean Hangul (a feature-based alphabetic script; see examples provided in Tzeng & Wang, 1983). Free recall and recognition were better when words were written with the logographs. Hatano et al. (1981) tested the degree to which Japanese readers could infer the meaning of unfamiliar words. The ability to do this was better for words written in Kanji than Kana, and the authors concluded that logographs produce efficient access to meaning since they are interpreted at the level of morphemes. This would explain the memorial superiority of the Kanji script: words presented with logographic script tend to be elaborated at the semantic level. Memory research over the last 10 years has provided convincing evidence to suggest the power of semantic elaboration within a level-of-processing framework.

To tap into the issue of different processing mechanisms across different orthographies, Turnage and McGinnies (1973) asked Chinese and American college students to study a 15-word list in a serial learning paradigm. They also manipulated the input modality of the stimulus presentation. It was found that Chinese students learned the logograph list faster when it was presented visually, whereas American students learned the word list faster when it was presented aloud. The finding on the Chinese logographs is opposite to the famous modality effect (Crowder, 1978), in which auditory presentation of English words results in better recall than does visual presentation for the terminal items in a serial list. The interpretation offered by Turnage and McGinnies is that Chinese logographs contain more symbols with similar sounds but different meaning than is the case for English, and this characteristic feature of the Chinese logographs may favor learning through the visual mode.

Turnage and McGinnies' (1973) study involved two different language populations: not only were the scripts different, there was also a difference in spoken language. Thus, the script may not be the determining factor; rather, the visual modality advantage could have been a result of difference in the two spoken languages. But this latter acount can be easily ruled out by the finding of Park and Arbuckle (1977) that the memorial superiority of Kanji is observed even when two different types of script (i.e., Chinese logographs vs. Hangul) actually map onto the same spoken language.

We decided to look into the memorial process more closely with a much finer analysis, by comparing serial recall performance of native English and Chinese speakers. We presented a series of nine items to subjects either orally with a tape recorder or visually by means of a slide projector. In the visual presentation, the items were either English words or Chinese logographs. The subjects were asked to recall the nine items according to their position in the series and the probability of correct recall was plotted according to the item's position. Previous findings from other laboratories indicate that it is usually easier to remember items at the end of a list if they are presented orally than visually (i.e., the modality effect). The data from both English and Chinese groups in our study showed that this effect occurred with the last two items. The interesting difference is that the Chinese subjects recalled the nonterminal items in the series consistently better when these were presented visually rather than orally, whereas no difference was found for the English subjects with the English words. Visual presentation was superior for Chinese readers regardless of whether they were asked to recall the items orally or in writing. In addition, this visual superiority effect on the long-term retention of Chinese logographs has also been observed with American students who are learning to read Chinese as a second language.

How can we explain these findings? Studies using visual nonverbal input have often shown very high levels of performance, without higher efficiency memory for the presented items (Shaffer & Shiffrin, 1972). Thus, our results could be explained as attributable to the very high level of uniqueness of logographic symbols. Spoken words, after all, are made up of only 40 or so phonemes in various combinations, and visually presented English words of only 26 elements. Unrestricted logographs could plausibly contain a much wider range of components. This would mean that each item could contain unique features, thus making it relatively easy to recall on a later memory test. Furthermore, it also means that later logographs from the same series could fall into a separate part of the sensory field, and thus would not overwrite the earlier items: i.e., no recency effect. This explanation is further supported by an impressive series of experiments by Phillips and colleagues (Phillips & Christie, 1977a, 1977b). Their results showed a low level of performance and the appearance of a recency effect when the visual items consisted of pattern of points chosen from a limited set of possibilities.

Taken together, these findings suggest that processing of a logographic script involves more visual/spatial memory than that of an alphabetic script. In a recent study carried out in the Salk Institute, Bellugi and associates (personal communication) compared the writing errors of American deaf children and Chinese deaf children when they are learning to read and write their respective scripts. From the analysis of the error patterns, they are able to show that the Chinese subjects tend to explore the spatial layout of each logograph while the American subjects tend to focus on the linear arrangement of the letter strings. As a consequence, deaf Chinese seem to have an easier transition from sign language to logographs than deaf American do in their transition from signing to the alphabetic script, such as reading English words.

Differing Patterns of Hemispheric Asymmetries

Throughout the history of hemispheric specialization research, there has been speculation about the possibility that the functional organization of a literate brain may be related to the type of written script one has learned to read. From Dejerine (1891) to Hinschelwood (1917) and, more recently, from Benson and Geschwind (1969), Luria (1970), and Hecaen and Kremin (1975), to Zaidel and Peters (1981), evidence has been provided to show a selective sparing of reading one type of script despite severe impairments in the reading of other scripts in bilingual aphasic patients (for a more detailed review, see Hasuike, Tzeng & Hung, 1986). Data from these bilingual studies are illuminating. However, they suffer from the lack of appropriate control of the degree of impairment of the spoken language. In this respect, recent findings of selective impairment in the reading of Kanji and Kana scripts by Japanese aphasic patients within a single spoken language have strengthened the hypothesis of the scriptal effect on cerebral organization (Hung & Tzeng, 1981; Sasanuma 1980).

It should be noted that the finding of selective impairment in the reading of the two types of Japanese script does not necessarily implicate a right hemispheric involvement for processing Kanji. In fact, Sasanuma and her associates (Sasanuma, 1975, 1980; Sasanuma & Fujimura, 1971; Tatsumi, Itoh, Konno, Sasanuma, & Fujisaki, 1982) have argued for a differential disruption of language due to localized lesions in the *left* hemisphere, rather than postulating a dichotomy of right and left hemispheric processing for Kanji and Kana scripts. According to Hasuike et al. (1986), before the mid-1970s, there seemed to be no disagreement about the role of the left hemisphere for processing Chinese logographs. However, in 1977 two papers attracted much attention because both showed some evidence for right hemispheric involvement in reading Chinese logographs.

The first study was by Hatta (1977), whose results showed that native Japanese readers identified singly presented Kanji characters better when they were presented in the left visual field than in the right visual field, implying a stronger right hemispheric involvement. In previous studies (Hirata & Osaka, 1967), native Japanese readers had showed the reverse lateralization pattern in identifying Kana symbols, implying a left hemispheric involvement in the processing of such sound-based script. Hatta's new finding was in accord with results obtained by Sasanuma, Itoh, Mori & Kobayashi

(1977), in which nonsensical two-character Kana and Kanji characters were presented to native Japanese readers for identification. They found a significant right visual field superiority for the recognition of Kana symbols and a non-significant left visual field superiority for Kanji characters. Results from these two studies have often been cited to give evidence for right hemispheric involvement in the processing of Kanji logographs.

However, the seemingly clear picture begins to look very messy when one examines data from studies using Chinese readers. Visual hemifield experiments with Chinese subjects (Hardyck, Tzeng & Wang, 1977, 1978; Kershner & Jeng, 1972) clearly showed a right visual field (left hemisphere) superiority for processing Chinese logographs. The discrepancy between the Japanese and Chinese results in these studies is curious. One possible interpretation is that Japanese readers process Kanji characters differently from the way Chinese readers process Chinese logographs, perhaps because of some unknown interaction between Kanji and Kana. Put another way, the Japanese not only borrowed the Chinese logographs, but also developed a different brain function in order to read them – hardly a plausible interpretation!

The major problem with visual hemifield experiments using a tachistoscopic procedure is the lack of control over the variables that could affect the results. Paradis, Hagiwara, and Hildebrandt (1985) discuss such factors related to the nature of the stimulus, the presentation conditions, the task demands, the response, and the subjects, and note that in most studies the familiarity, concreteness, and types of logographs are often not specified, let alone controlled. Thus, discrepancies could easily arise because of procedural differences. Tzeng et al. (1979) manipulated the number of logographs in two experiments, and found a left visual field superiority for recognition of single logographs and a right visual field superiority for two-logograph words. Hasuike et al. (1986) went one step further in carrying out an extensive comparison among all relevant experiments up to 1985. They identified the stimulus exposure duration as the key variable because superiority of the left visual field (right hemisphere) was found only in those studies in which exposure duration was less than 50 ms. This makes sense: short exposure duration produces an incomplete visual image with a very low spatial resolution, and the literature has shown that the right hemisphere is adept in perceiving the relationship between these fragmentary components and the whole configuration (Sergent, 1983). When the stimulus is presented for a longer exposure the spatial resolution is better, and under such conditions the left hemisphere seems to take over, especially when the task requires further linguistic analysis.

It should be concluded then that there is very little evidence, from either experimental or clinical studies, to suggest a stronger right hemispheric involvement in the linguistic analysis of Chinese logographs. Findings from two recent experiments in our laboratory on the writing and recognition of Chinese logographs by brain-damaged Chinese patients have shed much light on this point. Results of the first study revealed that while right-hemisphere-damaged Chinese readers showed typical left visual field neglect in copying geometric figures and in spontaneous drawing of familiar objects, they never missed any subcomponent in writing Chinese logographs. In contrast, left-hemisphere-damaged patients, while showing perfect performance in both figure copying and spontaneous drawings, did poorly in writing logographs, with numerous errors of omission and commission. Results of a further logograph-construction experiment with brain-damaged patients identified the left posterior region as the important site for integrating graphemic features into a whole logograph based upon a sequential organized graphomotoric schema. These data provide unequivocal evidence against any suggestion that Chinese logographs are processed in the right hemisphere. They also provide a possible answer for why the left hemisphere is responsible for processing Chinese logographs as well as other scripts.

Conclusion

The relation between written script and spoken languages seems so close that one would expect that anyone who is able to speak should be able to read. Nevertheless, this is not the case. Whereas all humans learn to speak effortlessly and naturally, indicating that there must be a significant influence from genetic facilitation, the situation is very different with writing. Many societies still do not have written languages, and in most literate societies there are some people who cannot read or write, either for social or organic reasons. Thus, for cognitive theorists and practitioners alike, the question becomes: Why do some children fail to learn to read? This question is particularly baffling when the reading failure is completely unexpected and defies commonsense explanations (Frith, 1979). For example, given that the child already has learned the spoken language, and that each letter on the printed array corresponds roughly to a visual analog of some known speech category, it seems that reading should be an easy deciphering task. Yet, this view is simply wrong. Decades of intensive research have revealed that the problem of reading may have something to do with the cognitive prerequisites to understanding one's own spoken language and to appreciating the script/speech relations embedded in a particular writing system (Hung & Tzeng, 1981; Tzeng & Singer, 1981).

The recognition that purely external linguistic factors may contribute to the incidence of reading disability immediately brings our research focus onto several directions of inquiry. First, what are the linguistic factors that affect the process of learning to read at the entry level? Are they language-specific? Second, what are the basic processing components in skillful reading? Again, are they language-specific? Third, what are the defining features of developmental and acquired dyslexia? What insight about the processes of normal reading can we gain from studying the similarities and differences of these two types of reading disorders? Finally, given the varieties of writing systems with different types of script/speech relations, how does the brain adapt to these orthographic variations?

From the literature review in the previous section, we have seen that orthographic variations affect basic visual information processing with respect to lexical access, code activation speed, memorial processes, and, to a lesser degree, visual lateralization patterns in normal readers. However, we have not seen any convincing evidence to suggest a modification of cerebral organization due to such orthographic variations. Moreover, we have presented strong evidence from the brain-damaged Chinese patients to suggest a strictly left hemispheric involvement in the writing and recognition of Chinese logographs. This discrepancy has an important message: Can different mental processes be driven by similar cortical functions? Or, do we have to entertain the possibility that our analysis of cortical functions has not been detailed enough to allow the manifestation of differences as shown in the higher cognitive processes? It is extremely important to be cautious about drawing conclusions from studies involving these two levels of analysis.

When group differences are obtained, there is a tendency to account for them in terms of the most salient differences between the groups, which in this case is the linguistic descriptions of the different types of written scripts. Methodologically speaking, nothing is wrong with this as long as the theorist stipulates his/her propositional account within the same level of descriptions, i.e., without attempting to stipulate underlying *psychological* or *neurological* mechanisms in order to account for the differences. In recent years, we have seen many model builders incorrectly assume that a linguistic description must have an implied knowledge of language structure, which then provides an independent rationale for the proposed specialized mechanism (or neurolinguistic pathway) in order to access such knowledge. For instance, many reading researchers propose that words are read by retrieving a single pronunciation from memory and that pseudowords are pronounced by using abstract context-sensitive GPC rules such as those developed by pure linguistic analysis (Venezky, 1970; Wijk, 1966). However, Glushko (1979) and Hung, Tzeng, Salzman, & Dreher (1984) were able to show that the complex linguistic description of such GPC rules, although satisfactorily formulated in pure linguistic terms, does not have any psychological reality at all.

Therefore, if we want to have a better understanding about the relations among orthography, reading, and cerebral functions, we need to pay more attention to research in the following four areas. First, we need to build a comprehensive theory of orthography. That theory should be capable of explaining the relationship between script and speech. Turvey (1984), consistent with the tradition of his associates at the Haskins Laboratories, employs the concept of "depth" of orthography to specify this relationship. Wang (1981) also discusses the concept of an "optimal orthography" based on the way in which the relationship between script and speech is captured in a two-dimensional array. Second, we need to build a theory of perceptual learning in which the perceptual and cognitive capacity of beginning readers can be specified, and the processes of their learning to deal with the cognitive demands imposed by the various orthographic structures can be outlined. Third, we need to have a comprehensive theory of reading that specifies its various components and explains the way in which those components interact with other conditions, such as the nature and presentation of the reading materials and the nature of the task. Finally, we need to develop a theory of neuronal organization in which the neural basis and mechanisms of reading can be detailed in both normal and aphasic populations. Each type of theory can be approached independently, and each can stand as a separate explanatory level of reading behavior. However, for any one type to be complete it will be necessary to understand the others, in order to gain an adequate understanding of itself. From a biological consideration, behaviors have been selected for their adaptation qualities, and the selection influences do their work on processes that are available at every level of the explanation.

Human beings stand alone in history as the sole creatures on earth who have invented written symbols and have benefited from these symbols. Because these symbols are arbitrary inventions external to our organismic structure, both accommodation and assimilation processes must have worked to extremes for us to achieve efficiency in manipulating them. It took a span of many thousands of years for our ancestors to come up with a system that worked for a particular language, and it takes a great deal of effort on the part of a modern learner to become a fluent reader. The diversity of orthographic structures provides excellent opportunities for investigators of human cognition to examine how children who speak different languages adjust themselves to meet various task demands imposed by their respective writing systems. Needless to say, there is still much to be learned in future research about the relations among orthography, reading, and cerebral functions.

References

Benson, D.F., & Geschwind, N. (1969). The alexias. In Vinkel, P.U., & Bruyn, G.W. (Eds.), *Handbook of clinical neurology*, vol 1. Amsterdam: North Holland.

Besner, D., & Coltheart, M. (1979). Ideographic and alphabetic processing in skilled reading of English. *Neuropsychologia, 17*, 467–472.

Besner, D., Davelaar, E., Alcott, D., & Parry, P. (1984). Wholistic reading of alphabetic print: evidence from the FDM and the FBI. In Henderson, L. (Ed.), *Orthographies and reading*, pp 121–135. Hillsdale: Erlbaum.

Biederman, I., & Tsao, Y.C. (1979). On processing Chinese ideographs and English words: Some implications from Stroop-test results. *Cognitive Psychology 11*, 125–132.

Coltheart, M. (1980). Deep dyslexia: a review of the syndrome. In Coltheart, M., Patterson, K. & Marshall, J.C. (Eds.), *Deep dyslexia*. London: Routledge & Paul.

Coltheart, M. Patterson, K., & Marshall, J. C. (1980). *Deep dyslexia.* London: Routledge & Kegan Paul.
Crowder, R. G. (1978). Memory for phonologically uniform lists. *Journal of Verbal Learning and Verbal Behavior, 17,* 73–89.
Dejerine, J. (1891). Sur un cas de cecite verbale avec agraphie suivi d'autopsie. *Mémoires de la Societé Biologie, 3,* 197–201.
Erickson, D., Mattingly, I. G., & Turvey, M. T. (1977). Phonetic activity and reading: an expert with Kanji. *Language and Speech, 20,* 384–403.
Fang, S. P., Tzeng, O. J. L., & Alva, E. (1981). Intra- versus inter-language Stroop interference effect in bilingual subjects. *Memory and Cognition, 9,* 609–617.
Fang, S. P., Horng, R. Y., & Tzeng, O. J. L. (1986). Consistency effects in the Chinese character and pseudo-character naming tasks. In Kao, H., & Hoosain, R. (Eds.), *Linguistics, psychology, and the Chinese language.* Hong Kong: University of Hong Kong Press.
Fotz, G. S., Poltrock, S. E., & Potts, G. R. (1984). Mental comparison of size and magnitude: size congruity effects. *Journal of Experimental Psychology: Learning, Memory, and Cognition, 40,* 442–453.
Frith, U. (1979). Reading by eye and writing by ear. In Kolers, P. A., Wrolstad, M., & Bouma, H. (Eds.), *Processing of visible language,* I. New York: Plenum Press.
Gleitman, L. R., & Rozin, P. (1977). The structure and acquisition of reading: Relation between orthographic and the structure of language. In Reber, A. S., & Scarborough, D. L. (Eds.), *Toward a psychology of reading: the proceedings of the CUNY conference.* Hillsdale: Erlbaum.
Glushko, R. (1979). The organization and activation of orthographic knowledge in reading aloud. *Journal of Experimental Psychology: Human Perception and Performance, 5,* 674–691.
Gough, P. B. (1972). One second of reading. In Kavanagh, J. P., & Mattingly, I. G. (Eds.), *Language by ear and by eye.* Cambridge, MA: MIT.
Gough, P. B., & Cosky, M. J. (1977). One second of reading again. In Castellan, I. G., Pisoni, D. B., & Potts, G. R. (Eds.), *Cognitive theory,* vol. 2. Hillsdale: Erlbaum.
Hardyck, C., Tzeng, O. J. L., & Wang, W. S.-Y. (1977). Cerebral lateralization effects in visual half-field experiments. *Nature, 269,* 705–707.
Hardyck, C., Tzeng, O. J. L., & Wang, W. S.-Y. (1978). Cerebral lateralization of function and bilingual decision processes: is thinking lateralized? *Brain and Language, 5,* 56–71.
Hasuike, R., Tzeng, O. J. L., & Hung, D. (1986). Script effects and cerebral lateralization. In Vaid, J. (Ed.), *Language processing in bilinguals: psycholinguistic and neuropsychological perspectives,* pp 275–288. Hillsdale: Erlbaum.
Hatano, G., Kuhara, K., & Akiyama, M. (1981). Kanji help readers of Japanese infer the meaning of unfamiliar words. *Quarterly Newsletter of the Laboratory of Comparative Human Cognition, 3,* 30–33.
Hatta, T. (1977). Recognition of Japanese Kanji in the left and right visual field. *Neuropsychologia, 15,* 685–688.
Hatta, T. (1981). Differential processing of kanji and kana stimuli in Japanese people: some implication from Stroop-test results. *Neuropsychologia, 19,* 87–93.
Hecaen, H., & Kremin, H. (1975). Neurolinguistic research on reading disorders resulting from left hemisphere lesions: aphasia and "pure" alexias. In H. Whitaker & H. A. Whitaker (Eds.) *Studies in neurolinguistics,* vol. 2, pp. 263–269. New York: Academic.
Henderson, L. (1982). *Orthography and word recognition in reading.* New York: Academic.
Hinschelwood, J. (1917). *Congenital word-blindness.* London: Lewis.
Hirata, K., & Osaka, R. (1967). Tachistoscopic recognition of Japanese letter materials in left and right visual fields. *Psychologia, 10,* 17–18.
Hung, D. L., & Tzeng, O. J. L. (1981). Orthographic variations and visual information processing. *Psychological Bulletin, 90,* 377–414.
Hung, D. L., Tzeng, O. J. L., Salzman, B., & Dreher, J. (1984). A critical evaluation of the horse-racing model of skilled reading. In Kao, H. S. R., & Hoosain, R. (Eds.), *Psycho-*

logical studies of the Chinese language. Hong Kong: Chinese Language Society of Hong Kong.

Kay, J., & Marcel, A. (1981). One process, not two, in reading aloud: lexical analogies do the work of non-lexical rules. *Quarterly Journal of Experimental Psychology, 33A,* 397–413.

Kershner, J.R., & Jeng, A.G.R. (1972). Dual functional hemispheric asymmetry in visual perception: effects of ocular dominance and post exposural processes. *Neuropsychology, 10,* 437–445.

Kintsch, W., & Van Dijk, T.A. (1978). Toward a model of text comprehension and production. *Psychological Review, 85,* 363–416.

Kroeber, A.L. (1948). *Anthropology: Race, language, culture, psychology prehistory.* New York: Harcourt, Brace.

Liberman, I.Y., Liberman, A.M., Mattingly, I.G., & Shankweiler, D. (1980). Orthography and the beginning reader. In Kavanagh, J.F., & Venezky, R.L. (Eds.) *Orthography, reading and dyslexia.* Baltimore: University Park Press.

Lien, Y.W. (1985). *The reading process of Chinese character: the activation of phonetic information in the phonogram.* Master's thesis. Taipei: National Taiwan University.

Luria, A.R. (1970). *Traumatic aphasia.* The Hague: Mouton.

McCusker, L.X., Hillinger, M.L., & Bias, R.G. (1981). Phonological recoding and reading. *Psychological Bulletin, 89,* 214–245.

Marshall, J.C., & Newcombe, F. (1973). Patterns of paraplexia: psycholinguistic approach. *Journal of Psycholinguistic Research, 2,* 175–199.

Miller, R.A. (1980). *Origins of the Japanese language.* Seattle: University of Washington Press.

Paradis, M., Hagiwara, H., & Hildebrandt, N. (1985). *Neurolinguistic aspects of the Japanese writing system.* New York: Academic.

Park, S., & Arbuckle, T.Y. (1977). Ideograms versus alphabets: effects of script on memory in "biscriptual" Korean subjects. *Journal of Experimental Psychology: Human Learning and Memory, 3,* 631–642.

Peereman, R., & Holender, D. (1984). Relation entre taille physique et taille numerique dans la comparaison de chiffres ecrits alphabetiquement ou ideographiquement. *Psychologica Belgica, 24,* 147–164.

Phillips, W.A., & Christie, D.F.M. (1977a). Components of visual memory. *Quarterly Journal of Experimental Psychology, 29,* 117–134.

Phillips, W.A., & Christie, D.F.M. (1977b). Interference with visualization. *Quarterly Journal of Experimental Psychology, 29,* 637–650.

Rubenstein, H., Lewis, S.S., & Rubenstein, M.A. (1971). Evidence for phonemic recoding in visual word recognition. *Journal of Verbal Learning and Verbal Behavior, 10,* 645–657.

Rumelhart, D. (1985). Toward an interactive model of reading. In Singer, H., & Ruddell, R.B. (Eds.), *Theoretical models and processes of reading,* pp 722–750. Delaware: International Reading Association.

Saffran, E.M., & Marin, O.S.M. (1977). Reading without phonology: evidence from aphasia. *Quarterly Journal of Experimental Psychology, 29,* 515–525.

Sasanuma, S. (1975). Kana and kanji processing Japanese aphasics. *Brain and Language, 2,* 369–383.

Sasanuma, S. (1980). Acquired dyslexia in Japanese: clinical features and underlying mechanism. In Coltheart, M., Patterson, K., & Marshall, J.C. (Eds.), *Deep Dyslexia.* London: Routledge and Kegan Paul.

Sasanuma, S., & Fujimura, O. (1971). Selective impairment of phonetic and nonphonetic transcription of words in Japanese aphasic patients: Kana vs. Kanji in visual recognition and writing.*Cortex, 7,* 1–18.

Sasanuma, S., Itoh, M., Mori, K., & Kobayashi, Y. (1977). Tachistoscopic recognition of Kana and Kanji words. *Neuropsychologia, 15,* 547–553.

Scribner, S., & Cole, M. (1978). Unpacking literacy. *Social Science Information, 17,* 19–40.

Seidenberg, M. S. (1985). The time course of phonological code activation in two writing systems. *Cognition, 19,* 1–30.
Sergent, J. (1983). Role of the input in visual hemispheric asymmetries. *Psychological Bulletin, 93,* 481–512.
Shaffer, W. O., & Shiffrin, R. M. (1972). Rehearsal and storage of visual information. *Journal of Experimental Psychology, 92,* 292–296.
Shimamura, A. P. (1987). Word comprehension and naming: an analysis of English and Japanese orthographies. *American Journal of Psychology, 100,* 15–40.
Smith, M. C., & Kirsner, K. (1982). On the nature of bilingual representation: Implications from interference studies. *Quarterly Journal of Experimental Psychology, 34,* 153–170.
Stroop, J. R. (1935). Studies of interference in serial verbal reaction. *Journal of Experimental Psychology, 18,* 643–662.
Takahashi, A., & Green, D. (1983). Numerical judgments with kanji and kana. *Neuropsychologia, 21,* 259–263.
Tatsumi, I. F., Itoh, M., Konno, K., Sasanuma, S., & Fujisaku, H. (1982). Identification of speech, kana and kanji, and the span of short-term memory for auditorily and visually presented stimuli in aphasic patients. *Annual Bulletin of R.I.L.P. 16,* 205–218.
Thibadeau, R., Just, M. A., & Carpenter, P. A. (1982). A model of the time course and content of reading. *Cognitive Science, 6,* 157–203.
Turnage, T. W., & McGinnies, E. (1973). A cross-cultural comparison of the effects of presentation mode and meaningfulness in short-term recall. *American Journal of Psychology, 86,* 369–381.
Turvey, M. T. (1984). Investigations in a phonologically shallow orthography. Paper presented at the third international symposium on psychological aspects of the Chinese language, Hong Kong.
Tzeng, O. J. L. (1981). Relevancy of experimental psychology to reading instruction. In Tzeng, O. J. L., & Singer, H. (Eds.), *Perception of print: reading research in experimental psychology.* New Jersey: Erlbaum.
Tzeng, O. J. L., & Hung, D. L. (1981). Linguistic determinism: a written language perspective. In Tzeng, O. J. L., & Singer, H. (Eds.), *Perception of print: reading research in experimental psychology.* Hillsdale: Erlbaum.
Tzeng, O. J. L., & Singer, H. (1981). *Perception of print: reading research in experimental psychology.* Hillsdale: Erlbaum.
Tzeng, O. J. L., & Wang, W. S.-Y. (1983). The first two R's. *American Scientist, 71,* 238–243.
Tzeng, O. J. L., Hung, D. L., & Wang, W. S.-Y. (1977). Speech recording in reading Chinese characters. *Journal of Experimental Psychology: Human Learning and Memory, 3,* no. 6, 621–630.
Vaid, J. (1985). Numerical size comparisons in a phonologically transparent script. *Perception and Psychophysics, 37,* 592–595.
Venezky, R. L. (1970). Regularity in reading and spelling. In Levin, H., & Williams, J. P. (Eds.), *Basic studies on reading.* New York: Basic.
Wang, W. S.-Y. (1981). Language structure and optimal orthography. In Tzeng, O. J. L., & Singer, H. (Eds.), *Perception of print: reading research in experimental psychology.* Hillsdale: Erlbaum.
Wijk, A. (1966). *Rules of pronunciation for the English language.* Oxford: Oxford University Press.
Zaidel, E., & Peters, A. M. (1981). Phonological encoding and ideographic reading by disconnected right hemisphere: two case studies. *Brain and Language, 14,* 205–234.
Zhou, Y. G. (1978). To what degree are the "phonetics" of present-day Chinese characters still phonetic? *Zhongguo Yuwen, 146,* 172–177.

CHAPTER 15

Literacy and the Brain

ANDRÉ ROCH LECOURS[1], JACQUES MEHLER[2], MARIA-ALICE PARENTE[3], and ALAIN VADEBONCOEUR[1]

Neuroscientists who are interested in language and linguists, psychologists, and philosophers who are interested in the brain all deal with the relations that exist between mind and language. Two fundamental postulates of modern neurolinguistics are (1) that brain-language relationships are determined by a genetic program specific to the human species (Broca, 1865), and (2) that this innate genetic predisposition produces a functional lateralization in the governing of linguistic behavior (Broca, 1863, 1865; Dax, 1836, 1865). Specifically, except among some ambidextrous and left-handed individuals, there is usually a dominance of the left cerebral hemisphere. The first of these postulates has been derived from studies of the cerebral cortex during gestation, and the second from the clinical observation that aphasia frequently occurs following lesions to the left hemisphere, but almost never following lesions to the right. Each of these points merits a moment of reflection.

It was on the basis of the neuroembryological observations of Gratiolet (1854) that Broca (1865) inferred the innateness of left cerebral lateralization for language. (This was somewhat ironic, as Gratiolet was one of the most articulate adversaries of the localizationist theories that Broca propounded.) Spectacular as this breakthrough was, conclusive evidence for it was in fact quite slim, and remained so up to and including the studies carried out more than a century later by Geschwind and Levitsky (1968), Tezner, Tsavaras, Gruner, and Hécaen (1972), Witelson and Pallie (1973), Wada, Clarke, and Hamm (1975), and many others. The problem was that the anatomical asymmetries that Broca and his successors considered to be the foundation of their arguments could conceivably been unrelated to the functional asymmetries of language disorders. Early supporting evidence came from embryological studies, which showed that asymmetries are visible by the 7th month of gestation; systematically favor the left cerebral hemisphere; and affect structures that are recognized as belonging to the *language zone.* In our opinion, a relatively convincing demonstration was made by Ratcliff, Dila, Taylor, and Milner (1980). This group established a firm relationship be-

[1] Université de Montreal, Centre Hospitalier Côte-des-Neiges, 4565, Queen Mary, Montreal, H3W 1W5, Prov. Québec, Canada.
[2] Ecole des hautes études en sciences sociales, F-Paris, France.
[3] Pontifica Universidade Catolica, Sao Paulo, Brazil.

tween anatomical and functional asymmetries by correlating particular brain regions (made visible through radiological techniques) with a transitory aphasia induced by injection of a barbiturate substance into the left carotid artery.

The postulate of an innate predisposition for language has also gained strong support in recent years, in the domain of neonatal cognition (Mehler & Fox, 1984). Within a few days of birth the left hemisphere seems to react more to linguistic information than the right (Entus, 1977; Glanville, Best, & Levenson, 1977; Molfese & Molfese, 1979; Segalowitz & Chapman, 1980). It must be recognized, of course, that even if it is established that a form of linguistic lateralization exists at birth, this does not necessarily prove that it is "innate," in the sense, say, as the behavior of a plover chick that scurries off after breaking its shell. It is hard to imagine what prenatal experience could have been relevant for the bird's motor behavior, whereas we do know that the human maternal environment lets through some sounds, many of which would be linguistic in nature. As the auditory pathways of the human fetus are well developed by the 7th month of gestation (Lecours, 1975; Yakovlev & Lecours, 1967), it could be that the lateralization seen at birth has been influenced by some rudimentary prenatal linguistic experience. On the other hand, it could be argued (although difficult to test) that prenatal auditory experience acquired in a liquid environment is unrelated to experience in the open-air environment.

Whatever the outcome of this debate may be, it is clear that the realization of the genetic program related to language – a maturation process which, if one believes the aphasiological data (Lecours & Lhermitte, 1979), could, in the final analysis, be subject to variations according to the individual – is very largely postnatal, and is only attained after several years of exposure to a linguistic environment. In other words, an innate predisposition does not in this case eliminate the need for learning. Analyses of the effects of cerebral lesions in children have provided particularly strong evidence that biological and environmental maturation work together in the achievement of left lateralization for language (Basser, 1962; Branco-Lefèvre, 1950; Lenneberg, 1967; Woods & Teuber, 1978).

Functional Lateralization in Illiterates

Without questioning the importance of genetic considerations, we will be dealing in this chapter with linguistic abilities that are completely acquired: specifically, neurological aspects of literacy. This topic could be addressed from a number of angles. For instance, there is the observation that lexic disorders are frequently associated with a personal or family background of left-handedness or ambidextrousness (Critchley, 1964), a fact which in itself bears witness to the existence of exceptions to the standard genetic programs that we alluded to earlier (Galaburda & Kemper, 1979). It is generally left cerebral lesions that (in literate subjects) disturb in an immediately observable fashion the encoding and decoding of both alphabetic (Dejerine, 1914) and syllabary (Sasanuma, 1975) writing systems. Furthermore, sometimes such deficits may persist for the decoding of syllabaries, such as Kannada, while alphabetic writing returns to normal or almost normal (Karanth, 1981). It has been speculated that the right brain (of right-handers) may play an important although not a dominant role in the encoding and decoding of ideographical writing (Sasanuma & Fujimura, 1971, 1972). Finally, there are cases in which a cerebral lesion has disrupted the phonographemological decoding of an alphabetic writing system, while sparing at least partial retention of the meaning of an "open list" of words (nouns, roots of verbs, etc.) (Coltheart, Patterson & Marshall, 1980).

We have chosen to address this topic by discussing the cerebral representation of language in illiterates (Lecours & Parente, 1982). Our discourse will concern itself with the adult nonreader, with particular reference to a question formulated in 1904 by Ernst Weber: Does the acquisition of literacy play a role, as does the acquisition of oral speech, in the process of left lateralization for language?

To the best of our knowledge, the research undertaken on this issue has been limited to the paradigms of dichotic listening tests and aphasiology.

Dichotic Listening Data

Damasio, Damasio, Castro-Caldas, and Hamsher (1979) presented three dichotic listening tasks to 47 adult subjects (14 men, 33 women) who were all right-handed and native speakers of Portuguese. The first test consisted of pairs of digits; the second, of pairs of words that were phonologically dissimilar (e.g., *colher* and *arvore, meia* and *livro*); and the third, of pairs of words that were phonologically related, differing only in the first consonantal phoneme (e.g., *ponte* and *fonte, caneta* and *maneta*). The population was initially subdivided into three groups: 16 subjects who had never been to school and were completely illiterate; 10 semiliterate subjects who had had 4 years of schooling with subsequent abandonment of all reading and writing; and 21 subjects who had completed secondary or university level studies and had maintained regular reading habits. On the basis of virtually identical results, the first two groups were merged (26 subjects between 26 and 67 years of age, mean age 47.9 years) and compared with the more educated group (21 subjects between 20 and 69 years of age, mean age 45.8 years). Both groups showed a right-ear preference on the first two tests. On the test with the phonologically similar words, right-ear preference persisted for the more educated subjects, but the subjects of the illiterate and semiliterate group showed a left-ear advantage.

In Epeiros, Tzavaras, Kaprinis, and Gatzoyas (1981) administered a dichotic listening task to 111 subjects (36 men and 75 women), all of whom were right-handed and native speakers of Greek. This test consisted only of digit pairs. Subjects were divided into two groups: 60 illiterate individuals (10 men, 50 women) and 51 who had completed at least 12 years of schooling and had kept regular reading habits (26 men, 25 women). The illiterate women were between 23 and 72 years of age (mean age 48.4 years); the illiterate men were between 29 and 80 years of age (mean age 60 years); the educated women were between 18 and 38 years of age (mean age 25.8 years); and the educated men were between 18 and 48 years of age (mean age 31.7 years). As with Damasio et al., both groups showed a right-ear preference on this task; however, the extent of this preference was more marked in illiterate subjects.

Aphasiology

It was observation of a number of illiterate and semiliterate patients with left cerebral lesions who did not present the predicted linguistic deficits that led Weber (1904) to consider whether the acquisition of reading and writing might not contribute to the degree of lateralization for language. More than 50 years later, Critchley (1956) reexamined Weber's question and recognized its pertinence, but claimed, without really specifying what he meant, that the problem was not as simple as the German neurologist had thought. Gorlitzer von Mundy (1957), a military doctor working in India, described the case of a right-handed, illiterate "boy" who, following a large left hemisphere lesion, showed no signs of linguistic disturbance despite the persistence of a severe right hemiplegia. He further claimed to have noted numerous other instances in which a left cerebral lesion had occurred in illiterates without causing aphasia, or had caused a discrete and transitory disturbance at most. Eisenson (1964) made similar observations on the effects of left cerebral lesions on illiterate and semiliterate American soldiers (the "low-level military population"), noting that aphasia in these cases was relatively rare and transitory in nature. On the other hand, Jakobson (1964) claimed to have observed illiterate World War I veterans who were severely aphasic due to left cerebral lesions, and who showed no evidence of recovery. More recently, Wechsler (1976), and Métellus, Cathala, Issartier, and Bodak (1981) separately reported atypical occurrences of aphasia in illiterate right-handers: in one case, linguistic disturbance followed a *right* hemispheric lesion [crossed aphasia (Joanette, Puel, Nespoulous, Rascol, & Lecours, 1982)], and in the other, a large infarct of the left sylvian fissure was accompanied by only a discrete and transitory aphasia.

Aside from anecdotal and single-case studies, there have been two systematic aphasiological investigations of the cerebral representation of language in illiterates, one in Mississippi (Cameron, Currier, & Haerer, 1971) and the other in Portugal (Damasio, Castro-Caldas, Grosso, & Ferro, 1976a).

Cameron et al. reviewed the cases of 62 right-handed and three left-handed adults suffering from right hemiparesis or hemiplegia following a cerebrovascular accident in the area of the left sylvian fissure. Thirty-seven subjects were defined as illiterate, with a mean of only 2.5 years of schooling; 14 as semiliterate, with a mean of 5.6 years; and 14 as literate, with a mean of 10.6 years. Transitory or persistent aphasia was observed in 78% of the literate subjects, in 64% of the semiliterates, and in 36% of the illiterates. Chi-square analysis yielded a significant difference between the first and third groups ($p = 0.02$).

Damasio et al. reviewed for their part the cases of 247 adults with unilateral focal lesions of various etiologies. This population included 209 educated subjects and 38 illiterates. Amongst the former, 114 (54.4%) exhibited aphasia: they are presumed to have been right-handed individuals with left hemisphere lesions, but these points were not specified by the authors. Amongst the illiterate subjects, 21 (55.2%) exhibited aphasia: 20 of these were right-handed with left-sided lesions, and one was left-handed with a right-sided lesion. Four of the 21 had global aphasia, ten had Broca's aphasia, and seven had some form of fluent aphasia. Proportionate diagnoses were observed in the group of educated aphasics. Two groups were then established, matched for age, sex, clinical form of aphasia, and localization of lesion, consisting of the 20 right-handed illiterate aphasics and an equal number of the educated aphasics. In contrast to the findings of Cameron et al., performance on the Token Test (de Renzi & Vignolo, 1962) did not differ significantly between the two groups.

Discussion

Dichotic listening tasks using pairs of digits show a right-ear preference in right-handed children between 8 and 12 years of age (Vargha-Khadem, Genesee, Seitz, & Lambert, 1977) and in right-handed adults regardless of literate abilities (Damasio et al., 1979; Tzavaras et al. 1981). For Tzavaras et al., this preference was found to be more marked in the illiterate than the educated group: however, since the illiterate subjects were also considerably older, a possible interpretation is simply that right-ear preference increases with age. That is, it is possible that certain functional lateralizations show a progressive realization extending well beyond childhood and adolescence, perhaps even into old age (Brown & Jaffe, 1975).

Whatever the real significance of the age-confounded data, it is questionable whether digit-pair tasks should be considered to be viable indicators of the lateralization of language. The dissociation observed by Damasio et al.

between pairs of phonologically dissimilar and phonologically related words seems to us to be more interesting. These researchers conclude that the cybernetics of language are on the whole the same in illiterates and educated subjects, but the former show "less mature [cerebral] dominance, calling for particular perceptual strategies in specific circumstances." These proposals might come across as a contradiction in terms: for our part, we would emphasize the latter point. The reversal by illiterates of ear preference when presented with a more subtle phonological distinction seems to us to constitute the "specific circumstances" that Damasio et al. speak of: that is, these subjects may be more dependent on mechanisms governed by the right cerebral hemisphere. This phenomenon would correspond exactly to the more specifically linguistic aspects of dichotic listening. This is consistent with observations that adult illiterates have great difficulty with tasks requiring phonemic segmentation, whereas syllabic segmentation is readily accomplished. It might be concluded that the former ability does not develop spontaneously with general cognitive maturation, but depends upon the formal acquisition of alphabetic reading skills (Morais, Cary, Alegria, & Bertelson, 1979; Cary & Morais, 1980).

With regard to the aphasiological studies, Damasio et al. (1976) claim that if any genuine differences existed between illiterate and educated subjects in the effects of cerebral lesions, they would have been observed very early on, since such studies began at a time when illiteracy was still common in Western Europe. This argument seems to us to be quite weak, for, along with all the early publications that supported the localizationist theory, there were in fact many that ran counter to it. For example, Moutier (1908) presented 430 cases of Broca's aphasia, including 43 personal observations, in an attempt to present precisely those cases that refuted Broca's doctrine. When the relevant documentation was available (which was far from being the rule) he mentioned the educational history of his patients. Although this was not done with the intention of relating the cerebral representation of language to literacy, it is clear that it was not uncommon to find illiterate subjects amongst those cases that were apparent exceptions to the anatomoclinical rules set forth by Broca.

Over and above the historical aspects, it is evident that modern research on this question has led to contradictory conclusions. Of the anecdotal reports, those of Critchley (1956) and Eisenson (1964) are supportive of Weber's hypothesis that literacy enhances lateralization, whereas those of Jakobson (1964) and Tikofsky (personal communication to Cameron et al., 1971) contradict it. In fact, Tikofsky suggests that aphasia in illiterates may merely be escaping clinical detection, on account of their low premorbid language abilities (which appears to us to be both unjustified and absurd). With regard to clinical observations, both the individual case reports (Gorlitzer von Mundy, 1957; Métellus et al., 1981; Weber, 1904; Wechsler, 1976) and the systematic research carried out by Cameron et al. (1971) indicate an influence of literacy on the cerebral representation of language, but Damasio and his Portuguese colleagues (1976) unequivocally conclude the opposite (as emphasized in the title of their publication, "Brain Specialization for Language Does Not Depend on Literacy").

In a subsequent exchange of letters published in *Archives of Neurology*, Currier, Haerer, and Farmer (1976) point out the disproportion of left and right lesions in Damasio et al.'s illiterate population, as well as the lack of mention of which side the lesions were on for their educated population. They furthermore suggest that these samples may have been biased, due to their having been taken from an institution known for its interest in aphasia. Damasio, Hamsher, Castro-Caldas, Ferro, and Grosso (1976b) reply (in what seems to us an enigmatic fashion) that even if such a bias existed, it would affect both the illiterate and the educated subjects equally and thus could be discounted; and they produce new data on 191 cases of left-sided and 34 right-sided lesions in right-handed patients that support their original conclusions.

Assuming that the Portuguese subjects were exposed to as expert and detailed an analysis as were the subjects from Mississippi, we cannot conclude on the basis of these reports that one claim is right and the other wrong. Possibly the same criticism applies to both studies. In both, the conclusions were based on the *observed frequency* of aphasia in groups that had been categorized by educational background and side of cerebral lesion, but the definition of aphasia is relative. A subjective diagnosis can be made on the basis of an investigator's clinical expertise (as seems to have been the case in the Mississippi study), and it is obvious that a given patient may appear slightly aphasic to one clinician but normal to another. Alternatively, a diagnosis can be based on a series of objectively scored tests (as in the Portuguese study), in which a subject scoring below a predetermined criterion is defined as aphasic. These tests, however, do not necessarily evaluate patients on all aspects of "normal" linguistic behavior. It is well known, for example, that right-handed subjects with right cerebral lesions, for whom an official diagnosis of aphasia is rare, may still have noticeable linguistic abnormalities, which may be correlated with educational background (Joanette, Lecours, Lepage, & Lamoureux, 1983). A more serious criticism is that all modern test batteries apparently use a single scale, and in the case of studies that compare two populations with strongly contrasting cultural backgrounds, this could artificially bias the relative percentage of cases that reach the conventional threshold for aphasia. Finally, neither study takes the *evolution* of the disorders into account. If a more diffuse representation of language indeed exists in illiterates, one should look for eventual differences in the modes of evolution of the disorders.

Thanks to a grant from the Harry Frank Guggenheim Foundation in New York, we have recently undertaken a study to address this question ourselves. In it, we have tried to take into account what we consider to be the mistakes of our predecessors (which may not have prevented us from making our own). In this study, 269 right-handers of 40 years of age or over were examined on a test battery that we believe to be highly valid. The population consisted of both illiterate and educated subjects, with three subgroups within each set: control subjects who were neurologically healthy, patients with left-sided cerebral lesions, and patients with right-sided lesions. In all the pathological cases the lesion had been caused by a cerebrovascular accident. Subjects were tested first at least 2 months after their stroke, and then again 6 months later provided there had been no subsequent neurological illness or neurosurgical intervention.

This research is still in progress and it would be premature to discuss its findings here. However, initial results indicate that the scores obtained by the illiterate controls differ significantly from those of their educated counterparts on all the tests employed. (In this type of research it is necessary to compare illiterate and educated subjects on more than one scale.) Moreover, the notion of left cerebral dominance for language continues to be valid for both illiterate and educated subjects, but for certain aspects of language function, such as access to the lexicon, this dominance is less absolute for illiterates.

Another important question is left unanswered. Since literacy and education often correlate to a very great extent, not to mention illiteracy and malnutrition, how legitimate is it to attribute any observed neurological differences to literacy per se? One cannot exclude the possibility that general cognitive maturation linked to schooling, rather than the specific phenomena of reading and writing, may be responsible for the realization of functional cerebral lateralization – and certainly, one cannot disregard the possibility that malnutrition may interfere with the development of genetic programs.

References

Basser, L. S. (1962). Hemiplegia of early onset and the faculty of speech with special reference to the effects of hemispherectomy. *Brain, 85,* 427–460.

Branco-Lefèvre, A. F. (1950). Contribuçao para o estudo de psicopatologia da afasia em rianças. *Archivos de Neuro-Psiquiatria, 8,* 345–393.

Broca, P. (1863). *Exposé des titres et travaux scientifiques de M. Paul Broca,* Paris, April 1863. Cited by Quercy, Annales médico-psychologiques, 101, 1943.

Broca, P. (1865). Sur le siège de la faculté du langage articulé. *Bulletin de la Société d'anthropologie, 6,* 337–393.

Brown, J., & Jaffe, J. (1975). Hypothesis on cerebral dominance. *Neuropsychologia, 13,* 107–110.

Cameron, R. F., & Currier, R. D., Haerer, A. F. (1971). Aphasia and literacy. *British Journal of Disorders of Communication 6,* 161–163.

Cary, L., & Morais, J. (1980). A aprendizagem da leitura e a consciencia da estrutura fonetica da fala. *Revista Portuguesa de Psichologia, 4,* 97–106.

Coltheart, M., Patterson, K., & Marshall, J. C. (Eds.), (1980). *Deep dyslexia.* London: Henley, Routledge & Kegan Paul.

Critchley, M. (1956). Premorbid literacy and the pattern of subsequent aphasia. *Proceedings of the Society of Medicine, 49,* 335–336.

Critchley, M. (1964). *Developmental dyslexia.* London: Heinemann.
Currier, R.D., Haerer, A.F., & Farmer, L.J. (1976). Letter to the editor. *Archives of Neurology, 33,* 662.
Damasio, A.R., Castro-Caldas, A., Grosso, J.T., & Ferro, J.M. (1976a). Brain specialization for language does not depend on literacy. *Archives of Neurology, 33,* 300–301.
Damasio, A.R., Hamsher, K. de S., Castro-Caldas, A., Ferro, J.M., & Grosso, J.T. (1976b). Letter to the editor. *Archives of Neurology, 33,* 662.
Damasio, H., Damasio, A.R., Castro-Caldas, A., & Hamsher, K. de S. (1979). Reversal of ear advantage for phonetically similar words in illiterates. *Journal of Clinical Neuropsychology, 1,* 331–338.
Dax, M. (1836). Lésions de la moitié gauche de l'encéphale coïncidant avec l'oubli des signes de la pensée. Communication presented to the Congrès méridional de médecine, Montpellier.
Dax, M. (1865). Lésions de la moitié gauche de l'encéphale coïncidant avec l'oubli des signes de la pensée. *Gazette hebdomadaire de médecine et de chirurgie, 33,* 259–262.
Dejerine, J.J. (1914). *Séméiologie des affections du système nerveux.* Paris: Masson.
de Renzi, E., Vignolo, L.A. (1962). The Token Test: a sensitive test to detect receptive disturbances in aphasics. *Brain, 85,* 665–678.
Eisenson, J. (1964). Discussion. In de Reuck, A.V.S., & O'Connor, M. (Eds.), *Disorders of Language,* p. 259. London: Churchill.
Entus, A.K. (1977). Hemispheric asymmetry in processing of dichotically presented speech and nonspeech stimuli by infants. In Segalowitz, S.J., & Gruber, F.A. (Eds.), *Language Development and Neurological Theory,* pp. 63–73. New York: Academic.
Galaburda, A.M., & Kemper, T.L. (1979). Cytoarchitectonic abnormalities in developmental dyslexia: a case study. *Annals of Neurology, 6,* 94–100.
Geschwind, N., & Levitsky, W. (1968). Human brain: left-right asymmetries in the temporal speech region. *Science, 161,* 186–187.
Glanville, B., Best, C., & Levenson, R. (1977). A cardiac measure of cerebral asymmetries in infant auditory perception. *Developmental Psychology, 13,* 54–59.
Gorlitzer von Mundy, V. (1957). Zur Frage der paarig veranlagten Sprachzentren. *Der Nervenarzt, 28,* 212–216.
Gratiolet, L.P. (1854). *Mémoire sur les plis cérébraux de l'homme et des primates.* Paris: Bertrand.
Jakobson, R. (1964). Discussion. In de Reuck, A.V.S., & O'Connor, M. (Eds.), *Disorders of Language,* p. 259, London: Churchill.
Joanette, Y., Puel, M., Nespoulous, J.-L., Rascol, A., & Lecours, A.R. (1982). Aphasie croisée chez les droitiers: 1. Revue de la littérature. *La Revue Neurologique, 138,* 575–586.
Joanette, Y., Lecours, A.R., Lepage, Y., & Lamoureux, M. (1983). Language in right-handers with right-hemisphere lesions: a preliminary study including anatomical, genetic and social factors. *Brain and Language, 20,* 217–248.
Karanth, P. (1981). Pure alexia in a Kannada-English bilingual. *Cortex, 17,* 187–198.
Lecours, A.R. (1975). Myelogenetic correlates of the development of speech and language. In Lenneberg, E.H., & Lenneberg, E. (Eds.), *Foundations of Language Development* vol. 1, pp 121–135. New York: Academic.
Lecours, A.R., & Lhermitte, F. (1979). *L'aphasie.* Paris: Flammarion.
Lecours, A.R., & Parente, M.A. (1982). Alfabetizaçao como fator determinante na fisiologia do cerebro humano. *Seara Medica Neurocirurgica, 11,* 1–14.
Lenneberg, E.H. (1967). *Biological foundations of language.* New York: Wiley.
Mehler, J., & Fox, R. (Eds.) (1984). *Neonate cognition: beyond the blooming buzzing confusion.* Hillsdale, Erlbaum.
Métellus, J., Cathala, H.P., Issartier, A., & Bodak, A. (1981). Une étude d'aphasie chez une illettrée (analphabète): réflexions critiques sur les fonctions cérébrales concourant au langage. *Annales médico-psychologiques, 139,* 992–1001.

Molfese, D. L., & Molfese, V. J. (1979). Hemisphere and stimulus differences as reflected in the cortical response of newborn infants to speech stimuli. *Developmental Psychology, 15,* 505–511.

Morais, J., Cary, L., Alegria, J., & Bertelson, P. (1979). Does awareness of speech as a sequence of phones arise spontaneously? *Cognition, 7,* 323–331.

Moutier, F. (1908). *L'asphasie de Broca.* Paris: Steinheil.

Ratcliff, G., Dila, C., Taylor, L., & Milner, B. (1980). The morphological asymmetry of the hemispheres and cerebral dominance for speech: a possible relationship. *Brain and Language, 11,* 87–98.

Sasanuma, S. (1975). Kana and kanji processing in Japanese aphasics. *Brain and Language, 2,* 369–383.

Sasanuma, S., & Fujimura, O. (1971). Selective impairment of phonetic and nonphonetic transcription of words in Japanese aphasic patients: kana vs. kanji visual recognition and writing. *Cortex, 7,* 1–18.

Sasanuma, S., Fujimura, O. (1972). An analysis in writing errors in Japanese aphasic patients: kanji vs. kana words. *Cortex, 8,* 265–282.

Segalowitz, S. J., & Chapman, J. S. (1980). Cerebral asymmetry for speech in neonates: a behavioral measure. *Brain and Language, 9,* 281–288.

Tezner, D., Tzavaras, A., Gruner, J., & Hécaen, H. (1972). L'asymétrie droite-gauche du planum temporale: à propos de l'étude anatomique de 100 cerveaux. *La revue neurologique, 126,* 444–449.

Tzavaras, A., Kaprinis, G., & Gatzoyas, A. (1981) Literacy and hemispheric specialization for language: digit dichotic listening in illiterates. *Neuropsychologia, 19,* 565–570.

Vargha-Khadem, F., Genesee, F., Seitz, M. M., & Lambert, W. E. (1977). Cerebral asymmetry for verbal and nonverbal sounds in normal literate and illiterate children. Communication, *Biennial Meeting of the Society for Research in Child Development,* New Orleans.

Wada, J. A., Clarke, R., & Hamm, A. (1975). Cerebral hemispheric asymmetry in humans. *Archives of Neurology, 32,* 239–246.

Weber, E. (1904). Das Schreiben als Ursache der einseitigen Lage des Sprachzentrums. *Zentralblatt für Physiologie, 18,* 341–347.

Wechsler, A. F. (1976). Crossed aphasia in an illiterate dextral. *Brain and Language, 3,* 164–172.

Witelson, S. F., & Pallie, W. (1973). Left hemisphere specialization for language in the newborn. *Brain, 96,* 641–646.

Woods, B. T., & Teuber, H.-L. (1978). Changing patterns of childhood aphasia. *Annals of Neurology, 3,* 65–70.

Yakovlev, P. I., & Lecours, A. R. (1967). The myelogenetic cycles of regional maturation of the brain. In Minkowski, A. (Ed.), *Regional development of the brain in early life,* pp. 3–70. Oxford. Blackwell.

CHAPTER 16

The Processing of Japanese Kana and Kanji Characters

EDWARD A. JONES[1] and CHISATO AOKI[2]

Japanese can be written in either phonetic (Kana) or logographic (Kanji) characters (see Fig. 1). Each of these scripts can express almost any word in the Japanese lexicon. Kanji are of Chinese origin, having been derived from archaic pictograms (as shown in Fig. 2) while Kana are a purely Japanese invention derived from Kanji (see Fig. 3), as a means of representing words and parts of speech having no direct Chinese equivalent. During the early period of Japanese literary history, female writers used Kana almost exclusively, while male writers used Kanji.

Research on written Japanese suggests that there might be two different hemispheric systems for processing the two character types (Hatta, 1976, 1977a, 1978; Hirata & Osaka, 1967; Sasanuma, Itoh, Mori, & Kobayashi, 1977). This research has indicated that native speakers of Japanese process Kana characters better in their left hemispheres, while they tend to process Kanji characters better in their right hemispheres, in cases where simple recognition tasks are used. Stroop test results indicate an interference effect for Kanji in the left visual field, while Kana show no difference in effects between visual fields (Hatta, 1981a; Morikawa, 1981). This pattern is consistent with the idea that both Kanji and nonverbal visual information are predominantly processed in the right hemisphere, while the processing of Kana

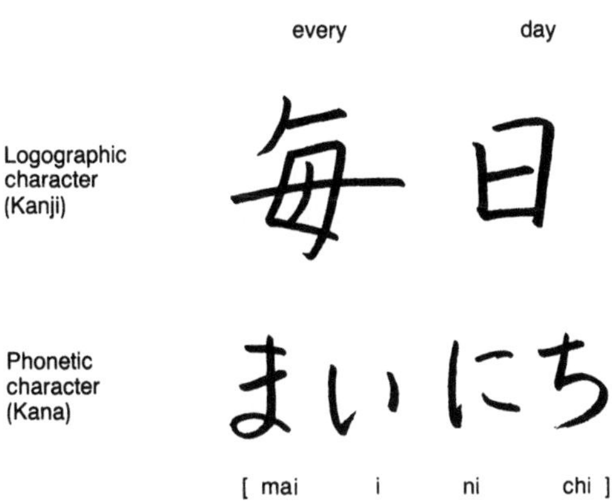

Fig. 1. Examples of logographic (Kanji) and phonetic (Kana) characters meaning "every day"

[1] Department of Psychology, Labouré College, Boston, Massachusetts 02124, USA.
[2] Department of Psychology, Northeastern University, Boston, MA, 02138, USA.

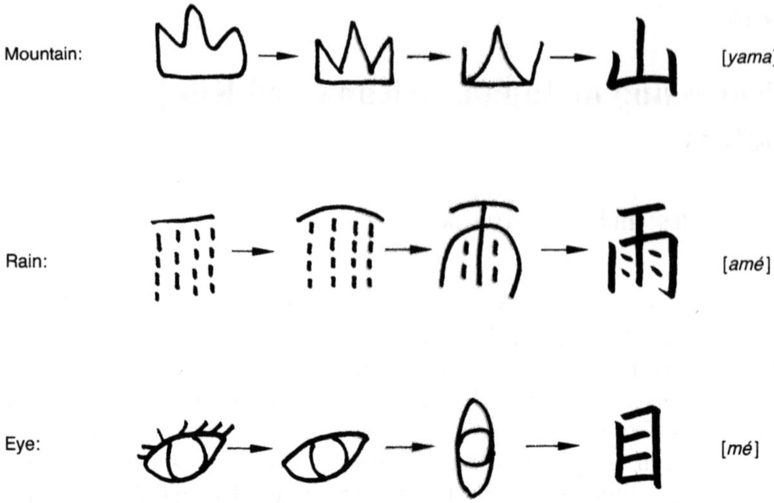

Fig. 2. Derivation of Kanji from archaic pictograms

takes place in areas that are independent of the processing of nonverbal visual information.

Hatta (1979, 1981b) has suggested that Kanji processing involves left hemisphere functioning when higher level tasks are performed. His studies indicate a right hemisphere superiority in tasks involving a simple physical recognition of Kanji form; no significant difference between the hemispheres in tasks involving a lexical recognition; and a left hemisphere superiority in those involving semantic congruence decisions. These outcomes suggest that there is a progressive shift toward the use of the left hemisphere in Kanji processing as the task moves from being one of simple stimulus recognition to one of making linguistic decisions.

The superiority of the right hemisphere in recognition tasks may be the result of characters being treated as whole figures. The lexical decision tasks appear to require a finer attention to detail, thereby reducing the role of this hemisphere. Finally, performance in the semantic congruence tasks appears to require even greater left hemispheric functioning, since the Kanji were no longer being treated as simple visual input, but were being processed with a view toward particular pronunciation and meaning.

The reading of Kanji in a phonological form yields left hemisphere superiority (Sasanuma, Itoh, Kobayashi, & Mori, 1980). Under these circumstances, the reader is converting Kanji information into the same sort of information that is provided by Kana; consequently, the processing shifts into the same hemisphere as that for Kana. This phonological reading of Kanji is commonplace in Japan: in fact, native speakers tend not to use Kana in writing whole sentences but rely instead on Kanji. Interestingly, most Japanese report that sentences written completely in Kana verge on be-

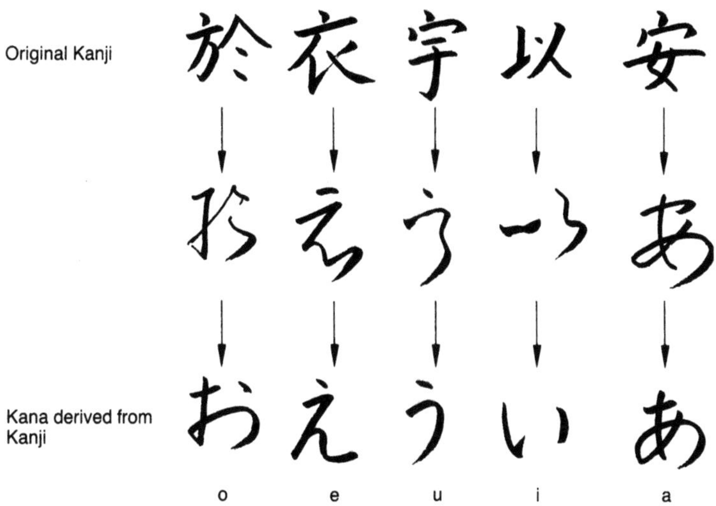

Fig. 3. Derivation of Kana from Kanji. The original kanji forms were used because they represented particular Japanese phonemes, and were employed as phonemic markers without regard to their actual meanings. Later they were broken down to the forms shown here

ing totally unintelligible. This suggests that the Kanji carry somewhat more information than just the phonetic patterns associated with them.

Sasanuma and Fujimura's (1972) observations on transcription task errors among aphasic and non-aphasic hemiplegics support the idea that the processing of Kana and Kanji involve different information systems. In this study, the researchers found that nonaphasics had no difficulty in representing words with their Kana equivalents, while they had a 22% error rate on Kanji words. The aphasic subjects had errors in over 50% of the Kana tasks and in 38% of the Kanji tasks. Their Kanji task performance was significantly worse than that of the nonaphasics ($p < 0.05$), but the types of errors followed the same pattern, chiefly involving either wrongly formed compounds or the addition or omission of strokes from characters.

The vast difference in performance in Kana processing appears to come from impaired phonological abilities on the part of the aphasics. This is supported by a comparison of the types of errors made by aphasics on Kana and Kanji transcriptions. Phonological confusion accounted for 59% of the errors made on Kana, but did not occur in Kanji. "Graphical confusions" accounted for almost 50% of the errors in Kanji transcription, but only 0.4% of the errors in Kana transcription. The overall error rates in Kana and Kanji were also quite different among aphasics, with Kana having a rate of 55.1% and Kanji having a rate of 37.6%. This indicates that the degree and types of impairments involved were quite different.

While the difference in error rates between aphasics and nonaphasics on Kanji processing was significant, it is important to note that the patterns of errors were similar, and that nonaphasics showed no errors in Kana transcription. This supports the idea that different processing mechanisms are involved in each. A recently published summary of studies of alexia and aphasia combined with alexia (Paradis, Hagiwara, & Hildebrandt, 1985) tends to support this point of view. Pure alexia affects both Kana and Kanji processing to equal degrees, but alexia involving aphasia was associated with either Kana processing superiority or Kanji processing superiority, depending on the types of lesions involved. In cases of Broca's aphasia, Kanji performance was superior, while in cases of transcortical sensory aphasia, Kana performance was superior. Iwata's (1984) findings confirmed these patterns as well.

Simple comparisons of Kana and Kanji are somewhat misleading, since they give the impression that all Kanji are equivalent in terms of their abstractness, concreteness, and pictographic dimensions. Kitao, Hatta, Ishida, Babazono, & Kondo (1977) addressed this problem by developing a classification of Kanji that describes them in terms of hieroglyphicity, familiarity, and concreteness. Examples of this classification system are presented in Fig. 4. Each character was rated by a group of judges for its ranking in each of these three categories, so that it can be described in terms of a "loading" for each characteristic.

Hatta (1977b) and Ohnishi and Hatta (1980) examined the effects of concreteness and abstractness in a lateralization study. They found that concrete Kanji are more often correctly recognized in the left visual field than abstract Kanji. This finding is consistent with recent work indicating that concrete English words tend to be lateralized more to the right hemisphere in processing (e.g., Kroll, 1985, personal communication). The impact of hieroglyphicity and familiarity on lateralization was not investigated by either Hatta (1977b) or Ohnishi and Hatta (1980).

Considerations of concreteness, hieroglyphicity, and familiarity are quite important, since the processing of Kanji having different degrees of each

		Concreteness	**Hieroglyphicity**	**Familiarity**
River:	川	High	High	High
Ocean:	海	High	Low	High
Left:	左	Low	Low	High

Fig. 4. Examples of classification of Kanji from Kitao et al.'s (1977) study. Each Kanji was rated for concreteness, familiarity, and hieroglyphicity

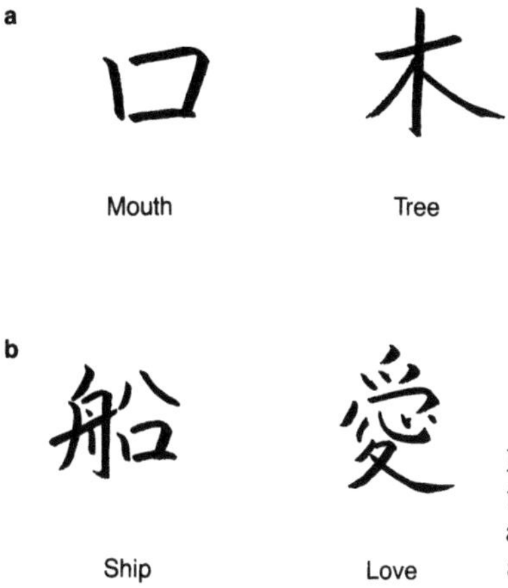

Fig. 5. Hieroglyphic Kanji close to the original pictograms for "mouth" (*Kuchi*) and "*Tree*" (ki). **b** Examples of nonhieroglyphic Kanji for "ship" (*Fune*) and "Love" (*ai*)

could conceivably be quite different. For example, highly hieroglyphic words tend to be very similar to the original pictograms from which Kanji were derived, as can be seen in Fig. 5a. In contrast, other characters, such as those in Fig. 5b, can be very far removed from any immediate visual equivalence to the concept that they represent. Under these circumstances, one could reasonably expect that the more hieroglyphic Kanji would tend to be more right hemispherically lateralized, and the less hieroglyphic ones would tend more toward left hemisphere lateralization.

Studies of hemispheric dominance in the reading of Kanji often treat the characters as graphemes representing spoken words. Under these circumstances, the tasks used require subjects to employ phonetic means of representing the information encoded by the Kanji at some point during the task. Kroll (1985, personal communication) has noted that the use of pictures rather than words in semantic tasks would yield a better right hemisphere performance. Kroll (1985) and Kroll and Potter (1984) have suggested that there are two different semantic systems involved in the processing of language, based on independent phonological and visual reading mechanisms. This is an important consideration in dealing with Kanji, since, as Iwata (1976) has noted, Kanji have both phonemic and ideographic elements, each of which contributes to their overall meaning. He has suggested that these elements are somewhat independent of each other and that they are capable of being selectively impaired.

Kroll's (1985) and Kroll and Potter's (1984) work on picture-matching tasks indicates that both concrete and pictorial processing may have their own mechanisms that might be located in the right hemisphere (Kroll, 1985, personal communication). Such mechanisms would presumably also function in the processing of logographs, in which case concrete Kanji should show a greater right hemisphere lateralization (e.g., Hatta 1977b). This effect would be heightened if concrete Kanji were tested in a picture-pairing task, rather that in a word-pairing task. Consideration of this point suggested the development of an experiment comparing Kanji and Kana processing in a picture-word task involving the use of concrete words.

In undertaking this research, the present authors also decided to control other factors for the purpose of intensifying any observed effect. These factors were familiarity and hieroglyphicity. By selecting highly concrete, highly familiar, and highly hieroglyphic Kanji, the authors determined that they would be most likely to observe a right hemisphere dominance on a Kanji-picture task involving semantic processing. If such dominance did not occur under these conditions, then it would be highly unlikely to occur under any conditions, since the variables selected were those most likely to produce right hemisphere dominance.

Experimental Design

Subjects used in the experiment were native Japanese who had been residing in Boston for about 2 years at the time of testing. They included ten females whose mean age was 26 years and nine males whose mean age was 27 years. All of the subjects regularly read Japanese during their stay in the United States, and all were college-educated. All were right-handed according to the standards of the modified Edinburgh Handedness Inventory (see Bryden, 1977; Oldfield, 1971; Raczkowski & Kalat, 1974 for a description of the technique). In addition, no members of their immediate families were known to be left-handed. All of the subjects had at least 20/20 corrected vision, as tested prior to the experiment.

The apparatus used in the experiment was a Gerbrands G1171 two-channel projection tachistoscope, composed of two Kodak Ektagraphic slide projectors and electric shutters controlled by a four-channel digital millisecond timer. Stimuli were rear-projected onto a translucent screen placed on a table in front of the subjects. The projection field subtended 24° of visual angle horizontally and 16° of visual angle vertically. Subjects were seated with their heads resting on a chin-rest 66 cm from the screen. A set of reaction keys was located by the right hand of each subject. The chin-rest was adjusted for each subject so that it would be comfortable, and the center point of the image was then adjusted to be at the subject's eye level. The stimuli used in the experiment projected to a size of 1.5° by 1.5° of visual angle for Kanji and pictures, and 1.5° of visual angle width for Kana, with a height that could reach 3.5° of visual angle.

The stimuli used in the experiment were developed from a series of Kanji words taken from Kitao et al.'s (1977) list of 881 basic Kanji characters. The items selected were 40 words having high hieroglyphicity, high concreteness, and high familiarity, as determined by Kitao et al. (1977). The primary considerations in character selection were concreteness and hieroglyphicity, with concreteness being rated more strongly. Once these 40 words were selected, their Kana equivalents were written out, and associated pictures representing each word were drawn. These were then photographed as 35-mm slides, and the pictures were tested by raters for recognizability. Once the Kanji, Kana, and pictures were set, a series of nonsense Kanji and nonsense Kana combinations was developed.

Forty picture-word pairs for each character type were developed for each visual field. Half of the characters in each of these series were either nonsense characters or characters that were mismatched with the picture, but resembled the appropriate word. Examples of the characters used in these combinations are shown in Fig. 6.

Subjects were instructed to sit with their chins on the chin-rest facing the screen, and the fixation point was adjusted to an appropriate height. They were told that their task was to determine if a word and picture were appropriately or inappropriately paired. Each subject was first given a practice trial, using 20 picture-word pairs equivalent to those which were to be used in the actual test session. Eye movement was controlled for by having a small cross or one of the numerals from 2 to 9 presented at the fixation point of the stimulus slide during the trial. If a subject was unable to correctly identify the numeral being presented, the result from that trial were omitted, on the assumption that the subject was not fixating. This occurred in only 2% of the trials.

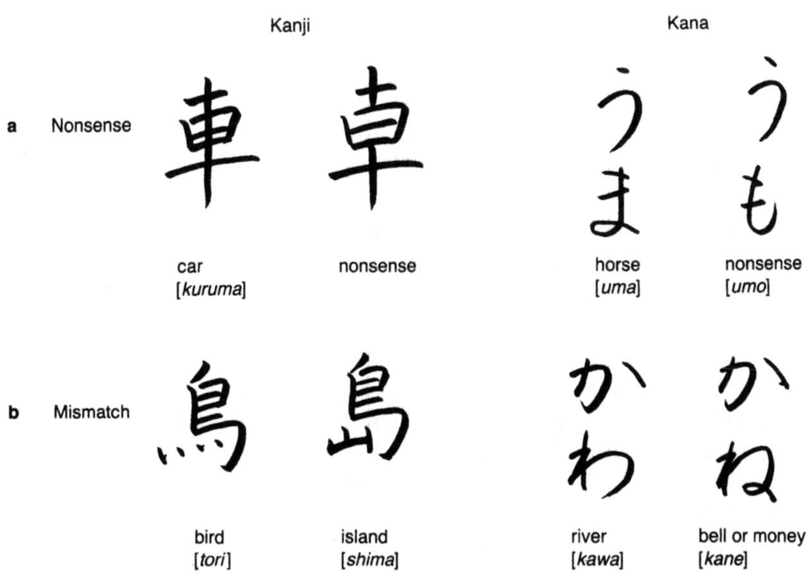

Fig. 6. Stimulus forms used in the experiment. Half were real, appropriate characters, while the others were either nonsense characters (**a**) or characters that appeared similar to the appropriate ones, but were mismatched (**b**). Real characters are shown on the *left* in each pair, their mismatched or nonsense counterparts on the *right*

Subjects were tested on a sequence of 40 trials for each type of character for each eye. A single trial consisted of the presentation of a fixation point followed by the presentation of the first stimulus; after an interval of 1 s, the second stimulus was presented. Once the two stimuli were presented, the subjects were required to indicate whether the match was correct or incorrect by pressing a response key.

Subjects' performances were evaluated by means of a modified threshold testing technique. Picture-Kanji pairs were initially presented to subjects for a duration of 170 ms, while picture-Kana pairs had an initial duration of 100 ms. These times were set on the basis of pilot research. An incorrect response would bring a 10-ms increase in presentation time, while two successive correct responses would bring a 10-ms decrease in presentation time.

The tester presented the subjects with either a right visual field or left visual field sequence on a random basis to control for order of presentation effects. The sequences contained the same series of trials for each mode. Subjects were tested in two sessions, with both Kanji and Kana being presented in each. During all of these sessions, the tester kept a record of reaction time, error rates, and stimulus duration, as well as checking the correctness of reported fixation numbers.

Results

A check of reaction time measurements showed no significant difference between visual fields. A slight difference did show up in the raw data, indicating that the reaction times for the right visual field were quicker. This nonsignificant difference can probably be attributed to right-hand, right-eye reaction times being better than left-eye, right-hand reaction times – as would be expected, since left-eye, right-hand reactions would involve interhemispheric coordination, while right-hand, right-eye coordination would involve only one hemisphere. This is consistent with the idea that an interhemispheric activity can cancel lateralization effects.

Stimulus durations for each hemisphere should not have been affected in the way reaction times could have been, since they did not require the use of both hemispheres. The results obtained on stimulus durations showed a very strong lateralization by character type. These results were determined by obtaining a mean stimulus duration for each character type for each visual field for each individual, for both correct and incorrect responses. These results were then analyzed by means of a mixed model analysis of variance (BMDP3V). The analysis was carried out for duration differences for visual fields, for character types, and for the interactions between character types and visual fields.

No overall difference was found for visual fields ($F = 0.89$, $p > 0.05$). This indicates no overall hemispheric superiority for the semantic task involved in the experiment. A significant difference in durations was found for character types ($F = 26.22$, $p < 0.001$). This difference results from Kana durations being much shorter than those of Kanji, and might be partially accounted for by the difference in initial times, which, however, was set on the basis of performance differentials found during the pilot study. The character difference by visual field interaction was also significant ($F = 3.90$, $p < 0.001$). This difference supports the idea that character types might be associated with different processing systems that are lateralized differently (Table 1, Fig. 7).

The results obtained for Kanji and Kana durations are interesting, given that most of the words used are not normally represented by means of Kana. It would seem that at this simple task level the ability to decide an appropriate picture-word match is enhanced by the use of Kana. Matching at this level is relatively easy, since the subjects were not required to do anything other than determine the identity of a given word from a picture. As noted earlier, Japanese typically report having great difficulty in reading long sequences of Kana, and find Kanji much easier to read under such circumstances. The task used here did not lend itself to testing this sort of performance.

Table 1. A comparison of the mean stimulus durations shows that there is a significant difference between Kana and Kanji ($p < 0.001$, indicated by*), and a significant interaction effect for character type and visual field ($p < 0.001$, indicated by**)

Mean stimulus duration	Right visual field	Left visual field	Average
Kanji	229**	207**	218*
Kana	90**	104**	97*
Average	159.5	155.5	

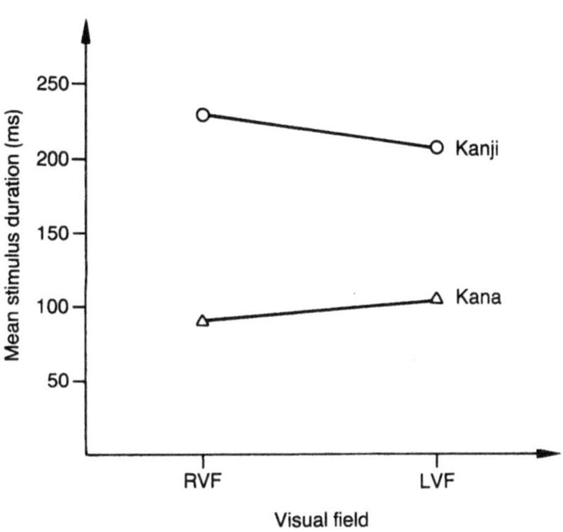

Fig. 7. Mean stimulus durations for Kana and Kanji by visual field. Visual field interactions with figure type are significant as shown ($p < 0.001$), with Kana faster for the right and Kanji faster for the left

Another limitation on the task is the fact that it relied on the use of familiar words, and hence familiar objects. All of the objects pictured in the test were readily identifiable, so the decision involved did not require information from the word to determine what was in the picture. If nonsense figures, such as those used by Kroll (1985) and Kroll and Potter (1984), were employed, then the information provided by the words themselves would become much more important in decision making, and this might in turn cause a different outcome in the comparison of Kanji and Kana durations.

Kanji and Kana processing for each hemisphere had comparable error rates, once the presentation duration reached the threshold level. This situation would be expected, since all of the subjects were normal in terms of brain functions and language processing. The question does arise as to whether this type of task would elicit different patterns of responses in split-brain subjects. Since this issue has yet to be investigated, it is best to assume that each hemisphere is capable of performing such tasks as was seen here, and that the right hemisphere is superior for Kanji, while the left hemisphere is superior for Kana.

Observations

Iwata (1976, 1984) has suggested that Kana and Kanji processing differences might involve the use of different neural pathways within the left hemisphere. This phenomenon is associated with the use of dictation tasks, which would cause left hemisphere dominance. If a purely visual task were used, it is possible that the right hemisphere would come into play. The research presented here supports this idea, and taken in combination with Iwata's findings suggests that both Kanji and Kana processing might involve both hemispheres in multiple-path functions.

A multiple-path bihemispheric system would have a certain amount of redundancy, and so the concept might not appeal to those who seek a very simple model to explain reading and writing. It is certainly far from being established as a concept, and requires a great deal more investigation. There is a good deal of observational evidence to support the idea that Kanji and Kana processing involve more than just a recognition of the phonetic or visual elements associated with given words. This is most evident in situations where individuals are presented with "broken" Kanji, such as those found in poetic "grass writing."

During the history of Japanese calligraphy, many styles for writing characters have been developed. These usually involve a "breaking" of the character, in which its overall pattern is simplified but its essential identity is preserved (see Fig. 8a). Kanji breaking involves the production of a somewhat abstract representation of a character that may have no immediately obvious visual relationship to its original formal version. It is based on the

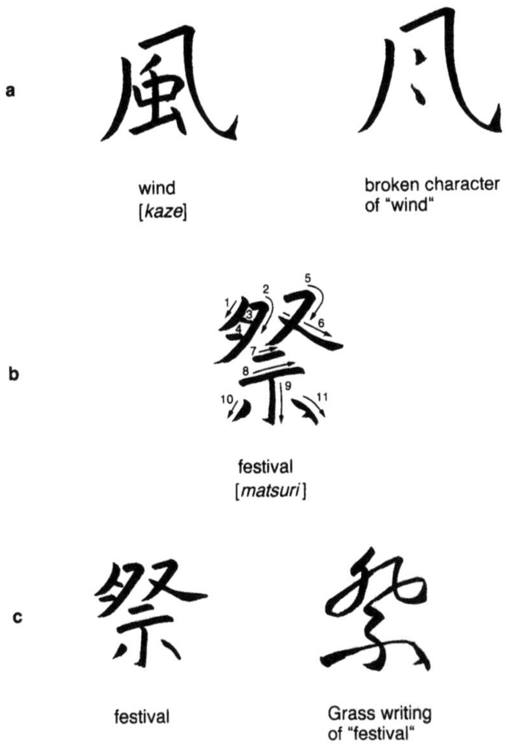

Fig. 8. **a** A broken character on the *right* compared with the original on the *left:* both mean "wind." **b** The stroke sequence for "festival" (*matsuri*). **c** "Festival" in grass writing. The original character on the left looks quite different, but the kinesthetic identity is preserved in the broken character on the right

fact that Kanji involve a certain number of strokes which must be written in a traditionally established sequence, as shown in Fig. 8b. This combination of a set number of strokes and a specific sequence for writing them makes it possible to identify a character by images that encode these features of the character, rather than simply copying them, as is shown in Fig. 8c.

Individuals who are presented with unfamiliar Kanji will often try to decipher them by tracing them in the air. This method of identifying characters is similar to the kinesthetic facilitation used by alexic patients (e.g., Ohashi, 1965; Torii, Fukuta, & Koyama, 1972; Yamadori, Nagashima, & Tamaki, 1983). It represents another way in which Kanji can be perceived and processed. Foreign students of Japanese, including one of the present authors, can quickly become aware of the kinesthetic dimension of Japanese and of another interesting aspect of the language.

Lovers of Japanese calligraphy are often wont to exclaim over the beauty of a particular example of the art and then admit that they are unable to read it. The beauty in such cases lies in the abstract images created by a particular pattern of strokes. If pressed to read the character, an individual can often do so through representing it kinesthetically. This observation led the present authors to undertake observations on individuals presented with difficult-to-read Kanji forms. Since subjects typically tended to resort to a kinesthetic representation of characters, a quasi-experiment was set up to determine the probability of this method being used as well as its effectiveness. Three female and two male subjects were presented with broken Kanji such as those shown in Fig. 8 and were asked to read them. In each case, the individual was not able to make an immediate positive visual identification of the Kanji, but then, without being instructed to do so, resorted to kinesthetically representing the Kanji, which led to correct responses in every case.

Kinesthetic cueing is also effective in the case of Kana (e.g., Ohashi, 1965) among alexic patients. It does not play a great role in the functioning of normal subjects, since Kana tend to be written in standard forms that are not broken. However, there are some stylized forms of Kana developed by sign painters and others interested in artistic effects that occasionally require tracing out for comprehension. This fact, in combination with observations on the treatment of alexic patients, supports the idea that both Kana and Kanji might be processed by more than one mechanism. This observation is consistent with Iwata's (1976, 1984) observations on Kanji and Kana processing.

The existence of parallel writing systems involving more than one reading mechanism might seem to be excessively complex to users of Greek alphabets. It is certainly quite different from the situation that exists in the case of Greek alphabet-based languages. One explanation for the multiple systems lies in the idea that the spoken words represented by Kanji are often homophones (Iwata 1984). A series of 30 characters representing the homophones of [ʃ in] is presented by Iwata (1984): some of these are shown in Fig. 9. Kates (1952) noted a similar situation in Chinese, where he found 164 different characters representing homophones for [ʃ i].

Homophones of [ʃ in]

新 : new

心 : heart

針 : needle

森 : forest

Fig. 9. Some of the homophones for (ʃ in) taken from Iwata's (1984) list of 30 such homophones

hen *tsukuri*

晴

清

精

Fig. 10. The *keisei* Kanji have left (*hen*) elements typically encoding semantic values, and right (*tsukuri*) elements typically encoding phonetic values. The *hen* element for each differs, indicating either "day," "water," or "rice," while the *tsukuri* element is the same, giving the pronunciation "sei." The meanings of the characters are "clear day," "clear water," and "polished rice" respectively

Iwata (1984) has observed that 80% of the Kanji in daily use are composed of two elements; a *hen* element indicating its semantic category, and a *tsukuri* element indicating its phonetic form. He further noted that these elements are usually arranged with the semantic element on the left and the phonological element on the right, and added that 70% of writing errors in Kanji are restricted to the phonological element. His results showed that such phonological processing errors in Kanji writing occur basically in cases of pure alexia, but not in cases compounded by aphasia.

The present authors' survey of commonly used *keisei* Kanji (those having a *hen/tsukuri* division) indicates that 76.5% of them follow the pattern described by Iwata; 8.17% of them follow the opposite pattern; 10% have the phonemic element on the top and the semantic element on the bottom; 8.83% reverse this pattern; and 0.5% have the semantic element in the middle (see Fig. 10). The preponderance of phonemic left, semantic right is significant both in statistical and logical terms. It indicates that the processing of phonological elements is probably easier in the right visual field (left hemisphere), while the processing of semantic visual elements is probably easier in the left visual field (right hemisphere). This is consistent with the idea that there are important hemispheric differences in the processing systems for Kana and Kanji.

Implications

The reading of Kanji obviously involves very different elements than the reading of Kana. However, this difference in processing does not appear to be simply a matter of hemispheric lateralization, nor one of different processing mechanisms within a single hemisphere. Rather, it would seem to involve more than one mechanism in both hemispheres. The lateralization that does occur appears to be linked to the type of reading or recognition task involved. Thus, phonological tasks elicit better left hemisphere processing, while purely visual tasks elicit better right hemisphere processing. This difference in task effects on lateralization gives rise to greater left hemisphere efficiency in Kana processing and greater right hemisphere efficiency in Kanji processing.

All of this suggests that Japanese speakers might be less lateralized than English speakers, in whom both speech and reading functions appear to be located in the left hemisphere. It should be borne in mind, however, that the results of the work by Kroll (1985) and Kroll and Potter (1984) suggest that the right hemisphere might play an important role in semantic decision making in English. If this is the case, then the differences between Japanese and English processing are less than they are currently thought to be, but it seems reasonable to assume that the logographic characters used must have some effect on the role of the right hemisphere in the reading of Japanese.

The patterns of *hen* and *tsukuri* in Kanji indicate that logographic characters are structured in ways that are most readily processed in a brain having a hemispheric division of labor in reading. They are also very strong evidence for the idea that the reading of Japanese involves more than one mode of simultaneous processing. One of the common assumptions made in research on the reading of Japanese is the idea that reading must be lateralized into one hemisphere or the other (see Paradis et al.'s 1985 review of the literature for examples of this). This assumption might reflect an ethnocentric bias created by the fact that most research on reading has been done on the highly lateralized, Greek-alphabet-based systems of European languages.

It is interesting to note in regard to the last point that research on thought and memory in psychology has been bascially dominated by language recognition and recall tasks. This pattern reflects an Occidental belief in the idea that thought must of necessity involve language processing. This is, of course, one of the fundamental bases of Greek philosophy, which appears to have grown out of the type of linguistic and logical systems associated with the left hemisphere. Educational systems in Western countries tend to be conducted in the context of the Greek philosophical framework that stresses the importance of verbal logic. Informal acquisition of skills has typically been considered to be inferior to verbal learning, with the net result that formal language learning has come to be seen as the only true form of learning.

The emphasis placed on formal language learning has tended to obscure the importance of skill learning. Consequently, kinesthetic elements of learning are overlooked in favor of purely verbal ones, in cases where unitized models of memory are used. Such unitized models could be said to be the consequence of the study of memory in cultures using highly lateralized Greek-style scripts. Models of multiple forms of memory appear to be more appropriate for an analysis of Kanji and Kana processing, since these tasks involve kinesthetic as well as visual and phonological elements in their storage in memory.

Schacter (1987) and Graf and Schacter (1985) have developed a model of multiple forms of memory that is consistent with an analysis of Kanji and Kana processing and differences in that processing. This model posits separate implicit and explicit memory processes. Implicit memory involves fairly automatic tasks, typically having sequential relationships. Explicit memory involves the retrieval of small units of information, requiring a focusing on elements rather than on whole patterns or sequences. The empirical distinction between the two types of memory functions is based on the types of materials used in a particular task. Schacter describes phrases as being either *unitized*, e.g., "sour grapes," "small potatoes," or *nonunitized*, e.g., "sour potatoes," "small grapes." The difference between these phrases lies in the fact that the first element of a unitized phrase serves as a predictor of the next element, while the first element of a nonunitized phrase does not.

Unitized phrases are predictable because subjects have had practice with them, while they have not had practice with nonunitized phrases. Practice causes the unitized phrases to develop as sequences that are in some ways equivalent to the sequences involved in kinesthetic skill learning, and this is one of the arguments that Schacter uses to support the idea of a sequential implicit memory system. The size and complexity of sequences in kinesthetic learning and in verbal learning can differ tremendously, and consequently the implicit or explicit nature of a task is determined by the scale of the sequences involved. The importance of this to Kanji and Kana processing lies in the fact that the processing of characters representing single words would be different than the processing of sequences of characters composing a sentence.

A string of Kana representing the name of an object in a picture would form an implicit memory structure, since the first character would serve as a predictor for the others. This was quite evident in our pilot study in which nonsense Kana were not used. Subjects in this study were able quickly to discriminate inappropriate picture-word matches on the basis of the first Kana in a given string. This led to an extremely rapid response rate and to very short stimulus durations. The introduction of nonsensical Kana combinations decreased response rates and raised stimulus durations, but responses to Kana were still much quicker than responses to Kanji. Apparently the sequential aspect of the task was still affecting responses, as subjects reported.

Kanji word pairs did not involve a simple sequential judgement. Instead, subjects had to attend to elements of the Kanji that could be modified in nonpredictable patterns, or which could involve having to attend to the correctness or incorrectness of a particular picture-Kanji pairing. In either case the subject did not have any predictive information to base decisions on, and so the task involved here was an explicit one. This difference between Kanji task types and Kana task types probably accounts for the difference in performance seen on the two types of characters.

The task used in the experiment conducted by the present authors involved only the processing of single word units. Kana superiority under these conditions has been attributed to the predictability of the structures of certain sequences of characters. This superiority would probably be lost in a more complex task involving sequences of words, since the context for interpreting a given set of Kana would no longer be a matter of matching a set of sounds with a given picture. Japanese informants typically report difficulty in reading sequences of words that are written exclusively in Kana, because of the problems of multiple homophones, and because the characters themselves do not have an implicit meaning. By contrast, they find the reading of sequences of Kanji to be much easier, because the meanings of the characters involved serve as predictors of successive sequences of characters. Thus, the shift from a single word to a multiple word context would probably cause processing of both Kanji and Kana to move into a different mode.

A shift from implicit to explicit memory functions would involve something quite different than a shift in hemispheres. It would more likely be a matter of changing processing systems within a given hemisphere. This means that the multiple processing systems would resemble the sort described by Iwata (1976), in which the processing of characters within a given hemisphere is the function of multiple systems. If this is the case, then Japanese writing is not lateralized in any particular hemispheric direction, and the processing of Japanese is not the product of a single processing mechanism. Instead, it would appear that Japanese is somewhat generalized in processing throughout a series of systems in both hemispheres.

Possible Cultural Significance of Processing Systems

Greek alphabet writing follows a sequential predictive pattern quite different from the patterns occurring in Japanese. It has been related to the use of space in the theater (de Kerckhove, 1981) and to philosophy and education (by the present authors). The literary and philosophical traditions that stem from Greek culture do a great deal to shape thought in the contemporary Western world. They differ tremendously from those of Japanese culture, in that they stress the definition of events in terms of predictive sequences of words, rather than in terms of subjective experience.

Japanese philosophy, especially Zen, differs radically from philosophies originating in ancient Greece. It is more a philosophy of action and experience than one of definitions and rules. Zen is supposed to transcend words, to reach a purely experiential plane in which judgement is suspended. Literature and art developed in this context are quite different from those of Western culture, which rely on formalism and canons of proportion. It is a form of expression that is strongly influenced by the desire to evoke feelings and to bring the viewer or reader into a particular state of awareness, rather than to elicit some intellectual reaction or critique.

Chinese art shares many of the traits seen in Japanese art, and both seem to have been strongly influenced by a common mystical philosophy and the use of logographic calligraphy. Brush writing has a very strong kinesthetic component, and, as Kates (1952) has noted, a very sensual aspect in the texture of the paper and the movement of brush and ink over it. This sensuality is closely linked to the aesthetics of calligraphy, being apparent in the pattern of brush strokes on a given surface. The process is more akin to ink drawing than to the transcription of a spoken word. This is especially true in the case of artistically "broken" Kanji.

The reading of calligraphy is a multidimensional experience. It involves a response to the written characters, to their sounds, to their meanings, to the textures of the writing surface, and to the abstract patterns made by the ink against the background of the writing surface. Furthermore, it involves a sensitivity to the movements of the brush and to the character of the writer as expressed in them. Part of the reading experience is very kinesthetic, particularly in the case of well-written artistic calligraphy, where readers often tend to trace out the flow of the brush strokes by moving their hands through the air in imitation of them. This is something that the present authors have yet to see as response to words written in a Greek-type alphabet.

The aesthetic response to calligraphy is an important element in the learning of the skills involved in good artistic writing. It is obviously a subjective reaction, of the sort commonly associated with the learning of art forms, and so may be discounted as poor support for the idea of cultural differences in learning processes. It would indeed be poor support if calligraphy were the only area in which such an approach to learning were used; however, this is far from being the case.

Learning in Japanese culture is premised on a very different concept of knowledge than formal Western education uses. It assumes that learning involves the attaining of some sort of insight into the subject matter or activity that defies objective identification. This feature of learning is embodied in the idea of different "paths" of experience as opposed to curricula. Traditional crafts, the martial arts, music, flower arranging, literature, and personal growth are all seen as having the potential of interacting in a process that brings the person to fuller awareness and sensitivity to the world. These interactions are based on a Zen approach to life which has made its way into Western industrial society by way of Maslow's concept of self-actualization.

The nature of learning in Japanese society is a direct expression of traditional concepts of the individual as a being having a very special subjective relationship to the universe, best understood by subjective experience. It is quite different from the Greek sense of the individual as an objective observer, who has a relationship with the outer world established by way of definition and the analysis of experience in objective terms. This difference between the traditional Japanese concept of the universe and the traditional Western concept has long been thought to account for the emergence of science in the West and its limited development in the East.

The failure of Chinese-influenced cultures to develop science is extraordinary in light of the high levels of technology that they attained. Westerners remark on this because they tend to see technology as the application of science. However, technology's relationship to science is not one of dependence; rather, technology can be enhanced through the application of scientific principles. Technology can develop without this enhancement through the use of "common sense" and insight, as appears to have happened in China and Japan. Much of the technological achievement attained in these ways was transmitted by means of an oral tradition involving masters and apprentices.

Traditional transmission of information through master–apprentice relationships often has a subjective mystical element to it. This means that it is not open to critical analysis, and is instead treated as a compendium of secrets or tricks of the trade. This is definitely the case in traditional Japanese crafts and arts which often involve combinations of surface methods and hidden methods passed to select disciples, who will succeed to the mastership of the craft. This type of treatment of knowledge is antithetical to the Western scientific tradition.

Science has traditionally been linked to the open publication of ideas in an objective form. Secrecy is anathema in scientific communication, and science could not exist without an objective means of representing ideas. Classical science developed through free exchanges of ideas that grew out of Greek philosophy, and the collapse of literate society with the fall of Rome clearly brought an end to it. The rise of universities in the late Middle Ages and the invention of the printing press both contributed strongly to the revival of scientific thought during the Renaissance.

Dry objectivity is in some ways the essence of science. It is something hard to attain in the classical literary forms used in Japanese and Chinese writing that emphasize the poetic and subjective aspects of experience. This suggests that the development of Western scientific and philosophical traditions might reflect a way of experiencing ideas that is the product of the Greek alphabetic system. Japan's development of science during the late nineteenth century grew along with the acquisition of scientific vocabularies from foreign countries, and in many cases scientific concepts are still expressed in foreign words, even though there are Japanese equivalents for them.

In summary, it can be said that Japanese language processing differs greatly from that of English, and presumably from that of other Greek-alphabet-style languages. Differences in hemispheric involvement as well as differences in procedures of processing and in grapheme forms are involved. These differences seem to be related to some elements of Japanese culture, particularly the development of aesthetics and traditional arts and crafts. This unique inclusion of arts and crafts in traditional aesthetics would seem to involve a much greater subjective component than Western science and technology, and this may account in part for the limited development of science and technology in Chinese-influenced cultures despite the high level of technology developed within them. If this is indeed the case, then it is reasonable to suggest that the invention of the Greek alphabet was a pivotal point in the emergence of the philosophy, science, and technology of Western culture, which is now being assimilated in Japan and China along with Western language skills.

References

Bryden, M.P. (1977). Measuring handedness with questionnaires. *Neuropsychologia, 15,* 617–624.

de Kerckhove, D. (1981). A theory of Greek tragedy. *Sub-Stance, 29,* 23–36.

Graf, P., & Schacter, D.L. (1985). Implicit and explicit memory for new associations in normal and amnesic subjects. *Journal of Experimental Psychology: Learning, Memory, & Cognition, 11,* (3), 501–518.

Hatta, T. (1976). Asynchrony of lateral onset as a factor in difference in visual field. *Perceptual and Motor Skill, 42,* 163–166.

Hatta, T. (1977a). Recognition of Japanese Kanji in the left and right visual field. *Neuropsychologia, 15,* 685–688.

Hatta, T. (1977b). Lateral recognition of abstract and concrete kanji in Japanese. *Perceptual and Motor Skills, 45,* 731–734.

Hatta, T. (1978). Recognition of Japanese kanji and hiragana in the left and right visual fields. *Japanese Psychological Research, 20,* 51–59.

Hatta, T. (1979). Hemisphere asymmetries for physical and semantic congruency matching of visually presented kanji stimuli. *The Japanese Journal of Psychology, 50* (5), 273–278.

Hatta, T. (1981a). Differential processing of kanji and kana stimuli in Japanese people: some implications from Stroop-test results. *Neuropsychologia, 19,* 87–93.

Hatta, T. (1981b). Different stages of kanji processing and their relations to functional hemisphere asymmetries. *Japanese Psychological Research, 23* (1), 27–36.

Hirata, K., & Osaka, R. (1967). Tachistoscopic recognition of Japanese letter materials in left and right visual fields. *Psychologia, 10,* 17–18.

Iwata, M. (1976). Yomukoto to kakukoto – moji no shinkeigaku [Reading and writing: neurology of literal activity]. *Kagaku [Sience], 46,* 405–410.

Iwata, M. (1984). Kanji versus kana: neuropsychological correlates of the Japanese writing system. *Trends in Neurosciences,* August, 290–293.

Kates, G. (1952). *The years that were fat.* Cambridge: MIT.

Kitao, N. Hatta, T., Ishida, M., Babazono, Y., & Kondo, Y. (1977). Concreteness, hieroglyphicity and familiarity of kanji. *The Japanese Journal of Psychology, 48,* 105–111.

Kroll, J.F. (1985). Colloquium presentation. Cambridge: MIT.

Kroll, J.F., & Potter, M.C. (1984). Recognizing words, pictures, and concepts: a comparison of lexical, object, and reality decisions. *Journal of Verbal Learning and Verbal Behavior, 23,* 39–66.

Morikawa, Y. (1981) Stroop phenomena in the Japanese language: the case of ideographic characters (kanji) and syllabic characters (kana). *Perceptual and Motor Skills, 53,* 67–77.

Ohashi, H. (1965). *Rinshoo noo byoorigaku [Clinical cerebral pathology].* Tokyo: Igaku Shoin.

Ohnishi, H., & Hatta, T. (1980). Lateral differences in tachistoscopic recognition of kanji-pairs with mixed image values. *Psychologia, 23,* 233–239.

Oldfield, R.C. (1971). The assessment and analysis of handedness: the Edinburgh Inventory. *Neuropsychologia, 9,* 97–113.

Paradis, M., Hagiwara, H., & Hildebrandt, N. (1985). *Neurolinguistic aspects of the Japanese writing system.* Orlando: Academic.

Raczkowski, D., & Kalat, J.W. (1974). Reliability and validity of some handedness questionnaire items. *Neuropsychologia 12,* 43–47.

Sasanuma, S., & Fujimura, O. (1972). An analysis of writing errors in Japanese aphasic patients, kanji vs kana. *Cortex, 8,* 268–282.

Sasanuma, S., Itoh, M., Mori, K., & Kobayashi, Y. (1977). Tachistoscopic recognition of kana and kanji words. *Neuropsychologia, 15,* 547–553.

Sasanuma, S., Itoh, M., Kobayashi, Y., & Mori, K. (1980). The nature of the task-stimulus interaction in the tachistoscopic recognition of kana and kanji words. *Brain and Language, 9,* 298–306.

Schacter, D.L. (1987). Multiple forms of memory in humans and animals. In Weinberger, N.M., McGaugh, J.L., & Lynch, G. (Eds.), *Memory systems of the brain: animal and human cognitive processes.* New York: Guilford Press.

Torii, H., Fukuta, T., & Koyama, Y. (1972). Junsui shitsudoku no shookooron ni tsuite: nookekkan shoogai no san rei o chuushin ni [Alexia without agraphia due to cerebrovascular disease: report of three cases]. *Seishin Shinkeigaku Zasshi [Psychiatria et Neurologia Japonica], 74,* 546–576.

Yamadori, A., Nagashima, T., & Tamaki, N. (1983). Ideogram writing in a disconnection syndrome. *Brain and Language, 19,*346–356.

Part 5 Brain, Lateralization, and Writing: Initial Models

Introductory Remarks

All the observations and arguments presented in the preceding sections lead to this one, which presents the model hypotheses. Four approaches are offered, including a cognitive one by Martin Taylor which serves as a general framework for the other three. Taylor's Bilateral Cooperative Model supports, by suggestions concerning reading processes at the level of cognitive programming, the kind of speculations about interhemispheric interaction delineated by Skoyles, de Kerckhove, and Jurdant. It reflects, like the other three models, much of the information contained in Part 4. John Skoyles' approach is historical, thus echoing the papers contained in Part 2, while Derrick de Kerckhove's theory looks for support in the logical and structural considerations presented in Part 3. Baudouin Jurdant's work presents a great deal of its own supporting evidence to provide for his model of cortical category differenciation of vowels and consonants. Finally, David Olson's paper sets the stage for further investigations of the effects of reading and writing, at the level of epistemic and cognitive functions.

CHAPTER 17

The Bilateral Cooperative Model of Reading

M. MARTIN TAYLOR [1]

Introduction

The Bilateral Cooperative Model of reading (BLC model) provides a descriptive framework within which not only reading, but also many of the processes used in language and symbolic thought may be considered. It is not a mathematical model of cognitive processes, and may not be easily amenable to computer simulation. Nevertheless, its use has proved stimulating to the author in considering many issues of perception, cognition, and scientific thought. In this chapter, the BLC model is described and its descriptive value illustrated through examples from the experimental psychology literature, dealing with levels of abstraction from recognizing letters to learning complex cognitive structures.

A long, though not unchallenged, tradition of research in reading suggests that there are two routes to word recognition. The studies sketched below seem to indicate that this duple characteristic occurs in other symbolic tasks, and that the bias of individual subjects may be consistent from one task to another. The BLC model claims that there are many routes to understanding the nature of the world, winding through two *tracks* of cooperating processes called LEFT and RIGHT. The model provides a framework within which many of the phenomena of symbolic understanding can be described. In the context of this book, it lends plausibility to the speculation that writing systems and cognitive styles could be related.

Experimental evidence relating to the reading aspects of the BLC model has been extensively discussed elsewhere (Taylor, 1981/1984; Taylor & Taylor, 1983). This chapter concentrates on describing the model and its implications in more detail than hitherto, and provides only illustrative examples from the experimental psychology literature. As previously, the emphasis will be on reading, but in view of the orientation of this book to "culture," occasional wider-ranging speculation may be in order. In particular, the two tracks of the model relate to possible differences of cognitive style; the LEFT track deals in abstract objects that have ordered and logical connections, whereas the RIGHT deals in the unordered associations among possibly

---

[1] Defence and Civil Institute of Environmental Medicine, Box 2000, Downsview, Ontario, M3M 3B9, Canada.

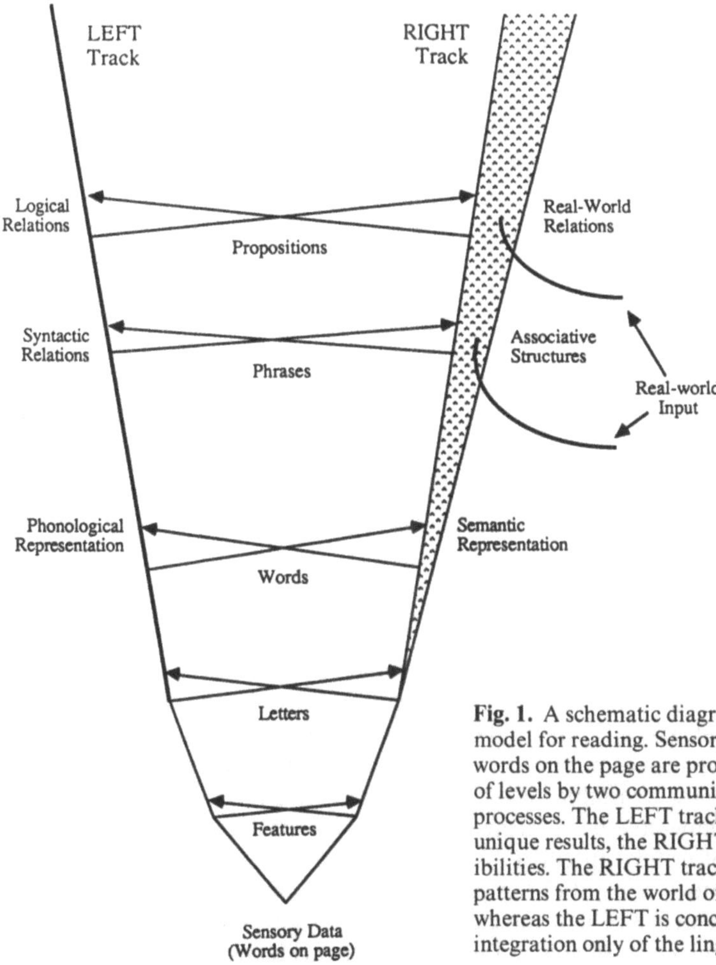

Fig. 1. A schematic diagram of the BLC model for reading. Sensory data from the words on the page are processed at a set of levels by two communicating tracks of processes. The LEFT track produces unique results, the RIGHT multiple possibilities. The RIGHT track also integrates patterns from the world of real objects, whereas the LEFT is concerned with the integration only of the linguistic objects

large numbers of concrete entities. The LEFT is concerned with precision, the RIGHT with inclusion.

Figure 1 shows a schematic representation of the relationship among processes of the BLC model associated with reading. This figure will be referenced implicitly throughout this chapter. It shows a structure of several layers. At each layer, information is available from the layers below about partially analyzed sensory data, and from the layers above about linguistic and real-world context at a suitable level of abstraction. The job of each layer is to bring the data to a new level of abstraction: from patterns of light and dark on the page to visual features, from features to letters, from letters to words, from words to phrases, from phrases to propositions, from propositions to larger scale concepts. Always the RIGHT track provides alternatives, multiple interpretations and various viewpoints, while the LEFT track checks the logical relations that preclude or favour various of these interpretations. The RIGHT track proposes, the LEFT disposes.

The RIGHT track connects with the real world, to provide the LEFT with symbols on which it can base its logical operations. A RIGHT track symbol can also enter into patterns at higher levels of its own track. Likewise, the symbolic results derived from LEFT track operations can themselves become elements of RIGHT track patterns. The RIGHT track makes sense of the world by integrating events from all sources, including the logical operations of the LEFT track. The LEFT makes sense of the world by determining how the symbols with which it is provided can be combined using coherent rules to produce a result on which effective action can be based. Intelligence may be casually defined as the ability to make sense of the world and to act appropriately. To be of high intelligence demands an effective RIGHT track that can provide a multitude of symbolic patterns, together with an effective LEFT track that can use a wide variety of rules for selecting and combining symbols.

Intellectual Roots of the BLC Model

The BLC model presumes that humans have not evolved any special kinds of process for handling language. This presumption is not required, but it enhances the elegance of the model while satisfying the demands of Occam's razor: "entities should not needlessly be multiplied." Any mechanisms used for language are available also for nonlinguistic perception or behavior, though they may have been refined and extended for use with language. The localization of critical linguistic functions in the left hemisphere argues for considerable development of the preexisting mechanisms, but it does not argue for the existence of new mechanisms specialized for language. The BLC model is conceived in this vein: the BLC mechanisms were evolved long before language, but have been enhanced substantially to deal with language. Accordingly, motivational justification for the model is to be found outside the realm of reading, in general studies of perception and behavior.

Evolutionary Necessity

Organisms must survive in a complex world. To do this, they must identify patterns that suggest appropriate behavior. Many different patterns can be extracted simultaneously from the stimuli available to the sensors, but only one behavior can be performed at any one moment in response to them, although the chosen behavior can advance many disparate goals simultaneously. Complex organisms must therefore have evolved to be able to do two things: to handle in parallel a large number of potentially important patterns in a large stimulus space, and to inhibit the deep processing of patterns that seem irrelevant to the goals of the organism. The BLC model is built around the interplay of these requirements.

The RIGHT track of the model represents the data-driven analysis that results in the extraction of many potentially useful patterns, especially patterns that are informationally independent. A familiar low-level example of informationally independent patterns is the set of components of a Fourier transform. The LEFT represents the goal-driven analysis that demands a unique result on which behavior can be based. The primary job of the RIGHT track is to provide possibilities consistent with a good portion of the data, that of the LEFT to inhibit possibilities inconsistent either with the goals of that level or with important elements of the data.

At a neural level, one may guess that both tracks use the same mechanisms. Neurons fire when the conditions are appropriate, which is to say when the right firing pattern of connected neurons has happened. Such a pattern-recognition mechanism is basic to the RIGHT track, but the LEFT track needs to build its logical operations on some kind of interlinked relationships among possibly many of these basic entities. Computers show the inverse relationship: a logical binary operation is the basic mechanism on which all else is built, including (after complex programming) approximate template-matching operations. One may expect a logical operation to be as difficult for a biological organism as an approximate template-match is for a computer. If this speculation is correct, each element of the LEFT track mechanisms must be very costly in both time and resources as compared to an element of the RIGHT. A ratio of three or four orders of magnitude seems not unreasonable.

Each processing level can be considered to provide another level of abstraction in the interpretation of the sensory data. Development of a level of abstraction is expensive in resources, especially if LEFT track operations are heavily involved. Humans seem to have reached a level at which logical relations can be abstracted, whereas it is doubtful whether any other animal has done so. Even if some animals have reached this level, they have not reached a level at which a conditional operation on propositions ("If X is true then Y is also true, otherwise Z is true") can be performed. Only at such a level can language as we know it be developed; language depends on the possibility of connecting propositions in a variety of ways that include at least two-way conditional branches.

Many mammals are, however, able to respond to language at an associative level, whether the effective stimulus be tone of voice or specific words from a small vocabulary. These animals can use RIGHT track language; J. Levy has been quoted as saying that the human right hemisphere has about as much language capability as a dog (K. Patterson, personal communication, 1982), which may be correct apart from the larger human vocabulary. Only humans (and perhaps clever and highly trained chimpanzees) can use LEFT track language effectively, and surely only humans can use language to build or to destroy.

J. G. Taylor's: *The Behavioural Basis of Perception*

In his marvellous book *The Behavioral Basis of Perception*, J. G. Taylor (1962) identified perception with the potential for behavior choices inherent in a stimulus array. According to him, it is impossible to perceive anything that does not have an implicit requirement for selection among behaviors. Following a thread of work derived from Pandemonium (Selfridge, 1959), the present author argued that data-driven pattern processing could develop in the absence of behavioral goals, and introduced the idea that the effects of lateral inhibition among developing feature detectors would necessarily produce sets of more or less mutually orthogonal feature-detectors (M. M. Taylor 1973). The output of these feature detectors would be fed to the behaviorally conditioned perceiving system. Recent simulation studies have verified and extended these ideas (e.g., Kohonen, 1982; Tattersal & Johnston, 1984). Neural arrays with lateral inhibition have been shown able to discriminate among spoken phonemes, even though no linguistic information was originally incorporated in the structure of the array.

The ideas on autonomous development of feature-detectors through lateral inhibition were soon applied to the perception and early learning of language, and used to account for some aspects of bilingualism (M. M. Taylor, 1974). A child would develop detector systems corresponding to regularities in the objects of the world. These regularities include features that occur in different objects as well as entire unique objects. Repeated relations among objects and events should also lead to the development of detectors. One important relation for which a detector pattern should develop is the link between an object or event regularity and a sound pattern (word) spoken when the object is in the field of attention. This relation would lead to the (perhaps unconscious) expectation that a regularity should have a word, and that a word should have a corresponding real-world regularity. A fallacy frequently observed in the history of philosophy has been the belief that a thing must exist because there is a word for it.

The coexistence of a regularity and a word for it should lead to more rapid learning of each, when a new word or a new regularity is encountered. As a specific example, it seems that it is easier to learn to segment the speech stream into words, syllables, or phonemes if at the same time one is learning a writing system that makes those distinctions (e.g., Alegria, Pignot, & Morais, 1982; Amano, 1970; Bradley & Bryant, 1983; Morais, Cary, Alegria, & Bertelson, 1979; Scribner & Cole, 1981).

If J. G. Taylor's theory were correct, the existence of detectors for patterns would not in itself be enough to allow the perception of those patterns in the stimulus array. For perception to occur, there must be a corresponding potential for differential behavior. One behavior is frequently incompatible with another, so that one member of the incompatible pair should inhibit the other. Accordingly, one might expect percepts to be unique, even in situations where mutually contradictory pattern detectors were simultaneously excited. The incompatible detectors need not inhibit one another, but the need for uniqueness among mutually incompatible percepts may cause the outputs of some of the detectors to be inhibited. As an example, consider a reversing figure such as the Necker Cube, which is seen as if from below or from above, but not both at the same time. This kind of inhibition is characteristic of the LEFT track of the BLC model. *The Behavioural Basis of Perception* (Taylor, 1962) thus not only prescribes the development of the LEFT track, but also identifies LEFT track function with conscious perception.

The rest of this chapter is organized as follows: The BLC model is described. Next, the reality of levels is discussed in context of the coding levels employed by different writing systems, then the reality of tracks is illustrated by studies of effects at various levels of abstraction. A few concluding words mention the possible implications of the BLC model for enquiring into the relations among writing, culture, and thought patterns.

The BLC Model

As the model now stands, collaboration is not between hemispheres, but between two "tracks" of processes, each of which can occur in either hemisphere, although each track is preferentially performed by one hemisphere rather than the other. These tracks are called, for mnemonic purposes, LEFT and RIGHT, to indicate the hemisphere that preferentially performs their processes. Because of this preference, much of the evidence for the model comes from studies of interhemispheric differences or cooperation. Hemispheric differences show the existence of two greatly different styles of processing sensory data, and can be taken to illustrate, in attenuated form, the characteristic differences between the RIGHT and LEFT process types.

Hemispheric Effects in Letter Recognition

Jones (1982) asked his subjects to identify uppercase letters presented briefly (90 ms) to the right or left of fixation, or to both sides at once. He found two distinct syndromes of behavior. If we call the groups "R" and "L," the following differences were characteristic:

- Group R recognized the letters better when they were presented to the right hemisphere (i.e., to the left of fixation); Group L showed the reverse preference.
- Group R improved performance when the same letter was presented on both sides of fixation; Group L showed no difference.
- Group R appeared to integrate information from both presentations before making any implicit recognition decision; Group L either ignored the right hemisphere information or were distracted by it.

- Group R confused O-Q 6 times more often than the next-worst confusion, C-G, which was itself more confusable than any confusion made by Group L. On the other hand, Group R found H to be very distinctive, whereas Group L confused it with several other letters, including U, N, K, M, D, and T.

The differences between the two groups were more marked for males than for females, but occurred in both sexes. Group R can be interpreted as those subjects who relied more on the RIGHT track than on the LEFT, and in doing so, preferred the right hemisphere (RH) for RIGHT track functions. Group L used the RIGHT track more in the left hemisphere (LH), but relied more on the LEFT track in any case (Taylor & Taylor, 1983, Chap. 11).

Each level of the BLC model has the same general characteristics. If the task is reading, data levels may be visual features, letters, words, phrases, propositions, situations, actions, and so forth. At a given processing level, operations from each of the two tracks are performed, using data derived from the results of processing at the lower levels. The RIGHT process is a fast, parallel template-based pattern-recognition system whose output is a possibly large number of identifications each of which is reasonably consistent with the data. The LEFT is a rule-based analytic recognition system that may take advantage of goals derived from earlier results or from the RIGHT recognitions. It proceeds serially and relatively slowly, but eventually obtains a result even if the input data form a pattern never before seen. The main job of the LEFT process is to inhibit the candidates from the RIGHT process that are inconsistent with the data, especially when there are a large number of plausible candidates. The LEFT results are (usually) unique, but slow, whereas the RIGHT results are (usually) multiple, but fast. The structure of a level is sketched in Fig. 2.

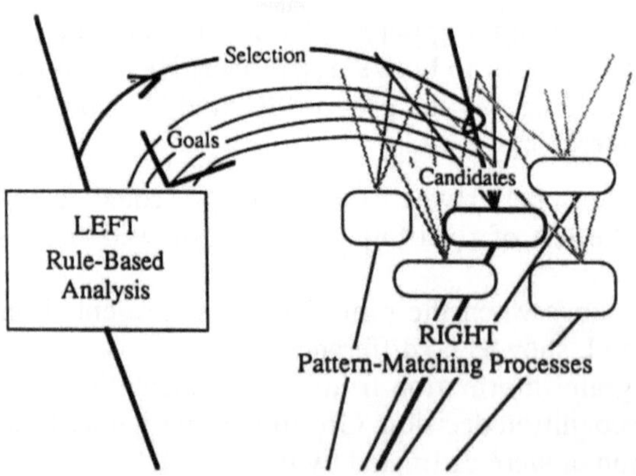

Fig. 2. An impression of a single level of the BLC model. Multiple pattern-matching operations in the RIGHT track feed to many associated patterns at higher levels, and also provide candidates that serve as goals for the rule-based operations of the LEFT track. The result of LEFT track success is to confirm one of the RIGHT track matches and inhibit the others

Hemispheric Effects in Character Matching

Nishikawa and Niina (1981) found interesting hemispheric differences in a category-matching task involving linguistic symbols as stimuli. They asked subjects to respond as fast as possible as to whether all symbols in a set of between two and five symbols were the same. Sometimes all the symbols were in fact the same, but sometimes one of them differed from the others. "Same" was sometimes the same name, like "A" and "a," sometimes the same shape, in which case "A" would differ from "a." The symbols were alphabetic characters, Kana (both Hiragana and Katakana, whose equivalences were treated like the equivalences of different-case alphabetics), and Kanji. Symbols might be presented normally or inverted, but only one orientation was used on a given trial. Subjects were both Japanese and French residing in Japan. The French were not given the Kanji task, and their results on the other tasks paralleled the results of the Japanese subjects in most respects.

Two distinctly different patterns of performance emerged. For some symbols, the reaction time increased with the number of items in the set presented. For these, presentation to the right visual field (left hemisphere) was faster by around 20 ms than presentation to the left visual field (right hemisphere). For other symbols, the reaction time was independent of the number of symbols displayed, and under these conditions, presentation to the left visual field (right hemisphere) was faster by the same amount for the Japanese subjects, but for the French, both visual fields gave about the same speed.

What symbols led to which result? The answer is simple: symbols that code phonetic information rather than meaning cause dependence on number, and are handled faster by the left hemisphere. This result is true whether the match is by name or by shape, though name is about 150 ms slower than shape. Symbols that do not code phonetic information (Kanji presented to the right hemisphere for Japanese subjects but no faster for French. Inverted symbols give faster reaction times than their normally oriented counterparts, except that for a set of two Hiragana, orientation made no difference. Kana were slower than alphabetics by 100 – 150 ms in each condition, and Kanji and inverted Kana were slower than inverted alphabetics. Finally, "Different" responses were slower than their "Same" counterparts, but displayed the same pattern of contrasts.

It is interesting and suggestive that the French gained no time advantage from presentation of nonphonetic symbols to the left visual field (right hemisphere), even though their pattern of performance was the same as that of the Japanese: parallel processing, independent of the number of distractor symbols. In this, they correspond to Jones's L-type subjects, who gained nothing from dual-hemisphere presentation of target letters. In both cases, the BLC implication is that when the L-type subjects used the RIGHT track processes, they were in the left hemisphere, whereas Jones's R-type subjects and Nishikawa and Niina's Japanese subjects were using the RIGHT track processes in the right hemisphere.

Description of One Stage

Each stage of the BLC model is responsible for interpretation of data partially analyzed at lower stages, and not necessarily only at the immediately lower stage. To keep matters concrete, this description will use as an example the stage dealing with word recognition in reading, but will not appeal to much experimental evidence at this point. The intention is to make clear exactly what the model claims to do.

The early processes of visual perception have recoded the incoming visual stimuli into various local features – textons (e.g., Beck, 1973, 1983; Julesz, 1981 a, 1981 b) such as line-ends, corners, crossing points, as well as variations in the densities of these textons – probably at a variety of scales. Very possibly, these features include some variety of local spatial spectral analysis, which identifies the existence of repetitive structures in the text, such as parallel lines, or repeated ascending or descending letter types.

The exact visual features used in letter and word recognition are not well known, but some indication may be found by comparing the various studies on letter discrimination in the Roman alphabet and among Kanji. In most studies, the outer shapes of letters were more important than inner features, symmetry of verticals was contrasted with outer diagonal components, elongated characters were contrasted with fat ones, and (for Kanji) dense characters were contrasted with open ones (Taylor & Taylor, 1983, Chap. 9). The contrast between circular components and rectilinear ones may be important, but one should note that rectilinear components imply the existence of line-end or corner textons, whereas circular components reduce the numbers of these textons. The texton density may be more important than the circularity. No matter what the features may be, the more important ones cluster around the borders of a word.

Pattern Based on Feature Data

In the BLC model for word recognition, the features are fed simultaneously to RIGHT track pattern-matching processes for words and for letters (Fig. 3). These processes are indifferent to the relative locations of the features, provided that they are not too far from their appropriate places. It is the associations of feature complexes that define the patterns, not rules relating their sequential connections. That such a sequence-free process based on coarse data can be effective has been shown by Marcus (1981) for speech recognition. Marcus used no local ordering information, but gave some weight to whether features occurred near the beginning, middle, or end of spoken words. Similarly, the RIGHT track detectors give no weight to rela-

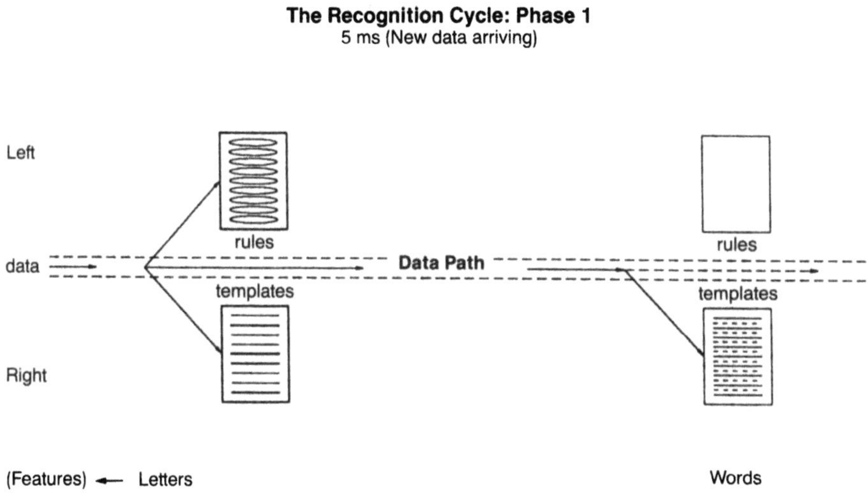

Fig. 3. The interaction of rules and templates at two levels, letters and words. Phase 1: feature-level information is available to letter templates, word templates, and letter rules. Word rules do not use feature information

tive location of features on a small scale, but do expect certain features in a given word to occur near the beginning and others near the end of the word.

The associations of features obtained from any one word are probably ill-discriminated from the associations that would have been derived from a few other words, but they are well discriminated from the majority of words. As a surrogate for the visual features, one can use a concept Taylor and Taylor (1983, Chap. 9) called "Bouma shape" after the letter-confusion groups described by Bouma (1971). Letters are segregated into seven groups of mutually confusible letters. Each letter in a group presumably contains much the same set of features. The groups are: (i) aszx, (ii) eoc, (iii), rvw, (iv) nmu, (v) dhkb, (vi) tilf, (vii) gjpqy. Among the words in a draft of their book, Taylor and Taylor found that the Bouma shape of a word (the sequence of Bouma shapes of the letters) was adequate to identify uniquely 6953, or 88.5% of the 7848 word types used. If multiply repeated letter shapes were collapsed down to pairs, 87.3% of the word shapes were still unique. These results suggest that much information about a word is available without explicit identification of its letters, although they do not indicate how much is available when the sequence information is also lacking.

Pattern Based on One Level Down

At the same time that the visual features are exciting the feature-based word detectors of the RIGHT track, the feature-based letter detectors of the same track are being excited by the same features (Fig. 4). After about 50 ms, both word and letter detectors have become excited in response to features

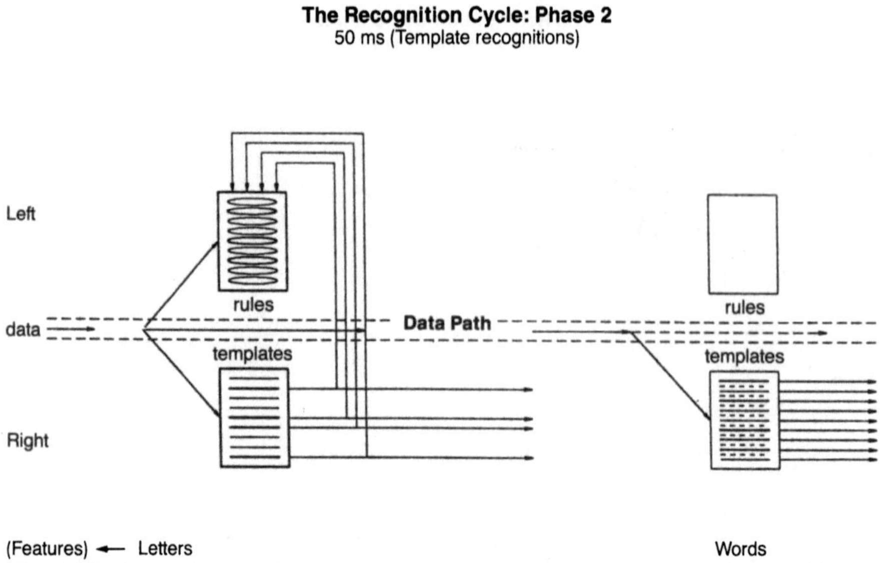

Fig. 4. The interaction of rules and templates at two levels. Phase 2: Candidates have been found for both letters and words, and are being made available to other processes. Letter candidates are available to letter rules and to word templates, word candidates are being sent up for higher-level processing

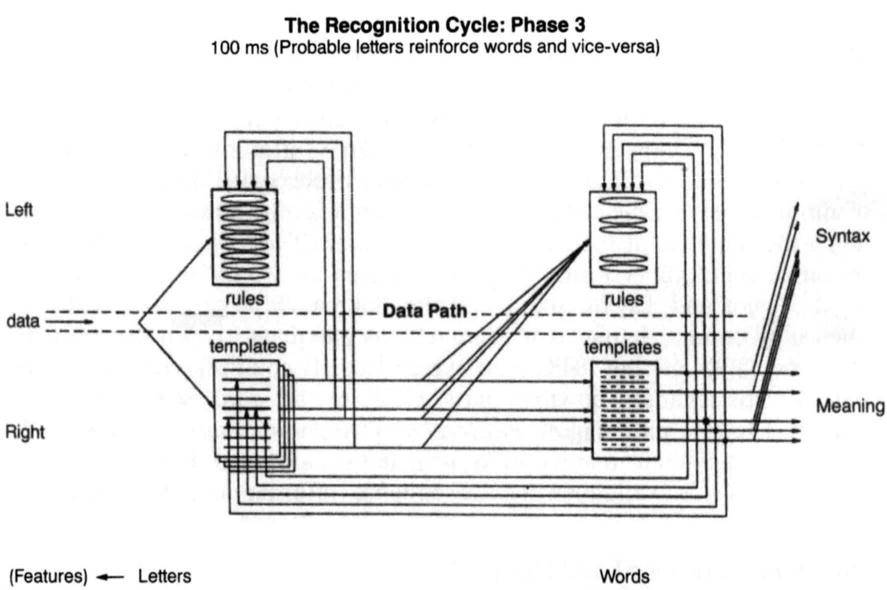

Fig. 5. The interaction of rules and templates at two levels. Phase 3: Word candidates are feeding back to letter templates and to word rules. Letter candidates are being refined and are still serving as goals for letter rules, but are now also being used as data for word rules whose goals have been supplied by the word candidates. Word candidates are being fed to both tracks at the next higher level

that belong to their patterns. Now, each level can begin to affect the other (Fig. 5). Excited word detectors feed back to letter detectors sensitive to letters that should occur if those words were correct. Simultaneously, excited letter detectors feed their letters to the word detectors, providing a more exact input than just the feature associations of the first stage. But just as in the first stage the precise locations of the features have little influence, so in this second whole-word stage do the sequential positions of the letters have little influence. Anagrams are almost as effective as the correctly ordered word, provided that the end letters are maintained (e.g., Rayner & Posnansky, 1978).

The feedback process between letters and words has its effect on both levels, confirming detections that are mutually consistent and depressing (but not inhibiting) those that are inconsistent with each other and with the feature data. At the same time, the earlier word detections are providing all their possible meanings to the next higher level detectors, and a similar mutual feedback happens there: meanings with associations in semantic fields consistent with prior context enhance the detection of those fields, and feedback enhances the outputs of the corresponding word detectors.

Rules

Neither stage of the RIGHT track recognition is very good at distinguishing words that are anagrams of one another and share the end letters (e.g., *goal* and *gaol*), but there are not many such pairs, and even fewer that share associations that would be appropriate in a particular context. Nevertheless, whatever the word, there is an exact test as to whether the lines on the page represent that word: are all the letters represented in the right order? To determine this is the job of the LEFT track detector. It can work either from the data up, or from goal-words down, or both. In either case, it can develop phonemic representations of the word as well as identifications that can lead to meaning.

Whereas the RIGHT track detector can work on the data from at least the two previous levels, the LEFT works normally on results from only the next level down. The LEFT track word recognizer uses only letter sequences as its data, ignoring the feature patterns (Fig. 6). Features are unlikely to be readily incorporated into rules that are more efficient than rules using the letters to which those features belong, but if they did, they might be used. Similarly, the LEFT track syntactic analyzer uses words (or rather, morphemes), not letters.

It is not possible to list the rules for a particular recognizer, but one may assume that they are similar in spirit to some of the "syntactic" methods that can be found in any issue of the journal *Pattern Recognition* (e.g., Isenor & Zaky, 1986, in the issue that was most recent at the time of writing). For example, "H" is more or less uniquely identified as "a crossbar connecting two verticals," although a poorly written "A" could have the same descrip-

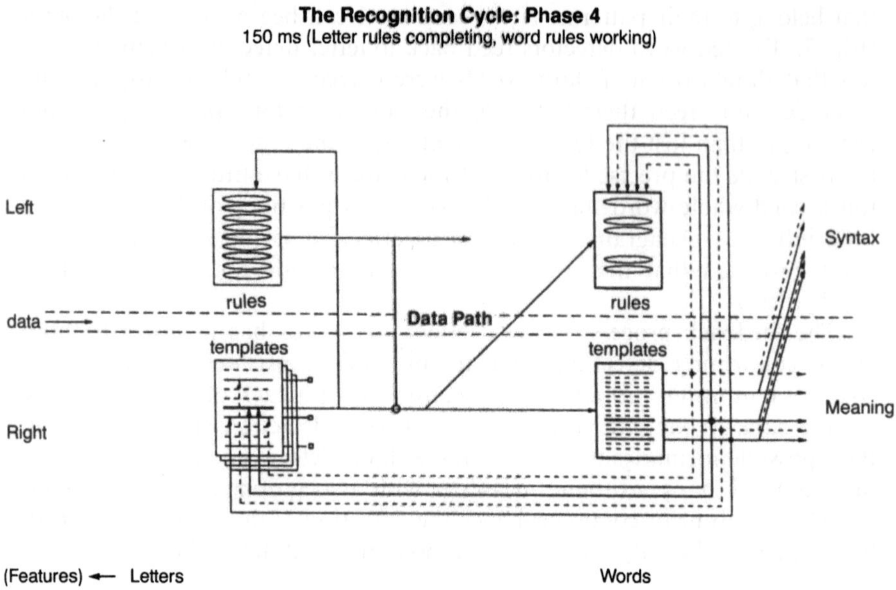

Fig. 6. The interaction of rules and templates at two levels. Phase 4: Feedback from word candidates and from letter rules has refined the letter candidates down probably to unique and correct recognitions. The correct letters define more precisely than hitherto the possible word candidates among the word templates, and specify more precisely the word rules. The increased precision of the word candidates also refines the precision of the goals for the word rules, and reduces the proliferation of possibilities at the next (syntactic or semantic) level

tion (and might look identical to a poorly written "H"). If the rules lead to two or more possible results, only one will be selected, the others being inhibited. On the other hand, if either context or the results of the RIGHT track lead to identifications that are a priori probable and as well can serve as goals for the rules, these goals will bias the detection process in such a way that a goal-free result will be obtained only if the data are seriously inconsistent with the goals. Usually, however, the data are consistent with the goals, and a full analysis is unnecessary.

Integration: The Three Phases of Word Recognition

What is meant by "word recognition"? In any discussion of reading, this must be an important issue. For one thing, word recognition is no more necessary for fluent reading than letter recognition is for word recognition. Even the concept of word may mean very different things in different languages. Consider the difficulties in identifying words that may be encountered by people who learn to read using a script that does not code word boundaries. Vai-script readers are poor at detecting word boundaries in

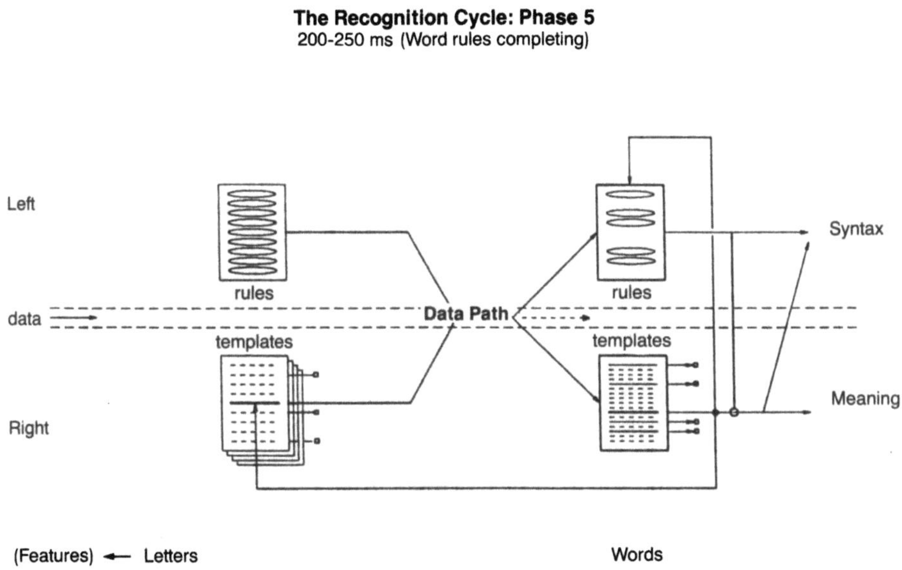

Fig. 7. The interaction of rules and templates at two levels. Phase 5: Word rules complete, and inhibit incorrect word candidates. Letters and words are now both precisely determined. In this sequence of figures, feedback and context from higher levels is ignored, but those interactions should occur in the same way as feedback from words to letters

speech (Scribner & Cole, 1981), and an educated but not linguistically trained Chinese speaker may deny that there is any difference between words and syllables (personal experience of the author). Although word recognition is not required for fluent reading, it may be required for exact reading, just as exact identification of letters is required for guaranteed identification of words in an alphabetic script.

According to the BLC model, three different processes are involved in word recognition, as shown in Figs. 3–7. Strictly speaking, there are four, but the rule-based letter recognition shown in Fig. 3–7 seldom has any significant effect on word recognition, though it might be useful in analyzing nonwords, or in reading familiar words in an unfamiliar script.

The first stage rapidly (50–100 ms) discovers one or more (usually many) candidates (Fig. 3, Fig. 4). (The timings presented here are justified in Taylor & Taylor, 1983.) While the second and third stages are progressing, these candidates are already being used as input to higher processing levels in the BLC model. At the same time that the first stage is discovering candidate words, candidate letters are being discovered, and these candidates are then used in a second wholistic recognition process (Fig. 5)[2].

---

[2] We prefer *wholistic* rather than *holistic* because it can retain the sense of using the whole pattern at once, without implying the metaphysical connotations that adhere to the term *holistic*. When we quote other authors who have used *holistic* we defer to their usage.

This second wholistic recognition phase (100–150 ms) reinforces some of the initial identifications, but not others (Fig. 6). The third recognition phase proceeds rather slower, taking 200–250 ms. The elements are combined according to rules, possibly using as goals the candidates discovered in the first wholistic recognition. This phase is almost guaranteed to finish with no more than one accepted candidate, and its results are used to inhibit the false candidates found by the wholistic recognition phases (Fig. 7). Often, however, there is no need for this third phase to run to completion, since strong candidates found by the quicker phases are adequate for the purposes of higher levels, and the analytic process can be used on subsequent words.

When a person reads at 600 words/min (faster than average, but by no means unusually fast), a process that takes 200–250 ms cannot be applied sequentially to every word, or even to more than half the words. Such a process must be applied judiciously to those words requiring its attention (the word *attention* is used with care). Uncertainty at the higher level might be able to invoke or inhibit operation of the LEFT track recognition process. There is evidence that LEFT track processes may be under some kind of strategic control (e.g., Tweedy & Lapinski, 1981), but this evidence applies only on a considerably longer time-scale than a few tens of milliseconds, which is the time scale within which this speculative invocation and inhibition operates.

The Integration of Meaning

The objective of reading is to acquire from the page the meanings that the writer intended to communicate. Each word must be integrated in two ways: its meaning must connect with the meanings of other words and probably with coherent features of the real world, and its relations with the other words must be consistent and make sense when converted to real-world meanings. The first integration is semantic, the second syntactic. Semantic integration is sometimes studied under the rubric of "semantic priming," although the experimental tasks under this heading sometimes seem far removed from the integration of meaning.

Semantic Priming

Present a subject with a letter string, and ask whether it is a real word. Next, do the same, but precede the target string with another, which might be a word and might be related to the target if that is in fact a word. The preceding string is called a "prime" and if it is a word (e.g., *doctor*) related to the target (e.g., *nurse*) the response is likely to be quicker than if it is not. If the prime is a word, but unrelated to the target, the response may be slower than if it is not a word, or is a word that has been repeated so often as to lose its effectiveness (e.g., *blank*). Such a word is a better neutral stimulus than the popular string XXXXX, which apparently produces some inhibition (de Groot, Thomassen & Hudson, 1982).

There are other kinds of priming experiment, such as measuring reactions related to a word that is consistent or inconsistent with the preceding phrase or sentence (e.g., Foss, 1982), or measuring event-related brain potentials associated with an anomalous word in a sentence (Kutas & Hillyard, 1980, 1982). Foss found that words in a sentence were processed faster if the sentence contained a prime some time before the target, but if the order of the words in the sentence had been scrambled, the prime was effective only when it immediately preceded the target. Kutas and Hillyard (1980) found a brain-potential anomaly starting about 250 ms after a semantically anomalous word was presented. Later, they found that brain potentials were larger in the left hemisphere for the normal part of a sentence, whereas a semantic anomaly produced a larger potential in the right hemisphere (Kutas & Hillyard, 1982). This latter finding suggests that the RIGHT track in the RH was doing a lot of work seeking areas of semantic association that would allow the anomalous word to be coordinated with the associations of the other words in the sentence, whereas when the semantic association occurred easily, the RH had relatively less work to do, and the resource-intensive syntactic processing of the LH dominated.

Semantic Features as Elements of a Quasi-Hologram

A visual feature may appear in many letters, a letter in many words, a word in many contexts. The "meaning" of the element changes, depending on what higher-level entity it turns out to be associated with. As a corollary, when the element is considered as data, it can evoke or be considered as evidence for many possible higher-level entities. In RIGHT track processing, the preferred entity is brought out when it is evoked by several lower-level elements: a letter by the conjunction of visual features, the meaning of a word by the related meanings of neighboring words. A RIGHT track element can also be evoked by the feedback from higher-level elements, especially if many higher-level elements evoked by the context are associated with the same lower-level one. Note that for the RIGHT track, there is no *right* choice; there is only a choice that has more supporting evidence than other choices. Furthermore, contradictory evidence has little or no weight.

In some respects, the operations of a RIGHT track level may be considered by analogy with a hologram. Low-level elements (e.g., visual features, words) combine in many ways to form higher-level patterns (e.g., letters, meanings), which in turn can be reconstructed into the originating elements. The more high-level patterns there are (the larger the piece of hologram) the more accurate the reconstruction can be. Unlike a hologram, however, both the imaged elements and the interference pattern can be sensibly interpreted: letters and words are both reasonable ways of describing a line of text.

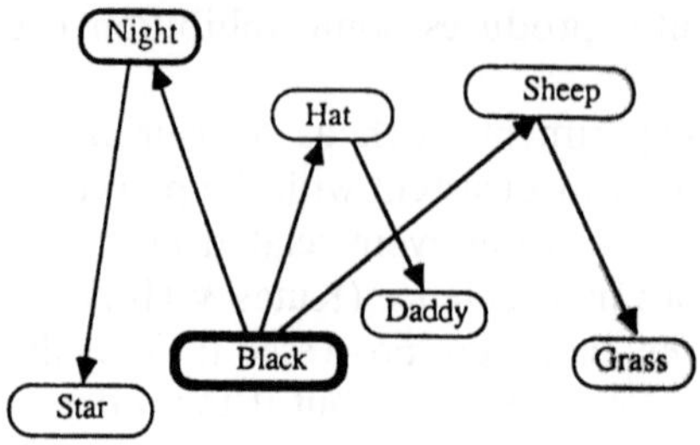

Child's Association Patterns

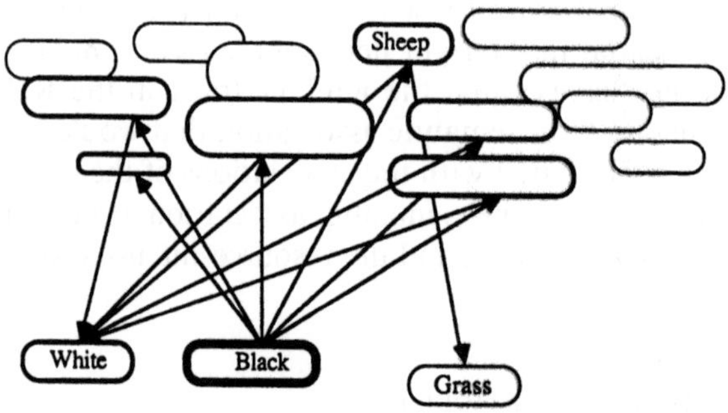

Adult: "Holographic Mirror" focused Association

Fig. 8. The difference between child and adult association patterns interpreted as a consequence of holographic imaging. The child has few primary associations to an initial word (*Black*), and the secondary associations of those are quite varied. No focusing occurs, and the strongest associations are those that normally occur along with the initial word. An adult has many more primary associations, many of which are associated also with words having the same syntactic and semantic features at the initial word. The strongest overall activation thus occurs not on a primary associations, such as *Sheep,* but on a secondary (*White*), which serves as a focus of the holographic mirror. Naturally, the inital word (*Black*) is the primary focus, but it is not permitted as a response in an association test. There are many secondary foci, but these have many different associations, which diffuses widely any further focusing, prohibiting a wider spread of semantic priming than is given by the main secondary association (*White*)

The holographic analogy proves fruitful when we consider the selective focusing that appropriate holographic filters and mirrors can perform. First, consider verbal association patterns. When a person is asked to say the first word that is evoked by a stimulus word, an adult is likely to provide a response that occurs in a similar semantic context (black – white), whereas a child may well say something that completes a context (black – night) (Anglin, 1977; Cramer, 1968; Entwisle, Forsyth, & Muus, 1964; Ervin, 1961; Petrey, 1977). Figure 8 illustrates what may be happening.

For the child, the stimulus word is strongly associated with the words with which it has co-occurred (as well as with semantic features of the real world). Each of these associated words is itself associated with further words, but the secondary associations are diffuse and unfocused, because the primary associations are few in number – just as the scene reconstructed from a small sliver of a hologram is defocused. The word most strongly evoked will thus be one of the primary associates. For the adult, the stimulus word is similarly strongly associated with words with which it has co-occurred, but there are more of them, and many of them have co-occurred also with words of a semantic (and syntactic) character similar to that of the stimulus. Furthermore, the associated semantic features of the real world are richer than they are for the child. The secondary associations focus strongly on one or two words semantically and syntactically like the stimulus, and one of these words is evoked more strongly than any of the primary associates. It becomes the overt primary associate that is recorded by the experimenter.

The holographic associations of the RIGHT track may account for the finding that semantic priming does not extend to secondary associates (de Groot, 1983). If word A has B as a strong overt associate, then presentation of A eases verbal tasks relating to B. If B has C as a strong overt associate, presentation of A has no apparent effect on performance with C, although a naive view of spreading activation would predict that it should. Considering the holographic analogy, one would not expect much effect of A on performance with C, because although the first "reflection" may focus on B most strongly, it focuses also on many other words, each of which focuses back not only on the original primary (covert) associates of A but on many others, creating a diffuse secondary reflection that is not especially focused on C. This argument predicts that if B were explicitly invoked in some way (thus bringing the LEFT track into play to inhibit the other foci of the first "reflection"), then C would be primed by A.

Consider now the refinement of meaning by way of context. A word evokes many associated semantic features, often in several different domains of meaning (e.g., *bank* could be where money is stored, or the edge of a river, or an expression of confidence, or the tilt of an aeroplane, or the sideways slope at a curve in a road). Almost all reasonably common words have some such ambiguity. If the word is presented in isolation, there is no way to tell which meaning is intended, but in a reasonable context, some of the associated semantic features have been evoked by preceding words, or will be evoked by following words. Appropriate meanings for each word is made more precise by the patterns of coherence among the features evoked by the others. Hirst (1983) uses essentially the same concept (he calls the idea "Polaroid words" because the words develop themselves).

Priming with Ambiguous Primes

Almost all common words in English have multiple meanings. One continuing research issue has been whether a reader has access to all these meanings, or whether only the contextually appropriate one is available. According to the BLC model, one would expect the RIGHT track processes to allow access to all the meanings of a word, with the LEFT inhibiting those that were inappropriate. In the absence of context, the more frequently encountered meanings of a word should dominate the rare meanings, in the sense that the more frequent meanings should more often be expressed in overt behavior (after transfer to LEFT track processes). With context, however, the semantic or syntactic features excited by the context should enhance the corresponding features of the appropriate meaning of a word, changing the balance between frequency and overt expression.

Holmes (1979) asked subjects either to detect possible ambiguities in sentences or to understand the same sentences. The ambiguities were due to the use of ambiguous words, and the contextually correct meaning might be due to a more frequent or a less frequent sense of the word. Subjects found it harder to detect the ambiguity, but easier to understand the sentence, if the more frequent meaning were the contextually correct one. As a control, Holmes found that there was no effect on comprehension time if high- or low-frequency synonymes of the contextually correct meaning were used instead of the ambiguous word. These results could be a consequence of RIGHT track operation alone, or they could be due to the RIGHT track's inability to inhibit one possibility when another exists, which would require the slow LEFT track to come into play.

When ambiguous words are used in a semantic priming experiment, a variety of results may be obtained. When the prime is an ambiguous word, both senses of the prime produce facilitation in a related target (Holley-Wilcox & Blank, 1980); but if that same ambiguous prime is preceded by a word related to one of its senses and not the other, only that sense facilitates the target, and the other sense inhibits as if it were an unrelated word (Schvaneveldt, Meyer, & Becker, 1976).

An effect hard to understand outside the BLC model was observed by Marcel (1980). Using an ambiguous prime with a preceding word related to one of its senses, he replicated the result of Schvaneveldt et al. when the prime was clearly visible. When he masked the prime, however, so that the subjects were unaware of its presence, both senses of the prime facilitated the target. This result is so paradoxical that some reviewers prefer to believe that it is an experimental artifact (e.g., Holender, 1986). It should be noted that being unaware of the prime is not the same as being unable to detect it, as many studies have shown (reviewed by Holender, 1986). When pressed, a subject can indicate that a word occurred at far greater than chance level, without being sufficiently certain of the fact to be conscious of the detection (a phenomenon well known to psychophysicists). Under stimulus conditions very like those used by Marcel (1980), other researchers have found subjects able to detect the existence of a word in a forced-choice experiment when they were subjectively unaware of its existence (see the discussion following Holender, 1986 for several examples). These objections, however, do not account for Marcel's finding of *qualitative* differences in performance when the prime was not consciously detected, a result since confirmed by others (e.g., Cheeseman & Merikle, 1985).

According to the BLC interpretation of Marcel's result, the masked prime did not provide enough information for the LEFT track to form a conscious percept of it, but it did provide enough information for the RIGHT track to access its meanings (and probably the meanings of several other confusable words as well). These meanings were all available to facilitate the target whereas when the ambiguous prime was clear, the LEFT track could inhibit the inappropriate meaning, which then became inhibitory for the following target.

Priming with Distorted Targets

Rather than requiring a lexical decision, Broadbent and Broadbent (1980) asked their subjects to name a target word made difficult by blurring or by deletion of letters. The experiment was arranged so that subjects were likely to report target words falsely when similarly shaped foil words were presented. Hence the experimenters could measure both the disciminability and criterion shift as a function of priming. They found that for blurred words, priming affected only the criterion: the willingness of the observer to report the word. Priming did not affect the ability of the subject to discriminate between one word and another. In another experiment, they found that the reliability of the primes (the probability that the prime really did relate to the target) affected blurred words differently from fragmented words. Reliable primes were better than unreliable ones if letters were deleted, but the reverse was true if the target was blurred.

Both effects can be interpreted as showing the influence of the LEFT track on RIGHT track operations. Broadbent and Broadbent (1980) interpret their first result as showing that priming affects a global, passive process (the RIGHT track), at least when the target is blurred. However, when the words are fragmented it is not easy (though it may be possible) for the RIGHT track to identify them, but they can be analyzed by the rule-based LEFT track processes. LEFT track operations tend to inhibit facilitative priming in the RIGHT track unless the prime-target relation is correct. When the prime-target relation is unreliable, the LEFT track operation may be inhibited, thus reducing the inhibition placed on the RIGHT, and allowing more responses (most of them correct) to be made. On the other hand, if the LEFT track must be strongly involved with the identification of the fragmented targets, false alarms are unlikely (Broadbent and Broadbent found none), and the depression of LEFT track function by unreliable primes can only lower the probability of a correct identification.

Syntax

Since the formal (LEFT track) approach to syntax is well known, this section concentrates on the RIGHT track's contribution. It is discussed largely through quotes from a rather inaccessible report by Taylor (1974), who described syntax as the reflection of frequent relationships that occur in the real world. This early paper did not take into account the rule-based functions the BLC model ascribes to the LEFT track, but nevertheless it was able to account for many of the phenomena observed in children's acquisition of one or more languages. It seems likely that the LEFT track is poorly developed, if it exists at all, in children under the age of perhaps 5 or 6, and its development and integration present a separate problem.

Syntactic Elements as Real-World Relational Patterns

The RIGHT track of the BLC model was foreshadowed by Taylor (1974), who built on the cognitive (semantic) network approach of Rumelhart, Lindsay, and Norman (1972) and Rumelhart and Norman (1973). Elements or *nodes* of a network detect and represent regularities in Nature, perhaps perceptual features, perhaps object types, perhaps recurrent relations such as "above." These nodes are built through repeated experience with similar associations of data; according to Taylor (1973), stochastic variability in the response of the nodes coupled with lateral inhibition among nodes responding to the overlapping inputs leads to statistical independence among the responses of groups of nodes. The importance of lateral inhibition in allowing data-directed development of features has recently been rediscovered (e.g., Kohonen, 1982; Tatersall & Johnston, 1984).

The network itself, like any network, consists of nodes connected by links. The nodes represent concepts, the links relationships among concepts. Some links represent class membership (a canary *isa* bird), some represent properties (a house *has* windows), some represent modifications of one concept by another (this roof *is* blue). ... Since the network contains programme as well as data, the flow of thought through it produces action and speech as well as mental operations. The programmes are not readily distinguished from the data on which they work. Vocabulary and syntax are both held in the network. ...

According to most theories of pattern recognition these [sensory processes] are coded in terms of features which suffice to describe the input patterns. Features are the simplest concepts in the network. A pattern which more or less matches some feature activates that feature to some degree, and the degrees of activation of the various features define the pattern. Under certain assumptions (Taylor, 1973) features will evolve so that the network of peripheral features will come to be an efficient description of the input.

Patterns activate features. These features themselves form patterns and enter into relationships, which can excite higher-level concepts; this process may continue until the activity reaches the level of concepts like *roof, house* and so forth. ... One should note carefully, though, that identification of the pattern as a *house* in no way prevents its simultaneous identification as something else, such as a concrete structure. ... Different properties of the item are effective in arousing different concepts. The chair in the living room may be a *chair,* but it may also be a *red thing,* or *a soft thing,* or a *grandfather thing,* or *a very valuable thing,* depending on what aspect of the world is salient at the moment. [Note the implications for semantic priming of and by ambiguous words.]

Concepts do not represent only object properties. A concept comes into being because a particular pattern of relationships recurs. Indeed, the notion of a particular relationship is itself a concept, which has grown because many pairs of objects were related in that particular way. Relationships such as *above, inside, behind,* are quite elementary relationships, which require only the segregation of objects in the world for their definition. ... A relationship such as *lives in* depends on a great deal of prior conceptual structure. But all relationships can be dealt with as concepts built because they recur in experience, and can be used as components in high-level concepts. (Taylor, 1974, pp. 70–76)

Repeated associations among aspects of the perceptual world define not only objects, but relations among parts of objects, and between separate objects. Furthermore, many independent objects or relations can be simultaneously activated, although related ones tend to be mutually inhibitory. The paper goes on to discuss critical periods in development, and points out that if the statistics of the environment do not change, there is no pressure for the distribution of commitment of nodes to features.

Another stabilizing feature is the feedback that is postulated to occur between higher and lower-level nodes (Taylor, 1973). When a higher-level node becomes active because most of its defining features are present, lower-level nodes representing features normally associated with that higher-level concept also tend to be activated. For example, if a word is detected despite illegibility of some of its letters, those letters also tend to be perceived. Something like this may account for the occasional misperceptions of the font or case in which barely detectable words are presented (Adams, 1979; Friedman, 1980).

Verbs as Relational Items

Much of syntax centers on the role of the verb. Verbs are usually considered as content words, and so they are. But verbs also are the primary linguistic means of expressing dynamic relationships among objects.

> Not all recurring patterns produce static concepts. Many, perhaps most, are dynamic or involve relationships among static patterns. The concept of someone *doing* something must be very strongly embedded in our cognitive structure. Most of the time we see people, they are actively engaged in affecting something else. This abstract relationship, which we may call "agent," is extremely common, and should be very early derived as a component of almost all patterns of events involving people or animals. Other relationships are almost as ubiquitous. When someone does something, the usually does it *to* something. The object of the action is a very common relationship. The action itself, or the concept of action, is also a very frequently occurring relationship among people and other people or things. We can demonstrate a whole abstract pattern of relationships, as follows: someone does something to something for someone with some instrument by some method for some reason. All of these relationships are common, and should occur as concepts in their own right within the cognitive network. When language develops, they should find some means of expression, regardless of which language it is. If universals of language exist, surely they should be found at this level.
>
> Generally, the concepts which define relationships tend to drive action programmes rather than invoke static images. My image of "gave," for example, is a dynamic picture of someone actually passing something over to someone else, not a static picture of "a giving." I have similar dynamic images for the more abstract relationships of *agent* – a multiplicity of people do all sorts of things; in some sense, *agent* is what is common to all these events.
>
> When it comes to [language], the relationships tend to drive transformations on the words with which we label concepts. They select word orders, affect number and gender agreements, select prepositions and postpositions, change the forms of words by declension and conjugation, and so forth. The more abstract the relationship, the more likely it is to be expressed in a "syntactic" manner, rather than as a content word. These concepts tend to be expressed as programmes rather than as data. (Taylor, 1974, pp. 80–81)

Syntactic relationships are relationships among words. From the viewpoint that the function of language is to transfer structure from one person's cognitive network to another's the only reason for these relationships to remain consistent is that they should signal consistent patterns within the cognitive network. In other words, *all* rules of syntax should find their counterpart in aspects of the network structure. If we turn this around, we should find that particular relationships within the cognitive network drive, as labels or programmes activated by those relationships, syntactic function words and syntactic transformations. In listening word patterns should drive transformation routines which activate the appropriate relational constructions in the network.

This position suggests that the complexity of individual syntactic rules in mature speech should match that of the structures they signal. Most relationships in the network are bilateral. The agent links the actor with the act, and so forth. Some relationships are three-way. A verb, such as "gave" may link three items, the agent, the recipient, and the object. Stevens (1973) attempted an exhaustive listing of English verb types, finding 25 different types, and discovered none with more than three relationships. ... A three-way relationship seems to be the most complex allowed in modern English. Schank (1972) also claims three-way relationships to be the most complex that occur in the related concept structures. (Taylor 1974, pp. 92–93)

The interrelation of LEFT and RIGHT tracks in syntactic analysis is complex. The RIGHT track presumably provides the symbols for the LEFT to work with; perhaps real-world relationships associated with the words being syntactically related. *Between,* for example, indicates that three things are being located in space or time; the pattern of the things is identified by the RIGHT track, while the role of each word in the syntax of the sentence is passed to the LEFT. The LEFT can handle the syntactic relation without any knowledge of the meaning of the words. It can interpret *Jabberwocky* sufficiently to constrain the kinds of things the nonsense words could mean. *The flogglewhopper is between the amindy and the exticase* becomes intelligible once one can guess the meanings of the three words that obviously represent visible objects. On the other hand, if one knows the meanings, the LEFT track analysis may be unnecessary: cup–table in an appropriate context probably means the cup is on the table, or should be put there. Much of everyday conversation lacks the syntactic markers required in academic writing, simply because context and RIGHT track appropriateness can supply the necessary relations.

The BLC Model in Practice: Some Examples

The BLC model depends on two related concepts: *levels* and *tracks.* Levels are important because it is only at discrete levels of abstraction that the two tracks communicate. This section considers some of the evidence for the existence of discrete levels of abstraction, by pointing out some of the different things that can be coded by writing systems. It finishes with a few illustrations that demonstrate that people actually use at least some of these levels in recognizing the letters, the names, and the meanings of printed words in English and French.

Written and spoken language each code concepts and the relationships among concepts. There is no logical necessity for writing to code speech, and in some writing systems it does not. Chinese is the most famous such writing system; far from coding speech, it does not even code the sounds of words, except in an occasional vague and suggestive way. No writing system codes all of speech, unless we count as a writing system the coding methods used by precise phoneticians. Even coding for sound is not a prominent feature of the writing systems of the world. According to the BLC model, decoding the sound correspondences of a written word is a LEFT track function, performed often in parallel with decoding the name or the meaning of the word using both tracks or the RIGHT track alone.

Sound is just one of the ways that a concept can be identified from marks on a page, and the use of sound in decoding those marks is effective only in languages whose writing systems are reasonably exact in coding sounds. In other writing systems, it is in principle possible, and in some it is necessary, that the meaning of a word be inferred from its visual form independently of its sound. Despite this obvious fact, a great deal of the research in reading has been devoted to whether and how people use sound coding. (The answer seems to be that readers use the sounds of words for short-term memory while decoding complex syntactic structures.) Much of the research on sound coding has been conducted by experimenters who speak English or another European language, using subjects who read the same language. The writing systems have therefore been mostly alphabetic ones which code phonemes with greater or lesser regularity. In an alphabet with highly regular symbol-sound correspondence, readers probably use sound intrinsically in decoding words (Feldman, 1981); in a logography, they probably do not, but still use sound for short-term memory (e.g., Chu-Chang & Loritz, 1977; Tzeng, Hung, & Wang, 1977; Yik, 1978).

Taylor and Taylor (1983, Interlude) tabulated a small sample of writing systems to show the variety of things that writing can code, such as phonemes, syllables, morphemes, words, meaning, and syntactic roles. Not all of these items represent different levels within the BLC model, some (such as phonemes and syntactic roles) being handled by the LEFT track exclusively, whereas others (morphemes and meaning) are handled by the RIGHT track. Taylor and Taylor omitted prosody, which conveys much information not contained in the words of speech, and which has its parallel in the punctuation marks that occur in some, but not all forms of writing systems.

A phonetically coded system does not have to code phonemes; it can code syllables or parts of syllables. As already pointed out, the Japanese Kana provides a pure representation of syllables (almost all of CV form), with no indication whatsoever of commonality among syllables with the same consonant or with the same vowel. The Vai script of Liberia is also a syllabary, but more elaborate than Kana both in the sounds represented and in the complexity of the signs. There are 220 sound signs and 23 auxiliary signs, including punctuation, according to Jensen (1970), in contrast to the approximately 100 Kana (including all the elaborated Kana). Both Vai and Kana can be used to represent the sounds of foreign words as accurately as their speakers can pronounce those words.

Most phonetically coded writing systems, however, do code phonemes, sometimes uniquely, as in Korean, Finnish, or Serbo-Croatian, sometimes approximately, as in English or French. When the correspondence between phoneme and symbol is close, it usually signifies that the writing system has been deliberately designed, rather than having evolved. In the case of Korean, the design was done in the fifteenth century, for Finnish and Serbo-Croatian, as late as the nineteenth century.

It is unlikely that the wide variety of writing systems in the world are all processed in the same way during reading. It is equally unlikely that any of them is processed by mechanisms unavailable to readers of other systems. It follows, almost necessarily, that reading is usually performed using a mix of processes, whose interactions and balance probable differ across writing systems. Different writing systems therefore offer different opportunities for the processes of the two BLC tracks to operate, and the mix differs with the level of abstraction.

Hieroglyphics

In the light of the BLC model, Egyptian hieroglyphics presents a particularly interesting writing system. In the history of the world, only Chinese has survived longer than hieroglyphics, with its derived cursive hieratic and demotic scripts. The long survival of these two systems argues that they are both very effective in representing their languages, even though phonetic coding plays a very small role in either script. The reasons for the demise of hieroglyphic writing may be as much political as technical, hieroglyphics having been largely used by a priestly class whose dominance was destroyed by foreign invaders who wrote with alphabets. It remains to be seen whether any alphabetic system will survive as long as either Chinese or hieroglyphics.

Why is the hieroglyphic system interesting? From our point of view, it is because it codes many aspects of language, but none of them uniquely. In this, it is like English or French, but balanced toward less phonetic coding and more semantic coding. The LEFT track would have a great deal of trouble reading it, but the RIGHT would perform well. The following discussion is based largely on Meltzer (1980).

The basic objective of hieroglyphic writing, as with all language, is to transfer meaning from the writer to the reader. Hieroglyphic does not represent sounds uniquely, although some of the symbols in a word provide clues to the sound of a word, and they may even represent all the consonants (sometimes twice). Hieroglyphic does not represent ideas in the way a logography does, although in many words there are symbols that give more or less explicit clues to the meaning. Hieroglyphic does not require separation among words; sometimes whole phrases are represented in a single written "word" when the phrase represents a coherent idea. The concept of "word" seems in any case to depend strongly on the writing system used for a language.

The uniqueness of a hieroglyphic word depends on the relation of its sound-carrying and meaning-carrying elements, and (as with English) its context. In essence, the reader finds "what word with this kind of meaning sounds kind-of like that and makes sense here." Meltzer (1980) uses the term "word-picture as a mnemonic unit" to describe the use of the written symbols. The whole complex is written about as consistently as was English before the lexicographic revolution of the 18th century. The writer did not have freedom to use any arbitrary group of symbols that might have a particular sound value, nor arbitrary symbols that might have a particular general meaning, but had to use the right ones, especially if there was available a symbol group that conveyed both sound and meaning.

Now, what did the hieroglyphic system actually code? Phonemes, or at least consonants, were often represented, but their ordering might vary. Typically, a complex (biliteral or triliteral) root such as *btp* might be used along with signs for "t" and "p" which helped to reinforce the recognition of the whole. In this particular case, the *btp* sign could be written above the "t" and the "p," and all three were symmetric. Since hieroglyphic characters could be written right to left or left to right, there was no way other than context to determine an ordering for the "t" and "p" if one did not know the *btp* sign itself, as shown in Fig. 9.

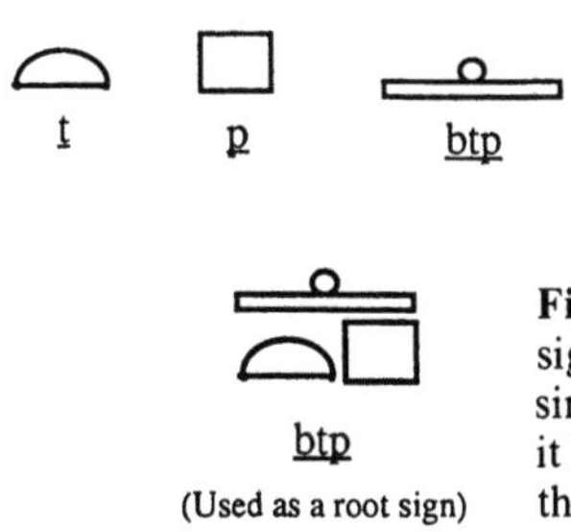

Fig. 9. Hieroglyphic writing: the combination of phonetic signs to provide redundancy. *btp* could be written with three single-consonant signs, but when it serves as the root of a word it has a triple-consonant sign together with signs for two of the three consonants

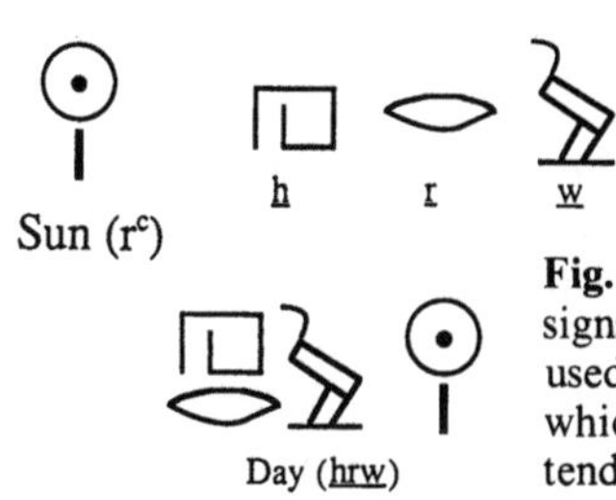

Fig. 10. Hieroglyphic writing: phonetic signs with semantic signs to provide redundancy in a different way. The *sun* sign is used with the consonants for *day* to show both the order in which to pronounce the consonants, and which *hrw* word is intended, by indicating the semantic field of the word

Semantic fields were coded. For example, the sign for "sun" (a circle with a central dot) is frequently found in words signifying time, such as *hrw* ("day"), which is written with the three symbols for "h," "r," and "w," plus the sun sign (Fig. 10). "Sun" itself is written with phonetic signs plus the circle-dot.

English

English is an alphabetic system with a complex set of added logographs that are used conventionally or for special purposes. Unlike Serbo-Croatian, English has no consistent mapping between graphemes and sounds, even in the number of symbols used to represent one phoneme, which can vary from less than one to two or three. Symbols can alter the sounds of other symbols without themselves representing a sound. Consider the difference between "us" and "use" (the verb) caused by the appended silent "e." The two words have no phonemes in common. Any given symbol and many of the symbol groups can represent a variety of different sounds, and most of the phonemes of the language can have a variety of symbolic representations. English may therefore be considered closer to Egyptian hieroglyphic in its processing requirements than to a fully phonetic alphabetic system such as Finnish or Serbo-Croatian. A reader must discover the meaning and (if necessary) the sound of a word by using clues from many sources: context, orthography, and, most particularly, direct knowledge of many of the words as individual entities.

A large proportion of the research in reading has been done in English, which is a mixed blessing. The variety of processing requirements leads to a variety of processing techniques, and this in turn leads to much disagreement in the research literature. Some researchers will develop evidence showing that readers use one particular processing technique, and will on that account criticize the work of others who find evidence for the use of a different technique. There is also a difference of style among the researchers, which seems to relate to the BLC model, in that some react as if only one processing technique could be true, whereas others accept the simultaneous use of a variety of techniques that complement one another.

Masking Effects in English

That there must be different recognition processes for words and for letters is attested by several studies on masking. If a word is presented for a short time and then followed by a flash of light or a pattern of shapes, the word is harder to see than if there is no following masker. Different kinds of maskers have different effects. For example, Johnston and McClelland (1973) found that a patterned masker might hardly affect the recognition of a letter in a target word, but the same masker might severely diminish the recognition of the same letter alone. Johnston and McClelland (1980) refined this result, finding that if the masker consisted of letters or words, there was little difference between detection of a letter in a target word as compared to detection of the letter alone; but if the masker were made of pseudo-letters consisting of parts of real letters, then target letters were identified as parts of words much better than alone. Jacobson (1974) had previously found that whereas words were much more strongly masked by letter strings than by pseudo-letters, strings of simple shapes the same size as letters were not. Presumably, the pseudo-letters masked the letters of the words as effectively as did strings of normal letters, but the pseudo-letters did not strongly mask the words themselves.

All of the studies mentioned above are consistent with the idea that letters and words can be differentially masked, but they do not prove the fact. Jacobson and Rhinelander (1978) asked subjects either to name a masked word or to spell it. When the masker was a string of letters shaped similarly to the letters of the word they masked, the word was relatively easy to read, as compared to when the masker was made of dissimilarly shaped letters. The reverse was the case when the task was to spell the target word rather than to name it. Anagrams of the target word also masked the word poorly, even though each of its letters might be masked by a geometrically dissimilar letter. The dissimilarity of the masking letters to the target letters they masked should have made the word hard to read, but the existence of the same letters in the mask as in the target countered this difficulty. This again suggests the existence of a word-recognition process that, if not independent of the letters, is at least independent of the sequence of letters in the word.

Jacobson (1976) studied masking at a yet more abstract level of similarity between target and masker. In this study, the masker was always a word, which might be an associate of the target, a homonym of the target, or unrelated to the target. Associates masked less than the control maskers, whereas homonyms masked more than the control maskers. Jacobsen concluded his paper with the suggestion "that reading involves at least two analyses, one for sound and one for meaning, and that they proceed at the same time upon word presentation. Models of reading in which one process is concluded before another is begun are precluded, and those involving parallel processing of different aspects of information are supported." Gekoski, Jacobson, and Frazao-Brown (1982) followed up this work by showing that with compound bilingual subjects, associates in the same language as the target masked less than associates in the other language, but both masked less than unassociated words in the same language. The most severe masking was caused by unassociated words in the other language. There seems to be not only parallel processing for meaning, but also a partial split between languages in performing this processing.

Reading Regular and Exception Words

What are the differences between good and poor normal readers? (Dyslexics are a class, or rather several classes, unto themselves, and are not the subject of this question.) Clearly, good readers read more quickly than do poor readers, and as a rule have a larger working vocabulary. But this is not all there is to the difference. Good readers are *slower* than poor readers in lexical decision but faster in word naming (saying a written word presented to them) (Butler and Hains, 1979).

Baron and Strawson (1976) characterized two groups of subjects, whom they labeled "Chinese" and "Phonecians." "Chinese" tend to use the visual forms of words (use the RIGHT track) more than do "Phonecians." To label a subject, Baron and Strawson used two tests: identifying homophones of English words ("Phonecians" should do better), and selecting the correctly spelled member of a homophone pair ("Chinese" should do better). Once the subjects had been identified, they were asked to read lists of 10 words or letterstrings. Three types of lists were used: regularly spelled words, exception words, and nonwords. Overall, the reading times were in the order given, but the "Chinese" subjects read the exception words faster than the regular words, provided that the words were written in lowercase. If the strings were written in mixed case, "Chinese," too, were faster on regular words, though not as much so as the "Phonecians."

Other researchers have not always replicated Baron and Strawson's finding that "Chinese" are faster on exception words than on regular words. The finding probably depends on the precise experimental conditions and the familiarity of the subjects with the words, among other factors. Nevertheless, the separation between the two groups seems clear. Two groups of people, who can be distinguished by their performance with homonyms, perform differently when reading words that are spelled regularly or irregularly. Using the notation used earlier, we may call the "Chinese" subjects R-type (RIGHT track preferred), the "Phonecians" L-Type.

Seidenberg and associates have studied the exception effect in lexical decision, sometimes finding it, sometimes not. In a lexical decision task, the subject is asked to indicate as fast as possible whether a string of letters (the target) is a word or not. Typically, reaction times are of interest only for the real words. According to Waters and Seidenberg (1986), the exception effect is found in lexical decision only when the set of real words includes some whose orthography is "strange" (i.e., unlike that of other legal words: *yacht* is an example). The inclusion of "strange" words has no effect on word-naming, since the exception effect usually appears in this task. In none of these experiments, however, have results been analyzed separately for "Chinese" and "Phonecian" subjects.

The main thrust of the various experiments of Seidenberg and associates has been to show that phonological coding is a slow process compared with visual word recognition, but that it will manifest itself if the visual process does not succeed rapidly enough. The exception effect is found only in slow namers, and it disappears when responses are forced to be given before a quick deadline. The exception effect is usually found only for low-frequency words, but in children and poor readers it is found in words of all frequencies (Waters, Seidenberg, and Bruck, 1984). In all cases, when the exception effect is found, the naming times are slow. Baron and Strawson's "Chinese" subjects behave like good readers, and their "Phonecian" subjects like poor readers.

The BLC interpretation of these results is that rapid naming probably is based on using the RIGHT track, which does not care whether a word is an exception or not, though it does care whether the word is of low familiarity. The phonological coding processes of the LEFT track can be used for all words, as well as for nonwords, but may fail with exception words if they are not recognized as being words. Good readers have a larger RIGHT track vocabulary, but this implies that they have a more difficult time discriminating nonword strings from as not belonging to this large vocabulary. A nonword string will partially match several templates from the vocabulary, and perhaps may match some of them more strongly than a low-frequency word matches its own template.

Brain Damage and Memory for Words

In presenting the illustrative data of this section, an attempt has been made to avoid studies involving brain damage, partly because such studies figured heavily in the earlier descriptions of the BLC model, and partly because each brain-damaged individual is unique in both lesion and behavior pattern. There is, however, one study that bears presentation here, because it seems to illustrate in fairly pure form a characteristic difference between LEFT track inhibition (manifest in LH function) and RIGHT track acceptance (manifest in right hemisphere function).

Rausch (1981) presented his subjects with a list of 60 words, in which the last 30 words each had some kind of relationship with words in the first half of the first. For each word, the subjects were asked whether it had appeared earlier in the list. Six of the last 30 words were repeated from the first half, and the other 24 were foils of different kinds: six were homophones of words in the first half of the list, six were from the same semantic category, six were close associates, and six were (apparently) unrelated to any word in the first half.

Using normal subjects as controls, Rausch found all groups detected the repeated words to about the same degree (70%–80%), but the three groups differed drastically in their false acceptance of the foils. The subjects who had a right temporal lesion almost never accepted a foil; their discrimination was almost perfect. The subjects with a left temporal lesion accepted a large number of foils, particularly words of the same semantic category (about 45%) and homophones (about 35%). They also accepted associates and unrelated words, but to a lesser degree (about 20% and 15% respectively). Normals fell between the two lesioned groups in performance, accepting many homophones (about 25%) and a few of the other foils (generally less than 5%).

The characteristic differences among the groups in Rausch's study were that one group (with left hemisphere lesions) accepted almost anything, one group (with right hemisphere lesions) accepted almost nothing unless it was exactly correct, and one group (normals) accepted a few false recognitions, especially homophones.

If the objective of verbal perception were to separate true repeats from new material, it would seem to be an advantage to have a lesion of the right temporal lobe! Obviously, it cannot be an advantage to be brain-damaged, so one must ask what benefit is given by the processes that reduce discrimination ability. The answer is probably flexibility: a normal person can readily see multiple possibilities in a verbal stimulus, so that metaphors, jokes, and creative opportunities are easily grasped. Gardner and coworkers have, for example, shown that patients with RH damage have difficulty with such linguistic and pragmatic tasks (Brownell, Michel, Powelson, & Gardner, 1983; Wapner, Hamby, & Gardner, 1981; Winner & Gardner, 1977). Normal people sit on a balanced seesaw, able to discriminate, but also able to accommodate.

Reading Speed and Semantic Priming

Eisenberg and Becker (1982) tested the effects of passage difficulty in skim-reading. They separated out the extreme subjects into two groups: one group (R) maintained almost the same speed in both passages, whereas the other group (L) slowed drastically when skimming the difficult passage. The selected subjects were then tested in a lexical decision task with semantic priming.

In a lexical decision task, subjects are asked to indicate as quickly as possible whether a target letter string is a real word. The "prime" consists of another word or letter string that is shown before the target. The prime may be a word related to the target (*doctor–nurse*), unrelated to the target (*butter–nurse*), or a neutral string of letters (such as "XXXXX"). After the related prime, the reaction time tends to be faster, and after the unrelated prime, slower, than after the neutral prime. Eisenberg and Becker found that their two extreme groups of subjects reacted differently to the primes: group R showed mainly the facilitation effect of related primes and little inhibition from unrelated primes, whereas group L showed the reverse, little facilitation but much inhibition.

In a second experiment, subjects were asked to indicate the truth of sentences such as *Summer is hot* following presentation of related or unrelated propositions (*Snow falls in winter,* or *Milk comes in bottles*) or a neutral stimulus ("XXXXX"). In this experiment also, group R showed mainly facilitation from the related proposition, with no inhibition from the unrelated one, whereas group L showed the reverse. The R-L bias holds across levels of abstraction.

Sentence-Picture Verification

A classic experiment is to present people with a sentence like *PLUS is (not) above/below STAR* followed by a picture of a plus-sign above or below an asterisk. Subjects are then asked to state as quickly as possible whether the sentence is true (Clark & Chase, 1972). A "linguistic" model for the task has been developed by Carpenter and Just (1975); it asserts that the subject makes a variety of linguistic transformation, such as negation, in order to map the picture onto the sentence. A certain number of operations, say *K*, are needed to create the base sentence corresponding to the picture, (e.g., "STAR is above PLUS"), and then between one and four more are required to convert it into the form of the presented sentence. One is required for a false affirmative ("PLUS is above STAR"), four for a false negative ("STAR is not above PLUS"), and five for a true negative ("PLUS is not above STAR"). The time taken for a correct reaction is asserted to be linearly dependent on the number of transformations required.

The linear assertion was tested for a set of 70 subjects by MacLeod, Hunt, and Mathews (1978). They found that it held well for 43 of the subjects, very badly for 16, and moderately well for 11. The 16 badly fit subjects showed no effect at all of sentence form, except for whether it was true or false (true sentences were around 200 ms faster than false). These 16 poorly fit subjects were substantially faster under all conditions than were the well-fit subjects (who took from 60% to 300% longer). The poorly fit subjects, however, took about 60% longer than the well-fit subjects to comprehend the sentence in the first place.

MacLeod et al. interpreted the well-fit subjects as supporting the Carpenter and Just model: they converted the picture to linguistic form and then compared the result with the original sentence. But the poorly fit subjects used another strategy: convert the sentence to pictorial form, and then compare pictures.

For many of the subjects, MacLeod et al. had access to tests of verbal and spatial ability taken at least 2 years beforehand. They found no difference between the groups on average in verbal ability, but a large difference in spatial ability in favor of the poorly fit group. When the effect of spatial ability was removed, reaction time correlated strongly (negatively) with verbal ability for the well-fit group but not at all for the poorly fit group, whereas when the effect of verbal ability was removed, spatial ability correlated with performance only for the poorly fit group. In no case, apparently, did both abilities contribute to performance (although this might possibly have occurred for the 11 subjects of intermediate fit, whose data were not further analyzed).

In context of the other studies mentioned above, we can designate the well-fit subjects as L-type; these use linguistic methods to solve the problem, and tend to have higher verbal than spatial abilities. The poorly fit, R-type subjects are unaffected by linguistic form, are faster in their responses, and have relatively high spatial abilities, which they use in solving the problem.

Learning and Conversation Theory

"Conversation Theory" has been developed over many years by Gordon Pask and his collaborators (Pask, 1984 gives an overview with 30 references to earlier work). It is a precise description, based on philosophical, mathematical, and practical considerations, of the process whereby a concept in one mind is transferred to another. Pask's own writings are hard for the uninitiated to interpret, but a simplified introduction to Conversation Theory has been provided by Ogborn and Johnson (1982).

A practical outgrowth of Conversation Theory is CASTE (Course Assembly System and Training Environment) (Pask & Scott, 1973). CASTE is an adaptive training environment, within which a student can select sequences of things to learn on the way to understanding a moderately complex topic such as the theory of resonant circuits or elementary algebra. Pask and Scott found that different students choose to learn in different ways, one they called "serialist" and one they called "holist." In the course of learning, students of either type work on topic relations, aiming toward some knowledge node in the network defined by the course developer. Students may have some ideas about an "aim node" before actually reaching it through the topic relations. Pask and Scott found:

> Serialist students do, in fact, appreciate topics ahead of those they understand (as we have found by probing in a pre-programmed fashion for correct belief measures). But the region in advance of the current topic relation where current beliefs are entertained is canalized and often rather short; beyond that region, no belief is offered. In contrast, a holist student is quite willing to express beliefs about many topic relations in advance of those that currently concern him and very often those are correct beliefs. Succinctly, the serialist proceeds from certainty to certainty: the holist also achieves certainty at points where explanation is required. But these points are embedded in a nexus of dimly perceived but often correctly perceived relations. (Pask & Scott, 1973, pp. 41–42)

Learning programmes can be devised that suit serialists or holists, but a programme suited for one class gives poor results when presented to a student of the other class (Pask & Scott, 1972). Serialists and holists were identified in an initial "free learning" session in which the students learned the taxonomy of a fictitious family of "Martian fauna." Training programmes were devised that provided information on a more complex, unrelated taxonomy, either in logical sequence using simple extensions of known information (serialist), or in small complexes of various types in no necessary sequence (holist). Each type of training was effective when used with the appropriate type of student, but ineffective when used with the other kind of student.

The type of learner transfers across knowledge domains. Pask and Scott (1972) taught the biochemical concept of the "operon" (Jacob & Monod, 1961) to some of the same serialist and holist students using matched and mismatched teaching strategies. On a final test, the matched students obtained scores ranging from 17 to 20 out of 20, whereas the mismatched students scored between 6 and 13 out of 20. In both the operon experiment and the taxonomy experiment, there was a suggestion that holists might be a little better with serialist training than serialists are with holist training.

Serialists and holists can be identified with the L-type and R-type of the studies described above. Serialists tend to emphasize LEFT track processes, where precision dominates, whereas holists emphasize RIGHT track processes, where partial information leads to rapid and possibly ambiguous assessment. As Pask and Scott (1972) say:

> Holists ... commit mistakes due to simple over-generalization (for example, that (β) always implies "Bushy Tail" which is true for only *some* subspecies) or systemic over-generalization to render the classification *scheme* more rational or symmetric than it actually is (p. 237).

On the other hand:

> Serialists fall into difficulties if they fail to distinguish the wood from the trees and consequently try to assimilate masses of sparsely related irrelevant information. ... But, even if relevance is determined systematically, the serialist strategy appears to impose an appreciable load on storage capability. Some students who accumulated all the relevant data failed, subsequently, to reconstruct the entire taxonomy. (p. 238)

In other words, RIGHT track patterns can be wrong by omitting detail, whereas LEFT track analyses can fail if provided with too much information at once.

Are There Two Types of People?

Throughout this section, R-type people are those who are relatively biased in favor of using the RIGHT track, or (in the case of the hemispheric bias studies) in favor of using the RH. The question that remains is: are people R-type or L-type as a general characteristic, or is the categorization specific to the task at hand, or perhaps even to the strategic choice of the moment? The case for the BLC model would be strengthened if R-typeness were characteristic of a person rather than of the performance of a person on a couple of closely related tasks. The case for an effect of alphabetic literacy on cultural phenomena such as philosophical style would be strengthened by the same kind of evidence. Unfortunately, I know of no such studies, although Pask and Scott's results on serialist and holist learners suggest that there is some long-term consistency and a resistance to change in style at the higher cognitive levels. Pask and Scott even claim to be able to recognize serialists and holists from their personalities and occupations, although they claim to "have only the most general concept of the cues [they] employ for this purpose." If people are consistent in style over time, the question remains as to whether they are consistent across levels of abstraction. An appropriate study would use tasks similar to those discussed in this section, correlating the results within subjects, and using a factor analysis to detect commonalities in performance style.

Conclusion: Tracks and Modes of Thought

As we have seen, different scripts make different use of RIGHT and LEFT track processes (Taylor & Taylor, 1983), and it is plausible to argue that reading in a script that demands the LEFT track could enhance the tendency to use the LEFT track in other symbolic processes: logical thinking, for example. In particular, LEFT track processing can be important for an alphabet that yields to phonetic processing of its symbols, whereas an allusive script such as Egyptian hieroglyphic, or Chinese, must make more demands on the RIGHT track. The development of the Greek alphabet coincided with the development of logical modes of philosophical thought. Seen from the viewpoint of the BLC model, the temporal coincidence could have been causally related; but it is hard to imagine how such a relation could ever be proved.

The suggestion of Kerckhove in this book is that three related changes occurred around 600 B.C. in Greece. The earlier Phoenician writing system acquired symbols for vowels and became the Greek alphabet; the direction of writing changed from preferentially leftward to preferentially rightward after a short period of boustrophedon writing; and logical, mathematical thinking became the dominant philosophical mode.

It seems quite plausible within the BLC model that a shift to alphabetic writing, with its concomitant increase of emphasis on the LEFT track, might have induced a corresponding increase of LEFT track operations in higher mental processes – a shift from the "appreciative" to the "scientific" manner of understanding the universe. The serialist proceeds like a mathematician, or a logician. Nothing is assumed except what can be proven; disconfirming evidence would block a line of inquiry. For the holist, on the other hand, pieces of evidence gradually conform to a new synthesis, which is taken to be correct. Disconfirming evidence has little weight if there is a great deal of confirming evidence.

Scientists, of necessity, rely on the analytic methods of the LEFT track, wherein evidence contradictory to an interpretation can destroy that theory, but these methods may provide only a precise view of small details of the world, contrasting with the diffuse, global, but equally valid view provided by Oriental philosophies such as the Tao, using the RIGHT track. In a RIGHT track approach, confirming evidence is of much more importance than disconforming evidence; it is the way most people operate much of the time, and even scientists are loath to discard a theory that makes many good predictions when a single piece of evidence goes the wrong way.

In practical life, the holist probably is right to depend on confirming evidence, since disconfirming evidence may well be extrinsic to the question at issue: if the context demands a letter "O," but a "Q" is seen, there is probably a flaw in the paper or a misprint, and "O" is correct. In science, however, an effective theory must (by convention) include the description of things that *must not* happen, as well as of things that should be expected. Scientists ask that a theory be falsifiable before they accept it. But useful, practical, theories may not thus be falsifiable. The Theory of Signal Detectability (e.g., Green & Swets, 1966), for example, has for over 30 years been tremendously useful in understanding psychophysical phenomena, yet is in principle unfalsifiable. Furthermore, in the development of a falsifiable theory, the holist approach should lead to descriptive possibilities that the theory can then predict in a logical, analytic way. The holist thus provides the goals for the serialist's logic. In just such a way does the BLC model provide for coherent symbolic perception: the RIGHT track provides the goals which the LEFT may attain or destroy.

The speculation is plausible that the three Grecian events – emergence of a true alphabet, rightward writing, and logical thinking – are related, but it is not scientifically provable. All one can say, in a holist RIGHT track manner, is that there is no evidence against the speculation that the change of writing system caused and was caused by a corresponding change in manner of thought. It is not at all clear what could constitute such evidence.

The evidence suggests that there is a difference between those who tend to use RIGHT track and right hemisphere processes more and those who use them less, and that Japanese may tend more to be RIGHT users than are Europeans. This evidence is, however, tenuous at best. Furthermore, Chinese, like Westerners, use phonetic coding, a LEFT track function, in short-term memory. To counter this point, it appears that deaf people who use Ameslan (American Sign Language) do not have a left hemisphere bias, whereas hearing interpreters of Ameslan do use their left hemisphere for short-term memory (Suter, 1982). Furthermore, Ameslan syntax does not depend much on word order, and deaf signers tend not to remember word order, which suggests they rely more on RIGHT track functions rather than LEFT for processing Ameslan (Hanson & Bellugi, 1982).

Whatever the reason for left hemisphere bias, it surely is not reading in an alphabetic script, since Japanese Kana is heavily left-biased. All we can say at the moment seems to be that Western analytic philosophies use methods appropriate to the LEFT track more than do Oriental philosophies such as the Tao.

The classical Chinese word was very different from an abstract sign representing a clearly delineated concept. It was rather a sound symbol which had strong suggestive powers, bringing to mind an indeterminate complex of pictorial images and emotions. The intention of the speaker was not so much to express an intellectual idea, but rather to affect and influence the listener. Correspondingly, the written character was not just an abstract sign, but was an organic pattern – a 'gestalt' – which preserved the full complex of images and the suggestive power of the word. ... The Chinese, like the Indians, believed that there is an ultimate reality which underlies and unifies the multiple things and events we observe: *There are three terms – 'complete', 'all-embracing', 'whole'. These names are different, but the reality sought in them is the same: referring to the One thing.* They called this reality the Tao, which originally meant 'the Way'. (Capra, 1976, p. 110)

The formal scientific model (if anyone uses it) is explicitly goal-oriented, looking for data that could deny prespecified hypotheses. That is almost a definition of the LEFT track process. In everyday life, however, people are more inclined to see what the data tell them, and how it fits into their already conceived context. Disconfirming data tend to be rejected or assigned to interfering processes, rather than being used to change ideas as to the truth of the world. Multiple, perhaps contradictory, ideas can be held simultaneously, and new ones can be built from sufficient new data, without necessarily displacing the old. These are characteristics of the RIGHT track.

RIGHT track processes seem to describe the manner most people, even scientists, deal with the world except when they are being deliberately analytic and thoughtful or critical. In order to make more definite statements about the impact of alphabetic writing systems as opposed to nonalphabetic systems, one would have to ask whether educated and literate people not exposed to an alphabetic system would normally use or appreciate "scientific" thought. Do such people exist? If not, we must rely on historical evidence and intuition, because it is unlikely that the question could ever be resolved scientifically. The RIGHT answer must be LEFT untold.

References

Adams, M. J. (1979). Models of word recognition. *Cognitive Psychology, 11,* 133–176.

Alegria, J., Pignot, E., & Morais, J. (1982). Phonetic analysis of speech and memory codes in beginning readers. *Memory & Cognition, 10,* 451–456.

Amano, K. (1970). Formation of the act of analyzing phonemic structure of words and its relation to learning Japanese syllabic characters (Kanamoji). *Japanese Journal of Educational Psychology,* 1970, *18,* 76–89 (in Japanese with English abstract).

Anglin, J. M. (1977). *Word, object and conceptual development.* New York: Norton.

Baron, J., & Strawson, C. (1976). Use of orthographic and word-specific mechanisms in reading words aloud. *Journal of Experimental Psychology: Human Perception and Performance, 2,* 386–393.

Beck, J. (1973). Similarity grouping of curves. *Perceptual and Motor Skills, 36,* 1331–1341.

Beck, J. (1983). Textural segmentation, second-order statistics, and textural elements. *Biological Cybernetics 48,* 125–130.

Bouma, H. (1971). Visual recognition of isolated lower-case letters. *Vision Research, 11,* 459–474.

Bradley, L., & Bryant, P. E. (1983). Categorizing sounds and learning to read: a causal connection. *Nature, 301,* 419–421.

Broadbent, D.E., & Broadbent, M.H.P. (1980). Priming and the active/passive model of word recognition. In Nickerson, R.S. (Ed), *Attention and performance,* vol 8. Hillsdale: Erlbaum.
Brownell, H.H., Michel, D., Powelson, J., & Gardner, H. (1983). Surprise but coherence: Sensitivity to verbal humor in right-hemisphere patients. *Brain and Language, 18,* 20–27.
Butler, B.E., & Hains, S. (1979). Individual differences in word recognition latency. *Memory & Cognition, 7,* 68–76.
Capra, F. (1976). *The Tao of physics.* Boulder: Stromhala.
Carpenter, P.A., & Just, M.A. (1975). Sentence comprehension: a psycholinguistic processing model of verification. *Psychological Review, 82,* 45–73.
Cheeseman, J., & Merikle, P. (1985). Word recognition and consciousness. In Besner, D., Waller, T.G., & MacKinnon, G.E. (Eds.), *Reading research: Advances in theory and practice, Vol. 5.* New York: Academic.
Chu-Chang, M., & Loritz, D.J. (1977). Phonological encoding of Chinese ideograms in short-term memory. *Language Learning, 27,* 344–352.
Clarke, H.H., & Chase, W.G. (1972). On the process of comparing sentences against pictures. *Cognitive Psychology, 3,* 472–517.
Cramer, P. (1968). *Word association.* New York: Academic.
de Groot, A.M.B. (1983). The range of automatic spreading activation in word priming. *Journal of Verbal Learning and Verbal Behaviour, 22,* 417–436.
de Groot, A.M.B., Thomassen, A.J.W.M., & Hudson, P.T.W. (1982). Associative facilitation of word recognition as measured from a neutral prime. *Memory & Cognition, 10,* 358–370.
Eisenberg, P., & Becker, C.A. (1982). Semantic context effects in visual word recognition, sentence processing, and reading: Evidence for semantic strategies. *Journal of Experimental Psychology: Human Perception and Performance, 8,* 739–756.
Entwisle, D.R., Forsyth, D.F., & Muus, R. (1964). The syntagmatic-paradigmatic shift in children's word association. *Journal of Verbal Learning and Verbal Behavior, 3,* 19–29.
Ervin, S.M. (1961). Changes with age in verbal determinants of word association. *American Journal of Psychology, 74,* 361–372.
Feldman, L.B. (1981). *Visual word recognition in Serbo-Croatian is necessarily phonological.* Status report on speech research SR-66. Haskins Laboratories, April–June.
Foss, D.J. (1982). A discourse on semantic priming. *Cognitive Psychology, 14,* 590–607.
Friedman, R.B. (1980). Identity without form: abstract representations of letters. *Perception & Psychophysics 28,* 53–60.
Gekoski, W.L., Jacobson, J.Z., & Frazao-Brown, A.P. (1982). Visual masking and linguistic independence in bilinguals. *Canadian Journal of Psychology, 36,* 108–116.
Green, D.M., & Swets, J.A. (1966). *Signal Detection Theory and psychophysics.* New York: Wiley.
Hanson, V.L., & Bellugi, U. (1982). On the role of sign order and morphological structure in memory for American Sign Language sentences. *Journal of Verbal Learning and Verbal Behavior, 21,* 621–633.
Hirst, G. (1983). *Semantic interpretation against ambiguity.* Technical report CS-83-25. Providence, RI: Brown University Dept of Computer Science.
Holender, D. (1986). Semantic activation without conscious identification in dichotic listening, parafoveal vision, and visual masking: a survey and appraisal. *The Behavioural and Brain Sciences, 9,* 1–66.
Holley-Wilcox, P., & Blank, M.A. (1980). Evidence for multiple access in the processing of isolated words. *Journal of Experimental Psychology: Human Perception and Performance, 6,* 75–84.
Holmes, V.M. (1979). Accessing ambiguous words during sentence comprehension. *Quarterly Journal of Experimental Psychology, 31,* 569–589.
Isenor, D.K., & Zaky, S.G. (1986). Fingerprint identification using graph matching. *Pattern Recognition, 19,* 113–122.

Jacob, F., & Monod, J. (1961). On the regulation of gene activity. In *Cellular regulatory mechanisms,* vol 26. Cold Spring Harbour symposia on quantitative biology.
Jacobson, J. Z. (1974). Interaction of similarity to words of visual masks and targets. *Journal of Experimental Psychology, 102,* 431–434.
Jacobson, J. Z. (1976) Visual masking by homonyms. *Canadian Journal of Psychology, 30,* 174–177.
Jacobson, J. Z., & Rhinelander, G. (1978). Gemoetric and semantic similarity in visual masking. *Journal of Experimental Psychology: Human Perception and Performance, 4,* 224–231.
Jensen, H. (1970) *Sign, aymbol and script.* London: Allen and Unwin.
Johnston, J. C., & McClelland, J. L. (1973). Visual factors in word perception. *Perception & Psychophysics, 14,* 365–370.
Johnston, J. C., & McClelland, J. L. (1980). Experimental tests of a hierarchical model of word identification. *Journal of Verbal Learning and Verbal Behavior, 19,* 503–524.
Jones, B. (1982). The integrative action of the cerebral hemispheres. *Perception & Psychophysics, 32,* 423–433.
Julesz, B. (1981 a). A theory of preattentive texture discrimination based on first-order statistics of textons. *Biological Cybernetica, 41,* 131–138.
Julesz, B. (1981 b). Textons, the elements of texture perception, and their interactions. *Nature, 290,* 91–97.
Kohonen, T. (1982). Self-organized formation of topologically correct feature maps. *Biological Cybernetics, 43,* 59–69.
Kutas, M., & Hillyard, S. A. (1980). Reading senseless sentences: brain potentials reflect semantic incongruity. *Science 207,* 203–205.
Kutas, M., & Hillyard, S. A. (1982). The lateral distribution of event-related potentials during sentence processing.*Neuropsychologia, 20,* 579–590.
MacLeod, C. M., Hunt, E. B., & Mathews, N. (1978). Individual differences in the verification of sentence–picture relationships. *Journal of Verbal Learning and Verbal Behavior, 17,* 493–507.
Marcel, A. J. (1980). Conscious and preconscious recognition of polysemous words: locating the selective effects of prior verbal context. In Nickerson, R. S. (Ed.), *Attention and Performance,* vol 8. Hillsdale: Erlbaum.
Marcus, S. M. (1981). ERIS-context-sensitive coding in speech perception. *Journal of Phonetics, 9,* 197–220.
Meltzer, E. S. (1980). Remarks on ancient Egyptian writing with emphasis on its mnemonic aspects. In Kolers, R. A., Wrolstad, M. E., & Bouma, H. (Eds). *Processing of visible language,* vol. 2. New York: Plenum.
Morais, J., Cary, L., Alegria, J., & Bertelson, P. (1979). Does awareness of speech as a sequence of phones arise spontaneously? *Cognition, 7,* 323–331.
Muraishi, S., & Amano, K. (1972). *Reading and writing abilities of preschoolers: a summary.* Tokyo: The National Language Research Institute (in Japanese).
Nishikawa, Y., & Niina, S. (1981). Modes of information processing underlying hemispheric functional differences. *Japanese Journal of Psychology, 51,* 335–341 (in Japanese with English abstract).
Ogborn, J. M., & Johnson, L. (1982). *Conversation Theory.* Technical Report MCSG/TR30. Div. of Cybernetics, Brunel University.
Pask, G. (1984). Review of Conversation Theory and a protologic (or protolanguage) Lp. *Education Communications and Technology Journal, 32,* 3–40.
Pask, G., & Scott, B. C. E. (1972). Learning strategies and individual competence. *International Journal of Man-Machine Studies, 4,* 217–253.
Pask, G., & Scott, B. C. E. (1973). CASTE: a system for exhibiting learning strategies and regulating uncertainties. *International Journal of Man-Machine Studies, 5,* 17–52.
Petrey, S. (1977). Word associations and the development of lexical memory. *Cognition, 5,* 57–71.

Rausch, R. (1981). Lateralization of temporal lobe dysfunction and verbal encoding. *Brain and Language, 12,* 92–100.
Rayner, K., & Posnansky, C. (1978). Stages of processing in word identification. *Journal of Experimental Psychology: General, 107,* 64–80.
Rumelhart, D.E., & Norman, D.A. (1973). *Active semantic networks as a model of human memory.* Report CHIP 33. Center for Human Information Processing. San Diego: University of California.
Rumelhart, D.E., Lindsay, P.H., & Norman, D.A. (1972). A process model for long-term memory. In Tulving, E., & Donaldson, W. (Eds.), *Organization of memory.* New York: Academic.
Schank, R.C. (1972). Conceptual dependency: a theory of natural language understanding. *Cognitive Psychology, 3,* 552–631.
Schvaneveldt, R.W., Meyer, D.E., & Becker, C.A. (1976). Lexical ambiguity, semantic context, and visual word recognition. *Journal of Experimental Psychology: Human Perception and Performance, 2,* 243–256.
Scribner, S., & Cole, M. (1981). *The psychology of literacy.* Cambridge: Harvard University Press.
Selfridge, O. (1959). Pandemonium: a paradigm for learning. In *Symposium on the mechanisation of thought processes.* London: HM Stationery Office.
Stevens, W.J. (1973). Verb types in modern English. *Language Sciences, 26,* 29–31.
Suter, S. (1982). Differences between deaf and hearing adults in task-related EEG asymmetries. *Psychophysiology, 19,* 124–128.
Tattersall, G.D., & Johnston, R.D. (1984). Self organizing arrays for speech recognition. *Proceedings of the Institute of Acoustics.*
Taylor, I., & Taylor, M.M. (1983). *The psychology of reading.* New York: Academic.
Taylor, J.G. (1962). *The behavioral basis of perception.* New Haven: Yale University Press.
Taylor, M.M. (1973). The problem of stimulus structure in behavioral theory of perception. *South African Journal of Psychology, 3,* 23–45.
Taylor, M.M. (1974). Speculations on bilingualism and the cognitive network. *Working Papers on Bilingualism, 2,* 68–124
Taylor, M.M. (1981/1984). The Bilateral Cooperative Model of reading: a human paradigm for artificial intelligence. In: Elithorn, A., & Banerji, R. (Eds.), *Artificial and Human Intelligence.* Amsterdam: Elsevier. Distributed as DCIEM Research Paper 81-P-4, 1981.
Tweedy, J.R., & Lapinski, R.H. (1981). Facilitating word recognition: evidence for strategic and automatic factors. *Quarterly Journal of Experimental Psychology, 33A,* 51–59.
Tzeng, O.J.L., Hung, D.L., & Wang, S.Y. (1977). Speech recoding in reading Chinese characters. *Journal of Experimental Psychology: Human Learning and Memory, 3,* 621–630.
Wapner, W., Hamby, S., & Gardner, H. (1981). The role of the right hemisphere in the apprehension of complex linguistic materials. *Brain and Language, 14,* 15–33.
Waters, G.S., & Seidenberg, M.S. (1986). Spelling-sound effects in reading: time-course and decision criteria. *Memory and Cognition, 13,* 557–572.
Waters, G.S., Seidenberg, M.S., & Bruck, M. (1984). Children's and adults' use of spelling-sound information in three reading tasks. *Memory and Cognition, 12,* 293–305.
Winner, E., & Gardner, H. (1977). The comprehension of metaphor in brain-damaged patients. *Brain, 100,* 717–729.
Yik, W.F. (1978). The effect of visual and acoustic similarity on short-term memory for Chinese words. *Quarterly Journal of Experimental Psychology, 30,* 487–494.

CHAPTER 18

Right Hemisphere Literacy in the Ancient World

JOHN ROBERT SKOYLES[1]

The Left Hemisphere Reading Paradigm

Is there such a thing as right hemisphere literacy? There is a paradigm in neuropsychology that the brain processes enabling right-handed people to read are restricted to the left hemisphere. The right hemisphere, according to this view, while involved in non-reading-related processes, lacks the neurological competence required for such tasks as the recognition and understanding of written words. Most paradigms do not arise out of thin air but have their foundation in some kind of fact. Our present understanding on reading lateralization has its roots in the effects of brain injuries, for it is well established that while injuries to the left hemisphere affect reading, sometimes destroying the ability completely, this is rarely the case with the right hemisphere. If reading abilities existed in this hemisphere, why should those individuals with a damaged left hemisphere, but an intact right one, not use the latter to take over their impaired reading abilities? This lack of compensation has been taken to indicate that the left hemisphere is specialized for reading, while the right hemisphere is illiterate.

There is a further reason for the paradigm. Writing is often viewed metaphorically as a kind of 'frozen speech', which has the unfortunate consequence of suggesting that many of the processes underlying reading are the same as those underlying oral language. Speech is strongly left-lateralized; thus, it could be intuitively felt that reading would be too. However, the relationship between speech and reading is more complex than is suggested by this metaphor. According to Levy's (1982) interpretation of Gloning, Gloning, Haub, and Quatember's data (1969), the lateralization of reading is independent, at least for left handed and ambidextrous individuals, of the lateralization of speech. Knowing which hemisphere speech resides in for such people does not inform us which hemisphere possesses reading abilities. In right-handers, speech and reading are normally both lateralized to the left hemisphere: However, this is probably accounted for not by some direct association between them, but by a common lateralization of both functions to the dominant hemisphere.

[1] Department of Psychology, University College London, Gower Street, London WC1E 6BT, UK.

Why question this orthodoxy? One reason is the problem of generality: we only have data on modern western readers. Is it also true for readers of other writing systems, and of other times? Although our understanding of brain lateralization is based on a very large sample of brain-injured people, the population studied has consisted mainly of readers of modern European phonetic alphabets: that is, alphabets descended from the Greek. What if reading lateralization were peculiar to European phonetic alphabetic writing or some aspects of reading in the western world?

It is only recently that such research has begun to be carried out upon modern non-European forms of writing. Its findings will to some degree remove the homogeneity of the data base upon which conclusions about reading lateralization rest. However, many forms of writing developed in the ancient world that are no longer used. Without knowing how these were processed, any generalizations about laterality are incomplete. That is, it may be some factor in the modern world, rather than the intrinsic nature of reading, has led it to be processed by the left hemisphere. Moreover, the lateralization of ancient readers is an important question in its own right. Reading and writing have been extremely important influences on the development of civilization, and anything that furthers our knowledge of them will aid in our understanding of how cultures developed in the ways they did. This would be particularly true if ancient readers were shown to have used a different hemisphere from that of modern people. The development of civilizations could then perhaps be related to changes in the hemisphere responsible for literacy (Skoyles, 1984).

Conclusions about contemporary reading do not necessarily apply retrospectively to ancient reading, as there are several ways in which the former may be unrepresentative. The role of literacy in the modern world is similar everywhere irrespective of orthography; for example, most readers regularly pick up printed newspapers and books. A consequence of the printing press has been that all individuals have convenient, everyday access, at little or no expense, to a wide variety of written material. This opportunity, however, is a modern phenomenon that did not exist in the ancient world. The average reader today may spend an hour or more daily in scanning newspapers, advertisements, books, and ephemera sent through the mail. What can we infer from this about ancient readers who encountered none of these uses of writing, and who probably in consequence had infrequent need or opportunity to read? Are the processes that developed to read printed matter on a regular basis the same as those used to read handwriting on an occasional one? Finally, we know from St. Augustine and other sources that prior to about 400 A.D. readers normally pronounced the words aloud. When researchers study reading processes, they are basing their conclusions on people who normally read silently. How far can such research be generalized to the ancient world?

There are many unknowns, then, that should caution us against the assertion that reading must necessarily be lateralized to the left hemisphere.

We cannot tell, from evidence based on modern phonetic readers, whether, the ancients, under very different circumstances used similar processes. My intention here is to present evidence that suggests that the right hemisphere played an important role in processing reading in other times. In particular, I shall argue that clues as to which hemisphere was used, are to be found in the remnants of ancient writing that have survived. Such clues are of course indirect, but this does not mean they cannot be taken seriously: many retrospective and scientifically respectable inferences about the past in astrophysics, geology, and the fossil sciences are made on equally indirect grounds.

On the basis of this indirect evidence I shall argue that early cultures were right-hemisphere literate, at least up until the period of Classical Greece. Since such arguments challenge the current assumptions about left hemisphere literacy, I shall first discuss some reasons why this paradigm need not necessarily have applied in the ancient world. I shall, as it were, clear the ground for my later arguments by presenting certain observations which, although they do not prove ancient reading was right hemispheric, at least open up the possibility that it could have been.

The Case Against the Left Hemisphere Literacy Paradigm

Various arguments can be advanced against the paradigm of left hemisphere literacy.

First, reading is a highly complex process requiring many years to learn, and it is unlikely that it could have arisen in its present form. No biologist believes that intricate forms of life arose spontaneously: rather, they developed from simpler forms through various stages of increasing complexity. Similarly, one should not assume that the neurological processes of modern reading, including lateralization, existed in a similar form in earlier times. In biology there is a saying by Haeckel that "ontogeny recapitulates phylogeny": the development an individual goes through as an embryo repeats the evolutionary stages of its species. It might be heuristic to apply this concept to reading. Reading develops from limited and simple cognitive strategies in young children to complex ones with greater ability in older children and adults. By studying these maturational stages, we might gain some clues as to the processes used in the early stages of the historical development of reading. Silverberg, Gordon, Pollack, and Bentin (1980) tested the lateralization of Israeli children in reading as they matured. They found that, at least for Hebrew readers, the early stages are not dominated by the left hemisphere, as in adults, but by the right. It is only after the age of 7 years that the left hemisphere gains dominance.

Second, neurological studies of nonphonetic writing systems suggest they may have a different hemispheric representation than that encountered with phonetic ones. Hatta (1977, 1981) and Hatta, Honjoh, and Mito (1983) have provided evidence that ideographic characters are recognized better by the right than by the left hemisphere. Sasanuma (1975) has shown that Kanji, the nonphonetic script of Japanese, is vulnerable to different left hemispheric injuries than are Kana, the phonetic characters which chiefly mark syntax. These studies suggest that the representation of writing in the brain may be related to the nature of the script as well as to general neurological reading processes.

Third, there is evidence for the existence of some direct right hemisphere reading capacities. Zaidel (1982) has shown that this hemisphere possesses some reading ability when tested separately from the left hemisphere, as can be done in split-brained individuals. However, it recognizes words visually (Zaidel & Peters, 1981), lacking the ability to make phonetic analyses. The reading abilities retained in a certain form of dyslexia caused by damage to the left hemisphere deep dyslexia, have also been suggested to derive from right hemisphere reading (Saffran, Bogyo, Schwartz, & Martin, 1980; Coltheart, 1983). It must be noted that this claim has been challenged (Patterson & Besner, 1984a; but also see comments on this paper: Rabinowicz & Moscovitch, 1984; Zaidel & Schweiger, 1984; and Patterson & Besner's reply, 1984b).

It is possible to show word images to the right hemisphere without doing so directly to the left, by means of tachistoscopic projection. Images projected to the left visual field pass directly to the right but not to the left side of the brain. Studies projecting words to the right hemisphere by such a technique have found that subjects are able to read them. However, there is little agreement as to whether this reveals the presence of real right hemisphere literacy. The problem is that although directly projected only to the right hemisphere, the images can pass through the corpus callosum to the left side, which thus might ultimately be responsible for their recognition. For this reason, studies on people without such connections to the left hemisphere or with damaged left hemisphere reading competences are judged to be the most important evidence for right hemisphere literacy. Such results have been inconclusive and are currently the subject of vigorous debate.

Fourth, there is clear evidence that something as yet not understood is going on in the right hemisphere during reading. Regional cerebral blood flow varies with the energy consumption of neurons in its various areas. This blood flow can be measured with modern techniques showing, in terms of energy consumption, those areas of the cortex whose neurons are silent and those where they are active. In effect, this gives a map of the areas of the cortex doing "brain work." Studies of people engaged in reading show particular areas of the cortex to be in states of brain work. These areas should correspond to those which affect reading upon injury. That is, it is reasonable to suppose that if damage to an area of the brain affects reading, then that area is involved in some aspect of literary processing, and thus will normally, consume extra energy during the act of reading. In the left hemisphere, as expected, the active regions correspond to those areas whose damage affects reading ability. The homologous areas in the right hemisphere would not be expected to show similar patterns of energy consumption, on the basis that their damage does *not* affect reading performance. However, blood flow maps during reading in fact show these areas to be consuming extra energy, and by inference doing brain work, just like their counterparts on the left side (Larsen, Skinhøj, & Lassen, 1979).

Such results have been treated as anomalies by many researchers, as they conflict so strongly with the established paradigm of reading lateralization. There are several ways of viewing this apparent contradiction. First, it could be that the right hemisphere *is* contributing to reading abilities, in spite of the evidence to the contrary. Alternatively, it has been suggested (Cook, 1984) that homologous areas of the two hemispheres are bilaterally activated by subcortical structures, in particular the reticular formation. This may activate the right hemispheric counterparts of areas on the left side that subserve reading processes. This activation is then *inhibited* by the left hemisphere through the corpus callosum, thus preventing the right hemisphere areas from being involved in reading. However, it is reasonable to suppose that the right hemisphere could interfere with left hemisphere reading only if it had some reading potential itself, though perhaps more limited.

It is possible, and indeed likely, that the truth is more complex than that given by either of these two explanations. However, any kind of explanation of the right hemisphere's energy consumption during reading is restricted to an account either suggesting that it does contribute to reading (i.e., that it is literate), or that its contribution to reading is inhibited (i.e., that it is potentially literate). While blood flow studies do not give clear support to the notion of right hemisphere literacy, they do question our present theories about complete left lateralization based upon the data from brain-injured people. The activation seen in the right hemisphere was not predicted from such theories.

This undermines our confidence in the assertion that reading is necessarily lateralized to the left hemisphere, especially with regard to circumstances for which we have no direct evidence from brain injuries. Consequently, the lateralization paradigm should not prevent our search for evidence of right hemisphere literacy in ancient readers: it is quite consistent with present evidence for left hemisphere laterality in modern readers that we should find evidence of right hemisphere laterality in ancient readers.

Writing Direction as Culturgen

The preceding arguments open up the possibility that the right hemisphere might have played some role in early reading, but are not evidence for its actuality, the proof of which would require some empirical facts. It might be felt that such evidence could not exist. Lateralization, after all, is a function of the brain, a soft tissue that is not preserved, and it is doubtful even if it were preserved that we could know much about its operations. Nevertheless, there is evidence as to which hemisphere was dominant in the reading processes of the ancients. Although the lateralization of the brain leaves no analyzable remains, it could have affected the nature of something which has been preserved: ancient writings.

There are three possible effects that provide evidence for right hemisphere lateralization: writing direction, mirror reversal of literals, and handwriting position. Each of the following arguments will depend upon auxiliary theories of how brain functions could influence these effects. Thus, although my claims concern events which occurred over 2500 years ago and of which we have no direct record, they are contestable on the basis of these auxiliary theories and are thus not above critical testing or falsification.

My first line of argument is that reading lateralization might affect writing direction. Obviously, if writing has a particular direction, the eye must follow it: for instance, leftward writing requires for its perception leftward eye scans. This connection is a strong one in the sense that whatever direction writing takes, so must eye scan. However, although eye movements are passively determined in this way by writing, the possibility exists that the reverse may also have occurred. That is, although the eye can scan equally in all directions, certain directions might be more effective than others for word perception.

Writing, like any activity, can be done in a way that is efficient or inefficient. The purpose of writing is to be read: legible writing is efficient, illegible writing inefficient. One way in which scripts will differ in legibility is in their *style,* a term I shall use in a broad sense to include not only the formation of letters and words but also writing direction.

The direction we read in is the one our script traditionally uses. However, people can adopt new directions. Both Leonardo da Vinci and Lewis Carroll were brought up to write conventionally, yet kept private notes in mirror image writing (Harris 1980). Whole civilizations have changed their writing directions: the Greeks in the sixth century B.C. from boustrophedon to rightward, and the Sumerians from downward to sideway.

Thus, writing direction need not necessarily be fixed, but can change in response to factors such as perceptual advantage in reading in one direction over another.

Some writing styles will be easier to read than others. Styles are cultural artifacts, in that the way a person writes will largely depend upon the style that he or she was brought up to use. Since writing styles have been propagated from generation to generation, and it is reasonable to suppose that individuals who wrote legibly would have been more encouraged to pass their skills on than those who did not, it is likely that those styles that were easiest to read were preferentially propagated. Style, then, is like a gene, or "culturgen," with ease of its reading being a kind of "fitness," that enables some styles to propagate in preference to others. Early writing direction is known to have been variable, some writers using one direction, some another, and yet others mixing several directions at once, within a common locality, period and writing system. Given this situation, it is possible that reading scan could have influenced the creation of a uniform writing direction, if one scan direction were more efficient than the others for reading.

The Evolution of Writing Direction

Historically, the scanning involved in reading went through three stages: the multidirectional, the vertical, and the horizontal.

The Multidirectional Stage

Much of the very earliest or the most primitive writing showed no consistent directional preference. This was especially the case with the early variants of the consonantal alphabet. This point is important, since the fact that these alphabets all settled for the same direction after the multidirectional stage could not be due to their using some initial direction passed on from the alphabets from which they originated.

Such indeterminacy of direction suggests that at this stage, writing had not developed to the point where eye scan or other factors could produce a stabilized direction through their influences.

The Vertical Direction

The ideographic precursors of the early phonetic cuneiform writing in Sumeria and hieroglyphics in Egypt were arranged vertically, as were the logographic writings in the Far East in China, Korea, and Japan until recently, although they have tended under western influence to become written horizontally. Why should these writing systems have been arranged originally in a vertical direction, and only with phonetic development have become horizontal? One possibility is that the reading of logographs is more efficient with vertical eye scans than with horizontal ones. This type of writing differs from phonetic systems in not using peripheral vision for its perception. Ideographs are highly complex visual patterns, often differing only in small details, which can best be distinguished when the character is at the center of foveal vision. This would have been even more important with ancient logographs, which contained many visual irregularities due to their being created by hand. This would have made them visually "noisy" in comparison with modern uniformly printed ideographs, thus making their visual identification even harder and more cognitively demanding. There has also been a simplification of logographs over the ages: ancient logographs were more visually complex than modern ones.

One conjecture that would explain the development of the vertical direction for writing logographs is that vertical positioning of the fixation point is more accurate than horizontal positioning. Some evidence in support of this view was presented by Shen (1927). However, more recent studies do not confirm this: the positioning of the fovea appears to be more accurate horizontally than vertically. Amongst western subjects, this might be attributed to the experience of horizontal reading, causing them to acquire superior skills in this direction compared to the less used vertical one. However, there also does not appear to be any superiority for vertical scanning amongst native readers of vertical writing. In a study by Tanaka, Iwasaki, and Miki, cited by Taylor and Taylor (1983), Japanese readers showed a superior discrimination of single Kana graphemes when they were horizontally rather than vertically arranged. There is, of course, a large difference between the Kana graphemes used in Japanese and visually complex (especially when hand-written) ancient logographs. Consequently, the scanning of simple graphemes may be unrepresentative of the problems of reading complex logographs. However, research is lacking on this issue at present.

The Horizontal Direction

In contrast, phonetic writing can be usefully identified outside the center of vision in the parafovea.

The words of phonetic writing, unlike ideographs, consist of a series of signs, each of which is a simple visual pattern. There is an advantage in reading horizontally, if percpetion of such simple signs is better when moving from the center of vision in a horizontal rather than in a vertical direction. Taylor and Taylor (1983) have reviewed the limited available evidence, and suggest that phonetic writing can be identified further horizontally than vertically from the center of vision. Such studies on the identification of letters confirm what is known about the extent of visual activity from perimetry (Aulmorn & Harms, 1972) and the representation of the visual field upon the cortex (Van Essen, Newsome, & Maunsell, 1984). However, retinal acuity is only slightly greater in a horizontal direction than a vertical one from the fovea, which suggests that the advantage of horizontal word perception might lie at higher levels of image analysis rather than simple perception. For instance, in such processes as visual memory and stimulus integration are involved in word identification. Whether and to what degree these processes differ when scanned horizontally as opposed to vertically is as yet largely unstudied.

Leftward Writing and the Right Hemisphere

There is evidence, as I shall outline below, that of the two possible horizontal directions, leftward scans are more efficient for right hemispheric processing of visual information than rightward ones, and vice versa. From this, it can be speculated that if early readers were right-hemisphere literate, they would have had a preference for leftward scans when reading, and consequently favored the development of leftward writing. This is supported by archeological records showing that virtually all early horizontal writing systems were written leftward. The one exception was a group that utilized clay as the writing medium: a practical factor that would overwhelm any scanning advantages of leftward writing, as discussed further on. For these people, there is additional evidence on hand position that suggests that in spite of their rightward reading direction they too were right-hemisphere literate. For the rest, there is strong evidence connecting leftward scans with right hemisphere processing of written words. To this I now turn.

Each side of the brain mainly controls the search and scanning of the contralateral space around the individual (Girotti, Casazza, Musicco, & Avanzini, 1983). Thus, what happens to our left is the concern of the right hemisphere, while that occurring to the right is mainly the concern of the left hemisphere, although it must be noted that the right hemisphere also has some ability to process stimuli on the right. One consequence of this is that the activation of a hemisphere produces eye scan movements into its contralateral space. People using their right hemispheres, such as when working out visuospatial problems, tend to gaze to the left, while when using their left hemispheres, as in working out verbal reasoning problems, gaze rightwards (Kinsbourne, 1972). This is also the result if the cortex is activated by electrical stimulation: leftward eye scans are produced by touching areas on the right hemisphere, while touching the corresponding areas on the left hemisphere produces rightward scans (Robinson & Fuchs, 1969). Injuries to both hemispheres impair scanning (Ogren, Mateer & Wyler, 1984): lesions to the right hemisphere disturb leftward eye movements, and left hemisphere lesions disturb rightward ones.

Thus, we would expect some relationship between leftward reading and the right hemisphere, and rightward reading and the left hemisphere. Two areas of research suggest that such a relationship exists.

Tachistoscopic presentation of pseudo-word Hebrew and English letter sequences (Tramer, Butler, & Mewhort, 1985) found that Hebrew letter sequences read leftward were perceived best by the right hemisphere, and English letter sequences read rightward were best perceived by the left hemisphere. This study, however, does not prove that eye scan in normal reading is associated with contralateral hemisphere perception. This has been more directly shown in a study using controlled perceptual masking of words on either side of the subject's fixation point during actual reading (Pollatsky, Bolozky, Well, & Rayner, 1981). Both eyes pass visual information equally to both hemispheres of the brain; for instance, although our right eye may be dominant, it does not send visual information just to the dominant left hemisphere but also to the right hemisphere. However, there is a lateralization of this information: the left half of our visual experience – the left visual field – is directly communicated to the right hemisphere, and what we see in our right visual field passes to the left hemisphere. Where these two fields join is a bilateral strip, which contains the fovea. Both this bilateral strip and the fovea send information directly to both sides of the brain. Words perceived to the left of the fixation point (fovea) project to the left hemisphere, while words to its right project to the right.

There is no apparent physiological reason why leftward and rightward writing should be better perceived to the left and right of the fixation point, respectively. For instance, a person reading leftward writing should at least in theory be free to fixate upon words so that they projected equally to the right and left hemispheres. Studies have been carried out that manipulate writing on a video monitor such that the reader can only perceive a limited number of letters to the left and right of the fixation point. In this way, the importance of letters on either side of the fixation point during actual reading can be determined. For normal reading of rightward script, an asymmetry is found, with letters to the right of the fixation point being more important for word recognition than letters to the left of it. This asymmetry is reversed for leftward writing. This suggests that reading direction, for some unknown psychophysical reason, affects which side of the brain words are projected to during reading, with the rightward scanning of words leading to their greater perception on the right visual field side of the fixation point, and vice versa for leftward scanning.

It is important to understand what this research implies. It does not prove that leftward writing is processed by the right hemisphere; merely that it is read within the left visual field, which is the field more closely connected to this hemisphere. It is quite possible for leftward writing to be perceived by the left visual field without entailing right hemisphere literacy, because what is seen in the left visual field is indirectly passed on to the left hemisphere through the corpus callosum. There is a further complication in that some of the information in the left visual field goes directly to the left hemisphere, via the connections of the fovea. It is reasonable to assume that perception of words by the left hemisphere would be more efficient if the information were transmitted directly, rather than exclusively via the right hemisphere and corpus callosum. I suggest that readers of modern Hebrew gain an important part of their access to writing through these direct connections of the left hemisphere from the fovea in the left visual field. If this is so, it could explain why the adoption of vowel representations into Hebrew writing was accomplished by representing them below the consonant letters, rather than in the same row as in Greek. A consequence of this is that Hebrew words are shorter than they would otherwise be if written like modern European phonetic writing. Such shortness is needed if words are to be read through that restricted part of the left visual field with direct connections to the left hemisphere.

The abilities of the left hemisphere to read leftward writing might appear to contradict my argument for right hemisphere literacy in the ancient world. What is suggested, however, is that early scripts favored the scan direction most efficient for right hemisphere processing of writing. In some cases, then lateralization changed, without a corresponding change in writing direction, to the hemisphere more efficient for perception of the writing concerned, as has been the case in modern Hebrew and Arabic.

Non-Right-Hemisphere Influences on Writing Direction

Writing direction is influenced by factors other than eye scan effects. For instance, it will be influenced by the ease with which the right-handed majority can write it, and also may be the product of influences which no longer exist but have been preserved through the inertia of tradition.

The former factor, that of the physically easiest direction in which to write, could well have influenced the direction of early scripts.

The Influence of Clay

Most people of ancient civilizations, like most people today, were right-handed (Coren & Porac, 1977). As the majority, any conventions that held an advantage for them would tend to be adopted into the culture at large. Right-handed individuals face specific problems when writing: for instance, they are more likely to smudge what they have just completed if they write upward or leftward, rather than downward or rightward. Such an influence on directionality would occur mainly when the commonly used writing materials could smudge. Some writing techniques, such as scratching letters on stone, slate, lead or way, would be very hard to smudge. Others such as using an ink pen or brush on paper, papyrus, or ostraca (pieces of pot or stone), would smudge with difficulty. Writing on damp clay with a stylus, however, such as is done in cuneiform, is very vulnerable. When using ink and pen, the ink dries before more than a few words have been written, but in cuneiform every word remains wet until the whole text is baked or placed in the sun to dry. This means that the line next to the one that is currently being written is still wet, which greatly increases the chances of accidental smudging. Because of this problem, those societies that used clay as a writing medium could be expected to try to reduce smudging by writing rightward.

As mentioned earlier, most ancient cultures employed leftward writing. One can thus distinguish two groups: those that produced leftward scripts on dry surfaces, such as hieroglyphics, hieratic, Etruscan, Phoenician, Proto-Canaanite, Hebrew, and Aramaic; and those that produced rightward scripts on clay, such as the Sumerian, Akkadian, Hittite cuneiforms, Ugarit, and the Greek Linear B. The former group is stylistically heterogenous, suggesting that the development of individual scripts was largely autonomous. In view of such independent evolution, their common leftward direction must have come about through some common intrinsic advantage for this orientation. There are, as far as I can see, no viable interpretations for leftward writing other than a role of the right hemisphere in literary processing. This is in contrast to the situation with rightward writing, which can be attributed either to its better perception by the left hemisphere, or to the practical aspect of its being the easiest form of writing for the right-handed majority.

The Influence of Cultural Inertia

By definition, the factor of cultural inertia could not have influenced the direction of writing used by early cultures. However, it is important in explaining why readers of Hebrew and Arabic read leftwards in the modern world. Based on the preceding arguments, one might conclude that all leftward scripts are evidence of right hemisphere processing. There is indeed an advantage to leftward scanning for right hemisphere reading. However, this does not rule out the possibility that the hemisphere used for reading could change without writing direction necessarily changing as well.

It is noted, that in those societies in which script direction changed, writing was not important in propagating religious beliefs. In other cultures it has taken on this role, largely through the belief that written religious works contain the "word of God." Since such writings are believed to be holy, it is central to the religion to propagate them unchanged – indeed, there may be injunctions to this effect. This could have the consequence of "freezing" the writing direction a society uses to that in which its religious works were originally written. My hypothesis that leftward writing relates to right hemisphere reading would seem to be immediately refuted by the leftward Hebrew and Arabic scripts, whose readers are known to use left hemispheric processes. However, both of these scripts have been used to propagate religious works. I suggest that the convention of leftwardness has been preserved due to the central importance of the Torah and the Koran in these societies, dating back to an earlier period when the leftward script used to write them reflected right hemispheric reading processes.

The Influence of the Left Hemisphere

The same arguments which suggest that right hemisphere literacy favored the development of leftward writing would apply to left hemisphere literacy favoring rightward writing. That is, there are advantages to the left hemisphere perceiving rightward script that correspond to those of the right hemisphere perceiving leftward script. The left hemisphere, when activated, makes eye scans to the right (Kinsbourne, 1972; Robinson & Fuchs, 1969), and gains a perceptual advantage during reading if upcoming letters appear in the right visual field (Pollatsky et al. 1981). Given that writing is flexible enough to change direction, an initially leftward script could become rightward if literacy changed from the right to the left hemisphere. Conversely, then, a change in writing direction, in the absence of any other factors to account for it, suggests a change in the hemisphere used for literacy.

There is an historical case where this occurred. Archaic Greek writing was originally written leftward, but by the fifth century B.C. (the Classical period of Greek civilization), had come to be written rightward. In between these periods, it was written in an intermediary form, *boustrophedon*, which entails writing both leftward and rightward with complete mirror reversal of individual letters on alternate lines. I propose that this indicates a bilateral literacy: the existence of dual right- and left-hemisphere literate individuals employing the most efficient system to read a single text. Such a situation would be predicted during the transitional period as literary processing shifted from the right hemisphere to the left. The modern European alphabets that subsequently developed from Greek retained the rightward direction, not just out of cultural inertia but because of the continued existence of left hemispheric literacy. Thus, this change within Greek society became the basis for modern lateralization.

Mirror Reversal of Letters and Hemisphere Reversal Among Ancient Alphabetic Readers

The arguments connecting right hemisphere literacy with leftward writing would gain in plausibility if such lateralization had produced other effects upon writing as well. Another such outcome, I suggest, was the mirror reversal of alphabetic letters. The letters of the Greek alphabet, a rightward script, are nearly identical but mirror-reversed from those of the Phoenician alphabet, the leftward script from which they developed. The previous arguments propose that the Greeks used a rightward script because of their left hemisphere processing, in contrast to earlier cultures who read leftward. Could the mirror reversal of Phoenician letters also be connected with this change in lateralization?

Each hemisphere can acquire a neurological representation of a motor skill or image through one of two means: either directly by learning it or indirectly through copying of a *representation* that the other hemisphere has already acquired. In the latter case, the hemisphere gains access to the other's learning through the corpus callosum, which connects them. There is some evidence on the nature of this indirect learning: one example concerns the transfer to the right hemisphere of the skill acquired by the left hemisphere in right-handed people for writing. Due to injury, right-handed adults often have to learn to write with their left hands. However, this hand is controlled by the right hemisphere, which has had no direct experience of writing. In consequence, the only representation it has of how to write is that which it has passively acquired from the left hemisphere through the corpus callosum. Some degree of right hemisphere writing ability is shown by the majority of people in the initial stages of learning to use their left hands. However, since this motor skill was acquired indirectly, it is not exactly like the skill of the left hemisphere. The left hand tends to produce mirror reversals: although such people are trying to write rightward, as they do with their right hand, they often produce leftward written letters, especially if not concentrating. According to one individual, it felt "natural to mirror-write and write right-to-left." (Schott, 1980).

Such mirror reversal of a learned representation when transferred from one hemisphere to the other is not confined to motor skills. Noble (1966) has shown that the corpus callosum mirror-reverses learned visual images. If the right hemisphere learns to recognize a certain pattern, this learning will be transferred through the corpus callosum. The left hemisphere will then be able to recognize the pattern too, but does so in mirror-reversed form. Studies (Harcum & Finkel, 1963; Bradshaw, Bradley, & Patterson, 1976) have shown that while the left hemisphere recognizes letters best when they are in their normal orientation, the right hemisphere recognizes mirror forms of letters better than the normal forms. The implication of this is that literacy acquired indirectly through the corpus callosum tends to be the mirror form of the active literacy in the other hemisphere. It is likely that this mirror-reversed literacy is inhibited in normal readers, but there is some evidence for its existence. Following certain injuries to the left hemisphere, individuals may develop, probably because of release from left hemispheric inhibition, the ability to read leftward, while losing that of reading rightward (Heilman, Howell, Valenstein, & Rothi, 1980).

This notion of hemisphere mirror reversal might explain the change in letter direction that accompanied the change in script direction. If right hemisphere literates used a leftward script with leftward orientated characters, the left hemisphere would passively acquire the ability to recognize these letters, but in mirror-image form: i.e., orientated to the right. When literacy changed to being left hemispheric, the left hemisphere would thus be most familiar with reading these letters in their reversed forms.

It is unlikely that there was a dramatic break from right to left hemisphere literacy. More probably, there was a period of transition during which some readers were right-hemisphere literate and others left-hemisphere literate. Alternatively, both sides of the brain may have possessed reading skills for some time, with the transition occurring gradually as the left hemisphere gained more importance. In the event of bilateral representation, there would have been a need for compromise between the left hemisphere favoring rightward script, and vice versa for the right hemisphere. This appears to have been what happened. In the sixth century B.C., Greek was written in boustrophedon. The first line of this script is rightward, the second is leftward, the third line is rightward again, and so on. Such a system of writing combines the advantages of leftward writing for the right hemisphere with those of rightward writing for the left hemisphere. Consistent with this, the letters of the leftward lines are mirror-reversed forms of those in the rightward lines. One can hypothesize that in reading boustrophedon, the left hemisphere read the rightward lines and the right hemisphere the leftward ones. Alternatively, ancient readers might have possessed a "main" reading hemisphere that processed all the information it received, and a "secondary" hemisphere that only processed the initial perceptual stages before communicating its information across to the main one. Thus, if the left side served as the main reading hemisphere, a individual would read rightward lines of boustrophedon normally, but for leftward ones, the right hemisphere would only process letters partially prior to passing this information through the corpus callosum to the left hemisphere to be completed.

Inverted Hand Position

There is a third line of evidence, independent of the two already outlined, for the existence of right hemisphere literacy amongst the ancients. The hand position which people use for writing is affected by the hemisphere used in reading, so examining records of this could give us information as to which hemisphere was dominant in the ancient world. People who write normally read with the hemisphere contralateral to their writing hand. However, individuals with a dominant hemisphere on the same side as their writing hand, e.g., demonstrating both left hemisphere representation of reading and left-handedness, use an inverted or "claw-writing" position (Levy & Reid, 1976; Herron, Galin, Johnstate, & Orstein, 1979). Normally, people hold the writing instrument between thumb and forefinger and use these to control its movement, with the writing end pointing away from them. Thus, the instrument as seen from above is below the line of writing. In the inverted position, the hand holds the writing instrument in a claw-like way, clasping it such that its writing end points toward the writer, and manipulating it by using wrist movements. Seen from above, the instrument is above the line of writing.

Ancient writers used their right hands (Coren & Porac, 1977); therefore it would be expected that if they used their right hemispheres in reading, they would use an inverted hand position or something closely resembling it when writing. This prediction must, however, be qualified. Present studies on inverted hand position in handwriting are based on people holding pens or pencils, but much of the writing in the ancient world was done with brushes. The "normal" hand position can control the application of pressure upon a brush, but this is more difficult with the wrist movements of the inverted position. Therefore, it is possible that observations of pen use would not apply to brush use, since a brush requires sensitive control not only of the direction in which it is moved, but also, unlike a pen, of the pressure exerted upon it.

Evidence as to the hand position of ancient writers comes from two sources: pictorial representations, and the marks that were produced in writing. Examples of the former show Assyrian and Egyptian scribes using pen and brush, as well as Assyrian scribes writing cuneiform. Such representations are naturally stylized; still, they appear to show users of ink-writing instruments (it is not known whether these were pens or brushes) using a normal hand position, and writers on clay (Driver, 1976) using something like an inverted hand position. These representations do not imply that ancient right- handed writers employing the normal hand position used their contralateral (left) hemispheres for reading. It is possible that the artists' stylizations did not accurately represent the actual position used; however, more likely, the writers using brushes found it difficult to use an inverted hand position and it is possible that the hand positions depicted relate to this.

Fortunately, there is more direct evidence of hand position than the stylized representations produced by artists: ancient writing itself. Both cuneiform and ink writing preserve clues as to the hand position used to write them. One analysis (Daniels, 1984) was carried out upon a script written in Aramaic, a consonantal alphabet from which modern Hebrew derives. Examination of letters written with a reed pen by a scribe named Haggai between 437 and 402 B.C., revealed them to be done in a way suggestive of the inverted hand position. Daniels' study is based up on examining very closely the edges of the pen strokes used to make up individual letters. Through careful analysis of these marks, he was able to conclude, that the pen was held with the convex side of its nib pointing toward the writer. This indicates that the writer positioned the pen slanting away from his body, which is characteristic of the inverted handwriting position – normally, people write with the convex side of the nib facing away from them and the pen slanted toward them.

The hand position used to write cuneiform has been more intensely studied. However, the only recent work is that of Powell (1981) which I shall briefly discuss. Cuneiform was written with a stylus cut from a reed. This stylus was then used to make variously orientated wedge-shaped impressions upon soft clay. Phonetic signs were constructed by arranging such wedges into recognizable patterns. From the shape of the wedge impressions, it is possible to infer the shape of the stylus' point as well as some information as to how it was held. Powell's investigations suggest that it was held in a kind of grip below the fingers, almost certainly to avoid the smudging of the damp clay that would otherwise occur when writing. The position he suggests is made by the thumb pressing the stylus against the tip of the middle finger, with the tips of the two fingers furthest from the thumb slightly curled around the stylus and the forefinger resting on top. Since cuneiform writers would have been expected to avoid smudging the surface of clay upon which they were working as much as possible, they would have avoided holding the stylus such that their fingers were below it, as is the case in both normal and inverted pen-writing positions. The question then is whether this grip is the equivalent of the inverted or the normal hand-position, taking into consideration its adaptions for writing on clay. There are several features which suggest that it is the equivalent of the inverted position.

First, from Powell's illustration it is clear that he interprets cuneiform writers as holding the bottom of their clay tablets to the left, with the top directed rightward. An important difference between normal and inverted hand positions is that right-handed people using the normal position hold their paper orientated with its top slightly toward the left and its bottom toward the right. In contrast, the inverted writer, like the cuneiform scribe, holds the writing medium with its top slightly rightward and its bottom directed leftward.

A second important distinction between normal and inverted hand position is whether the pen is held below or above the line of writing. Inverted hand-position is often defined as holding the pen above this line, and normal writing as holding it below it. The stylus in Powell's illustration is held in such a way that it is above the line of cuneiform being written. Thus, in this respect it more closely resembles an inverted hand position than a normal one.

Third, the cuneiform grip allows the stylus to be controlled only with wrist movements. It is believed that the inverted hand position derives from a restricted ability to control the fingers, thus preventing fine manipulation of the pen. This impairment is related to the control of the fingers residing in a different hemisphere from that which is reading what is being written (and presumably involved in some kind of visual coordination of word production): that is, the right-hand fingers are being controlled by the contralateral left hemisphere, while the reading hemisphere is on the right. This separation, although preventing a fine control of the pen through finger movements, allows a better wrist movement control: this movement is bilaterally controlled and so can be influenced by the hemisphere on the same side as the writing hand. Alternatively, it has been suggested that the crude movements of the wrist, unlike the fine ones of the fingers, can be coordinated between the hemispheres through the corpus callosum.

These three factors suggest similarities between the cuneiform grip and the inverted handwriting position. Whether this is actually the case may be revealed by further studies on the hand position phenomenon. It would be interesting to know how modern individuals would attempt cuneiform writing. Would inverted hand position users be at an advantage?

I have not attempted in this chapter to prove beyond a doubt that ancient readers were right-hemisphere literate in the period prior to the Classical Greeks. Rather, my aim has been to show that speculation on this issue is possible, and that the present evidence, insofar as it indicates a hemisphere for literacy, suggests the ancients were right- rather than left-hemisphere literate.

References

Aulmorn, E., & Harms, H. (1972). Visual perimetry. In, Tameson, D., & Hurrich, L.M. (Eds.), *Handbook of sensory physiology,* vol. VII/4. Berlin Heidelberg New York: Springer.

Bradshaw, J., Bradley, D., & Patterson, K. (1976). The perception and identification of mirror-reversed patterns. *Quarterly Journal of Experimental Psychology, 28,* 221–246.

Coltheart, M. (1983). The right hemisphere and disorders of reading. In, Young, A.G. (Ed.), *Functions of the right cerebral hemisphere.* London: Academic.

Cook, N. (1984). Homotopic callosal inhibition. *Brain and Language, 23,* 116–125.

Coren, S., & Porac, C. (1977). Fifty centuries of right-handedness. *Science, 198,* 631–632.

Daniels, P. (1984). A calligraphic approach to aramaic paleography. *Journal of Near Eastern Studies, 43,* 55–68.

Driver, G.R. (1976). *Semitic writing.* London: British Academy/Oxford University Press.

Girotti, F., Casazza, M., Musicco, M., & Avanzini, G. (1983). Oculomotor disorders in cortical lesions in man: the role of unilateral neglect. *Neuropsychologia, 21,* 543–553.

Gloning, I., Glonong, K., Haub, G., & Quatember, R. (1969). Comparison of verbal behaviour in right-handed and non-right handed patients with anatomically verified lesion of one hemisphere. *Cortex, 5,* 43–52.

Harcum, E.R., & Finkel, M.E. (1963). Explanation of Mishkin and Forgay's result as a direction reading conflict. *Canadian Journal of Psychology 17,* 224–235.

Harris, L.J. (1980). Left-handedness: early theories, facts and fancies. In, Herron, J. (Ed.), *Neuropsychology of left handedness.* New York: Academic.

Hatta, T. (1977). Recognition of Japanese *Kanji* in the left and right visual fields. *Neuropsychologia, 15,* 685–688.

Hatta, T. (1981). Differential processing of *Kanji* and *Kana* stimuli in Japanese people: some implications from Stroop-test results. *Neuropsychologia, 19,* 87–93.

Hatta, T., Honjoh, Y., & Mito, H. (1983). Event-related potentials and reaction times as measures of hemispheric differences for physical and semantic *Kanji* matching. *Cortex, 19,* 517–528.

Heilman, K.M., Howell, G., Valenstein, E., & Rothi, L. (1980). Mirror-reading and writing in association with right-left spatial disorientation. *Journal of Neurology, Neurosurgery, and Psychiatry, 43,* 774–780.

Herron, J., Galin, D., Johnstone, J., & Ornstein, R.E. (1979). Cerebral specialization, writing posture and motor control of writing in left-handers. *Science, 205,* 1285–1289.

Kinsbourne, M. (1972). Eye and head turning indicates cerebral lateralization. *Science, 176,* 539–541.

Larsen, B., Skinhøj, E., & Lassen, N. (1979). Cortical activity of left and right hemisphere provoked by reading and visual naming. A rCBF study. *Acta Neurologia Scandinavia* [Suppl.], *72,* 6–7.

Levy, J., (1982). Handwriting posture and cerebral organisation: how are they related? *Psychological Bulletin, 91,* 589–608.

Levy, J. & Reid, M. (1976). Variations in writing posture and cerebral organization. *Science, 194,* 337–339.

Noble, J. (1966). Mirror-images and the forebrain commissures of the monkey. *Nature, 211,* 1263–1265.
Ogren, M.P., Mateer, C.A., & Wyler, A.R. (1984). Alterations in visually related eye movements following left pulvinar damage in man. *Neuropsychologia, 22,* 187–196.
Patterson, K., & Besner, D. (1984a). Is the right hemisphere literate? *Cognitive Neuropsychology, 1,* 315–341.
Patterson, K., & Besner, D. (1984b). Reading from the left: a reply to Rabinowicz and Moscovitch and to Zaidel and Schweiger. *Cognitive Neuropsychology, 1,* 365–380.
Pollatsky, A., Bolozky, S., Well, A., & Rayner, K. (1981). Asymmetries in the perceptual span for Israeli readers. *Brain and Language, 14,* 174–180.
Powell, M.A. (1981). Three problems in the history of cuneiform writing: origins, direction of script, literacy. *Visible Language, 15,* 419–440.
Rabinowicz, B., & Moscovitch, M. (1984). Right hemisphere literacy. *Cognitive Neuropsychology, 1,* 343–350.
Robinson, D.A., & Fuchs, A.F. (1969). Eye movements evoked by stimulation of frontal eye fields. *Journal of Neurophysiology, 32,* 637–649.
Saffran, E.M., Bogyo, L.C., Schwartz, M.F., & Martin, C.S.M. (1980). Does deep dyslexia reflect right-hemispheric reading? In, Coltheart, M., Patterson, K., & Marshall, J. (Eds.), *Deep dyslexia.* London: Routledge and Kegan Paul.
Sasanuma, S. (1975). *Kana* and *Kanji* processing in Japanese aphasics. *Brain and Language, 2,* 369–383.
Schott, G.D. (1980). Mirror movements of the left arm following peripheral damage to the preferred right arm. *Journal of Neurology, Neurosurgery and Psychiatry, 43,* 768–773.
Shen, E. (1927). An analysis of eye movements in the reading of Chinese. *Journal of Experimental Psychology, 10,* 158–183.
Silverberg, R., Gordon, N.W., Pollack, S., & Bentin, S. (1980). Shift of visual field preference for Hebrew words in native speekers learning to read. *Brain and Language, 11,* 99–105.
Skoyles, J.R. (1984). Alphabet and the western mind. *Nature, 309,* 409–410.
Taylor, I., & Taylor, M.M. (1983). *The psychology of reading.* New York: Academic.
Tramer, O., Butler, B.E., & Mewhort, D.J.K. (1985). Evidence for scanning with unilateral visual presentation of letters. *Brain and Language, 25,* 1–18.
Van Essen, D.C., Newsome, W.T., & Maunsell, J.H.R. (1984). The visual field representation in striate cortex of the macaque monkey. *Vision Research, 24,* 429–448.
Zaidel, E. (1982). Reading by the disconnected right hemisphere: an aphasiological perspective. In, Zotterman, Y. (Ed.), *Dyslexia: Neuronal, Cognitive and Linguistic Aspects.* Oxford: Pergamon.
Zaidel, E., & Peters, A.M. (1981). Phonological encoding and ideographic reading by disconnected right hemisphere: two case studies. *Brain and Language, 14,* 205–234.
Zaidel, E. & Schweiger, A. (1984). On wrong hypotheses about the right hemisphere. *Cognitive Neuropsychology, 1,* 351–364.

CHAPTER 19

The Role of Vowels in Alphabetic Writing

BAUDOUIN JURDANT[1]

Introduction

Over the past few years a number of different writing systems have been the object of intensive investigation, and researchers from widely differing disciplines have begun to meet in order to share their thoughts and compare results. Philosophers, anthropologists, scholars of ancient history, psychologists, linguists, and neuroscientists are now combining their efforts in order to reach a better understanding of the role played by writing in the sociocultural construction of the modern world.

Among writing systems, the Greek alphabet has aroused particular interest. Its convenience is unequalled; the time required for learning it is minimal; its effectiveness is potentially universal. However, this alphabet is only slightly different from the consonantal alphabet previously developed by the Phoenicians, which the Greeks only borrowed. The difference between these two systems appears almost negligible, based on a few letters that were diverted from their initial consonantal function to allow the Greek writing system to include a new sensory dimension, already implemented in the human voice as vowels.

Most authors emphasize the graphic originality of the Greek system, and most also consider this system to have the mark of an alphabetic "authenticity" that cannot be attributed to the Phoenician consonantal system. However, few have sought to explain just why vowels are of such importance.

It is, however, obvious that the notation of vowels has a particular importance for the learning of reading and writing. At first glance, the learning process appears to be surprisingly simple, judging by the ease with which children gain access to alphabetic systems sometimes from a very early age (Cohen, 1982). On the other hand, the process is marked by difficulties that do not appear in learning other writing systems. These consequences of the notation of vowels have been demonstrated by numerous neurophysiological investigations on the cortical processing of language data.

This article proposes that the "alphabetic" particularities of cortical language processing are linked to the visual perceptual autonomy possessed

---

[1] Groupe d'Etudes et de Recherches sur les Sciences, Université Louis Pasteur, Strasbourg, 4, rue Blaise Pascal, F-67070 Strasbourg, France.

by vowels, which in turn reflects the importance of vowels during the earliest auditory experiences of the human being. The vowels of the written alphabet lead to bilateral cortical processing, because of hemispheric differences implied in processing by the distinct functional mechanisms in which vowels participate during reading.

In the first section below, we will examine the relationship between writing systems and the speech signals that they represent. Next, we will analyze the graphic status of the difference between consonants and vowels during the acquisition of literacy. We will then attempt to evaluate the auditory importance of vowels in terms of the early sensory experience they induce during the perinatal period, which will lead to a reconsideration of the impact of literacy on the functional lateralization of the cortex. This in turn will allow us better to understand the modalities of the specifically linguistic function of vowels during reading. We will then try to define the equivocal cortical status of vowels by discussing their processing modes in the right hemisphere, and will conclude with an examination of the cognitive consequences of the proposed hypotheses.

Writing Systems and Speech

Insofar as it was "perfected" by the Greeks, the alphabetic system differs from the earlier Phoenician system by the notation of vowels. This initiative of giving *an autonomous graphical status* to vowels was certainly not the result of a deliberate wish to innovate. As Février (1963) has stated, the Greeks simply made use of what was left over: certain consonants belonging to Semitic languages did not occur in Greek phonology and the signs thus made available were then used to note vowels. Unlike Semitic languages, where the internal vocalic dimension of the words was likely to vary depending on their grammatical function in a sentence, Greek possessed a lexicon linked to Indo-European roots, constructed around relatively stable vocalic nuclei (Lévеque 1964).

Many authors believe that the notation of vowels reinforced the faithfulness of writing to its oral model, speech. According to Havelock (1981), Greek writing succeeded in giving a more exact visual representation of speech than any preceding graphic system. Such an unequalled phonetic faithfulness meant an exhaustive and precise sensory adjustment between the graphic elements and the elementary sounds of the language in question. In this sense, a perfectly "faithful" writing system would be one capable of giving an exact and completely unambiguous reflection of the sounds of spoken language. But it is possible to consider this matter differently. For example, the best guarantee of the faithfulness of a written representation to its oral model could be the correlation between the segmentation of the graphic units offered to the reader's perception, and the segmentation that defines the structures of speech on either the acoustic or the conceptual level. Henceforth the problem is one of representational capacity. Graphic signs thus serve to represent the "entities" that already exist in speech, depending on its particular structural constraints.

According to this criterion of faithfulness, most non-alphabetic systems are extremely faithful to the speech they represent. Saussure (1972) has proposed that the linguistic sign is a double-sided psychological entity whose acoustic component (the signifier) echoes the syllabic structure of the language, and whose conceptual component (the signified) echoes semantic structures. Systems of notation can thus choose to represent one of these two components, on the basis of the segmentation (syllabic or conceptual) imposed by speech. Depending on which choice is made, writing systems are either syllabic, as are Indian Sanskrit, Japanese Katakana, and as was Linear B, which the Greeks used before they adopted the alphabet; ideographic, as is Chinese; or "lexicographic," as was the consonantal writing of the Phoenicians, which was adjusted to the lexical segmentation of their language. But each of these options obeys in its own way the principle that it is the structure of speech that defines how the graphic signs organize the visual representation of language.

How are the modalities of reading defined in these different systems? Tzeng, Hung, and Wang (1977) have shown that the reading of Chinese ideograms is facilitated by the mental restitution of a phonological support. This is explained by the conformity of the written structures to structures of conceptual representation, which in turn depend on semantic structures imposed by the Chinese language. Because of the indissociability of the acoustic and conceptual components of the linguistic sign, Chinese readers must provide for themselves a dimension that is missing from this visual representation but is necessary if the visual representation is to make sense. That is, in order to allow the *visual* structures to be considered in a linguistic framework, the reader must refer to the *acoustic* structures of the language which he or she in fact speaks.

From this point of view, the status of syllabaries is different, since their graphic elements exist as a visual reflection of the syllabic sounds of speech. Each graphic unit has a name which is the *sound* of a syllable of speech. Here the readers themselves give meaning to the text. Through this meaning, the graphic data allow the reader access to syllabic structures that give a representation to speech independent of that given to meaning. The faithfulness of the text thus depends directly on the strength of the link between the graphic elements and their exact sound interpretation, with the element binding these two modalities of linguistic representation together being the reader's previous knowledge of the possible meanings of the text being read. As Havelock (1981) well understood, a consequence of such a principle is an emphasis on the role of tradition in those societies that have opted for a syllabic system.

Learning To "See" the Vowels

The rise of the vowel to an autonomous graphic status – that is, to an existence apparently separate from the consonantal environment imposed by the syllabic function of speech – had significant consequences. Février (1963) considers that the Greek alphabet broke the unity of certain syllables. With the notation of vowels, the graphic system no longer obeyed the principle of syllabic segmentation that marks the oral acquisition of a language (Mehler, 1981).

It is of course during the acquisition of literacy that the graphic autonomy of the vowel makes its greatest impression; Liberman, Shankweiler, Liberman, Fowler, and Fisher (1976) observed that the need to distinguish vowels from consonants constitutes one of the major difficulties of learning alphabetic writing. This is not surprising. The graphic break-up of the syllable imposes a perceptual constraint on the reader that is not encountered on the acoustic level: one must learn to carry out consciously, on the basis of the visual impression of consonants and vowels, the syllabic synthesis that relates the written text to the structures organizing auditory perception of speech. This effort is not required in oral acquisition of speech: in speech, syllabic synthesis takes place spontaneously and without one being conscious of it.

In later stages, the active role of consciousness is no longer necessary; perceptual automatisms take over. But it is important to note that the capacity for literal decipherment will always remain available, because of its usefulness in situations that require one to read unfamiliar words (proper names, rare words, foreign words, etc.).

It might seem surprising that a categorical distinction along an *acoustic* dimension of linguistic data would establish a constraint on reading, whose internal integration requires *visual* training. However, the participation of visual systems in reading appears to be minimal: once graphic data reach the primary visual zone of the cortex (Brodmann's area 17), they apparently do not undergo further cortical processing that would lead to a more complete visual analysis, but are instead transmitted toward the auditory zones of the left hemisphere (Wernicke's area). This would explain why some readers have the impression of *hearing* what they are reading, or why patients suffering from certain reading disorders associated with alexia claim to have a subjective feeling of *deafness* (Lenneberg 1967), even though their peripheral system of auditory perception is functioning normally!

The syndrome of alexia is of particular interest for our discussion. This disorder is generally associated with lesions to certain auditory zones in the temporal region of the left hemisphere. Among the different forms of this syndrome, "literal alexia" refers to the inability to identify letters and words. Such patients attempt to guess words by means of an ideographic strategy that seeks to compensate for the absence of acoustic cues, an absence that is confirmed by their complete inability to read nonsense syllables or anagrams. The eyesight of alexics is in no way altered (Sugishita, Iwata, Toyokurma, Yoshioka, & Yamada, 1978). Their errors on letters and words thus seem to result from a breakdown of the auditory cortical processing of these data.

Interestingly, in the symptomatological test developed by Kremin (1976), the author indicates that the errors in the identification of letters always occurred either within the category of consonants or of vowels, but never resulted in a confusion between these two letter categories. This symptomatological particularity makes it clear that even in the case of lesions that seriously perturb the auditory cortical zones specialized for the processing of linguistic sounds, the ability to differentiate between consonants and vowels is preserved. One can conclude that there has been visual training of the distinction between these two categories of letters. Such training has been shown by the works of Morais, Cary, Alegria, and Bertelson (1979), in determining the acoustic sensitivity of literates – as opposed to illiterates – to the phonetic structures of language.

However, there is nothing in the actual form of consonants and vowels to account for this perceptual difference between them, and on the level of auditory acquisition it is impossible to find elements that reflect this underlying phonetic contrast. One can then legitimately ask whether this distinction is purely arbitrary. But despite the arbitrary nature of the distinction between consonants and vowels, it is apparently a necessary prerequisite for the acquisition of literacy. If this distinction does not have an auditory foundation and if its visual basis is not supported by any constraints on form, why is it necessary to learn it during the acquisition of literacy? Could it be that it results from certain peculiarities of cortical processing reserved for alphabetic writing?

If it is true that once the graphic data reach the primary visual area of the cortex they are redirected to the *auditory* cortex, it is possible to propose that it is only at this level that the distinction between consonants and vowels acquires an acoustic meaning. The problem is to discover the neurophysiological structures that underlie such a distinction, whatever the sensory modality.

Early Vocalic Differences

> A noir, E blanc, I rouge, U vert, O bleu: voyelles,
> Je dirai quelque jour vos naissances latentes . . .
>
> *Arthur Rimbaud, Voyelles*

By giving the vowel a perceptual status independent of its consonantal context, alphabetization called into question an important established fact of previous linguistic training: that is, the syllabic structuring of our perception of the sounds of speech. But one could ask: might this graphic autonomy of the vowel not be simply the echo of an early auditory sensitivity, which during the perinatal period attaches itself precisely and selectively to vowel sounds?

It is known that the auditory sensory experience of the fetus begins in the 5th month of gestation, in the course of the first stage of maturation of the auditory system (Lecours & Lhermitte, 1979). From the 6th month this experience is contrasted, with external auditory stimuli being distinguished from the internal background noise originating from the mother (Granier-Deferre, Lecanuet, Cohen, Querlen, Busnel, & Swean, 1984). Recordings carried out in utero show that, in terms of external stimuli, not only is the maternal voice distinctly perceptible, but its prosodic and vocalic components undergo the least amount of acoustic deformation on entering the uterus.

By the 9th month, the fetus reacts to stimuli that include vocalic differences (Granier-Deferre, personal communication). This discriminative ability remains for a few days after birth, and is still limited to vowels (Bertoncini, 1985). The selective nature of this sensitivity may be linked to the fact that the vocalic dimension is the principal foundation of the prosodic modulations to which the neonate pays particular attention in order to recognize the maternal voice (Mehler, 1985). These observations seem to confirm the notion that the vocalic components of the voice are able to trigger a sensory experience at a very early stage, independent of any reference to the linguistic context of the ultimate syllabic integration.

Of course this sensory experience ceases to be distinguished from other linguistic processes once the maturation of the central processes starts, at the beginning of the period of acquisition of language. The awakening of the baby's discriminatory sensibilities to consonantal differences follows its access to vowel differences very closely. It is only then that the perceptual conditions are assembled that allow one to syllabically structure the acquisition of language.

Between vowels and consonants, the first difference is that temporal gap which seems to separate the moments when the discriminatory sensibilities of the neonate come into play: first the vowels and then the consonants!

It is important to note the existence of a purely acoustic difference between these two categories of sounds. Spectral analyses show that vowels include a fairly low fundamental frequency and a harmonic structure of formants, of which the first two – F_1 and F_2 – are essential for the recognition of speech. For vowels, the value of the first formant varies between 250 Hz (i) and 750 Hz (a). The variations of the second formant are much clearer and range from 750 Hz (o) to 2200 Hz (e) or even 2500 Hz (i).

In contrast, the spectral distribution of consonants shows that they generally require two zones of spectral reinforcement: a first zone that is low and narrow and has an upper limit of 500 Hz, and a second zone, high and wide, that is situated between 1000 Hz and 1600 Hz (f), between 2000 Hz and 5000 Hz (n), or even between 70 Hz and 4500 Hz (l). As Versyp (1985) has stated, consonants, whose acoustic structure is more varied and complex than that of vowels, are characterized by rapid changes of formant frequency. These phonetic transitions are closely determined by the vocalic environment (Evans, 1982) and are important cues for the perception of individual consonants.

It seems that it is the second formant of vowels, F_2, that produces sufficiently clear contrasts to assure the specifically linguistic pertinence of vowel sounds and their localized cortical processing in the temporal region of the left hemisphere. This cortical activity begins very soon after birth (Molfese and Molfese 1979), and it is tempting to associate the beginnings of such hemispheric specialization with the requirements of syllabic structuring of speech. The aural acquisition of language would thus be based on a hemispheric preference for certain frequential zones that would facilitate the "acoustic welding" between consonants and vowels leading to the perceptual unity of the syllable.

It is, however, necessary to stipulate that this "acoustic welding" is carried out on the basis of an acoustic selection of certain harmonic components (F_1 but above all F_2) of the vowel. One can ask why selection of frequencies can occur: may be the constraint of extremely rapid – and unconscious – cortical processing of the units destined for syllabic synthesis. One of the major consequences of such a process is to inhibit cortical processing of vowels in any manner other than as imposed by language acquisition. What is involved here is the cortical processing of the sensory data associated with the melodic curve of the maternal voice.

Linguistic Sounds and Hemispheric Specialization

The long period of oral apprenticeship that follows the awakening of the consciousness to vocalic and consonantal differences is accompanied on the cortical level by the ever-increasing specialization of the left hemisphere for the neurophysiological processing of linguistic sounds. This functional lateralization corresponds to a fairly clear auditory superiority of the right ear for such sounds.

A large amount of research carried out in sometimes very diverse disciplinary contexts has yielded results that give a more nuanced picture to hemispheric specialization. Observations have brought two facts to light: the left hemisphere is not *only* linguistic, and the right hemisphere is *more* linguistic than had previously been believed!

Witelson (1983), for example, following a number of different studies leading to the same results (Schwartz & Tallal, 1980; Tallal & Newcombe, 1978) was able to show that the specialization of the left hemisphere seems to correspond to its ability to carry out a rapid analytico-temporal processing of the stimuli presented in the form of discrete units. It is this capacity that would be responsible for the left hemisphere's superiority in processing linguistic sounds.

For their part, Studdert-Kennedy and Shankweiler (1981), while confirming the advantage of the left hemisphere in the processing of consonants and syllables, showed that this superiority did not exist in the processing of sustained vowels: here, the right hemisphere performed even better. This capacity of the right hemisphere has been amply confirmed by different methods. For example, Molfese and Erwin (1981) arrived at a similar result with the aid of electrophysiological recordings of cortical activity in response to a variety of vocalic sounds.

According to Tsavaras, Kaprinis, and Gatzoyos (1981), the linguistic and cognitive capacities of the right hemisphere are not universally present. Illiterates, for example, show a clearly superior degree of auditory sensitivity for the right ear as compared to the left for all sounds, whether linguistic or not. This phenomenon is accompanied by the apparent absence of a specifically right hemispheric cognitive strategy. The authors suggest that it is the process of learning to read and write that triggers the increased linguistic and cognitive integration of the right hemisphere.

However, earlier research (Damasio, Castro-Caldos, Grosso & Terro, 1976) had led to the conclusion, on the basis of the clinical observation of 247 cases of lesional aphasia, that there was *no* effect of literacy on hemispheric lateralization. Such an effect had been proposed to explain the fact that the illiterate subjects (and, curiously, women also) who had suffered severe left hemispheric lesions had a greater facility than literate subjects (or men) in inducing the shift of linguistic functions from the left hemisphere to the right. This hemisphere seemed to be better prepared to take over from the left for both these categories of aphasics. Damasio's observations led to the proposal that there was either a lesser degree of lateralization among illiterates and women, or a greater degree of latent linguistic capacity in their right hemispheres. This finding stands in contradiction to the later research by Tsavaras.

Without seeking to call into question the validity of these clinical observations, one can certainly qualify their interpretation. It could be proposed, for example, that if illiterates and women are able to regain their faculty of language more quickly and more easily thanks to intervention of the right hemisphere, it might be because this hemisphere is more immediately available to function as a backup. This greater availability could result from the right hemisphere's relative absence of specialization, which in no way prejudices the degree of specialization that the left hemisphere shows *before* the lesion and the aphasia. In other words, the results of Tsavaras are perfectly compatible with those of Damasio. Are they also compatible with those of Tsunoda (1985)?

Tsunoda (1985) has worked for more than 20 years on the differences in hemispheric specialization between the Japanese and the Western brain, and claims to have discovered a clear auditory superiority of the right ear/left hemisphere in Japanese subjects. This superiority is not limited to the processing of linguistic sounds, including vowels, but extends to animal cries, insect buzzes, traditional Japanese music, emotional sounds, etc. Only mechanical noises (e.g., helicopters), Western instrumental music, and a variety of background noises are processed by the right brain of Japanese subjects. In 1975, Shimizu published identical results showing the greater sensitivity of the right ear of Japanese subjects to vowel sounds. The Western brain on the other hand, according to Tsunoda and other authors, shows a specialization of the left hemisphere for syllables and consonants, but most other sounds (mechanical noises, Western instrumental or Japanese traditional music, sounds of nature, sounds associated with emotions, and sustained vowels) are processed on the right. This auditory characteristic of the Western brain has been confirmed by clinical observations made by Assal and Aubert (1979), who discovered a right hemisphere auditory superiority for animal cries.

Since Tsunoda's tests reveal the same hemispheric specializations for subjects who have been blind from birth, the author has hastened to exclude the influence of writing as an explanation for this neurocultural contrast. His interpretation is based on certain particularities of the Japanese language, especially the important role that the vocalic dimension plays in it.

But what *is* the linguistic status of vowels in Japanese? Is this status not necessarily syllabic? Even in the case of utterances that are entirely syllabic, how could the vowels escape the syllabic role that linguistic functioning imposes upon them? The clear superiority of the right ear/left hemisphere that the Japanese show for almost all sounds – linguistic or not – must thus be related to the absence of autonomy of their vowels. One could see two complementary reasons for this: on the one hand, as Tsunoda strongly emphasizes, the phonological structures of the Japanese language are heavily marked by a clear dominance of the vocalic dimension. The fundamental importance of their linguistic role forces vowels to remain under the exclusive control of the syllabic structures of the language. Hécaen (1976) proposed a similar interpretation for the results obtained by Shimizu (1975).

On the other hand, the graphic representation of Japanese simultaneously obeys two principles; a syllabic principle for the Katakana and Hiragana scripts, and an ideographic principle for Kanji script. The advantage of this mixture is to effectively involve both cerebral hemispheres in reading and writing: a greater involvement of the left hemisphere for syllabic graphisms and a greater involvement of the right hemisphere for ideograms. But neither of these systems calls graphically into question the syllabic integration of the vowel.

What can we conclude from these various works on the role of linguistic sounds in hemispheric specialization? Some authors have claimed that literacy leads to an accentuation of cortical asymmetry; others, like Tsavaras, claim that it is the opposite that occurs and that reading and writing open new cognitive possibilities for illiterate subjects by mobilizing their "right brain"! If this were the case, then why would the Japanese, who have not just one but three writing systems, show a strong cerebral asymmetry analogous to the one that Tsavaras claims to have discovered in illiterates?

The Role of Vowels in Reading

When I read, I hear the words.

Einstein

In order better to understand the ultimate impact of literacy on hemispheric lateralization, it seems indispensable to consider alphabetic writing at the level of the vowels. This is what Adams (1981) has tried to do with regard to the perceptual problem posed by graphism. As he observes, the ability of the visual system to process, in terms of spatial position, the literal data of graphism, is in fact both "crude and sluggish". Basing his comments on the research carried out by Estes (1972), Adams underlines the fact that at the edges of the fovea, and further and further away from it, the receptive pathways are less dense, thereby leading to a significant degree of "positional uncertainty" in the perception of letters. This accounts for the perceptual strategy employed by readers in using the probabilities of the sequential assembly of letters (Mewhort and Beal, 1977).

What is the role of vowels in such a perceptual strategy? On their own, vowels represent a considerable degree of graphic redundancy, since roughly 40% of graphic signs serve to note the five principal vowel sounds of English in the majority of texts. In other words, vowels provide very little precise information about the probabilities of sequential assembly. It is also the vowels which, in English, are involved in most of the orthographic irregularities, which explains why they are most often responsible for reading errors during acquisition (Liberman et al., 1974). Thus, the redundancy of vowels is accompanied by considerable phonological imprecision. But if vowels are not irreplaceable for the phonemes that they represent, what is their purpose?

The answer suggested by Adams (1981) is that "the primary function of the vowels within our writing system is orthogonal to their phonological significance. Their primary function is that of preserving the syllable as a perceptual unit and as such derives directly from the redundancy they carry" (p. 211). Vowels allow each syllable to possess a vocalic "center" or "nucleus," which serves the sequential ordering required by the temporal linearity of speech. Vowels can thus be viewed as the perceptual tools that allow the reader to proceed to a syllabic segmentation of the graphic data.

If it is true that vowels play this role in the perception of graphic data, it is necessary to consider that their status as letters is very different from that of consonants. Consonants seem to be the foundation for sensorimotor connections between the visual perception of the graphic data and the latent articulation that underlies – and facilitates – each reading. Vowels, on the other hand, seem to be designed for a role of syllabic segmentation and spatiotemporal orientation.

Alphabetic graphism is thus at the origin of the role differentiation of consonants and vowels. As these roles are critical for structuring the visual perception of the graphic data, one can better understand why and how the difference between consonants and vowels becomes the object of categorical training during the acquisition of literacy. The consonants provide a visual representation of the articulatory movements at the origin of the vocalization of the written text, whereas the vowels serve as cues for the rapid identification of syllabic units, allowing the transformation of their order in space into a temporal order that reflects the linearity of speech.

This function of vowels associates them closely with the syllabic – and therefore linguistic – processing of graphic data. But it is not because of their "phonological value" that vowels are thus solicited by the reader. One can suggest that the specifically linguistic processing of graphic data, in the current state of our knowledge of the functions of the left hemisphere, "neutralizes" all reference to the acoustic sensory dimensions of vowels.

It is thus probable that because of their perceptual role in the temporal ordering of the linear succession of syllables, the vowels of Greek alphabetic writing confirmed and even accentuated the specialization of the left hemisphere for the processing of linguistic data (De Kerckhove, 1984). It is also probable that this system of writing had, and still has, specific effects on the cortical conditions of linguistic processing. For if it is true, as Harris (1986) has suggested based on observations of owls, that vision is a more dominant sensory modality than hearing, then the acquisition of the alphabet requires the reader to adjust any previous auditory knowledge of linguistic structures to the perceptual automatisms that are imposed by visual graphic data. Bertelson (1972) has shown the direct influence exerted by the orientation of writing – from left to right, as is the case for most Western scripts, or from right to left, as in Hebrew writing – on the mechanisms of auditory temporal identification. Moreover, it is within this same domain that Morais et al. (1979) also discovered the effects of literacy on awakening an auditory sensitivity to the phonetic structures of language.

It is well known that writing mobilizes certain cortical zones of the left hemisphere. Gillis and Sidlauskas (1978) succeeded in increasing the reading speed of dyslexic children by raising the intensity of auditory feedback of vocalization to the right ear. The effect disappeared when feedback was systematically oriented toward both ears, which shows the importance of the left hemisphere in the cortical processing of writing.

Presented in this way, the role of vowels in writing appears to be exclusively linguistic i.e., not sensory, and completely under the control of the left hemisphere. Unfortunately, this interpretation does not account for a large number of results that show right hemisphere cortical activity in response to vocalic stimulation.

The Ambiguous Nature of the Cortical Status of Vowels

In their recent review of the research on hemispheric specialization, Versace and Tiberghien (1985) reconsider certain hypotheses, previously presented by Semmes (1968), suggesting that "the two hemispheres would be differentiated by their representation of elementary functions, sensory and motor: a precise, focalized representation in the left hemisphere, and a diffuse representation in the right hemisphere" (p. 268).

According to the authors, this suggestion confers particular interest upon the research done by Sergent (1982) in the domain of differences of hemispheric sensitivity to spatial frequencies in the visual spectrum. The hypothesis, they write:

> could be used to explain hemispheric sensitivity to different spatial and auditory frequencies. As a matter of fact, a precise projection of sensory pathways in the left hemisphere could explain this hemisphere's ability to process high frequencies, auditory or visual. In the same way, a diffuse projection would explain the superiority of the right hemisphere for the processing of low frequencies. (Versace & Tiberghien, 1985, p. 268)

This frequency-dependent characterization of hemispheric specialization sheds an unexpected new light on the problem raised by the cortical processing of vowels. But in order to understand this new approach, it is necessary to return to the beginning of our reflection on alphabetic writing.

The originality of alphabetic writing lays in the graphic autonomy that the system accorded to vowels. During the acquisition of reading, vowels gain access to a perceptual existence separate from the syllabic context that controls their linguistic function. This "visual resurrection" of the sensory autonomy of vowels should be considered as an only temporary stage (putting into place provisional mechanisms), leading to the syllabic integration of vowels according to perceptual constraints imposed by visual access to linguistic data. It is during this stage that one must learn to dissociate vowels and consonants, so that each category can play its specific perceptual role in reading; with the understanding, however, that this dissociation calls into question neither previously acquired linguistic knowledge, nor the cortical modalities of the processing of such data in the left hemisphere. These elements require the direct participation of the right hemisphere using the learning process. The isolated vowel, which was not previously perceived, now appears to the learner isolated from any context.

Such a situation in all likelihood leads to the reactivation of pathways originally associated with the maternal voice, which both in utero and for a brief postnatal period depend mainly on vocalic sounds, the primary prosodic component of this voice recognizable among all others. The acoustic characterization of these sounds cannot be a result of their linguistic status, and as Versyp (1985) has shown, it is in the fundamental frequency and the first harmonic (F_0 and F_1) that the vocalic dimension of the voice is best preserved. In other words, if the suggestions of Versace and Tiberghien are correct, the fact that vowels range in the low frequencies would lead to their spontaneous orientation to the auditory zones of the right hemisphere.

The sensory autonomy of the vowels that the alphabetic system induces on the visual level is thus at the origin of the reactivation of this early auditory sensitivity. The possibility of this reactivation is all the more probable given the research carried out by Kuhl and Meltzoff (1982), who showed that 4- to 5-month-old infants were able to associate the auditory perception of verbal stimuli (in this case vowels) with the visual perception of a face that produced the perceived sounds. They did not observe a left or right hemisphere preference for the perception of the face producing the vowel sounds, but this absence of lateralization is not surprising. At this age, neither vocalic sounds nor facial recognition evoke clear behavioral responses that would indicate marked hemispheric specialization. The initial stages of right hemisphere specialization for facial recognition associated with vocalic and prosodic components of the voice cannot be excluded if one takes into account the proposals set forth by Masland (1968), who claims that the establishment of cross-model sensory connections reinforces hemispheric asymmetry. This initial specialization would then be interrupted during the oral acquisition of language.

It is in fact at this point that the early sensitivity of the neonate to vowels would be *inhibited*, in order to accentuate the left hemisphere's specialization for syllabic processing of speech. The long period of language acquisition would thus be responsible for the disappearance of vowels as the source of sensory experience separate from linguistic function. But this "disappearance" should not be thought of as absolute. The right hemisphere's initial cross-modal sensory processing associated with the impact of the maternal voice and face has certainly left a tangible effect in the neonate. Research carried out by Moscovitch (1976) argues in favor of this hypothesis: the right hemisphere is not without linguistic abilities, but is unable to use them because of the control exerted by the left hemisphere over this function. Alphabetic writing allows vowels to partially escape from this control and to reactivate certain associative pathways of the right cortex.

This phenomenon of reactivation explains the results obtained by Baker, Smink, and Reitsma (1973) concerning the emergence of cortical ambilaterality in young readers. This ambilaterality would then give way to the traditional right ear advantage for all linguistic data. Similar results were obtained by Sadick and Ginsburg (1978), and led Davidoff, Done and Scully (1981) to criticize the way in which Bakker et al. and Sadick and Ginsburg interpreted this temporary ambilaterality as one of the elements that characterizes the cortical strategy of "good readers." In any case, such ambilaterality argues in favor of the hypothesis of the reactivation of cortical pathways initially opened by the diffuse impact of low-frequency vowels in the right hemisphere. This would certainly explain why vowels, once they are isolated, are so comfortably handled by this hemisphere, as has been shown by a wide variety of research. Molfese and Erwin (1981) conclude their observation of electrophysiological recordings of cortical response to computer-synthesized vowels as follows:

Overall these findings suggest that vowels are not processed by specific mechanisms which are restricted to only one hemisphere. This is in agreement with most previous findings. No single scalp recording region was found to differentiate between vowel sounds either within or between hemispheres. Rather, many discrete regions over the temporal and parietal areas of both hemispheres reflect various differences between vowel sounds.

It is even possible that this cortical reactivation of certain pathways (initially opened by the diffuse impact on the right hemisphere of maternal vowels), triggered by the onset of literacy, is not simply a temporary phenomenon. It is well known that fluent reading implies perceptual strategies that are precisely focused on certain points of the text, and also more global and diffuse strategies involving the more or less wide zones around the fovea. Polich (1978) has shown that the right hemisphere is superior to the left for the visual perception of verbal material under certain conditions of presentation (specifically, when the angle of presentation to the fovea is increased, or when stimulus quality is mediocre!). Other research has already shown this particular superiority of the right hemisphere. As Hécaen (1976) reminds us:

In a number of experiments, the right hemisphere was shown to be superior for the processing of verbal material under certain conditions: judgement of the similarity of the physical form of letters (Cohen, 1982) or of their orientation, matching a word to a previously presented word, etc.

In other words, and as Neville (1985b) has shown with electrophysiological recordings, the right hemisphere "sees" the letters of the graphic alphabet, even if their specifically linguistic interpretation can only be carried out under the exclusive control of the left hemisphere.

These observations lead us to conclude that alphabetic writing provided the vowel with a bilateral cortical status: on the one hand, as an indispensable perceptual mark for left hemisphere pinpointing and recognition of the syllabic units of the text; on the other hand, as an autonomous perceptual unit dissociated from the syllabic context of linguistic function, and processed in the right hemisphere where they would reactivate cortical pathways originally traced by the primitive sensory contact of the fetus and neonate with the maternal voice.

A Neurocultural Indetermination

The graphic autonomy accorded to the vowel by the alphabetic system thus reactivates a specific cortical sensitivity of the right hemisphere. The consequences of such a process are all the more difficult to imagine, since the right hemisphere is unable to express verbally what it is "seeing."

Gazzaniga, Ledoux, and Wilson (1977) have proposed interesting hypotheses on the specific role of the right hemisphere in development of the mechanisms at the foundation of consciousness. These hypotheses were formulated on the basis of responses of a commissurotomized patient to certain tests designed to examine the role of each hemisphere in various linguistic tasks. The authors were able to show that the right hemisphere, even though it does not have the capacity for verbal expression, is able to respond appropriately to written requests. Faced with reactions whose origin it does not know, the left hemisphere rationalizes the behaviors thus induced by attributing invented "causes" to these behaviors:

When 'laugh', for example, was flashed to the right hemisphere, the subject commenced laughing, and when asked why, said, 'Oh, you guys are too much'. When the command 'rub' was flashed, the subject rubbed the back of his head with the left hand. When asked what the command was, he said 'itch' . . . (Gazzaniga et al., 1977, p. 1146).

The authors conclude their presentation as follows:

This process of attribution by the verbal system seems to be a major mechanism of consciousness. The verbal system is not always privy to the origins of our actions. It attributes cause to behavior as if it knew, but, in fact, it doesn't. One's belief system could arise as a consequence of this attribution process. We may build our sense of reality by considering what we do. It is as if self-consciousness involves verbal consideration of our actual sensorimotor activities. (Gazzaniga et al. p. 1147)

This concept, when applied to our hypothesis of a nonlinguistic "vision" of vowels in the right hemisphere, leads us to strange neurocultural speculations. Whatever the sensorimotor modality of the reaction of the right hemisphere to the stimuli triggered by the vocalic dimension of writing, this can only lead us to an enigma. Vowels in themselves do not have meaning. Their concrete reality is dependent upon a dimension of the human voice, below the level of syllabic segmentation which associates them irreversibly to the structures of speech. The "vision" of vowels, to which the right hemisphere has access because of the alphabet, would translate into an incomprehensible and diffuse request, carrying an infinite number of possible sensory and behavioral interpretations.

The alphabetic system is thus at the origin of a very peculiar cortical effect: the vowel has become the foundation of an indetermination on both the cognitive and cultural levels, which affects the left hemisphere all the more since it is the graphic breakup of the syllabic unit that produces the effect. This origin calls into question the principle according to which the syllabic structuring of speech is to be carried out in the left hemisphere. The left hemisphere is forcibly required to comment on the sensorimotor effect triggered by the vocalic data perceived by the right hemisphere.

According to the suggestions made by Gazzaniga et al. (1977), one finds oneself in the presence of the very mechanisms that presided over the emergence of consciousness as produced by a quasi-permanent commentary by the left hemisphere on our sensorimotor reactions to vocalic graphic data. This hypothesis would explain why the Greek version of the alphabet led to completely new writing practices, which deeply modified the cultural space of the Mediterranean world.

As Finley (1983) has observed, this writing system, which was initially used by the Greek bards and rhapsodists for the transcription of stories in the oral tradition, very quickly gave birth to new texts that were not written in the Homeric or Hesiodic epic style. These were the texts of *authors*, designed for the poetic evocation of intimate emotions and personal feelings. Instead of being used for the recording of important events or solemn declarations, writing was used for the individualized expressions of the *internal* psychic life of the poet. Such texts spotlighted a new dimension of consciousness: a *private* dimension.

Acknowledgement. The author would like to thank C. Granier-Deferre for her judicious remarks and counsel. The author retains sole responsibility for the imperfections and errors that may have escaped his notice.

References

Adams, M.J. (1981). What good is orthographic redundancy? In, Tzeng, O., & Singer H. (Eds.) *Perception of print: reading research in experimental psychology*, pp. 197–221. Hillsdale: Erlbaum.

Assal, G., & Aubert, C. (1979). La reconnaissance des onomatopées et des cris d'animaux lors de lésions focalisées du cortex cérébral. *Revue Neurologique, 135*, 65–73.

Assal, G., Aubert, C., & Buttet, J. (1981). Asymétrie cérébrale et reconnaissance de la voix. *Revue Neurologique, 137*, 4, 255–268.

Baker, E., Blumstein, S.E., & Goodglass, H. (1981). Interaction between phonological and semantic factors in auditory comprehension. *Neuropsychologia, 19*, 1–15.

Bakker, D.J., Smink, T., & Reitsma, P. (1973). Ear dominance and reading ability. *Cortex, 9*, 301–319.

Bakker, D.J., Hoefkens, M., & Van Der Vlugt, H. (1979). Hemispheric specialization in children as reflected in the longitudinal development of ear symmetry. *Cortex, 15*, 619–625.

Baty, C.W., & McConnel, J.K. (1976). Two sides of the brain in language and art. *Educational Research, 18*, (3), 201–207.

Benowitz, L.I., Bear, D.M., Rosenthal, R., Mesulam, M.M., Zaidel, E., & Sperry, R.W. (1983). Hemispheric specialization in nonverbal communication. *Cortex, 19*, 5–11.

Bentin, S., & Carmon, A. (1983). The relative involvement of the left and right cerebral hemispheres in sequential vs. holistic reading: electrophysiological evidence. Paper presented at the *Annual BABBLE* Meeting. Niagara Falls, Canada, March 1983.

Berlin, C.I., & McNeil, M.R. (1976). Dichotic listening. In, *Contemporary issues in experimental phonetics*, pp. 327–387. New York: Academic.

Bertelson, P. (1972). Listening from left to right versus right to left. *Perception, 1*, 161–165.

Bertoncini, J. (1984). L'équipement initial pour la perception de la parole. In, Moscato, M., & Pieraut-Le Bonniec, G. (Eds.), *Le langage. Construction et actualisation*, pp. 39–51. Rouen: Publications de l'Université de Rouen.

Bertoncini, J. (1985). Infants sensitivity to onset spectra of CV syllables. In, Mehler & Fox (Eds.) *Neonate Cognition*. Hillsdale: Erlbaum.

Blanc, M. (1976). L'idéographie cérébrale. *La Recherche, 7*, (71), 878–881.

Bradshaw, J.L., & Nettleton, N.C. (1981). The nature of hemispheric specialization in man. *The Behavioral and Brain Sciences 4*, 51–91.

Bryden, M.P., & Sprott, D.A. (1981). Statistical determination of degree of laterality. *Neuropsychologia, 19,* (4), 571–581.

Carmon, A. (1981). Temporal processing and the left hemisphere. *The Behavioral and Brain Sciences, 4,* 183–228.

Changeux, J.P. (1983). *L'homme neuronal.* Paris: Fayard.

Clifton, R.K., Morrongiello, B.A., & Dowd, J.M. (1984). A developmental look at an auditory illusion: the precedence effect. *Developmental Psychobiology, 17,* (5), 519–536.

Cohen, R. (1982). Plaidoyer pour les apprentissages précoces. Paris: Presses Universitaires de France.

Coltheart, M. (1980). Deep dyslexia: a right-hemisphere hypothesis. In, Coltheart, M., Patterson, K., & Marshall, J.C. (Eds.) *Deep dyslexia,* pp. 326–380. London: Routledge Kegan Paul.

Cutting, J.E. (1974). Two left hemisphere mechanisms in speech perception. *Perception and Psychophysics, 16,* 601–612.

Damasio, A.R., Castro-Caldas, A., Grosso, J.T., & Ferro, J.M. (1976). Brain specialization for language does not depend on literacy. *Archives of Neurology, 33,* 300–301.

Davidoff, J.B., Done, J., & Scully, J. (1981). What does the lateral ear advantage relate to? *Brain and Language, 12,* 332–346.

De Kerckhove, D. (1984). Effets cognitifs de l'alphabet. In, de Kerckhove, D., & Tutras, D. (Eds.) *Pour comprendre 1984,* pp. 112–129. Ottawa: UNESCO, occasional pages N49.

Deutsch, D. (1978). Pitch memory: an advantage for the left-handed. *Science, 199,* (4328) 559–560.

Eimas, P.D. (1985). Some constraints on a model of infant speech perception. In, Mehler, J., & Fox, R. (Eds.), *Neonate Cognition. Beyond the Blooming Buzzing Confusion.* Hillsdale: Erlbaum.

Estes, W.K. (1972). Interactions of signal and background variables in visual processing. *Perception and Psychophysics, 12,* 278–286.

Evans, E.F. (1982). Functions of the auditory system. In, Barlow, H.B., & Mollon, J.D. (Eds.), *The senses,*pp. 307–332. Cambridge: Cambridge University Press.

Fevrier, J. (1963). Les sémites et l'alphabet. In Centre International de Synthèse (Ed.) *L'écriture et la psychologie des peuples,* pp. 117–129. Paris: Colin.

Friedman, R.B. (1980). Identity without form: abstract representations of letters. *Perception and Psychophysics, 28,* (1), 53–60.

Funnell, E. (1983). Phonological process in reading: new evidence from acquired dyslexia. *British Journal of Psychology, 74,* 159–180.

Finley, M. (1983). *Le monde d'Ulysse.* Paris: Maspéro.

Gazzaniga, M.S., & Smylie, C.S. (1984). Dissociation of language and cognition. *Brain, 107,* 145–153.

Gazzaniga, M.S., LeDoux, J.E., & Wilson, D.H. (1977). Language, praxis, and the right hemisphere: clues to some mechanisms of consciousness. *Neurology, 27,* 1144–1147.

Gazzaniga, M.S., Smylie, C.S., Baynes, K., Hirst, W., & McCleary, C. (1984). Profiles of right hemisphere language and speech following brain bisection. *Brain and Language, 22,* 206–220.

Geschwind, N., & Levitsky, W. (1968). Human brain: left–right asymmetries in temporal speech region. *Science, 161,* 186–187.

Gillis, J.S., & Sidlauskas, A.E. (1978). The influence of differential auditory feedback upon the reading of dyslexic children. *Neuropsychologia, 16,* 483–489.

Granier-Deferre, C., Lecanuet, J.-P., Cohen, H., Querleu, D., Busnel, M.-C., & Sureau, C. (1984). Le foetus entend. *Les Dossiers de l'Obstétrique, n° 107,* 27–31.

Havelock, E.A. (1982). *The literate revolution in Greece and its cultural consequences.* Princeton: Princeton University Press.

Hardyck, C., Tzeng, O.J.L., & Wang, W.S.-Y. (1978). Cerebral lateralization of functionand bilingual decision processes: is thinking lateralized? *Brain and Language, 5,* 56–71.

Harris, W.A. (1986). Learned topography: the eye instructs the ear. *TINS,* March, 97–99.

Hécaen, H. (1976). La contribution de l'hémisphere droit aux fonctions du langage. *Lyon Médical, 236,* 19, 699–715.

Hynd, G. W., Teeter, A., & Stewart, J. (1980). Acculturation and the lateralization of speech in the bilingual native American. *International Journal of Neuroscience, 11,* 1–7.

Jaynes, J. (1982). *The origin of consciousness in the breakdown of the bicameral mind.* Toronto: Toronto University Press.

Joanette, Y. (1985). Hémisphere droit et langage: la quatrième dimension. *Annales de la Fondation Fyssen, 1,* 25–32.

Jurdant, B. (1984a). Ecriture alphabétique et stratégies cognitives. *Cahiers STS, 5,* 92–102.

Jurdant, B. (1984b). *Ecriture, monnaie et connaissance,* Thèse de Doctorat d'Etat. Strasbourg: Université Louis Pasteur.

Jusczyk, P. W., Pisoni, D. B., Reed, M. A., Fernald, A., & Myers, M. (1983) Infants' discrimination of the duration of a rapid spectrum change in nonspeech signals. *Science, 222,* 1275–1276.

Katz, A. N. (1986). Meaning conveyed by vowels: some reanalyses of word norm data. *Bulletin of the Psychonomic Society, 24-1,* 15–17.

Kellar, L. A. (1978). Words on the right sound louder than words on the left in free field listening. *Neuropsychologia, 16,* 221–223.

Kelly, R. R., & Tomlinson-Keasey, C. (1981). The effect of auditory input on cerebral laterality. *Brain and Language, 13,* 67–77.

Kolers, P. A. (1983). Polarization of reading performances. In, Coulmas & Ehlich (Eds.), *Writing in focus,* pp. 371–391. Berlin: Mouton.

Kremin, H.(1976). Les problèmes de l'alexie pure. *Langages, 44,* 82–124.

Kuhl, P. K., & Meltzoff, A. N. (1982). The bimodal perception of speech in infancy. *Science, 218,* 1138–1141.

Lafont, R. (Ed.) (1984). *Anthropologie de l'écriture.* Paris: Centre Georges Pompidou.

Lackner, J. R., & Teuber, H. L. (1973). Alterations in auditory fusion thresholds after cerebral injury in man. *Neuropsychologia, 11,* 409–416.

Lazorthes, G. (1982). *Le cerveau et l'esprit.* Paris: Flammarion.

Lecours, A. R., & Joanette, Y. (1985). Keeping your brain in mind. In, Mehler & Fox (Eds.). *Neonate cognition,* pp. 327–348. Hillsdale: Erlbaum.

Lecours, A. R., & Lhermitte, F. (1979). *L'aphasie.* Paris: Flammarion.

Lenneberg, E. H. (1967). *The biological foundations of language.* New York: Wiley.

Leveque, P. (1964). *L'aventure grecque.* Paris: Colin.

Levy, J., & Trevarthen, C. (1977). Perceptual, semantic and phonetic aspects of elementary language processes in split-brain patients. *Brain, 100,* 105–118.

Liberman, I. Y., Shankweiler, D., Fischer, F. W., & Carter, B. (1974). Explicit syllable and phoneme segmentation in the young child. *Journal of Experimental Child Psychology, 18,* 201–212.

Liberman, I. Y., Shankweiler, D., Liberman, A. M., Fowler, C., & Fischer, F. W. (1976). Phonetic segmentation and recoding in the beginning reader. In, Reber, A. S., & Scarborough, D. L. (Eds.), *Toward a psychology of reading.* Hillsdale: Erlbaum.

Masland, R. L. (1968). Some neurological processes underlying language. *Annals of Otology, Rhinology and Laryngology, 77,* (4), 787–804.

McKain, K., Studdert-Kennedy, M., Spieker, S., & Stern, D. (1983). Infant intermodal speech perception is a left-hemisphere function. *Science, 219,* 1347–1348.

Mehler, J. (1981). La connaissance avant l'apprentissage. *Symposium de l'Association de psychologie scientifique de langue française,* 129–155.

Mehler, J. (1985). Language related dispositions in early infancy. In: Mehler, J., & Fox, R. (Eds.), *Neonate Cognition. Beyond the Blooming Buzzing Confusion,* pp. 7–28. Hillsdale: Erlbaum.

Mehler, J., Segui, J., & Frauenfelder, U. (1981). The role of the syllable in language acquisition and perception. In, Myers, T., Laver, J., & Anderson, J. (Eds.), *The cognitive representation of speech,* pp. 295–305. Amsterdam: Morth-Holland.

Mewhort, D.J.K., & Beal, A.L. (1977). Mechanisms of word identification. *Journal of Experimental Psychology, 3,* 629–640.
Molfese, D.L., & Erwin, R.J. (1981). Intrahemispheric differentiation of vowels: principal component analysis of auditory evoked responses to computer-synthesized vowel sounds. *Brain and Language, 13,* 333–344.
Molfese, D.L., & Molfese, V.J. (1979). Hemisphere and Stimulus differences as reflected in the cortical responses of newborn infants to speech stimuli. *Developmental Psychology, 15,* 5, 505–511.
Morais, J., & Bertelson, P. (1973). Laterality effects in diotic listening. *Perception, 2,* 107–111.
Morais, J., Cary, L., Alegria, J., & Bertelson, P. (1979). Does awareness of speech as a sequence of phones arise spontaneously? *Cognition, 7,* 323–331.
Moscovitch, M. (1976). On the representation of language in the right hemisphere of right-handed people. *Brain and Language, 3,* 47–71.
Neville, H.J. (1985a). Brain potentials reflect meaning in language. *Trends in Neurosciences, 8,* (3) 91–92.
Neville, H.J. (1985a). Effects of early sensory and language experience on the development of the human brain. In, Mehler & Fox (Eds.), *Neonate Cognition,* pp. 349–369. Hillsdale, Erlbaum.
Nickerson, R.S. (1981). Speech understanding and reading: some differences and similarities. In: Tzeng, O., & Singer, H. (Eds.), *Perception of print: reading research in experimental psychology,* pp. 257–289. Hillsdale: Erlbaum.
Polich, J.M. (1978). Hemispheric differences in stimulus identification. *Perception and Psychophysics, 24,* 49–57.
Rakerd, B. (1984). Vowels in consonantal context are perceived more linguistically than are isolated vowels: evidence from an individual differences scaling study. *Perception and Psychophysics, 35,* (2), 123–136.
Rivers, D.L., & Love, R.J. (1980). Language performance on visual processing tasks in right hemisphere lesion cases. *Brain and Language, 10,* 348–366.
Rosen, G.D., & Galaburda, A.M. (1985). Development of language: a question of asymmetry and deviation. In, Mehler & Fox (Eds.), *Neonate Cognition,* pp. 307–325. Hillsdale, Erlbaum.
Ross, E.D. (1982). The divided self. *The Sciences,* February, 8–12.
Rozin, P., Poritsky, S., & Sotsky, R. (1971). American children with reading problems can easily learn to read English represented by Chinese characters. *Science, 171,* n° 3977, 1264–1267.
Sadick, T.L., & Ginsburg, B.E. (1978). The development of lateral functions and reading ability. *Cortex, 14,* 3–11.
Satz, P. (1975). Cerebral dominance and reading disability: an old problem revisited. In: Knights, R., & Bakker, D.J. (Eds.) *The Neuropsychology of Learning Disorders: Theoretical Approaches.* Baltimore: University of Park Press.
Scholes, R.J., & Fischler, I. (1979). Hemispheric function and linguistic skill in the deaf. In, *Brain and Language, 7,* 336–350.
Schwartz, J., & Tallal, P. (1980). Rate of acoustic change may underlie hemispheric specialization for speech perception. *Science, 207,* 4437, 1380–1381.
Segalowitz, S.J., & Bryden, M.P. (1983). Individual differences in hemispheric representation. In, Segalowitz, S.J. (Ed.), *Language Functions and Brain Organization,* pp. 271–272. New York: Academic.
Segalowitz, S.J., & Chapman, J.S. (1980). Cerebral asymmetry for speech in neonates: a behavioral measure. *Brain and Language, 9,* 281–288.
Saussure, F. de (1972). *Cours de linguistique générale.* Paris: Payot.
Semmes, J. (1968). Hemispheric specialization: a possible clue to mechanism. *Neuropsychologica, 6,* 11–26.
Sergent, J. (1982). Theoretical and methodological consequences of variations in exposure duration in visual laterality studies. *Perception and Psychophysics, 31,* 451–461.

Shankweiler, D., & Studdert-Kennedy, M. (1967). Identification of consonants and vowels presented to left and right ears. *Quarterly Journal of Experimental Psychology 19,* 1, 59–63.

Shimizu, K. (1975). A comparative study of hemispheric specialization for speech perception in Japanese and English speakers. *Studia Phonologica, 9,* 13–24.

Sidtis, J.J. (1980). On the nature of the cortical function underlying right hemisphere auditory perception. *Neuropsychologia, 18,* 321–330.

Sidtis, J.J., & Bryden, M.P. (1978). Asymmetrical perception of language and music: evidence for independent processing strategies. *Neuropsychologia, 16,* 627–632.

Sidtis, J.J., Volpe, B.T., Wilson, D.H., Rayport, M., & Gazzaniga, M.S. (1981). Variability in right hemisphere language function after callosal section: evidence for a continuum of generative capacity. *The Journal of Neuroscience, 1,* (3) 323–331.

Sperry, R. (1982). Some effects of disconnecting the cerebral hemispheres. *Science, 217,* 1223–1226.

Studdert-Kennedy, M., & Shankweiler, D. (1981). Hemispheric specialization for language processes. *Science, 211,* 960–861.

Studdert-Kennedy, M., Shankweiler, D., & Pisoni, D. (1972). Auditory and phonetic processes in speech perception: evidence from a dichotic study. *Cognitive Psychology, 3,* 455–466.

Sugishita, M., Iwata, M., Toyokura, Y., Yoshioka, M., & Yamada, R. (1978). Reading of ideograms and phonograms in Japanese patients after partial commissurotomy. *Neuropsychologia, 16,* 417–426.

Tallal, P., & Newcombe, F. (1978). Impairment of auditory perception and language comprehension in dysphasia. *Brain and Language, 5,* 13–24.

Tallal, P., Stark, R.E., & Kallman, C. (1980). Developmental dysphasia: relation between acoustic processing deficits and verbal processing. *Neuropsychologia, 18,* 3, 273.

Tsunoda, T. (1985). *The Japanese Brain.* Tokyo: Taishukan.

Tsavaras, A., Kaprinis, G., & Gatzoyas, A. (1981). Literacy and hemispheric specialization for language: digit dichotic listening in illiterates. *Neuropsychologia, 19,* 565–570.

Tzeng, O.J.L., & Hung, D.L. (1984). Orthography, reading and cerebral lateralization. In, Eing, C., Stephenson, H. (Eds.) *Current issues in cognition,* pp. 179–200. National Academy of Sciences and American Psychological Association.

Tzeng, O.J.L., Hung, D.L., & Wang, W.S.-Y. (1977). Speech recoding in reading Chinese characters. *Journal of Experimental Psychology: Human Learning and Memory, 3,* 621–630.

Tzeng, O.J.L., Hung, D.L., Cotton, B., & Wang, W.S.-Y. (1979). Visual lateralisation effect in reading Chinese characters. *Nature, 282,* 499–501.

Versace, R., & Tiberghien, G. (1985). Spécialisation hémisphérique et fréquences spatiales. *L'année psychologique, 85,* 249–273.

Versyp, F. (1985). *Transmission intra-amniotique des sons et des voix humaines.* Doctoral thesis, University of Lille.

Witelson, S.F. (1983). Bumps on the brain: right-left anatomic asymmetry as a key to functional lateralization. In: Segalowitz, S.J. (Ed.), *Language functions and brain organization,* pp. 117–144. New York: Academic.

Witelson, S.F., Pallie, W. (1973). Left specialization for language in the newborn. *Brain, 96,* 641–646.

Zaidel, E. (1976). Auditory vocabulary of the right hemisphere following brain bisection of hemidecortication. *Cortex, 12,* 191–216.

Zaidel, E. (1983). The split brain as a model for functional recovery from language deficits. Paper presented at *Annual Meeting of Claremont Conference on Applied Cognitive Psychology,* April.

Zurif, E.B., & Carson, G. (1970). Dyslexia in relation to cerebral dominance and temporal analysis. *Neuropsychologia, 8,* 351–361.

CHAPTER 20

Critical Brain Processes Involved in Deciphering the Greek Alphabet

DERRICK DE KERCKHOVE[1]

Introduction

The object of this chapter is to present a hypothesis concerning the underpinnings of Western culture. Did the fully phonetic alphabet developed by the Greeks and still used today in Greece (and in the rest of the West in its Latin and Cyrillic variations), have a conditioning impact on the biases of specialized brain processes? The hypothesis is that when the Greeks introduced vowels to adapt the Phoenician alphabet to suit the needs of their own Indo-European language, they changed the nature of the reading process from a context-based to a sequence-based decipherment. This change in turn may have been responsible for the reorganization of brain strategies, and this may explain why the direction of writing changed from the leftward orientation of Phoenician to rightward. The implications of such a change may have had far-reaching consequences on the biases of Western cognition.

Although Ovid Tzeng and Daisy Hung conclude their chapter in this volume (Part 4) by suggesting that more detailed work must be conducted in the analysis of different levels of reading processes before one can describe how orthographies determine differentiated patterns of processing, they have suggested elsewhere quite convincingly that such differentiation is more than plausible:

> Depending on how meanings are represented in print (i.e. what type of writing system is used), a reader may have to develop different processing strategies in order to achieve reading proficiency. Hence, by comparing experimental results on reading behavior across languages as well as across different writing systems, we should be able to gain some insight into the various *intricate processes* involved in reading. (Tzeng & Hung, 1981).

Comparative neurological studies on the effects of brain lesions in different cultures have noted, for instance, that some Chinese and Japanese patients with specific brain injuries do not evidence the same reading and writing impairments as corresponding Western patients (Jones & Aoki, this volume; Sasanuma, 1975). The reason may be that since their scriptforms are structured differently from Western systems, this has affected the organization of their cognitive processes. A real contribution can be made by the neuro-

---

[1] McLuhan Program in Culture and Technology, and Department of French, University of Toronto, Toronto, Ontario, M5S 1A1, Canada.

sciences if it can be shown that specific features of orthographies, such as the orientation of the letters and the direction of writing, relate to neurophysiological constancies, such as visual field preferences and linear processing. The existence of simple rules of orthographic layout over several thousands of years could conceivably be explained not only by circumstantial evidence, but also by causal connections between neurophysiological selectivities in the brain and graphological features of the script.

There are so few exceptions to the general rule correlating consonantal alphabets with a leftward direction and vocalized scripts with a rightward one (see my chapter, Part 3, this volume), that I have been led to investigate whether the brain, rather than the culture, has been primarily responsible for these determinations (de Kerckhove, 1981, 1982, 1984b). This hypothesis, which is testable on a purely logical level, may have profound sociocultural implications if tested at the neurobiological level.

Contextual Vs Contiguous Relationships in Alphabets and Syllabaries

To understand what role the brain may play in coding and decoding orthographies, we must clarify some relationships between oral languages and their scriptforms. Several authors in this book (e.g., de Kerckhove, Hagège, and Lafont) and elsewhere (Jurdant, 1984; Lafont, 1984; Sampson, 1985) have stressed that the most important linguistic difference between Semitic and Indo-European languages is that the morphological structure of the former is based on a division of labor between vocalic and consonantal sounds, while the latter make use of both vocalic and consonantal sounds to mark lexical oppositions (Semitic morphology reserves the use of vocalic intervals to modulate the relationships of consonantal lexical morphemes). This difference would lead one to expect, even before investigating the role of the brain, that any writing systems used to represent these languages at the phonological level would elicit different recoding strategies.

To explore just this sort of thing for Hebrew, Shlomo Bentin and colleagues presented readers with string of characters to find out whether their access to lexical values was predominantly phonological or orthographical. They report that "in Hebrew, orthographic codes play a more important role in the process of word recognition than do phonemic codes, especially in comparison with the roles played in other languages." They suggest also that "many (but not all) Hebrew words with the same sequence of consonant characters can be pronounced in several ways, each one a different legal Hebrew word. In order to pronounce the word, the reader must assign one of these alternatives to the character string on the basis of context" (Bentin, Bargai, & Katz, 1984).

By comparison, orthographies that include vowels invite their readers to rely more readily on phonemic mediation, except in the case of high-frequency words, which tend to be read quasi-ideographically (Read, 1985; Seidenberg, 1985). Indeed, there is a growing body of literature on the so-called dual-route theory of reading (for a critical review, see Humphreys & Evett, 1985).

The following is a series of eight propositions that sets the stage for speculations concerning neurological processes involved in coding orthographies. The emphasis here is placed on the degree to which phonological orthographies depend on contextual cues.

1. Speaking is acting symbolically on auditory *representations.* One of the properties of any oral language is to enable humans to act symbolically over and above their ability to behave pragmatically. One can structure one's activity either by gesture or by speech, or both. Acting symbolically, however, implies an articulation of differentiated elements that have to be combined in an orderly fashion. The production of oral speech is field-dependent (Witkin, 1962) and therefore highly contextualized.

2. Reading/writing is acting symbolically on visual *representations.* Any kind of writing, by emphasizing the representation of human experience in articulated forms, further refines and strengthens this capacity to act symbolically. That is, writing enables one to act on representations. Except for Braille and other specialized systems, writing is usually a *visual* representation of speech or thought. However, it is two quite different things to act on speechforms and to act on the iconic representation of their semantic contents. Writing removes speech from the immediate context of its production.

3. Because Phonological *writing is based on a* visual *representation of* auditory *sequences, it is bound structurally to present characters in* sequences. *Phonological* writing, as opposed to ideographic or other pictographic forms, emphasizes the process of representation by redoubling it: that is, phonological writing represents language, which in turn represents experience. Phonological writing is a *visual* representation of an *auditory* representation. Because the auditory representation of language consists of an ordered succession of sounds, its visual representation must perforce also evidence the orderly succession of characters that address the individual sounds. This is the principle of sequencing or *linearity.*

4. The linguistic structure of Semitic languages requires that the emphasis of phonological representation be placed on the visual representation of consonantal sounds alone. However, the structure of Semitic languages, because it reserves the use of consonants to distinguish the lexical values of individual words from each other, led the inventors of Semitic orthographies to place an emphasis on consonantal characters to represent speech visually. This means that the vocalic sounds, which are not used for lexical oppositions (although they are necessary for the full expression of oral speech) are left in abeyance in these scripts. That is, the intervals between the consonants are missing and have to be supplied by the reader. This fact implies that the decipherment of Semitic scripts *must remain bound to the context of speech production,* whether in its oral mode (via phonemic mediation), or more frequently in its semantic values (via ideographic mediation).

5. The linguistic structure of Indo-European languages requires that both *vocalic and consonantal sounds be visually represented.* The structure of Indo-European languages requires that the vocalic components of linguistic sounds be included in the visual representation of the language because they are just as critical as the consonantal components for discrimination between different words at the lexical level. The inclusion of characters for vowels in the orthographies of Indo-European languages, that is, in alphabets and syllabaries, implies that not only the lexical values but also grammatical values are represented. Consequently, there is no need for the reader to supply missing elements or depend on context in order to decipher the written line. The semantic and phonological dimensions of the text are adequately represented by the sequence of visual characters.

6. To decipher consonantal *alphabets, it is necessary to combine the symbols by* contextual *sequence: that is, to supply the* auditory *component.* Because they are structurally bound to an equal degree to represent the *sequence* of the phonological articulations of oral speech, Semitic as well as Indo-European scripts must present their characters in succession, namely in a *linear* sequence. However, the sequence in Semitic scripts is not linear to the same extent as in Indo-European alphabets and syllabaries. In deciphering Semitic scripts, because the reader has to supply the missing vocalic intervals, he or she cannot directly combine the sequence of letters as they appear on the line of the script. The consonants are "written" and the vowels are "oral." By reading, the decipherer of Hebrew or Arabic "gives life" to the text. The reader relates the shape of the letters to sounds and/or meanings that are given not only by the succession of the letters themselves, but by the global *context* that these letters summon. Therefore, the exact decipherment of the script depends not primarily upon the sequential order of the characters, but, by priority, upon the contextual order of the words, which alone permits the reader to choose safely among different potential interpretations of any single group of letters. This is the principle of *contextuality.*

7. To decipher vocalic *alphabets, it is sufficient to combine the symbols by* contiguous *sequence; I.E., the auditory component* is not mandatory. To decipher vocalic alphabets, that is, to access the auditory representation of meaning independently of the semantic values of the text, it is sufficient to combine the shapes of each individual letter into syllabic units and then to further combine these units along the linear sequence of the letters. This implies that, for the purpose of decipherment at least, the reader does not need to rely on the meaning of the whole sentence or of its context, but merely on an abstract process of letter by letter recombination. This property enables the alphabet to preserve both the content and the phonological structure of "dead" languages such as Latin or Classical Greek, or, in the case of the exacting Indic syllabaries, to preserve even the complex phonological values of dialectal variations over millenia (Lafont, this volume). The reading process of Indo-European alphabets is ruled by the principle of sequential *contiguity*, which does not and cannot apply to decoding consonantal alphabets.

8. Likewise, to decipher syllabaries, *it is necessary first to combine the symbols by contiguous sequence, but also to supply the* auditory *component.* In reading syllabaries, the principle of *contiguity* is again predominant. However, because syllabaries are meant to represent, as exactly as possible, phonological sequences rather than the combined lexical and grammatical structures of speech (Lafont, this volume), they cannot be as clearly detached from the oral context as are alphabets, whether consonantal or vocalic. The principles of both contiguity and contextuality are brought into play, but contiguity is applied first, to enable the reader to combine the syllabic symbols into a sequence of sounds. The reason why these orthographies have remained faithful to the principle of syllabic division, in spite of the sophistication of the phonemic analysis implicit in some syllabaries (such as the Indic or the Korean Hangul, or even the Ojibway/Cree systems) is precisely because they are meant to emulate phonology, not merely to suggest it as is the case for European alphabets.

In this paradoxical way, one can understand why it has always been difficult to decide whether Semitic alphabets are really syllabaries distinguished by an optional representation of the vocalic components, or whether they are bona fide alphabets implying a phonemic analysis. The answer is that they are both. They share with syllabaries the need to occasionally resort to the syllabic principle to effect the proper reading of unfamiliar strings of characters (Bentin et al. 1984), but they also contain, as do many sophisticated syllabaries still in use today, an awareness of the separate status of the phoneme. Their intermediate position allows them to benefit from the economy of means that is characteristic of the Indo-European alphabets, and even to better it, making do with fewer characters (22 instead of 26 letters for English and French alphabets). However, this quasi-shorthand condition binds them to the use of context to somewhat the same extent as ideographies in order to be deciphered.

Indeed, all alphabets without fixed signs for the sound of vowels, whether old Phoenician, modern Arabic, or Hebrew, require a thorough knowledge of the language and some indication regarding the context of any given text. These writings cannot be fully decontextualized except for short and frequently used sentences. In contrast, the fully phonetic alphabet can be read without any previous knowledge of the context: in extreme cases, it does not even require knowledge of the language itself. There is no guesswork; the reading process is ruled primarily by the analysis of the shapes and sequence of letters, not by their relationship to a given meaning. Thus the partial dissociation between language and experience, common to all forms of writing, is carried one step further to a dissociation of coding from meaning in the Greek and Latin alphabets.

Assuming that reading phonological sequences that depend on context is different from reading those that rely on contiguous sequences, it is conceivable that the consistently opposite orientation patterns of the scripts are correlated with this difference. In order to discover how such correlations might be causally established, it is necessary to introduce neurological considerations.

The Optic Chiasm and Brain Specialization

Neurological investigations over the last 20 years, especially in aphasiology and amnesiology (Galin, 1974; Geschwind, 1972; Kinsbourne, 1982; Krashen, 1972; Marcel, Katz, & Smith, 1974), have shown that the lateralization of cognitive processes is a critical issue in distinguishing the differences between specific tasks performed in different areas of the brain. It is well known that each hemisphere largely controls the opposite side of the body (see Grant, Tzeng, & Hung, and Lecours et al., this volume). Specifically, the right hemisphere is better equipped to deal with information presented in the left visual field, while the left hemisphere processes faster what is presented in the right visual field (Kimura, 1966, 1969, 1973; Tzeng, 1982).

The reason for this is anatomical; the *optic chiasm* (Fig. 1) splits the visual field of each eye vertically, with the right visual fields of both eyes being processed by the left hemisphere of the brain, while both left visual fields project to the right. In addition, it appears that within the visual system of a normally lateralized – i.e., right-handed – subject, the right hemisphere is better at deciphering icons and images, while the left is better at analyzing sequences (Bogen, 1969; Milner, 1971; Moscovitch, 1983; M. Taylor, this volume).

These preliminary observations first led me to speculate that, all considerations of posture and writing material aside (for discussions on writing postures and neurological correlates, see Levy and Reid, 1978; Shanon, 1978), iconic and pictographic writings might favor a kind of neurological processing quite different from that which is elicited by phonological systems. This notion is already receiving support from a growing literature on Japanese aphasiology and alexia. The Japanese are ideally suited to provide evidence on matters related to writing, because they, and they alone today, share with the ancient Akkadians the privilege of using two systems of orthography, the iconic Kanji and the syllabic Kanas. Localized aphasias among the Japanese show evidence that if a given lesion destroys the ability

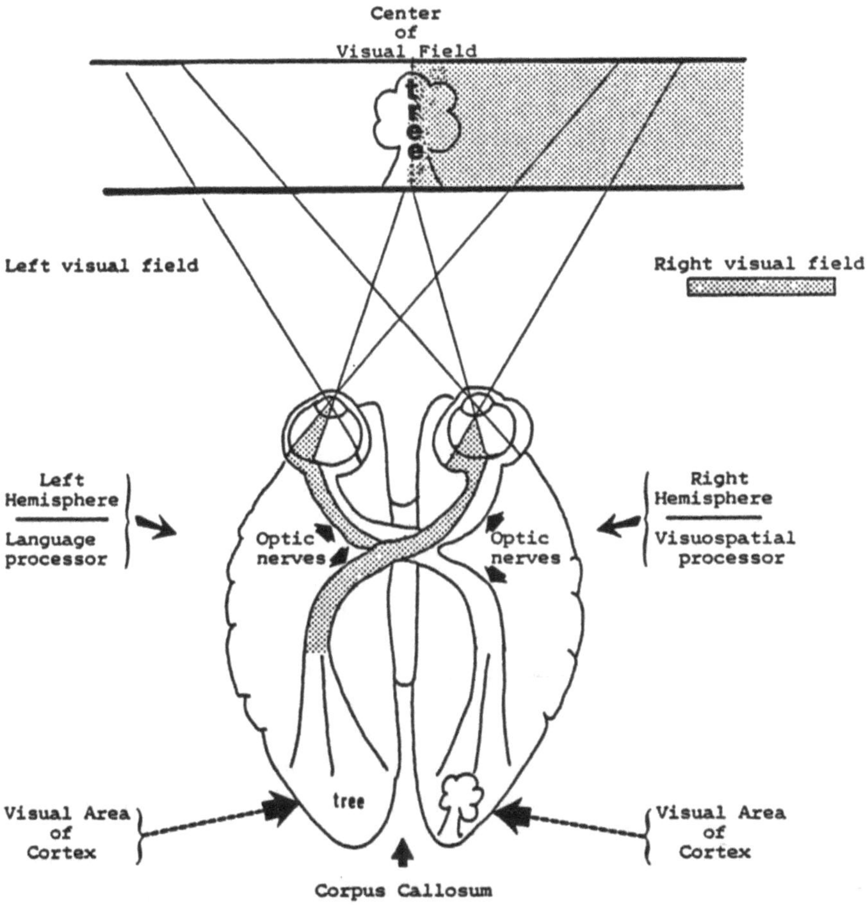

Fig. 1. The optic chiasm. (From Sinatra and Stahl-Gemake, 1983, reproduced with permission)

to decipher one system, it occasionally has little or no effect upon the ability to decipher the other (Sasanuma, 1975; Hatta, 1981; Tzeng & Wang, 1983). Investigations on normal subjects also show evidence of different lateralization patterns for both systems (Sasanuma, Itoh, Mori, & Kobayashi, 1977). However, it is becoming clear to investigators that both systems are predominantly processed in the left hemisphere (where language is usually processed in normal right-handed subjects), but that there are significantly different interhemispheric interactions for Kanji and Kanas (Iawata, 1984; Jones & Aoki, this volume).

There is more. Some 30 years ago the linguist Roman Jakobson, intrigued by conflicting reports on the effects of different types of aphasias, compiled enough evidence to regroup the major effects in two categories

(Jakobson & Halle, 1956). He classified the effects of brain lesions into disturbances of what he called relationships of *similarity* and relationships of *contiguity*. In the first instance, patients seemed to be unable to bring together different words, notions, or images into a coherent whole. However, while sometimes showing no evidence of comprehension, they could easily decipher whatever reading material they were presented with. In their conversations with the clinicians, they were unable to initiate any topic, but they were quite capable of carrying on a dialogue in a closely knit question and answer sequence. The other group could relate different objects or words by observing their correspondences and similarities, but they could neither read nor even put together a single sequence of simple letters. They also gave evidence of being incapable of structuring sentences according to syntactical rules.

Since that time, data have poured in from neurobiologists and, recently at the University of Toronto's Memory Disorders Unit, from amnesiologists (Schacter, Hanbluk, & McLachlan, 1984; Schacter, 1985), to show that indeed at least these two processes are needed to read, think, or speak, that they are presented in different configurations, and possibly located in different parts of the brain. The case of what Schacter and his colleagues call "source amnesia" is particularly relevant to the present discussion: this is the condition where a patient is able to recall, often with a high degree of precision, sequences of letters, words, and even actions, but cannot remember under what conditions these sequences have been learned, or even that they have been learned at all (Schacter et al., 1984). This condition would correspond to what Jakobson called the "disturbance of similarity" in one category of aphasics. There is no suggestion in Jakobson and Halle's paper as to what parts of the brain are involved.

To be sure, there is an unresolved controversy over whether these processes are in precisely isolated and localizable areas of the brain (for a review of the controversies over brain localization see Corballis, 1980, 1983). That indeed is another matter altogether. I do not pretend to resolve it here, but only attempt to present a workable suggestion. For the purpose of the present hypothesis, the question of lateralization is relevant only at the level of the visual field organization, independently of the precise configuration and lateralization of deeper processes in the brain.

Contextual Relationships and Left Visual Field Preference

The principal argument is that modern Arabic and Hebrew, which have to give precedence to the iconic features of groups of characters over their sequential order, would spontaneously favor the left visual field and be written leftward. Conversely, Greek, Latin, Cyrillic, and all other phonetic scripts, which can fully rely on the contiguous combination of the letters without the need to depend upon the proper separation and the order of groups of letters, would favor the right visual field and be written rightward.

Let us recall the standard clinical evidence that feature detection is best performed if objects are presented to the left visual field, and thus to the right hemisphere, while analytical processing abilities are enhanced if objects are presented in the right visual field (Bever, 1975; Kimura, 1966, 1969; Moscovitch, 1983; Tzeng & Hung, 1981; Tzeng & Singer, 1981; Tzeng & Wang, 1983). Using the tachistoscopic technique, Gloria Bradshaw and colleagues tested 46 right-handed college students to find which visual field was more suited to read words and nonwords. They found that the left visual field was superior for very short exposures (20 ms or less), but that the right visual field was unquestionably better suited for normal exposure time (Bradshaw, Hicks, & Rose, 1979). The value of this experiment is that it may indicate that feature recognition without decoding is better performed in the left visual field, while the right visual field is more appropriate for semantic decoding. Obviously, this experiment does not imply that the right hemisphere can "read," only that is is involved in feature recognition. However, although Rabinowicz and Moscovitch (1984) doubt that split-field experimental techniques can provide incontroversial evidence of right-hemisphere reading in neurologically intact people, they suggest that, regarding coding strategies, there is ample evidence that there are "right-field advantages emerging for verbally coded stimuli, and left-field advantages for identical stimuli that are visually imaged" (see Bersted, 1983). In any case, the right brain is far from incapable of recognizing letters: another experiment, involving tachistoscopic recognition trials for uppercase and lowercase letters in both visual fields separately, also indicated that reaction time was faster for left than for right visual field presentations (Hellige & Webster, 1981).

The suggestion is that with a consonantal alphabet, the reader requires rapid feature detection to facilitate contextual relationships. This is akin to establishing what Jakobson called the relationship of similarity, which, for the present purpose, I would like to rename the relationship of *contextuality*. From the clinical evidence, it follows that scripts that depend primarily on feature detection may require less reaction time by the brain if they are written toward the left. The reason is that, appearing first in the left visual field, they will first be addressing the part of the brain most appropriate for feature recognition.

Would this kind of specialization be a normal function of the anatomy of the brain, or does it correspond to a developmental condition? Trying to reproduce results obtained by Rizzolatti, Umilta, and Berlucchi (1971), Reynolds and Jeeves (1978) found that whereas there was no significant visual field lateralization among 7- to 8-year-olds, there was a definite "right visual field superiority in reaction time to letters in male adults." These results are consistent with the body of literature on the subject, notably with Luria's restatement of Karl Lashley's discovery that "the region responsible for sequential analysis [is] in the anterior region of the left hemisphere" (Luria, 1970). But there is an added feature to the Reynolds and Jeeves study, in that it emphasizes the developmental nature of the brain specialization.

The developmental nature of lateralization, as of many other aspects of brain specialization, is given support in this volume by the associated theories of Changeux and Finkel regarding the "selective stabilization of synapses" in the brain as it is exposed to environmental conditions. Learning to read any type of orthography is surely a powerful environmental condition. However, this suggestion brings up an old controversy over whether physiological or cultural determinants are responsible for the preferred direction of reading found in different orthographic environments.

Tzeng and Wang (1983) suggest that, as expected, a right visual field superiority is consistently found for readers of alphabetic scripts, such as English and Spanish. However, they add that "this RVF superiority obtains as well for scripts like Arabic and Hebrew, even though in these cases the letters run right to left across the page." Beside the fact that semantic analysis, irrespective of the direction of the orthography, is processed in the left hemisphere, and thus would tend to support a right visual field preference at the level of word interpretation, there may also be a cultural explanation for these results. Indeed, Martin Taylor (1983) cites studies by Nachshon, Shefler, and Samocha (1977) and Orbach (1967) to support this notion: "Hebrew readers' right-left scanning is not as strong as English readers' left-right scanning, perhaps partly because Hebrew readers at school must read left-right materials such as Arabic numerals and English books." A similar conclusion, attributing visual field preference to cultural indicators and acquired scanning habits rather than to hemispheric specialization, is found in Philip Bryden's review of the controversy (Bryden, 1978).

Perhaps the case for hemispheric involvement, especially at the developmental stage, is best made by Ruth Silverberg et al., who tested 72 Israeli children originally trained in Hebrew reading. They found that at different levels of training in English reading and comprehension, there was a progressive shift of visual field preference from the left to the right direction. Their data "suggest right hemisphere involvement in acquiring the reading skills of a new language" (Silverberg, Gordon, Pollack, & Bentin, 1980).

These last results are encouraging for the hypothesis at hand, in that they give evidence for right hemisphere involvement, but they do not clear the objection raised by Tzeng that even Hebrew readers eventually demonstrate a right visual field preference. The problem lies, in my opinion, not with the conclusions drawn from the experiments but with the level of processing that they address. A study by Zaidel and Peters (1981) on two commissurotomy patients may bring us closer to the answer: it appears that these subjects were able to "read" the words presented to their left visual field, that is, exclusively to their right hemisphere, but that they were not able to produce the "sound image" of the spelled words. Zaidel and Peters suggest that "these right hemispheres read 'ideographically' recognizing words directly as visual gestalts without intermediate phonetic recoding or grapheme-to-phoneme translation."

Needless to say, these results add further support to the notion of distinct and separate abilities in the hemispheres, but they also suggest that if the right brain is involved at all, it is not at the level of semantic decoding, but at that of feature detection. In this case, the work of reading is indeed shared by both hemispheres, with the right one providing the elements of feature recognition, and the left one providing the verbal interpretation (see Martin Taylor's theory of the Bilateral Cooperative Model, in this volume, for further clarification). It is the matter of differentiated reaction time which is relevant. By remaining faithful to the tradition of writing leftward, writers of Hebrew and other Semitic scripts were in fact adopting the direction that enhances feature recognition as the most valuable asset during the first few milliseconds of reading their kind of orthographies.

Contiguous Relationships and Right Visual Field Preference

The confusion regarding the levels of processing that are addressed by otherwise valid experiments probably originates in the fact that most controlled studies are performed with subjects who are trained to read English. For these people, the majority of whom are normal right-handers, the visual field preference coincides with the standard lateralization of language, the standard direction of writing, and the standard neurological conditions best suited to the analysis of sequences. It is not always easy to distinguish these levels from one another; hence the tendency, even among neurophysiologists and psychologists (e.g., Bryden, 1978; Taylor, 1983), to attribute preferred direction to cultural conditioning, rather than to conclude from the evidence that there is indeed a physiological reason to support the brain's choice of the right visual field for the layout of contiguous characters.

The suggestion of the present hypothesis is that, when one is reading a vocalized alphabet, one primarily calls on the brain's ability to process contiguous sequences. Ultimately, the rightward direction of the script must follow, because the most urgent task in this case is to combine the signs, a task probably best performed by the left hemisphere, and presented in optimal conditions if the line of writing is first viewed in the right visual field. This process is akin to establishing what Jakobson calls relationships of *contiguity*. However, although it is well suited to make distinctions among linguistic processes, the concept of contiguity may not be close enough to help identify the left brain's processing strategies.

To get down to what clearly differentiates the left from the right brain's contribution to reading, it may be necessary to distinguish between complementary processes of segmentation and combination. Indeed, in putting together a sequence both processes are needed almost at once: the recognition of individual characters, or parts of characters and parts of words for experienced readers, may still be a prerogative of the right hemisphere, while the analysis and combination of these features would be effected by the left hemisphere. This much is suggested by Martin Taylor in this volume and in a previous attempt at describing the collaboration (Taylor, 1983). However, Taylor understandably shies away from ascribing hemispheric localization to the processes he has identified. Another way to look at the question of contiguity would be to suggest that the line of characters is analyzed serially by the left hemisphere searching for grapheme-to-phoneme correspondences. This is Luria's position (1970), but it has received little support from recent investigations into the recoding strategies of the normal adult reader (Humphreys & Evett, 1985; Read, 1985; Seidenberg, 1985).

On the other hand, much of the discussion in the preceding section of this chapter, regarding the preferred (left) visual field for feature recognition, implied a contrast with the right field, which is deemed to elicit a better recognition of sequences. In a general way, based on their own investigations and on the work of Efron (1963), Carmon and Nachshon (1971), and others, Jerre Levy (1974), David Galin (1974), and Stephen Krashen (1972, 1975) have suggested that the left hemisphere specializes in abstract, temporal analyses of sequential information (for a discussion, see Sinatra and Stahl-Gemake, 1983). However, although there is a great deal of literature on the sequential character of *spoken* language in the brain, the same feature does not seem to have focused the attention of many neurological studies of *written* language. It is often assumed that the opposition between hemispheric abilities is best expressed in terms of verbal vs visuospatial material, not always in terms, more precisely, of the kinds of processing strategies involved. There are, to be sure, countless crude categorizations of left and right brain global strategies, from scientists such as Bogen (1975) and Ornstein (1976) to culturologists such as Jaynes (1976) and McLuhan (1978, unpublished paper), but these are not very helpful because they obscure by excessive generalization the finer processes at hand.

Among those who have explored the specific strategies of the left hemisphere for reading, Ovid Tzeng and Daisy Hung (1984) recently conducted an elegant experiment to demonstrate that the right visual field was indeed better suited for the sequential ordering of data. Inspired by Hammond's (1982) research in hemispheric differences in temporal resolution, Tzeng and Hung used a four-field tachistoscope to present 40 right-handed college students with letter sequences so arranged as to elicit alternate responses whether they were read at one glance or sequentially. In their own words: "The results are clear-cut. All subjects, regardless of sex, were better able to report the words according to the temporal order of the presented letters when the letters were presented in the RVF than when presented in the LVF."

The same experiment was later conducted with sequences of colored dots instead of letters. Although initially there was no evidence of visual field preference, on a second trial with the same subjects, now familiar with the possible permutations of the dotted sequences, a right visual field preference became evident in all subjects. Tzeng and Hung interpret these results as showing that "The left hemisphere's superiority in word recognition is due to its greater ability to track the sequences of segments, regardless of whether they are audible sounds or visible patterns." They add that "the temporally based mechanism is amodal as well as prelinguistic" and go on to cite a later study by Bentin and Carmon (1983), which used experimental paradigms similar to their own but with evoked potentials as the dependent measure; this experiment showed that "a greater number of brain activities occurred in the left hemisphere during reading, especially when the reader employed a sequential strategy to encode the input letters" (Tzeng & Hung, 1984)

This line of investigation would go a long way to explain why orthographies such as ancient Greek and Latin went through a *boustrophedon* stage (see de Kerckhove, this volume) after they generalized the use of vowels to complement the consonantal apparatus of their Phoenician and Etruscan models. In both cases, the initial tendency would naturally be to follow the example of the models and adopt their direction. But the next stage (Woodhead, 1981) required that the writing follow alternately the rightward and the leftward directions. The explanation for such a strange pattern of development could be that the inclusion of vowels was closing the gaps between the consonants. Then, the nature of the scriptform, and its dependent, though unconscious processing strategies, would change from a context-based to a sequence-based code. This would presumably generate a neurophysiological "pull" toward favoring the right visual field, and therefore, the rightward direction, because the sequences of contiguous letters would elicit a faster reaction from the brain in the right visual field. In fact, there are many examples of early rightward direction, even among the oldest Greek epigraphs (Jeffery, 1961).

The boustrophedon, then, would appear as a compromise based on the need to effect uninterrupted sequences of characters, and arising spontaneously from experimental writing conditions. It is also possible that the boustrophedon was related to abilities of mirror-writing, which is known to accompany the early stages of learning to write among children (for a development of this theme, see Cornell, 1985; also for the evidence of spontaneous mirror-writing in many children).

Regarding the determining nature of the principle of contiguity, as it applied to early vowel-based alphabetic orthographies, it is worth remembering that both Greek and many variations of Latin orthographies abandoned word and even sentence separations in the line of writing, which were mandatory within the Phoenician and Etruscan models (de Kerckhove, this volume). Cohen (1958) reports that even in the most ancient Greek manuscripts, evidencing a cursive type of writing, word separation was rarely indicated from the fifth century B.C. to the sixth century A.D.

For purposes quite unrelated to this issue, Karl Pribram once conducted experiments to identify what parts of the brain were responsible for parsing activities. He reports that:

It is remarkable that the same parts of the brain are responsible for the operations that determine context by way of pragmatic procedures and those that determine the pauses necessary to parsing utterances, i.e. expressions into words. This identity of neural substrate suggests that pauses in speech provide the contextual cues within which the content becomes related to the speaker's state. (Pribram, 1980)

If that were also to be the case for writing systems, then the adoption of *scriptio continua* (de Kerckhove, this volume; Saenger, 1982) was tantamount to eliminating context altogether, as if to reinforce the reliance on purely sequential decoding. The practice of lining up the letters one after the other, without any kind of parsing, either on stone or on lighter materials, seems to indicate that the deciphering process would henceforth give priority not to the context of individual words, but to the sequence of the letters themselves. One can assume that in order to read under such conditions, a process of phonemic mediation was mandatory, and therefore the need to combine phonemes to effect the syllabic synthesis was binding, rather than optional as is the case today for experienced readers presented with proper word separation.

Parenthetically, to dispel a possible misunderstanding, foveal vision is of course the operative visual mode for reading, and the scanning process of reading, irrespective of saccadic movements, depends on focusing, even for short periods of time, on the narrowest area covered by the line of writing. There is no evidence to indicate that foveal vision is susceptible to laterality preferences: on the contrary, standard neuroanatomy suggests that the information presented to the fovea of both eyes is distributed to both hemispheres at once. Bever (1980) suggests that "visual stimuli must be presented outside the fovea to be completely lateralized neuroanatomically." This, however, does not invalidate the above argument, because reading horizontally is a dynamic process requiring eye movements to scan the surface of the line in one or the other direction. To determine the speed and accuracy of such movements, an area surrounding the fovea, called the parafovea, is involved (Gould, 1967). It is the scanning process, which is controlled by what appears in the area of the visual field dependent on parafoveal vision, that is affected by preferential visual fields, rather than the visual system per se. Pollatsek, Bolozky, Well, and Rayner (1981) have shown that in Hebrew reading, the perceptual span covered by the parafovea is larger to the left, whereas for readers of English it is larger to the right.

Conclusions

Summing up the above observations, the major conclusions are as follows:

1. Different types of orthographies affect different processes of the human brain in differing proportions.
2. All orthographies elicit at least two fundamental responses from the brain: the recognition of the *shape* of the letters, and analysis of their *sequence*.
3. The aspects emphasizing the *shape* tend to be processed preferentially in the *left visual field*, while the aspects relating to the *sequence* tend to evoke a more accurate and faster response within the *right visual field*.
4. Thus, *contextual* relationships requiring speed of feature detection may involve the specific properties of the right hemisphere more readily.
5. Conversely, *contiguous* relationships requiring speed of feature connection would more readily involve the special abilities associated with the left hemisphere.
6. The Indo-European alphabets and syllabaries differ from Semitic consonantal alphabets by virtue of the fact that they attempt to present a visual analogue of the complete sequence of oral speech. They give precedence to contiguous over contextual relationships in the coding and decoding processes. It is this priority given to the sequencing over the contextualizing of the characters which, along with conditions of cultural and material reinforcement, *neurophysiologically* determines the direction of these orthographies to the right.
7. Conversely, it is the priority given to feature detection and contextualizing which *neurophysiologically* determines the direction of Semitic consonantal alphabets to the left.
8. Finally, syllabaries differ from the Indo-European alphabets, but only marginally, in that they require both contiguously and contextually bound decoding processes almost to the same degree in order to be deciphered. The fact that the priority ascribed to one or the other process is not clearly established explains why a small percentage of syllabaries have been, for a period of time, written to the left.

The implication behind these observations is that even though we can assume that both hemispheres always collaborate in the production of mental representations of the outside world, they contribute different and complementary processes in differing proportions according to the kind of training and development they have been subjected to (de Kerckhove, 1984b). Thus, it is conceivable that the brain will develop biases that are characteristic of the special abilities of one hemisphere, and which would be less pronounced under nonliterate conditions (Bogen, 1975; Segalowitz & Bryden, 1983). Specifically, if a developing brain has been exposed to the alphabet and to an environment strongly conditioned by a literate culture such as ours, the complex interactions normally engaging both hemispheres during instances of information processing are likely to be ruled predominantly by the left hemisphere. If this is true, it can also be suggested that the adoption of vocalized alphabets may, more than any other system, have promoted and reinforced reliance of left-hemispheric strategies for other aspects of psychological and social information processing. Segalowitz and Bryden (1983) suggest that "Written language imposes a certain form on communication: an increase in propositionalizing and an emphasis on grammar and form, with a relative decrease in emphasis on global and emotional communication, which of course is reintroduced in poetry." They go on to cite a study by Cameron, Currier, & Haerer (1971) to the effect that there is relatively less left hemisphere dominance for language among illiterates (see also Lecours et al., this volume).

Why is this selection of laterality so important? Precisely because we are dealing with the processing of language. The mental organization and representation of language implies the structuring of thought itself. Language cannot be separated from cognition, whether it evokes images and attitudes directly or is analyzed in the mind. It has been mentioned earlier that just as visuospatial relationships would require a greater involvement on the part of the right hemisphere, temporal sequences are processed mainly by the left. This is a useful way of distinguishing the properties attributed to each hemisphere. Assuming that, in spite of some reservations about making too clear-cut differentiations (Kinsbourne, 1982), there is some validity to this distinction, the dominating consequence of the hypothesis presented above would be that vocalized alphabets may have brought reading and writing in line with speaking as effects of the timing properties of the left hemisphere. Some neurobiologists (Changeux, 1983; Kinsbourne & Lempert, 1979; Krashen, 1975; Nebes, 1975) suggest that speech finds its place in the left hemisphere, not ontogenetically, but because the left brain's timing processes reflect and accommodate the serial nature of the production and reception of linguistic sounds. Tzeng and Hung's own theory makes a case precisely to that effect, by suggesting that it is the differential rate of processing that conditions the localizing of language in the left hemisphere in normal right-handed subjects (Tzeng & Hung, 1984).

Thus, for those people learning to use fully phonetic alphabets, the serial and timing properties associated with the left hemisphere in normally right-handed people may be given a special emphasis as the ground rule of processing. Those properties that favor bit-by-bit analyses of items or chunks of information could eventually be deemed to rule the coding and decoding operations involved not only in reading and writing, but also in "thinking." It is a matter of whether language is processed as "oral," that is, as evoking immediate and direct responses, or as "written," that is, as an object of mental scrutiny and interpretation.

In a complete alphabetic sequence, the reading brain can rely on the succession of letters without having to check its interpretation against the oral rendition of the text, the immediate context of the statement, or the situation of the reader. It is this level of abstraction which enables language to be processed in the mind as "written" rather than "oral" (see Olson, this volume). Because of this release from the need to contextualize, the reader can understand the text through a purely conceptual use of language. The written word refers not to a reality, nor to an image of reality, nor even to an idea of reality, but first of all to a mental image of a sound which itself can eventually yield a idea or an image of reality.

Furthermore, the principle of combining letters to form syllables and syllables to form words enables the reader/writer to perceive and use each level as a separate unit. As Lafont has suggested in this volume, the invention of the Greek alphabet opened the era of grammar. Just as letters constitute a finite set of modular structures that can be combined to form a higher level of modules, each alphabetically written word automatically acquires the status of a separate concept, and such concepts, if need be, can be ordered together in abstract sequences before they are put to the task of describing a given reality.

In Western philosophy, which is predicated on a written rather than an oral tradition, what most people call "thinking" is a predominantly conceptual and sequential activity (Jones & Aoki, this volume). It is the ability to organize concepts in chains and sequences. "Speed" reading, i.e., registering written information at high speed, appears to be more relevant to our culture than "deep" reading, which is restricted to hermeneutics and literary criticism. The order and succession of concepts generated by reading are more important, more "significant," than their full elaboration in the imaging processes. As Plato's philosophical investigations into the nature of discourse (*Cratylus, Phædrus, Seventh Letter*) and, later, Descartes' *Méthode* for scientific enquiry amply demonstrate, the literate bias has been to break information down into parts and then to order such parts in a proper sequence.

Metaphorically, one could say that this was the beginning of artificial intelligence. There is not much that is "natural" about Western intelligence. Indeed, I am considering the possibility that the adoption of the alphabet by Western cultures has had a reordering effect on the brain and the whole nervous system of literate people, including their sensory modes (de Kerckhove, 1981, 1982), an effect comparable to changing the program of a computer. With full phonetization, writing seems to have acquired a precision, a flexibility, and a paradoxical meaninglessness that is comparable to computer programming codes. I do not mean by this that alphabetic writing has turned people into computerized automatons, but that it made language available for a kind of information processing which is, technically, and especially in scientific investigations, very close to a mathematical model.

In evolutionary terms, with the advent of the Greek alphabet, the development of writing moved further and further away from the context of immediate experience, and took up its place as the abstract code of reality. It became possible to read meaningfully strings of visual speechforms that contained radically new ideas, concepts, or notions, some of which could even be completely foreign to the reader as he or she did not have to depend upon previous knowledge to decipher them. Hence, the origin of the first truly comprehensive scientific investigations was dependent upon a system of archival recording that was not bound to the traditional usages of oral speech, but only to the specialization of reliable written documents based on progressively more reliable empirical observations. This conclusion has intuitively and tentatively been reached by many scientists and cultural observers, and its consequences for the reinterpretation of cultural differences and historical developments may require a paradigmatic shift in scientific and scholarly investigations.

References

Bentin, S., & Carmon, A. (1983). The relative involvement of the left and right cerebral hemispheres in sequential vs. holistic reading: electrophysiological evidence. Paper presented at the annual BABBLE meeting, Niagara Falls, Canada, March.

Bentin, S., Bargai, N., & Katz, L. (1984). Orthographic and phonemic coding for lexical access: evidence from Hebrew. *Journal of Experimental Psychology Learning, Memory and Cognition, 10,* 353–368.

Bersted, C.T. (1983). Memory scanning of described images and undescribed images: hemispheric differences. *Memory and Cognition, 11,* 127–136.

Bever, T.G. (1975). Cerebral asymmetries in humans are due to the differentiation of two incompatible processes: holistic and analytic. In, Aaronson, D., & Rieber, R. (Eds.), *Developmental psycholinguistics and communication.* New York: Academy of Sciences.

Bever, T.G. (1980). Broca and Lashley were right: cerebral dominance is an accident of growth. In, Caplan, D. (Ed.) *Biological Studies of Mental Processes,* pp. 186–230. Boston: MIT.

Bogen, J.E. (1969). The other side of the brain. *Bulletin of the Los Angeles Neurological Society, 34,* 191–220.

Bogen, J. (1975). Educational aspects of hemispheric specialization. *UCLA Education, 17,* 24–33.

Bradshaw, G.J., Hicks, R.E., & Rose, B. (1979). Lexical discrimination and letter-string identification in the two visual fields. *Brain and Language, 8,* 10–18.

Bryden, M.P. (1978). Strategy effects in the assessment of hemispheric asymmetry. *Strategies of Information Processing,* pp. 117–149. London: Academic.

Cameron, R.F., Currier, R.D., & Haerer, A.F. (1971). Aphasia and literacy. *British Journal of Disorders of Communication, 6,* 161–163.

Carmon, A., & Nachshon, I. (1971). Effects of unilateral brain damage on perception of temporal order. *Cortex, 7,* 410–18

Changeux, J.P. (1983). *L'Homme neuronal.* Paris: Fayard.

Cohen, M. (1958). *La grande invention de l'écriture et son évolution.* Paris: Imprimerie Nationale.

Corballis, M. (1980). Laterality and myth. *American Psychologist, 35,* 284–295.

Corballis, M. (1983). *On human laterality.* New York: Academic.

Cornell, J.M. (1985). Spontaneous mirror-writing in children. *Canadian Journal of Psychology, 39,* 174–179.

de Kerckhove, D. (1981). A theory of Greek tragedy. *Sub-Stance, 29,* 23–36.

de Kerckhove, D. (1982). Ecriture, théâtre et neurologie. *Etudes françaises, 18,* 109–128.

de Kerckhove, D. (1984a). Introduction à la recherche neuroculturelle. In, de Kerckhove, D., & Iannucci, A. (Eds.), *McLuhan e la metamorfosi dell'uomo,* pp. 147–189. Roma: Bulzoni.

de Kerckhove, D. (1984b). Effects cognitifs de l'alphabet. In, de Kerckhove, D., & Jutras, D. (Eds.), *Pour comprendre 1984*, pp. 112–129. Ottawa: UNESCO. (Occasional paper N 49).
Efron, R. (1963). Temporal perception, aphasia and déjà vu. *Brain, 86*, 285–294.
Galin, D. (1974). Implications for psychiatry of left and right cerebral specialization. *Archives of General Psychiatry, 31*, 572–583.
Geshwind, N. (1972). Language and the brain. *Scientific American, 226*, 76–83.
Gould, J.D. (1967). Pattern recognition and eye movement parameters. *Perception and Psychophysics, 2*, 399–407.
Guarducci, M. (1967). *Epigrafia greca*, VI, *Caratteri e storia della disciplina, la scrittura greca dalle origini all'età imperiale.* Roma: Libreria dello Stato.
Hammond, G. (1982). Hemispheric differences in temporal resolution. *Brain and Cognition, 1*, 95–118.
Hatta, Takeshi (1981). Differential processing of Kanji and Kana stimuli in Japanese people: some implications from Stroop-test results. *Neuropsychologia, 19*, (1), 87–93.
Hellige, J.B., & Webster, R. (1981). Case effects in letter-name matching: qualitative visual field difference. *Bulletin of the Psychonomic Society, 17*, (4), 179–182.
Humphreys, G.W., & Evett, L.J. (1985). Are there independent lexical and non-lexical routes in word processing? An evaluation of the dual-route theory of reading. *The Behavioral and Brain Sciences, 8*, 689–740.
Iwata, M. (1984). Kanji versus Kana: neuropsychological correlates of the Japanese writing system. *Trends in Neurosciences, 54* (12), 290–293.
Jakobson, R., & Halle, M. (1956). *Fundamentals of language.* La Haye: Mouton.
Jaynes, J. (1976). *The origin of consciousness in the break-down of the bi-cameral mind.* Toronto: University of Toronto Press.
Jeffery, L.H. (1961). *The local scripts of archaic Greece.* Oxford. Clarendon.
Jurdant, B. (1984). Ecriture, monnaie, et connaissance. University of Strasbourg: unpublished thesis.
Kimura, D. (1966). Dual function asymmetry of the brain in visual perception. *Neuropsychologia, 4*, 275–285.
Kimura, D. (1969). Spatial localization in left and right visual fields. *Canadian Journal of Psychology, 23*, 445–458.
Kimura, D. (1973). The asymmetry of the human brain. *Scientific American, 228*, 70–78.
Kinsbourne, M. (1977). The evolution of language in relation to lateral action. In, Kinsbourne M. (Eds.), *The asymmetrical function of the brain*, New York: Cambridge University Press.
Kinsbourne, M. (1982). Hemisphere specialization and the growth of human understanding. *American Psychologist, 37*, 411–420.
Kinsbourne, M., & Lempert, H. (1979). Does left brain lateralization of speech arise from right-biased orienting to salient percepts? *Human Development, 22*, 270–275.
Krashen, S.D. (1972). Language and the left hemisphere. *Working Papers in Phonetics, 24.*
Krashen, S.D. (1975). The major hemisphere. *UCLA Educator, 17*, 17–24.
Lafont, R. (Ed.). (1984). *Anthropologie de l'écriture.* Paris: Centre Georges Pompidou.
Levy, J. (1974). Cerebral asymmetries as manifested in split-brain man. In, Kinsbourne, M. & Smith, W.L. (Eds.), *Hemispheric disconnection and cerebral function. Springfield: Thomas.*
Levy, J., & Reid, M. (1978). Variations in cerebral organization as a function of handedness, hand posture in writing, and sex. *Journal of Experimental Psychology: General, 107*, 119–144.
Luria, A.R. (1970). The functional organization of the brain. *Scientific American, 222*, 66–73.
Marcel, T., Katz, L., & Smith, M. (1974). Laterality and reading proficiency, *Neuropsychologia, 12*, 131–139.
McLuhan, H.M. (1978). *The hemispheres and the media.* Unpublished paper.
Milner, B. (1971). Interhemispheric differences in the localization of psychological processes in man. *British Medical Bulletin, 27*, 272–277.

Moscovitch, M. (1983). Stages of processing and hemispheric differences in language in the normal subject. In, Studdert-Kennedy, M. (Ed.), *Psychobiology of language*, pp. 88–104. Cambridge: MIT.

Nachshon, I., Shefler, G.E., & Samocha, D. (1977). Directional scanning as a function of stimulus characteristics, reading habits, and directional set. *Journal of Cross-cultural Psychology, 8,* 83–99.

Nebes, R.D. (1975). Man's so-called 'minor' hemisphere. *UCLA Educator, 17,* (2), 13–17.

Orbach, J. (1967). Differential recognition of Hebrew and English words in right and left visual fields as a function of cerebral dominance and reading habits. *Neuropsychologia, 5,* 127–134.

Ornstein, R. (1976). *The psychology of consciousness.* New York: Grossman.

Pollatsek, A., Bolozky, S., Well, A.D., & Rayner, K. (1981). Asymmetries in the perceptual span for Israeli readers. *Brain and Language, 14,* 174–180.

Pribram, K.H. (1980). The place of pragmatics in the syntactic and semantic organization of language. In, Linguarum, J. (Ed.), *Temporal variables in speech: studies in honour of Frieda Goldman-Eisler,* pp. 13–19. The Hague: Mouton.

Rabinowicz, B., & Moscovitch, M. (1984). Right hemisphere literacy: a critique of some recent approaches. *Cognitive Neuropsychology, 1,* (4), 343–350.

Read, C. (1985). Effects of phonology on beginning spelling: some cross-linguistic evidence. In, Olson, D.R., Torrance, N. & Hildyard, A. (Eds.), *Literacy, Language and Learning.* Cambridge: Cambridge University Press.

Reynolds, D.M., & Jeeves, M.A. (1978). A developmental study of hemisphere specialization for alphabetical stimuli. *Cortex, 14,* (2), 259–267.

Rizzolatti, G., Umilta, C., & Berlucchi, G. (1971). Opposite superiorities of the right and left cerebral hemispheres in discriminative reaction time to physiognomical and alphabetical material. *Brain, 94,* 431–442.

Saenger, P. (1982). Silent reading: its impact on late medieval script and society. *Viator, 13,* 369–414.

Sampson, G. (1985). *Writing systems: a linguistic introduction.* Stanford: Stanford University Press.

Sasanuma, S. (1975). Kana and Kanji processing in Japanese aphasics. *Brain and Language, 2,* 369–383.

Sasanuma, S., Itoh, M., Mori, K., & Kobayashi, Y. (1977). Tachistoscopic recognition of Kana and Kanji words. *Neuropsychologia, 15,* 547–553.

Schacter, D. (1985). Multiple forms of memory in humans and animals. In, Weiberger, N.M., McGaugh, J.L., & Lynch, G. (Eds.), *Memory systems of the brain: animal and human cognitive systems.* New York: Guilford.

Schacter, D.L., Hanbluk, T.A., & McLachlan, D.R. (1984). Retrieval without recollection: an experimental analysis of source amnesia. *Journal of Verbal Learning and Verbal Behaviour, 23,* 593–611.

Segalowitz, S.J., & Bryden, M.P. (1983). Individual differences in hemispheric representation of language. In, Segalowitz, S. (Ed.), *Language functions and brain organization,* pp. 341–372. New York: Academic.

Seidenberg, M.S. (1985). The time course of phonological code activation in two writing systems. *Cognition, 19,* 1–30.

Shanon, B. (1978). Writing positions in Americans and Israelis. *Neuropsychologia, 16,* 587–591.

Silverberg, R., Gordon, N.W., Pollack, S., & Bentin, S. (1980). Shift of visual field preference for English words in native Hebrew speakers. *Brain and Language, 11,* 99–105.

Sinatra, R., & Stahl-Gemake, J. (1983). *Using the right brain in the language arts. Springfield: Thomas.*

Taylor, I., & Taylor, M. (Eds.) (1983). *The psychology of reading.* New York: Academic Press.

Taylor, M. (1983). The bilateral cooperative model of reading. In, Taylor, I., & Taylor, M. (Eds.), *The psychology of reading,* pp. 233–267. New York: Academic.

Tzeng, O.J.L. (1982). Cognitive processing in various orthographies. In, Chu-Chang, M. (Ed.), *Asian and Pacific-American perspectives in bilingual education: comparative research.* New York: Columbia University Press.

Tzeng, O.J.L., & Hung, D.L. (1981). Linguistic determinism: a written language perspective. In, Tzeng, O., & Singer, H. (Eds.), *Perception of print: reading research in experimental psychology.* Hillsdale: Erlbaum.

Tzeng, O.J.L., & Hung, D.L. (1984). Orthography, reading, and cerebral lateralization. In, Ging, C., & Stephenson, H. (Eds.), *Current issues in cognition,* pp. 179–200. National Academy of Sciences and American Psychological Association.

Tzeng, O.J.L., & Singer, H. (Eds.), (1981). *Perception of print: reading research in experimental psychology.* Hillsdalle: Erlbaum.

Tzeng, O.J.L., & Wang, W. (1983). The first two R's. *American Scientist, 71,* 238–243.

Witkin, H.A. (1962). *Psychological differentiation,* New York: Wiley.

Woodhead, A.G. (1981). *The study of Greek Inscriptions,* 2nd edn. Cambridge: Cambridge University Press.

Zaidel, E., & Peters, A.M. (1981). Phonological encoding and ideographic reading by the disconnected right hemisphere: two case studies. *Brain and Language, 14,* (2), 205–234.

CHAPTER 21

Mind, Media, and Memory: The Archival and Epistemic Functions of Written Text *

DAVID R. OLSON [1]

We have confused reason with literacy, and rationalism with a single technology.

McLuhan, *Understanding Media,* 1964, p. 15.

A change in the means of communication between human beings . . . became the means of introducing a new state of mind – the alphabetic mind.

Havelock, *The Literate Revolution in Greece and its Cultural Consequences,* 1982, p. 7.

This paper is concerned with a question that originally took form in the writings of McLuhan (1962, 1964), Havelock (1963), and Goody and Watt (1963). The question is this: Has our new understanding of the media of communication provided a basis for a new understanding of mental life – the structure of thought, the organization and uses of memory, the elaboration of our mental states?

Even in advance of any new understanding of the consequences of the communications revolutions that have occurred, a number of promising theories have been advanced to link alternative "modes of thought" to social, linguistic, cultural, and historical factors, theories associated with such writers as Whorf, Levy-Bruhl, Vygotsky, Luria, and Jaynes. These theories have implicated language and literacy in a variety of ways, but, partly because of their categorical nature, they have not stood up to detailed criticism (cf. Gardner, 1985; Scribner & Cole, 1981). On the other hand, even in the absence of an understanding of the cognitive processes, several promising theories of the consequences of the media of communication have been advanced – theories associated with such writers as Innis, McLuhan, Havelock, Goody, and Ong. Yet theories that relate the structure of mental life to the forms of communication have been advanced – theories associated with such writers as Innis, McLuhan, Havelock, Goody, and Ong. Yet theories that relate the structure of mental life to the forms of communication are at an early stage of development. This paper is intended as a contribution to such a theory.

---

* Writing of this paper was supported in part by a commission from the Annenberg Scholars Program, Annenberg School of Communications, University of Southern California, and the paper was delivered at the Scholars conference, "Communication and Collective Memory," March 1986. A version of this article appeared in *Journal of Communication,* Spring, 1988. I am grateful to the Spencer Foundation and to the Social Sciences and Humanities Research Council for their support.

[1] Ontario Institute for Studies in Education, and McLuhan Program in Culture and Technology, University of Toronto, 39A Queen's Park Crescent, Toronto, Ontario, M5S 1A1, Canada.

One way, I suggest, to link the media of communication to the structure of the mind is through the concepts of representation and interpretation. This move requires that we focus primarily on the media of communication, that is, the symbolic form of the information, rather than on the technologies, whether speech, writing systems, print, or television, in which those forms of representation are "instantiated." And second, it requires that we analyze both the structure of information that is explicitly represented in that medium and the interpretive procedures that are required to use those representations. There is no representation without interpretation[2]. These changing representations and their schemas for interpretation for different media such as oral poetry, written prose, television programs, and computer programming will be our more specific concern.

Media of communication are important devices for conveying or exchanging information between people. Indeed, early models of communication reflected exclusively that function (cf. Katz & Lazarsfeld, 1955). Media were analyzed using a simple communication model of who says what to whom by what means. Consequently, media of communication were seen simply as channels of information rather than as forms of representation. Hence, their intellectual and social implications were largely overlooked. The media were seen largely as "extensions" of experience and memory. Print, television, and computing could store and retrieve more information and do so more quickly than ordinary memory, but they were not seen as changing those functions in any important way.

However recent developments in communication theory have suggested the possibility that the media of communication, including writing, do not simply extend the existing structure of knowledge; they alter it. The media impart a systematic bias to the information they communicate, as Innis (1951) and McLuhan (1962, 1964) put it. These writers, together with Havelock (1963), Goody and Watt (1963), and more recently Eisenstein (1979), Ong (1982), and Stock (1983), have all made the case, in quite different ways, that literacy did not simply extend the structure and uses of oral language and oral memory, but altered the content and form in important ways. But what was the nature of that alteration in memory and in mind more generally?

This literature is too rich to summarize briefly, but the arguments tend to run in one of two directions. One group interprets cultural changes associated with changes in the forms of communication in terms of changed social and institutional practices while assuming that the cognitive processes of individuals remain much the same. In this camp I would put such writers as Scribner and Cole (1981), Leach (1982), Douglas (1980), and Eisenstein (1979). The other group interprets these same cultural changes in terms of psychological changes, altered forms of representation and forms of consciousness. In this camp I would put McLuhan (1962, 1964), Havelock

[2] I am indebted to Derrick de Kerckhove for this way of putting it.

(1963), Goody (1977), Ong (1982), Stock (1983), and myself (Olson, 1970; 1977).

One should not make too much of this difference, but it may be illustrated by contrasting the work of Eisenstein (1979), who examined the importance of printing to the rise of Protestantism and to the rise of Modern Science, with that of Stock (1983), who examined the role that literacy played in setting the stage for the changes described by Eisenstein. In the first case, the emphasis falls upon changing technologies and their uses; in the second, on newly evolving forms of literate competence. Even in this case, however, is it not clear just what such competencies involve.

Some writers, particularly McLuhan (1962), Havelock (1963), Goody (1977), Yates (1966), and Ong (1982), have emphasized the changing form of the cognitive processes involved in the shift away from a reliance on oral tradition to written record, of from "orality" to "literacy." The differences are particularly noteworthy in regard to archival forms, oral epics, oral ritual, and the like, that is, to a society's way of preserving culturally important information, and less applicable to ordinary, everyday discourse. Yet the altered uses of memory are remarkable. In an oral tradition, such as that of the Homeric Greeks, culturally important information was biased in the direction of "poetized" speech, speech dependent upon rhyme and rhythm, and employing a cast of gods and heroes involved in dramatic action in order to be memorable. Such language coded a "panorama of happenings," not a "program of principles" (Havelock, 1982, p. 223). Written texts removed this constraint of memorability; information could be preserved without being memorable. But as Yates (1966, 1979) has shown, such prose was rendered memorable through the invention of remarkable "arts of memory," the method of loci, being the most famous, and rote verbatim memorization, being the most enduring. (The only people in the western world who still do verbatim memorization are the subjects in psychological experiments.) These arts permitted the memorization of vast amounts of information, information which was often already in books but which, through these arts, could be stored in memory.

But others, including Ong (1982) and Havelock (1963) have pointed out that when writing began to serve the memory function, mind could be redeployed to carry out more analytic activities such as examining contradictions and deriving logical implications. It is the visible artifact which, as Havelock says, "is to release the mind" from its burden of memory and free it for analytic, logical thought. Just why the release from memorization would lead to critical thought, rather than, say, increased leisure, is not clear. And just how quickly this transformation took place is open to dispute (Morrison, 1986). Nonetheless, literacy, particularly literacy in an alphabetic writing system, is seen as contributing to a specialized, analytic, formal mode of thought. Put generally, it is the availability of an explicit written record and its use for respresenting thought that impart to literacy its distinctive properties.

While these theories are correct in pointing to the distinctive properties of writing as a representational medium for culture and cognition – its permanence through time and across space, its invariance, its explicitness – they more or less uniformly overlook the problem literacy created. While writing solved the problem of storage of information, it created what I shall call the "meaning" or the "interpretation" problem. While writing preserves the very words of a text, it does not preserve the meaning; the text has to be interpreted. It is the progressive solutions to the problem of interpretation that, I shall argue, provide the motor for conceptual shifts and hence, for the altered uses of mind and memory.

One promising way, then, of relating culture and cognition is to suggest that the media of communication are directly related to the language of internal mental representation. Changes in the media of communication correspond to changes in the structure of internal, mental representations. But more importantly, new forms of representation require new modes of interpretation and new canons for interpretation. The conceptual changes we are interested in are tied to these new canons of interpretation.

But before we can begin to look for these relations between media, mind, and memory, it is necessary to say something further about the conception of mind that we require to understand these relations. This view of mind is quite widely held in cognitive science circles and is based largely on Searle's (1983) theory of Intentional States, Fodor's (1981) analysis of propositional attitudes, and, generally, a computational view of mind (cf. Pylyshyn, 1984; Gardner, 1985). The mind is viewed as a system for constructing mental representations of events and operating on those representations. These mental representations can be analyzed into two basic constituents, a psychological mode (or propositional attitude) and a propositional content. An intentional state is a relation, as Searle says, between a knowing subject and the world. So a typical mental state is, for example, my belief that this is Los Angeles. Formally we may represent this as:

Psychological mode (agent, propositional content);
e.g., believe (I, this is Los Angeles).

Such a structure would allow us to differentiate the features we require for such a representational theory of mind. The propositional contents can be any piece of knowledge, the agent can be any person, myself or another, and the psychological mode can be any attitude including such things as *believe, doubt, suspect, think, know, infer,* and the like. So we may differentiate my *believing* that this is Los Angeles from your *knowing* that this is Los Angeles from someone else's *doubting* that this is Los Angeles and so on. I introduce this fragment of an account of mind because what we shall require for our analysis of interpretation is just such variability not only in the propositions involved, but more importantly in the attitudes or modes qualifying those propositions. Now we can begin.

Literacy and Modernity

As mentioned, in their analyses of the consequences of a shift in the technology of communication, both Eisenstein (1979) and Stock (1983) emphasize the new uses to which written language was put in law, government, theology, philosophy, and science toward the end of the Middle Ages and at the beginning of the modern period. Stock, for example, shows how the legal system changed when courts began to use written records rather than oral testimony as evidence, how theology changed when it came to be text- rather than church-centered, and how these changes set the stage for the great modern reformations, the Protestant Reformation and the rise of Modern Science.

The changing role of written texts in the administration of justice is representative. Until the twelfth century complaints were delivered orally; the breach of law was stated and compensation was demanded. The defendant replied to the charge and the local "doomsman" indicated the type of validation to be used to decide the case. This decision was not a matter of weighing the evidence in the attempt to arrive at an abstract "truth." Rather it was a matter of fairness, of allowing some clue to indicate the defendant's innocence or guilt. This, of course, is trial by ordeal. The innocent, it was assumed, could survive some horrible ordeal; the guilty would perish by ordeal, lose the duel, or whatever. A physical sign, losing the duel, was a sign of guilt.

In the twelfth and thirteenth centuries written documents began to replace oral memory and oral testimony. Stock (1983) and Clanchy (1979) have both detailed how the scrutiny of written documents and records came to provide the evidential base permitting legally competent judges to pronounce on the innocence or guilt of the accused. Stock shows that changes in understanding of scripture, of the sacraments, and of nature underwent a corresponding transformation under the impact of literacy. The fundamental tenet of the later Middle Ages, Stock (1984–1985, p 24) points out, was "the identification of objectivity with a text. As a consequence . . . questions also began to be asked about the validity of hearsay testimony, oral family record, and collective memory."

Eisenstein (1979) was primarily concerned with the role of the printing press as an agent of change. As her account is extremely rich, it is difficult to state her resulting theory briefly. Yet central to her account is her demonstration of the role of the printing press in both the Reformation and the rise of Modern Science. She summarizes as follows: "Intellectual and spiritual life were profoundly transformed by the multiplication of new tools for duplicating books in fifteenth-century Europe. The communications shift altered the way Western Christians viewed their sacred book and the natural world" (p. 704). For Reformation theology, the printing press placed a copy of scripture in the hands of every reader and thereby circumvented the role of the Church. One could encounter God through the simple practice of reading for oneself without the mediation of the priest. An analogy is the way that U.S. Presidents now appeal to the public directly via radio and television rather than through the elected representatives in Congress. Second, the printing press provided an important means for spreading the gospel to a rapidly growing reading public.

For Modern Science, Eisenstein suggests that printing was primarily responsible for placing an "original" copy of a text, free from copyist errors, into the hands of hundreds of scholars who could study them, compare them, criticize and update them. New discoveries could be incorporated into newer editions. In this way printing contributed to the development of an accumulative research tradition. Hence, the developments in both science and religion were produced more by exploiting the new opportunities afforded by printed materials, whether books, diagrams, charts, or maps, than by any particular alteration in modes or forms of thought.

However, because Eisenstein focuses her attention on printing and on the different roles played by printing in science and religion, she overlooks an important aspect of the relation between the Reformation and the rise of Modern Science. The relation she overlooks, I suggest, may be crucial to detecting a way in which a change in the medium of communication could produce a genuine change in the structure of cognition. She argues that until the Reformation, science and religion were closely related, whereas after the rise of Modern Science, they went their own separate ways. They did so because "the effect of printing on Bible study was in marked contrast with its effect on nature study" (p. 701). Science, she suggests, used printing for the consensual validation of observations, the rise of objectivity, whereas religion used it primarily for the spread of glad tidings. Because of their different uses, "the changes wrought by printing provide the most plausible point of departure for explaining how confidence shifted from divine revelation to mathematical reasoning and man-made maps" (p. 701).

Indeed, the apparently different ways that religious and scientific traditions were affected by the communications revolution, suggested to Eisenstein (1979, p. 701) "the futility of trying to encapsulate its consequences in any one formula." Printing neither led from words to images – McLuhan's eye for ear formula – nor from images to words. The effects call for "a multivariable explanation even while stressing the significance of the single innovation" (p. 702).

Admittedly, printing (and writing) may have served different purposes in religion and in science, yet a second look reveals a deeper relation between them than Eisenstein makes out. To do this we must distinguish skill in the medium of writing, that is literacy, from the technology of printing (Postman, 1985). Printing may indeed have been used in quite different ways by science and religion as Eisenstein suggests. Yet writing as a medium of communication and the required competence with that medium, literacy, played much the same fundamental role in the Protestant Reformation as it did in the rise of Modern Science. In both cases it permitted the clear differentiation of the "given" from the "interpreted." Literacy generally, and printing in particular, fixed the written record as the given against which interpretations could be compared. Writing created a fixed, original, objective "text," as Stock pointed out. (It also created the problem of interpretation, but we shall return to that.) Eisenstein quotes Sprat as making precisely this point in his joint defense of the Anglican Church, of which he was a bishop, and the Royal Society, of which he was the historian. Both, he claimed, had achieved a Reformation:

> Both have taken a like course to bring this [Reformation] about; each of them passing by the corrupt copies and referring themselves to the perfect originals for instruction; the one to Scripture, the other to the huge Volume of Creatures. They are both accused unjustly by their enemies of the same crimes, of having forsaken the Ancient Traditions and ventured on Novelties. They both suppose alike that their Ancestors might err; and yet retain a sufficient reverence for them (Sprat, 1966).

Why this search for the perfect originals? Because the original would be the "given" against which any interpretation would be compared. The given in religion was the *word* of God, the given in Nature was the *work* of God, as Bacon would say. This was the conceptual shift that underlay both Reformation hermeneutics and the rise of scientific epistemology. Both were centered on the conceptual distinction between that given by God, whether in scripture or in nature, and those interpretations made by humans, only some of which strictly accorded with the given. Changing means of communication, writing and printing, accompanied by developing skills of literacy, could therefore be used as a single explanation for developments in two apparently different traditions – religion and science. It is this common role for written texts which Eisenstein, so far as I can tell, overlooks.

We may see the operation of this principle, the distinction between the given and the interpretation, in a variety of contexts. Luther spoke for the whole of the Reformation when he claimed that the "Scripture needed no interpreter" and again "the meaning of scripture lies not in the dogma of the church but a deeper reading of the text" (Gadamer, 1975; Ozment, 1980). Bacon spoke for the whole of Modern Science when he stated: "God forbid that we should give out a dream of the imagination for a pattern in the world" (1965). And Hoogstraten spoke for seventeenth century Dutch art when he criticized style in art and "chides those who read *meanings* into the clouds of the sky" (Alpers, 1983). He urged them to use their eyes to see clouds as clouds, not as symbols of the heavens. In all of these cases we see the strict imposition of a categorical distinction between the given and the interpretations made of it, which, of course, an emphasis on the former.

Written language, printing, and an appropriate degree of literacy provided the basis for an important conceptual distinction, that between something given in a text or in nature, and all the rest which could be seen as interpretation, a "dream of the imagination." Writing invited the distinction by providing an artifact for language, a fixed text. It provided the first, prototypical, "given." Moreover, the given/interpretation distinction reworked[3] to cope with the problem of interpreting scripture, what we now think of as Reformation hermeneutics, provided a model for the interpretation of nature, as Bacon called empirical, and we call Modern, Science. Now this distinction is not only fundamental to religion and science, it is a conceptual one and hence may simultaneously reorganize the structure of mind. Just how, we shall consider presently. For now, it is important to note that literacy, competence in reading and writing texts, provided a new foundation for a given/interpretation distinction; that distinction was the basis of Reformation hermeneutics, the adoption of a literal interpretation as being given in the text. All the rest – moral meaning, allegorical meaning – were products of the imagination and of tradition. Concurrently, the distinction was the foundation for the development of Modern Science in which the given was the product of the observation of nature, "the statement of observed facts" as Bacon put it; all the rest was a "dream of the imagination." His epistemology reflected his hermeneutics, as Stock said of the medieval philosophers. I develop this argument more fully in Olson (in preparation).

The evolution of a literate tradition, then, involves more than the accumulation of knowledge or the development of an accumulative research tradition. It involves a new way of classifying and organizing knowledge. First, it involves the systematic distinction between what a text says and what it means, between a text and its interpretation, and hence, between facts and theories, observations and inferences. But second, it sets up the possibility of collecting and organizing only the former. That is what authoritative texts are: compilations of the given. "Knowledge is learned as an 'ideal text', something which so far as the learners are concerned is given not created" (Hoskin, 1982). Seen as repositories of a society's valid and objective knowledge, textbooks are above criticism (Olson, 1980).

If knowledge in a society is organized in a different way, it may require a shift in the organization of knowledge in the mind and a change in the uses of memory. That is our next concern.

Literacy and the Making of the Modern Mind

As I have already mentioned, there are two primary ways of accounting for the role and consequences of literacy, through a change in institutions or through a change in modes of thought. As we have just seen, the former seems impossible to deny but the latter seems difficult to establish. A theory of modern as opposed to primitive thought was advanced by Levy-Bruhl

[3] Stock (1984–1985, p. 17) concludes that there was no generalized medieval hermeneutics but from about 1050 on, there were an increasing number of hermeneutic models competing with each other. The Reformation seized upon one of these models.

(1966) to explain the magical thinking of members of traditional societies. He later abandoned the argument because of the persistent criticism offered by anthropologists such as Evans-Pritchard (1937) and others, and claimed rather that while beliefs may vary from culture to culture it is difficult if not impossible to distinguish forms or modes of thought in these cultures (Leach, 1982). The notion of "primitive" thought has not survived for a variety of reasons. There is almost as much diversity in forms of thought within a literate society as there is between literate and nonliterate societies. Further, the commitment of "primitive" thought to magical thinking is not notably greater than in modern thought. Mary Douglas (cited by Lloyd, 1979) recounts one such case: "Once, when a band of Kung Bushmen had performed their rain rituals, a small cloud appeared on the horizon, grew and darkened. Then rain fell. But the anthropologists who asked if the Bushmen reckoned the rite had produced the rain, were laughed out of court." Finally, some anthropologists such as Douglas (1980) and Gellner (1973) argue that modern science is an institution rather than a mode of thought. It is science which keeps options open even if each individual scientist is doggedly devoted to a pet scheme. Consequently, "the majority of contemporary social anthropologists regards the distinction between 'primitive' man and man in general as anachronistic and untenable" (Leach, 1982, p. 55).

But even if there is considerable criticism of the notion of "primitive" or "oral" modes of thought, part of the problem lies in the theories of mind that such categories presuppose. If the argument I have laid out above is correct, what we must look for is not some grand difference between eye and ear, between left and right hemisphere or between oral and literate, although all of those categorical theories may overlap, but rather for the presence or absence of a given/interpretation distinction and for its manifestation in language and thought.

In the past few years my colleagues and I (Olson & Torrance, 1987; Olson & Astington, 1985) have been engaged in a program of research that is directed to examining the given/interpretation distinction and its development in a variety of contexts. If the distinction is a literate one, one invented for interpreting written texts, we should expect to find that members of traditional, that is nonliterate, societies will tend to conflate what is said with what is meant by it; that is, that they fail to distinguish the given from the interpretation. We read the anthropological literature and we ourselves conducted experiments with preliterate children. Of course it is dangerous to compare those two groups for two reasons. First, young children are not only nonliterate, they are immature. Second, by comparing nonliterate adults with young children we are perhaps inadvertently encouraging the myth that nonliterates are childlike. With those limitations in mind we may proceed, cautiously.

What we find varies from one traditional society to another somewhat but several researchers have reported the relevant points. Evans-Pritchard (1937) was surprised to find that what a suspected witch actually intended by his speech made no difference to his accusers, who interpreted it to their own ends. There was no sharp distinction between the text, what was said, and the interpretation. Similarly, Duranti (1985) in a recent study of the interpretive strategies of the Samoans reported that there is no sharp, perhaps even detectable, distinction between what is said, the "given," and the interpretation that is assigned: "A certain meaning is possible because others accept it within a particular context" (p. 47). Hence, a Samoan speaker will not "reclaim the meaning of his words by saying 'I didn't mean it'" (p. 49). Generalizing, there is no given/interpretation distinction. Moreover, not only is there no distinction pertaining to meaning, there are no ways for referring to intentions or for treating intentions and meanings as internal mental events. Consequently, "Samoans, as perhaps members of Polynesian cultures in general, don't seem to have the western notion of 'self'" (p. 48). Rosaldo's (1982) study of speech acts in Ilongot, a traditional Philippine society, found a corresponding absence of concepts of intentionality and of self. McKellin (1986), too, discusses how Papuan speakers in making offers and complaints allow that what is uttered means only what the speaker takes it to mean; the spoken "text," and the intention that it expresses are not distinguished.

Yet it is possible to overinterpret these differences. The absence of certain intentional states and of their corresponding linguistic expression invites the inference that at least some nonliterate societies have no ways of marking intentionality. That is, mental states marked by psychological mode such as *think, feel, believe, suspect* and the like appear to be missing in at least some nonliterate societies. Literacy, we might infer, is tied to the development of Cartesian mental states. Expressed formally, rather than having mental states of the form I discussed earlier which is typical of adults in western societies, namely:

psychological mode (subject, propositional content)

it is tempting to think that their mental representations fail to represent intentional states and rather simply have the form:

propositional content

In fact, if I understand him correctly, Jaynes (1976) in an influential book, suggested that the origins of consciousness, which he associates at least in part with the evolution of writing; began with the development of awareness of just these intentional states, thinking, believing, doubting, and the like. However, such a claim may be too strong. The complete absence of such awareness would not characterize nonliterate humans; rather its complete absence would be symptomatic of autism (Leslie, 1985). A person without some range of intentional mental states and some awareness of them would not be fully human. Further, it is difficult to imagine linguistic creatures who had no sensitivity to the intention of others. Hence, it seems essential to acknowledge that all human beings have some range of intentional states. What may vary from culture to culture is the range and degree of elaboration of those states, and, more importantly, whether or not the culture has a concept of that intentional state and can represent it in language. This appears to be the particular advantage of the literate subjects.

All linguistic groups talk, tease, lie, tell secrets, and appear to have some devices for referring to talk (Schieffelin, 1979), and they all have some ways of referring to the status of pieces of knowledge through the use of "evidentials" (Chafe, 1985). These devices are somewhat equivalent to speech act terms, that is, verbs of saying, and mental state terms, that is, verbs of thinking in English. Hence, it appears that it is not that intentional states are absent in nonliterate societies; rather, it is the range and explicitness of these concepts which is limited. Literacy, I suggest, gives these implicit distinctions conceptual status and turns them into basic categories of thought (cf. Fodor, 1975; Havelock, 1982, p. 290).

In our empirical studies of young children we can detect the beginnings of children's recognition of these intentional states and their acquisition of the metalinguistic and metacognitive verbs that mark these states. Hence, we can observe more directly the consequences of the presence or absence of these concepts. In a typical study, my colleague Nancy Torrance read stories or acted out events in which an ambiguous utterance is misinterpreted by a listener. She noted how children responded to the misinterpretation. Here is a sample:

One Saturday night, Lucy and Charlie Brown were going to a party. Lucy was all dressed in her brand new red party dress, but she didn't have her shoes on. She wanted to wear her *new* red shoes to go with her party dress. Linus was upstairs so she called up to him, "Linus, bring me my red shoes." Linus went to Lucy's closet where she kept her shoes. Now Linus picked up the *old* red shoes and rushed down the stairs with them. He said, "Here are your red shoes," and gave the shoes to Lucy. "Good grief," said Lucy, "how can you be so stupid?" and she gave him a whack on the head.

We follow such stories with a series of questions. Here is how a typical 5-year-old responds:

Experimenter: Did Linus bring the shoes that Lucy wanted?
Child: No.
E: Did he do what Lucy said to do?
C: No.
E: What did Lucy tell him to bring?
C: The red party shoes.
E: What were the exact words that Lucy said? She said, "Linus, bring me. . ."
C: My new red shoes.

Of course, the new red shoes is what Lucy meant, we would say, not what she said. All she said was "red shoes." The child conflated the "given" with the "interpretation" and consequently, when asked what was said, reported, instead, the interpretation. By the time they are 8 most children make this distinction spontaneously and all of them can make it when it is pointed out to them.

At about the same time that they are sorting out the difference between a text and its interpretation, they are also sorting out the difference between what is in their own minds, their prior knowledge and beliefs, and what is present and visible in the world. My colleague Janet Astington recently posed a series of questions to children as to what they see when they look at ambiguous displays and what other people, holding different beliefs, see when they look at the same display. To illustrate, young children assume that if they know that a cat-picture is red, and if they see a red patch through a window as the cat, they assume that someone else, not knowing that the cat is red, would also see the red patch as a cat. The striking feature is that until they are about 5 years old, children tend not to acknowledge the role of beliefs, of a private mental life which differs from person to person, in their perception. This is related to what Piaget called "ego-centrism." Only when they are 6 or 7 do they begin to acknowledge that another person, holding different beliefs, would see things differently than they themselves do. About this time they begin to mark the difference with mental verbs in such sentences as "He doesn't *know* that the cat is red" or that "She *thinks* that it is something else." Taylor and Flavell (1985) and Perner and Wimmer (1985) have reported similar findings.

Both of these given/interpretation distinctions may exist in embryonic form in preliterate children and in nonliterate adults. We need to check out these possibilities more exhaustively. But what is clearly the case is that these interpretive categories become the focus of a great deal of elaboration and development in a literate society during the middle school years. Janet Astington has taken a sample of some 30 of the principle verbs, technically speech act and mental state verbs, which mark and elaborate the distinctions in question. There are three points to note about these verbs. First, they are verbs that make up the "literate standard" language; they are not a part of the lexicon of many fluent speakers of English. Second, they were largely borrowed into English in the fifteenth and sixteenth centuries as English became the literate language of religion and government (cf. Olson and Astington, 1986; Traugott, 1985). And third, and I think this is the most promising aspect of this work, these verbs are not completely discontinuous from the simple verbs of saying and thinking that children acquire when they are 2 or 3 years old. Astington has shown that they can often be analyzed in terms of the more basic, and presumably universal, terms *say, mean,* and *think.* For example, the verb *interpret* can be analyzed into something like "to think that it means;" *understand* can be analyzed into "to know what something means."

The significance of such analyses is that they provide a solution to several paradoxes. Their distinctiveness would account for the particular properties of the "modern" Cartesian mind, namely, complex, explicit mental states which may serve as objects of thought. Second, they allow us to see the continuity between literate and nonliterate societies by showing where these literate distinctions come from: they are complex predicates assembled from the set of simpler, presumably universally available predicates. This, too, may provide the explanation of the observations that gave rise to the theory of "primitive" mentality, namely, the absence of concepts marking the given/interpretation and the observation/interference distinctions in traditional societies. All speakers presumably represent their experience in terms of some set of basic predicates including *say, mean,* and *think.* What literacy involves is the specialization of these distinctions and the representation of them in the form of explicit linguistic concepts. What the Inuit do for concepts of snow, literates do for concepts of speech acts and mental states.

Literate Memory and Thought

Let us summarize the resulting conception of mind and review the cognitive implications of literacy. A theory of mind that is to be sensitive to language and literacy has to specify both the structures of knowledge possessed by an individual and a set of optional attitudes or modes to that knowledge. The former specifies the propositional content of mind, the second the intentional stance, including believing, doubting, wanting, intending, and the like. Knowledge develops by increasing the propositional content of memory and by constructing new forms of organization for that knowledge in memory. While the content differs from individual to individual and society to society, the structure of this knowledge is, presumably, universal. But, more importantly, specifying alternative "attitudes" to that knowledge, and operating on those relations, is what is primarily responsible for thought. All linguistic creatures, we must assume, have some "attitudes." Even prelinguistic children have some states such as *wanting that p* or *expecting that p.*

Elaborating these attitudes and turning them into objects of thought is what accounts for literate, "modern," thought. Consider how this could occur. Rozeboom (1972) pointed out that the classical distinction between signs and symbols could be more profitably viewed as indicating different relations between the attitude and the propositional content. A sign, such as thunder, specifies both the propositional content and the psychological mode in one and the same operation. The proposition *it will rain* and the attitude *believing* are both specified by the event. Rozeboom argued that language changes that; we may entertain the propositional content of an utterance and, independently, choose whether or not to believe it. We will believe it if we trust the speaker and doubt it if we do not. The propositional content and the attitude come apart with language while they remain indissoluble for nonlinguistic signs, and nonlinguistic creatures.

My argument has been that they come apart in more ways and in systematically different ways when one is dealing with written texts. This is what I meant when I said that writing created the problem of meaning; it created the problem of attempting to specify just how a text should be taken. When one attempts to specify that, one gets, first of all, hermeneutic assumptions, and second, an elaborated set of speech act and mental state terms. And finally, it is that elaborated set that is responsible for the subjectivity and reflectiveness of modern thought.

We may take the argument one step further. Rozeboom may overstate the case when he suggests that language is responsible for the differentiation between the attitude or mode and the propositional content. In oral language contexts, as in oral societies, we noted that there is little or no systematic distinction between what is said and what is meant, or between "texts" and their interpretation. In hearing an utterance, its content and mode are just as clearly specified as if one had heard thunder. There is little or no problem of interpretation in speech; the context that is in place on hearing an utterance determines the meaning. What one means depends upon how it is taken by the listener. Meaning and interpretation become a problem primarily in dealing with decontextualized, written text. To deal with the problem one distinguishes the propositional content from the attitude assigned. Such a move makes possible the recognition that the same proposition may at different times be an expression of belief, or doubt, or possibility, or of necessary inference. Modern, scientific thinking is little more than assigning the appropriate propositional attitude to various propositions, some as hypotheses, some as assumptions, some as inferences, and some as conclusions. It is this move, I would suggest, which makes thinking rather than memory the primary cognitive activity in a literate society.

All human beings, and presumably other creatures as well, represent their experience in terms of knowledge and attitudes to that knowledge. Ordinarily both the attitude and knowledge are specified by the natural or linguistic event. It is largely, though not exclusively, in coping with written language, and in developing a tradition for dealing with written language, that one begins to differentiate the two. Once differentiated, one can hear the propositional content and then decide whether the believe or doubt it, or one can hear an utterance and then think whether it was meant literally or metaphorically. One can, as we say, begin to reflect on text rather than merely remembering it.

So what are we to say of memory? As we noted at the outset, writing extends memory. But the burden of my argument is that by creating the problem of interpretation, writing was responsible for elaborating thought. It did so by splitting the propositional content of memory from the attitudes that one could hold to that remembered content. Operation on those attitudes, that is, shifting from one attitude to another while holding the content relatively constant, is what we mean by thought. Thus to go from the suspicion that *p*, to the hypothesis that *p*, to the inference that *p*, to the conclusion that *p*, is what we mean by reflective thought. So it is not only by providing an artifact for speech that the resources of mind could be applied to thinking rather than to remembering, as Havelock suggested; it is also that writing provided the tools, the conceptual distinctions, for that thought. A symptom of that modern thought is the appropriate use of the speech act and mental state terms that I mentioned earlier.

Now here we have a possible explanation of how media of communication, literacy in particular, influence memory. They do so primarily not through changing the content but through marking the attitude to the content more precisely. Consider the interesting work by Loftus (1979) on eyewitness testimony. She has provided convincing evidence of the fact that memory is reconstructive, that recall cues interact with whatever is stored in memory to yield different patterns of recall. Further, she has shown that people's ability to remember events is not improved by practice. The kind of argument I have been developing would suggest that while one cannot improve one's memory for the propositional content, one may become more skilled in marking the attitude to that content more precisely. That is, one's reliability as a witness may be improved by marking some propositions as known or seen, namely those that others with different interpretive biases would also see, and marking others as inferred or interpreted, namely, those that could be open to reconstrual. Indeed, this is apparently what trained observers, such as policemen, are taught to do. Hence, their "objective" descriptions, their measurements, and their written notes. These are practices that began in the Middle Ages when literacy came to be the standard of objectivity.

So what is stored in memory? Ever since Bartlett (1932) it has been clear that people store their knowledge of events in terms of schemata from which particular events are reconstructed. All memory, therefore, involves not objective events but one's interpretation of those events. As I mentioned in our "red shoes" story, people ordinarily, and children always, remember the intention or the meaning or the gist of a story rather than the particular wordings used. But literacy plays into this pattern, as we have seen. The written record provides a verbatim account of the "given," not of its interpretation, and the meaning has to be reconstructed. It is this preserved text that makes possible the distinction between the given and the interpretation. And the process of interpretation is one of learning to assign the appropriate attitude to the content expressed. Learning to handle interpretations is therefore, learning to think.

The Given/Interpretation Distinction and the Media of Communication

We have seen how one medium of communication, writing, and particularly the processes involved in the interpretation of writing, led to the specialization of certain conceptual distinctions, in this case a distinction between the text and its interpretation. And these distinctions have been fundamental in the rise of Modernity, both in science and religion. Correspondingly, an educated mind is one that can systematically construe events in terms of those categories. That is what it is to be literate and to be educated.

Written texts, and, even more so, printed books, provided the basis for the, by modern standards, naive view of the objectivity of the text. Texts led to the given/interpretation distinction and were identified with the given. It was that assumption that gave, and continues to give, texts and textbooks their authority. That assumption determined that they would be written "objectively," from no one person's viewpoint. They simply set out how things truly and objectively are. As repositories of the given, the appropriate attitude to them is one of respect, study, and until recently memorization. In this century we have seen this assumption fall. Texts are no more given than eyewitness testimony is given. In the natural sciences, it is now common to think of facts, not as *a priori* given, but as theory-relative (Kuhn, 1962). "What is given," Barthes (1977) has said, "is no more than a way of taking." A less contentious way of putting this point is that what is given is relative to a way of interpreting. In spite of the fact that literacy led us to expectancies of objectivity and givenness that are indefensible, literacy, perhaps appropriately, continues to set the standard against which all other media of communication and expression are evaluated.

It is possible that visual media such as film and television cannot systematically honor that distinction between the given and the interpreted and as a result make critical response difficult. The literate given/interpretation distinction took as given the visual image and used it as the basis for interpretation. Television presents those interpretations as part of the visual image. Hence, viewers of film and television are tempted to take the visual image as "given," that is, as objective and free from bias. Failing to distinguish the given from the interpretation, they are as credulous when viewing television as they are when reading the *Encyclopaedia Britannica.* Because it makes criticism difficult, television is a threat to literacy and rationality.

Computing, on the other hand, is a rather direct extension of literacy. Computers insist on unambiguous messages; there is no scope for interpretation. Computer programming requires both the explicit representation of knowledge and explicit representation of procedures (Larsen, 1986). Indeed, there is every indication that the traditional competences in literacy and mathematics are the prime competencies for interacting with computers (Olson, 1985). Whether children find it useful to operate in terms of ascribing mental states to the computer ("It thinks I want to add,") or to recognize that it doesn't think, that is, interpret, at all, is of some interest. It is possible that in ordinary discourse children fail to distinguish what was said from the interpretation a listener assigns – indeed, that was the upshot of the studies I mentioned earlier – whereas with a computer, one can tell immediately whether the message was understood. But even here it is not clear whether children will simply note their success or failure or whether they will explain such failure by ascribing intentions to themselves or the computer. Turkle's (1984) findings of a heightened metalanguage in child programmers suggest the latter. It may depend upon the metalanguage that the children bring to the task.

For all new media new categories for interpretation and criticism must be developed and taught to children so that they will have the interpretive skills that are as powerful as those we have developed for the literate tradition. But at the same time we must entertain the possibility that the literate distinctions have run their course. There is no ontological distinction between the given and the interpreted. The closest we can get to events is through our construals, perceptions, and interpretations of them. Yet some interpretations are better, more exhaustive, more general, better tested than others. This is to adopt the new media as veridical representations of both mind and reality and to use them to explore the alternative ways of representing events.

Mind, Media and Memory

So what is the relation between mind and the media of expression and communication? Do media create new ways of representing the world? Yes. They do so by creating distinctions of the sort we examined in regard to literacy. Writing preserved an aspect of language that could be seen as fixed, given, and objective. It is this fixity that makes possible the cumulative research tradition discussed by Eisenstein; the rise of Modern Science was, in part, the attempt to pack knowledge into those given, objective forms. Textbooks to this day attempt to preserve and transmit that objectively given description of reality, free from interpretation and subjectivity. Even if we now realize that this is a goal that can never be reached, it establishes an important norm, the norm of objectivity, for judging representations.

But, at the same time, writing created the problem of interpretation. Its solution came through the invention of devices to specify just how the propositional content was to be taken, that is, to mark the proposition in such a way as to guide its interpretation – whether as a statement, a claim, a conjecture, or a conclusion. Literacy is responsible for creating texts and, subsequently, for creating an interpretive tradition. The visual media establish new norms of objectivity, of the given. Yet it is not easy to distinguish what is given from the interpretations made, and a new visual hermeneutics has not yet developed.

And do the media of communication create new ways of representing events in the mind? Partly. The human mind, and presumably other mammalian minds and computers, already have means for representing both propositional contents and holding them in different attitudes – believing as opposed to desiring, for example. Ordinarily, in memory the content is not distinguished from the attitude to that content. Content and attitude are specified by the structure of the event. Language may, as Rozeboom suggested, make it possible to distinguish content from attitude.

But the distinctions become elaborate and explicit in literate uses of language; for reading, interpreting, and commenting on written texts it is important to distinguish what a text says, from its meaning, from the author's intention, from its interpretation. Speech act and mental state verbs are devices for explicitly marking the attitudes to propositions and, hence, for making those attitudes into objects of thought. Literate competence is a matter of learning to assign attitudes to contents – whether to represent a proposition as a hypothesis, a belief, an inference, or a conclusion. Their use runs in two directions. The contents in memory come to be specified in terms of this increasingly sophisticated set of attitudes; one knows what one knows, believes, suspects, and doubts. Second, competence in operating on these attitudes while holding the contents invariant is what is required for reflective thought. So literacy, through the elaboration and explication of attitudes, alters memory and enhances thought.

Acknowledgement. I am grateful to Janet Astington for her help in the preparation of this paper.

References

Alpers, S. (1983). *The art of describing: Dutch art in the 17th Century,* p. 77. Chicago: University of Chicago Press.
Bacon, F. (1965). Francis Bacon. A selection of his works. In, Northrop Frye, H. (Ed.), *College classics in English.* Toronto: University of Toronto Press. Chapter edited by S. Warhaft.
Barthes, R. (1977). *Image, music, text.* London: Fontana.
Bartlett, F. (1932). *Remembering.* Cambridge: Cambridge University Press.
Chafe, W. (1985). Linguistic differences produced by differences between speaking and writing. In, Olson, D. R., Torrance, N., & Hildyard, A. (Eds)., *Literacy, language, and learning: the nature and consequences of reading and writing.* Cambridge: Cambridge University Press.
Clanchy, M. T. (1979). *From memory to written record.* London: Arnold.
Douglas, M. (1980). *Edward Evans-Pritchard.* Harmondsworth: Penguin.
Duranti, A. (1985). Famous theories and local theories: the Samoans and Wittgenstein. *The Quarterly Newsletter of the Laboratory of Comparative Human Cognition, 7,* 46–51.
Eisenstein, E. (1979). *The printing press as an agent of change. Cambridge: Cambridge University Press.*
Evans-Pritchard, E. (1937). *Witchcraft, oracles and magic among the Azande. Oxford: Oxford University Press.*
Fodor, J. (1975). *The language of thought,* p. 172. New York: Crowell.
Fodor, J. (1981). *Representations.* Cambridge: Bradford.
Gadamer, H. (1975). *Truth and method.* London: Sheed and Ward.
Gardner, H. (1985). *The mind's science.* New York: Basic.
Gellner, E. (1973). The savage and the modern mind. In, Horton, P., & Finnegan, R. (Eds.), *Modes of thought.* London: Faber and Faber.
Goody, J. (1977). *The domestication of the savage mind.* Cambridge: Cambridge University Press.

Goody, J., & Watt, I. (1963). The consequences of literacy. *Comparative Studies in Society and History, 5,* 304–345.
Havelock, E.A. (1963). *Preface to Plato. Cambridge: Harvard University Press.*
Havelock, E.A. (1982). The literate revolution in Greece and its cultural consequences. Princeton: Princeton University Press.
Hoskin, K. (1982). Examinations and the schooling of science. In, MacLeod, R. (Ed.), *Days of judgement: science, examinations and the organization of knowledge in late Victorian England,* p. 217. Duffield: Nafferton.
Innis, H. (1951). *The bias of communication.* Toronto: University of Toronto Press.
Jaynes, J. (1976). *The origin of consciousness in the breakdown of the bicameral mind.* Toronto: University of Toronto Press.
Katz, E., & Lazarsfeld, P. (1955). *Personal influence.* New York: Free Press.
Kuhn, T. (1962). *The structure of scientific revolutions.* Chicago: University of Chicago Press.
Larsen, S. (1986). Procedural thinking, programming, and computer use. In, Hollnagel, E., Mancini, G., & Woods, D. (Eds.), *Intelligent decision aids in process environments.* Berlin, Heidelberg, New York, Tokyo: Springer.
Leach, E.P. (1982). *Social anthropology.* Oxford: Oxford University Press.
Leslie, A. (1985). *Pretense and representation in infancy. Part 1: A cognitivist approach.* Medical research council, cognitive development unit. London: Mimeo.
Levy-Bruhl, L. (1966). *How natives think.* New York: Washington Square Press. Originally published in 1910.
Lloyd, G. (1979). *Magic, reason and experience,* p 3. Cambridge: Cambridge University Press.
Loftus, E. (1979). *Eyewitness testimony.* Cambridge: Harvard University Press.
McKellin, W. (1986). *Intentional ambiguity in Mangalese negotiations.* University of Toronto, Toronto: Mimeo.
McLuhan, M. (1962). *The Gutenberg galaxy.* Toronto: University of Toronto Press.
McLuhan, M. (1964). *Understanding media: the extensions of man.* Toronto: McGraw-Hill.
Morrison, K. (1986). *Greek literacy and textual practice: a re-examination.* York University. Toronto: Mimeo.
Olson, D.R. (1970). *Cognitive development.* New York: Academic.
Olson, D.R. (1977). From utterance to text: the bias of language in speech and writing. *Harvard Educational Review, 47,* 257–281.
Olson, D.R. (1980). The language and authority of textbooks. *Journal of Communication, 30,* 186–196.
Olson, D.R. (1985). Computers as tools of the intellect. *Educational Researcher,* May, 5–8.
Olson, D.R. (in preparation). *The world on paper.*
Olson, D.R., & Astington, J.W. (1985). *Seeing and knowing: on the ascription of mental states to young children.* Paper presented at the annual meeting of the Society for Philosophy and Psychology, Toronto, May.
Olson, D.R., & Astington, J.W. (1986). Children's acquisition of metalinguistic and metacognitive verbs. In, Demopoulos, W., & Marras, A. (Eds.), *Language learning and concept acquisition,* pp. 184–199. Norwood: Ablex.
Olson, D.R., & Torrance, N. (1987). Language, literacy and mental states. *Discourse Processes* (in press).
Ong, W. (1982). *Orality and literacy: the technologizing of the word.* London: Methuen.
Ozment, S. (1980). *The age of reform 1250–1550: an intellectual and religious history of late Medieval and Reformation Europe.* New Haven: Yale University Press.
Perner, J., & Wimmer, H. (1985). *Ignorance versus false belief: a developmental lag in epistemic state attribution.* Paper presented at the Annual Meeting of the Society for Philosophy and Psychology, Toronto, May.
Postman, N. (1985). Media and technology as educators. In, Fantini, M.D., & Sinclair, R.L. (Eds.), *Education in school and non-school settings: 84th yearbook of the National Society for the Study of Education.* Chicago: University of Chicago Press.

Pylyshyn, Z. W. (1984). *Computation and cognition: toward a foundation for cognitive science.* Cambridge: MIT.

Rosaldo, M. Z. (1982). The things we do with words: Ilongot speech acts and speech act theory in philosophy. *Language in Society, 11,* 203–237.

Rozeboom, W. (1972). Problems in the psycho-philosophy of knowledge. In, Royce, J., & Rozeboom, W. (Eds.), *The Psychology of Knowing.* New York: Gordon and Breach.

Schieffelin, B. (1979). Getting it together: an ethnographic approach to the study of communicative competence. In, Ochs, E., & Schieffelin, B. (Eds.), *Developmental pragmatics.* New York: Academic.

Scribner, S., & Cole, M. (1981). *The psychology of literacy.* Cambridge: Cambridge University Press.

Searle, J. R. (1983). *Intentionality: an essay in the philosophy of mind.* Cambridge: Cambridge University Press.

Sprat, T. (1966). *History of the Royal Society of London for the Improving of natural Knowledge,* part 3, section 23, p 371, Cope & Jones (Eds), St. Louis, Washington University Press. Originally published in London in 1667.

Stock, B. (1983). *The implications of literacy.* Princeton: Princeton University Press.

Stock, B. (1984–1985). Medieval history, linguistic theory, and social organization. *New Literary History, 16,* 13–29.

Taylor, M., & Flavell, J. H. (1985). *The development of children's ability to distinguish what they know from what they see.* Paper presented at the Annual Meeting of the Society for Philosophy and Psychology, Toronto, May.

Traugott, E. C. (1985). *Literacy and language change: the special case of speech act verbs.* Working papers of the McLuhan Program in Culture and Technology, no. 8, University of Toronto.

Turkle, S. (1984). *The second self: computers and the human spirit.* New York: Simon and Schuster.

Yates, F. (1966). *The art of memory.* Chicago: University of Chicago Press.

Yates, F. (1979). Print culture. *Encounter, 52,* 59–64.

General Conclusion

At the end of this book, the biggest question remains: Can one assume that as we are dealing with language, a building block of cognition, there might be a sort of spillover effect from the processing strategies required for reading to the processing strategies required for organizing knowledge and thought in the mind? The suggestion has been made by neurobiologist Joseph Bogen:

It is likely that some anatomical asymmetry underlies the potential for hemisphere specialization; but it is also clear that the extent to which capacities are developed is dependent upon environmental exposure. Although humans of any culture, so far as we know, have the potential for reading and writing, many remain nonliterate and thus fall short of acquiring the most special of left-hemisphere functions. Conversely, we can readily comprehend the concept of a society in which 'right-hemisphere illiteracy' is the rule. Indeed, our own society (admittedly complex) seems to be, in some respect, a good example: a scholastized, post-Gutenberg-industrialized, computer happy exaggeration of the Graeco-Roman penchant for propositionizing. *(Bogen, 1975, p. 29)*

Assuming that correlations between recognized structural features of comparable orthographies and patterns of lateralization were established in statistically reliable ways, what would this imply? First, it would raise the possibility that irrespective of the similarities and/or differences in cognitively attributable effects, such orthographies would require different processing strategies at the level of neurological organization, including the direction of scanning across the visual field.

Second, the kind of processing strategies required to decipher a string of characters sequentially from left to right could also be applied to the processing of the cognitive content of the reading/writing material. The reason is that such processing strategies probably depend on special and well-differentiated properties of the brain. As we have seen, this possibility is cautiously explored in Martin Taylor's paper on the Bilateral Cooperative Model of reading. The hypothesis is that by being taught to decode alphabetic texts, which are linear and sequential, the brain is encouraged to adopt strategies of sequential analysis for other cognitive operations too. The tendency to favor sequential processing of cognitive material might remain for the duration of adult life, even though we know that after the initial learning period many children can read without having to recode strings of letters phonetically except for rare or unusual words (Seidenberg, 1985; Read, 1985).

This possibility is especially relevant when one considers that two of the founders of modern science, Plato and Descartes, 2000 years apart, expressed almost the same sequential order of mental operations with similar concepts. In Plato's *Phœdrus,* we find suggestions about the art of speech writing that bear a striking similarity to the four rules of investigation (examination, division, order, and enumeration) that are enunciated as the substance of Descartes' *Discourse on Method:*

First, you must know the truth about the subject that you speak or write about; that is to say, you must be able to isolate it in definition, and having so defined it you must next understand how to divide it into kinds, until you reach the limit of division; secondly, you must have a corresponding discernment of the nature of the soul, discover the type of speech appropriate to each nature, and order and arrange your discourse accordingly.

Plato, *Phœdrus,* 277 b, c (Hamilton & Cairns, 1980)

Although it is very likely that Descartes, who was acquainted with Plato's writings, derived his own inspiration from this and other Platonic suggestions, it is worth noting that both philosophers came to articulate similar cognitive strategies within a comparable time-span after important advances in literacy in their respective situations; the alphabet itself for Plato, and the printing press for Descartes. The wider implication, of course, remains open for investigation, but there is, we wish to suggest, enough evidence to warrant continued study.

There are good reasons to believe that there are causal relationships between the nature of writing systems and specific cultural and cognitive consequences, but we cannot hope to answer this question completely at the level of neuropsychology right now. Our intent is to define the question and to show both its relevance and its urgency. This book does not claim to arrive at definite conclusions. Its main ambition is to set the stage for consideration and international discussion on the theme of the specific relationships between orthographies and the brain's hidden structure.

DERRICK DE KERCKHOVE
CHARLES J. LUMSDEN

References

Bogen, J. E. (1975). Educational aspects of hemispheric specialization. *UCLA, 17,* 24–33.

Descartes, R. (1967). *Le discours de la méthode.* Pataud, J. M. (Ed.), Paris: Bordas.

Hamilton, E., & Cairns, H. (Eds.), (1980). *Plato: The Collected Dialogues.* Bollingen Series. Princeton: Princeton University Press.

Read, C. (1985). Effects of phonology on beginning spelling: some cross-linguistic evidence. In, Olson, D. R., Torrance, N., & Hildyard, A. (Eds.), *Literacy, Language and Learning.* Cambridge: Cambridge University Press.

Seidenberg, M. S. (1985). The time course of phonological code activation in two writing systems. *Cognition, 19,* 1–30.

Index

MIX
Papier aus verantwortungsvollen Quellen
Paper from responsible sources
FSC® C105338
FSC
www.fsc.org

If you have any concerns about our products,
you can contact us on
ProductSafety@springernature.com

In case Publisher is established outside the EU,
the EU authorized representative is:
Springer Nature Customer Service Center GmbH
Europaplatz 3, 69115 Heidelberg, Germany

Printed by Libri Plureos GmbH
in Hamburg, Germany